TABLE T-3 Table Values for Confidence Intervals for σ. Entries Are Values of $T3L$ and $T3U$.

	0.90		0.95		0.99			0.90		0.95		0.99	
DF	***T3L***	***T3U***	***T3L***	***T3U***	***T3L***	***T3U***	**DF**	***T3L***	***T3U***	***T3L***	***T3U***	***T3L***	***T3U***
3	0.620	2.92	0.566	3.73	0.483	6.47	32	0.832	1.26	0.804	1.32	0.754	1.45
4	0.649	2.37	0.599	2.87	0.519	4.40	33	0.834	1.26	0.807	1.32	0.757	1.44
5	0.672	2.09	0.624	2.45	0.546	3.48	34	0.836	1.25	0.809	1.31	0.759	1.44
6	0.690	1.92	0.644	2.20	0.569	2.98	35	0.838	1.25	0.811	1.30	0.762	1.43
7	0.705	1.80	0.661	2.04	0.588	2.66	36	0.840	1.24	0.813	1.30	0.765	1.42
8	0.718	1.71	0.675	1.92	0.604	2.44	37	0.842	1.24	0.815	1.29	0.767	1.41
9	0.729	1.65	0.688	1.83	0.618	2.28	38	0.844	1.24	0.817	1.29	0.769	1.40
10	0.739	1.59	0.699	1.75	0.630	2.15	39	0.845	1.23	0.819	1.28	0.772	1.40
11	0.748	1.55	0.708	1.70	0.641	2.06	40	0.847	1.23	0.821	1.28	0.774	1.39
12	0.755	1.52	0.717	1.65	0.651	1.98	41	0.849	1.22	0.823	1.28	0.776	1.38
13	0.762	1.49	0.725	1.61	0.660	1.91	42	0.850	1.22	0.825	1.27	0.778	1.38
14	0.769	1.46	0.732	1.58	0.669	1.85	43	0.852	1.22	0.826	1.27	0.780	1.37
15	0.775	1.44	0.739	1.55	0.676	1.81	44	0.853	1.22	0.828	1.26	0.782	1.37
16	0.780	1.42	0.745	1.52	0.683	1.76	45	0.854	1.21	0.829	1.26	0.784	1.36
17	0.785	1.40	0.750	1.50	0.690	1.73	46	0.856	1.21	0.831	1.26	0.786	1.36
18	0.790	1.38	0.756	1.48	0.696	1.70	47	0.857	1.21	0.832	1.25	0.788	1.35
19	0.794	1.37	0.760	1.46	0.702	1.67	48	0.858	1.20	0.834	1.25	0.790	1.35
20	0.798	1.36	0.765	1.44	0.707	1.64	49	0.859	1.20	0.835	1.25	0.791	1.34
21	0.802	1.35	0.769	1.43	0.712	1.62	50	0.861	1.20	0.837	1.24	0.793	1.34
22	0.805	1.34	0.773	1.42	0.717	1.60	55	0.866	1.19	0.843	1.23	0.801	1.32
23	0.809	1.33	0.777	1.40	0.722	1.58	60	0.871	1.18	0.849	1.22	0.808	1.30
24	0.812	1.32	0.781	1.39	0.726	1.56	65	0.875	1.17	0.854	1.21	0.814	1.28
25	0.815	1.31	0.784	1.38	0.730	1.54	70	0.879	1.16	0.858	1.20	0.820	1.27
26	0.818	1.30	0.788	1.37	0.734	1.53	75	0.883	1.16	0.862	1.19	0.825	1.26
27	0.820	1.29	0.791	1.36	0.737	1.51	80	0.886	1.15	0.866	1.18	0.829	1.25
28	0.823	1.29	0.794	1.35	0.741	1.50	85	0.889	1.15	0.870	1.18	0.834	1.24
29	0.825	1.28	0.796	1.34	0.744	1.49	90	0.892	1.14	0.873	1.17	0.838	1.23
30	0.828	1.27	0.799	1.34	0.748	1.48	95	0.894	1.14	0.876	1.17	0.841	1.23
31	0.830	1.27	0.802	1.33	0.751	1.46	100	0.897	1.13	0.879	1.16	0.845	1.22

APPLIED STATISTICS

A FIRST COURSE IN INFERENCE

APPLIED STATISTICS

A FIRST COURSE IN INFERENCE

FRANKLIN A. GRAYBILL

Colorado State University

HARIHARAN K. IYER

Colorado State University

RICHARD K. BURDICK

Arizona State University

PRENTICE HALL
Upper Saddle River, New Jersey 07458

Library of Congress Cataloging-in-Publication Data

Graybill, Franklin A.
Applied statistics : a first course in inference / Franklin A. Graybill, Hariharan K. Iyer, Richard K. Burdick.
p. cm.
Includes index.
ISBN 0-13-621467-3
1. Statistics. I. Iyer, Hariharan K. II. Burdick, Richard K. III. Title.
QA276.12.G72 1998
519.5--dc21 97-11588
CIP

Executive Editor: Ann Heath
Editorial Director: Tim Bozik
Editor-in-Chief: Jerome Grant
Assistant Vice President of Production and Manufacturing: David W. Riccardi
Art Director: Maureen Eide
Editorial/Production Supervision: Judith Winthrop
Composition: Techsetters, Inc.
Managing Editor: Linda Mihatov Behrens
Executive Managing Editor: Kathleen Schiaparelli
Marketing Manager: Melody Marcus
Director of Marketing: John Tweeddale
Marketing Assistants: Diana Penha, Jennifer Pan
Creative Director: Paula Maylahn
Art Manager: Gus Vibal
Cover Designer: Lorraine Castellano
Manufacturing Buyer: Alan Fischer
Manufacturing Manager: Trudy Pisciotti
Editorial Assistant/Supplements Editor: Mindy Ince
Cover Photo: Jeremy Walker/Tony Stone Images.
Pair of industrial chimneys emitting smoke, sunset, silhouette and USA, Arizona, view over Grand Canyon at dusk.
Design and Layout: Lorraine Castellano

Printed in the United States of America
10 9 8 7 6 5 4 3 2 1

ISBN 0-13-621467-3

Prentice-Hall International (UK) Limited, *London*
Prentice-Hall of Australia Pty., Limited, *Sydney*
Prentice-Hall of Canada, Inc., *Toronto*
Prentice-Hall of India Private Limited, *New Delhi*
Prentice-Hall of Japan, Incc., *Tokyo*
Simon & Schuster Asia Pte. Ltd., *Singapore*
Editora Prentice-Hall do Brasil, Ltda., *Rio de Janeiro*

To my wife, Jeanne, for more than 50 years of support. To my children, Dan and Kathy, for more than 45 years of delight. To my grandchildren, Tonia, Jules, and Emily, for the happy hours you have given me. Thanks for all the good times.

Franklin A. Graybill

To my parents, for teaching me right from wrong; to my teachers, for showing me the way; to Pam, Matthew, Kristin, Kevin, and Geoffrey, for your encouragement and support; to Frank, for allowing me to work with you, and to Jeanne, for allowing Frank to work with Rick and me.

Hariharan K. Iyer

To my family: Pat, Stacy, Mickey, and Stephanie. Thanks for sharing me with Frank and Hari.

Richard K. Burdick

CONTENTS

CHAPTER 7 INFERENCE FOR A POPULATION MEAN AND MEDIAN FOR ANY POPULATION 155

CHAPTER 8 STATISTICAL TESTS 179

CHAPTER 9 SIMPLE LINEAR REGRESSION 209

PREFACE

APPROACH

This book is written for a first course in applied statistics that emphasizes statistical inference. It is not intended to be a compendium of every procedure in statistics. Rather, we focus on the basic ideas that form the foundation of statistical inference. We explain how statistical inference can be used to solve real problems. We have written this book in a manner that allows students to focus on statistical concepts and not become overly involved with mathematical details. We believe that students do not need to prove statistical concepts mathematically in order to understand them.

AUDIENCE

We have tried to present a clear and accurate account of statistical inference that can be understood by a student with a knowledge of college algebra. Some acquaintance with computers such as that typically obtained in high school is beneficial but not required. The book can be used for several different audiences. For example, students who desire an understanding of statistical results reported in the popular press, radio, and television should cover the following sections.

- Chapters 1 and 2 (all sections)
- Chapter 3 (sections 1 and 2)
- Chapters 4 and 5 (all sections)
- Chapter 6 (sections 1, 2, and 3)
- Chapter 8 (sections 1 and 2)
- Chapter 9 (sections 1, 2, 3, and 4)
- Interlude

This coverage will teach students the notation, terminology, and general concepts of statistical inference. It will allow them to understand why statistical inference is of fundamental importance in scientific investigations as well as everyday life where data are used. After learning these sections, students will be able to know what type of questions to ask professional statisticians and be able to understand the answers.

A more detailed course for an audience that intends to apply statistical procedures should cover all of the first 9 chapters, the Interlude, and sections of interest in the remaining chapters. Chapters 10 through 14 can be covered in any order. This course will allow students to study more advanced specialized material in statistical inference.

ORGANIZATION

The order of material is somewhat different than in other statistics textbooks at this level. In particular, we introduce and discuss statistical inference in Chapters 1 and 2 using a population proportion as the parameter of interest and do not encounter the mean and the normal population until Chapters 5 and 6, respectively. The reason for this ordering is that we want the student to learn the basic ideas of statistical inference early in the course. This requires an understanding of populations, variables, parameters, samples, point estimation and confidence intervals. We believe these concepts can best be learned using population proportions.

Population Proportions

We introduce statistical inference using population proportions in Chapter 1. Proportions are not only easy to understand but they are also extremely useful and the inference procedures are appropriate for any population. Since students are familiar with proportions, this leaves them free to concentrate on the concepts of inference without having to simultaneously learn the myriad of other concepts involved if statistical inference is introduced using the mean of a normal population.

Sample Size

The importance of determining sample size prior to collecting a sample is emphasized in this book. We believe that a user of statistics must always take the sample size into consideration when using statistical inference.

The Mean and Median

We believe that the median is under-used in statistics. We treat the median on a comparable level with the mean in Chapters 5 through 7 and provide an explanation to help the student determine when each should be used.

Confidence Intervals

We take the view that point estimation and interval estimation (margin of error) are the principal procedures in statistical inference. We use confidence intervals throughout the book. Statistical tests, which are discussed in Chapter 8, are introduced as an application of confidence intervals. This approach helps the student realize that interval estimation, rather than statistical testing, is by far the more useful inference procedure.

Nonparametric Procedures

There is no separate chapter devoted to nonparametric statistics (distribution-free statistics). We take the view that nonparametric procedures should be discussed as the need arises rather than deferring these procedures to a separate chapter. The Mann-Whitney and Wilcoxon procedures are used in Chapter 11 to construct confidence intervals for the difference between two population means.

FEATURES

A Systematic Approach

We introduce statistical inference in the first chapter using examples that students might encounter in their daily lives. Many of these examples, denoted by a signature illustration, are revisited in later chapters and help to establish a systematic approach for dealing with uncertainty using statistical inference. This approach requires students to understand the importance of defining the population of interest and to learn how to use appropriate samples from the population to measure uncertainty. Each topic in succeeding chapters is presented within the systematic approach introduced in Chapter 1.

Data

All of the data presented in this book are based on real or "real appearing" studies consistent with our experiences as professional statisticians. Because of the complexity of real studies, the data presented in the book have often been simplified to allow students to focus on the statistical concepts rather than the details of the study. We have used examples that do not require a knowledge of any special discipline. Rather, the examples involve familiar topics such as television ratings, household incomes, miles per gallon of gasoline for automobiles, performance in college, and so on. We have used problems that can be introduced simply, can be understood by all students, and relate to real-life situations. Most of the data sets used in the book are available on the Data Disk that is packaged with the book.

Student Study Tools

A number of special features have been built into the book to motivate students, to help them master concepts, prepare for tests, and apply their knowledge to real problems.

Explorations

Nine explorations are included in the book which engage the student in immediate practice of a skill or concept. Questions are posed for students to investigate, and these problems sometimes require computations. Reinforcement is provided by placing the answer (in italics) immediately after each question or problem statement.

Thinking Statistically: Self-Test Problems

Most sections end with a set of Self-Test Problems. These problems cover the material discussed in the section and provide students with an opportunity to test their comprehension of the material. The complete solution of each Self-Test Problem is located at the end of the chapter in which it appears.

Exercises

At the end of each chapter are exercises that cover all material in the chapter. Answers to odd exercises are in Appendix A at the end of the book.

Chapter Summary

At the end of each chapter is a summary of the material covered in that chapter.

In Practice: Interlude

The interlude (following Chapter 9) is a unique feature that provides students with practical guidelines for evaluating statistical studies. It also includes a discussion of an actual study used to examine the cause of air pollution in the Grand Canyon National Park. This discussion demonstrates some of the difficulties in detecting causal relationships and describes the difference between observational and experimental studies. Falling at the end of the first nine chapters that make up the conceptual core of the book, the Interlude provides a nice, practical capstone for the course. Some instructors may choose to discuss elements of the Interlude earlier in the course and revisit it again at the end.

Examples

Every technique discussed in this book is introduced with one or more examples. After a technique is discussed, more examples are provided to give the student an opportunity to see how the technique can be used.

Appendices

Some chapters include one or more *Appendices* that contain supplemental material. The appendices are more advanced and provide more complete explanations of some concepts. It is not necessary to read any appendix in order to understand other sections of the book or other appendices.

Computing

Appendix C provides a short introduction to the computing package Minitab. We use Minitab throughout the book to perform most of the required computations. However, this book is *not* dependent on the use of Minitab or any other statistical computing package. Exhibits are displayed throughout the book that show the results of statistical computations using Minitab. However, exhibits similar to those produced using Minitab can be produced with virtually any statistical package. In addition, we have provided necessary formulas so that the book can be used without a software package. There are some computations for which no simple Minitab command exists. For these situations, we have written Minitab programs, called macros, which are included on the floppy disk that accompanies the book. All of the data sets that require computations are also on the disk in two forms: (1) Minitab portable files that have the extension **mtp** and (2) ASCII files that have the extension **dat**.

SUPPLEMENTS

Instructor's Solutions Manual

(ISBN 0-13-621483-5) Written by the authors and independently checked, the Instructor's Solutions Manual contains solutions to all of the exercises in the book.

The Excel Companion

(ISBN 0-13-676487-8) Developed by the text authors, the Excel Companion to *Applied Statistics: A First Course in Inference* teaches students familiar with Excel how to use this package to perform statistical analyses described in the text.

ACKNOWLEDGMENTS

We owe a debt of thanks to the reviewers and to others who provided many helpful comments during the development of this book.

First we owe a great debt to Ann Heath, the Executive Editor of Statistics at Prentice Hall for her faith in the book and for her insights, patience, vision, and help.

To those who read the manuscript and made many important comments we also owe a debt of gratitude. These include

- Greg Davis, University of Wisconsin-Green Bay
- Chris Franklin, University of Georgia
- Joan Garfield, University of Minnesota
- Albyn Jones, Reed College
- Mohammed Kazemi, UNC Charlotte
- Sheldon Kovitz, University of Massachusetts-Harbor Campus
- Jack Leddon, Dundalk Community College
- Peter Matthews, University of Maryland-Baltimore
- Scott Preston, SUNY Oswego
- Robert Raymond, University of St. Thomas
- Galen Shorak, University of Washington
- Bruce Sisko, Belleville Area College
- Martha Van Cleave, Linfield College.

We also want to thank Don Gecewicz for his development efforts of this book, Judith Winthrop for her management efforts, and Sarah Streett for performing accuracy checks of all exercises.

Franklin A. Graybill
Hariharan K. Iyer
Richard K. Burdick

CHAPTER 1

DATA AND STATISTICAL METHODS

1.1 INTRODUCTION

To overcome the challenges of life, humans are continually exploring nature in an attempt to understand and control it. New challenges lead to new questions, and our ability to answer these questions and make correct decisions determines our future. Superstition and guesswork have given way to scientific methods where problems are solved, questions are answered, decisions are made, and actions are taken based on information.

No doubt you have heard it said that we live in the "information age." We are constantly bombarded with information from a myriad of sources—books, magazines, newspapers, computer output, the World Wide Web, radio, television, telephone, movies, tape and compact disc players, and gossip, among others. The Internet provides a continual flow of new information. The question that comes to mind is this: "What shall we do with all this information?" Although we spend vast amounts of time, money, and energy collecting information, it will do us little good unless that information can be processed, interpreted, and properly used to help make decisions.

In this book we are principally concerned with information in the form of data. **Data** is information expressed in a form that facilitates calculations and descriptions. The subject called **statistics**, which is the content of this book, can be viewed in a narrow sense as the science of collecting, processing (organizing, displaying, summarizing, and analyzing), and interpreting data. In any investigation the first thing you should do is specify the objectives of the study. If the study requires decisions or actions based on data, then statistical methods are needed to accomplish the study objectives. Box 1.1.1 lists three areas in which statistical methods can assist you when data are used in decision making.

BOX 1.1.1: THREE WAYS TO USE STATISTICAL METHODS

When you plan to make decisions or take actions based on data, statistical methods can be used to help in

1. data collection,
2. data processing, which includes organizing, displaying, summarizing, and analyzing data, and
3. data interpretation.

We use data in making simple decisions every day of our lives. For example, the selection of our driving route to work each morning often depends on data arising from radio traffic reports, weather, and our historical knowledge of traffic flow on that particular day of the week. In order to decide which route to take on our drive to work, we must process and interpret this information. Although this might not result in a formal statistical study (many of us hardly have time to grab the cup of coffee and get out the door), the logic that we employ is consistent with the theory of statistical methods to be discussed in this book.

Statistical methods are employed not only to help make simple decisions every day, but also to help answer profound questions that face all of society. In the early 1950s a new vaccine that was claimed to prevent the insidious disease polio was developed. In 1954 a large statistical experiment was conducted in the United States involving 740,000 children in the second grade to test the vaccine's effectiveness. From this experiment it was concluded that the vaccine was indeed effective in preventing polio. [For further information, see "The Biggest Public Health Experiment Ever: The 1954 Field Trial of the Salk Poliomyelitis Vaccine" by Paul Meier (1978) in *Statistics: A Guide to the Unknown*, Judith Tanur, ed., Wadsworth, Belmont, CA.] The beauty of the subject called "statistics" is that it can be applied to any field of interest. Statistical methods are used in such diverse fields as engineering, law, business, politics, psychology, agriculture, medicine, and geology. In the remainder of this chapter we provide several examples that demonstrate how statistical procedures can be used to help solve problems of interest in a variety of settings.

THINKING STATISTICALLY: SELF-TEST PROBLEMS

1.1.1 Find a newspaper article that discusses a result based on data. Summarize the contents of the article in your own words.

1.1.2 What are the three areas where statistical methods can assist you when using data to answer a question?

1.1.3 When you cross a busy street, what data do you use to help you decide when to cross?

1.1.4 When you are driving a car and approaching a traffic light, what data do you use to decide whether you should stop or proceed?

1.1.5 How many students in this statistics class are over six feet tall? How did you collect data to answer this question?

1.1.6 Do you think that cigarette smoking is harmful to a person's health? On what data do you base your conclusion?

1.1.7 The world population is growing. An article in the July 18, 1994, issue of *USA Today* stated that the United Nations Population Reference Bureau projected that the world population will exceed 10 billion people by the year 2050. What kinds of data do you think are useful in making such a projection?

1.2 USING DATA TO HELP ANSWER QUESTIONS

In this section we provide a wide variety of examples to show how data can be employed to help answer questions. In Examples 1.2.1 and 1.2.2 we show how data are used to answer very simple questions, similar to ones we confront many times each day. In these examples we also introduce some charts that are commonly used to display data. The charts are discussed in detail in Chapter 4.

EXAMPLE 1.2.1

PIZZA PARTY. You are planning a pizza party and must purchase 10 medium-sized pepperoni pizzas. Three restaurants serve pizzas of comparable quality; you want to determine which one will give you the best price. The question you want to answer is the following:

Question 1.2.1: Which of the three restaurants will sell 10 medium-sized pepperoni pizzas at the lowest price?

SOLUTION Now let's see how the three areas in Box 1.1.1 are used to answer Question 1.2.1.

1. **Data collection:** You call each restaurant and collect the following data.
   Restaurant 1: The cost of a medium-sized pepperoni pizza is \$10.00, and the second one is half-price.
   Restaurant 2: The cost of a medium-sized pepperoni pizza is \$8.00, with a 5% discount on the total bill if it is more than \$50.00.
   Restaurant 3: The cost of a medium-sized pepperoni pizza is \$9.00, but there is a 30% discount on every third pizza.

2. **Data processing:** You analyze the data you have collected by determining the cost of 10 pizzas at each of the three restaurants.
   The total price for 10 pizzas from restaurant 1 is $\$10.00(5) + \$5.00(5) = \$75.00$.
   The total price for 10 pizzas from restaurant 2 is $\$8.00(10)(0.95) = \$76.00$.
   The total price for 10 pizzas from restaurant 3 is $\$9.00(7) + \$6.30(3) = \$81.90$.
   The prices at each of the restaurants are displayed in Figure 1.2.1. This figure is called a **bar chart**. Each restaurant is represented by a bar (rectangle). The height of each bar represents the cost of 10 pizzas.

3. **Data interpretation:** The 10 pizzas cost the least at restaurant 1. The answer to Question 1.2.1 is as follows:

▶**Answer to Question 1.2.1**: The lowest total cost for 10 pizzas can be obtained from restaurant 1 and is \$75.00.

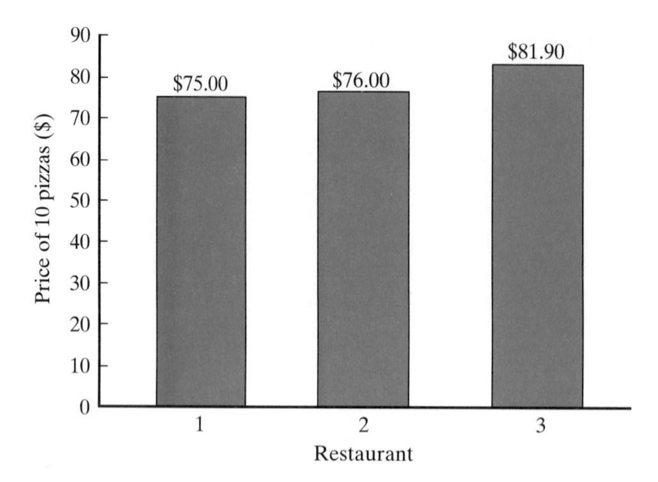

FIGURE 1.2.1
Price of 10 Pizzas at Three Restaurants

The answer to Question 1.2.1 can now be used to help decide where to purchase the 10 pizzas. You may want to consider other information, such as the distance to the restaurant, before making a final decision. This is a very simple example of collecting, processing, and interpreting data to answer a question. The question is not "earth shaking," but each day most of us collect data to help answer similar questions.

EXAMPLE 1.2.2

TRUCK ACCIDENT. A truck carrying 5,000 cartons of paper had an accident on Interstate 25 near Denver, Colorado, in which many of the cartons were split open and the paper destroyed. The truck driver must determine how many of the cartons were damaged and report the total cost of the damaged cartons to the owners. The questions that are to be answered are the following:

Questions 1.2.2: How many of the 5,000 cartons of paper were damaged? What is the total cost of the damaged cartons?

SOLUTION

1. **Data collection:** The truck driver investigated the accident scene and collected all damaged and undamaged cartons. She counted 263 damaged cartons.
2. **Data processing:** The number of damaged cartons is displayed with a bar graph in Figure 1.2.2. Each carton cost \$58, so the total cost of the damaged cartons is $\$58(263) = \$15,254$.

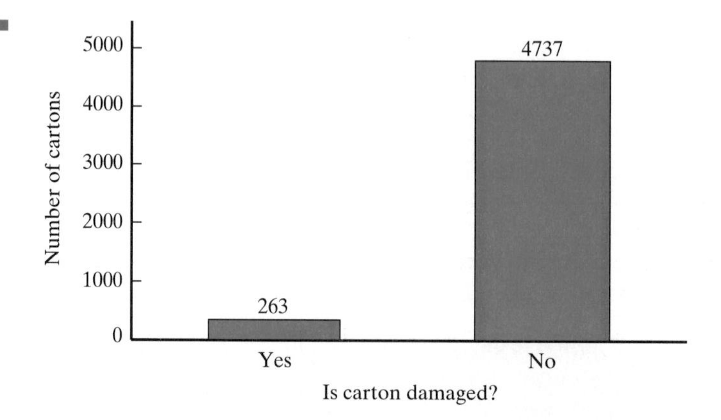

FIGURE 1.2.2
Number of Damaged Cartons

3. **Data interpretation:** The analysis in step (2) provides the following answers to Questions 1.2.2.

▶**Answers to Questions 1.2.2:** Exactly 263 cartons were damaged. The total cost of the damaged cartons is $15,254.

While Questions 1.2.1 and 1.2.2 may not be "vitally important," statistical methods play a key role in helping decision makers deal with problems that affect all of society. These problems include world hunger, domestic violence, and combating disease. Example 1.2.3 demonstrates how statistical methods are used to address a problem that concerns the safety of our environment. This example requires the computation of a **proportion** and a **percentage**. Before we present the example, we define these terms.

DEFINITION 1.2.1: PROPORTION AND PERCENTAGE

In a set of items, the **proportion** of items with a specific characteristic is a number between 0 and 1 (inclusive) and is defined by the fraction

$$\textit{proportion} = \frac{\text{number of items in the set with a specific characteristic}}{\text{total number of items in the set}}.$$

In a set of items, the **percentage** of items with a specific characteristic is a number between 0 and 100 (inclusive) and is defined by

$$\text{percentage} = \text{proportion} \times 100.$$

Thus,

$$\text{proportion} = \frac{\text{percentage}}{100}.$$

For example, consider a set of 20 people (items). If 12 of the people are married (married is the specific characteristic of interest), then the proportion of married people in the set of 20 people is $12/20 = 0.60$. A percentage is obtained from a proportion by multiplying the proportion by 100. The symbol "%" is used for the word "percent." For example, the proportion 0.60 is expressed in percentage form as 60%, where $60\% = 0.60(100)\%$. You can think of the symbol "%" as meaning "multiply by $1/100$." Thus, $60\% = 60(1/100) = 0.60$.

EXAMPLE 1.2.3

RADON GAS EMISSIONS. Much research has been conducted during the past few years in an attempt to understand the health effects of radon gas in dwellings. Radon is a radioactive gas that is generally present in harmless amounts in nature. However, in certain dwellings radon gas is known to be present in quantities that may be harmful to humans, particularly in basements of buildings where the air is stagnant. The Environmental Protection Agency (EPA) sets standards for environmental emissions from hazardous substances. According to the July 1995 issue of *Consumer Reports*, the EPA has suggested that radon levels exceeding 4.0 pc/l (*picocurie per liter*, which is a measure of radioactive emissions) are associated with an increased risk of lung cancer. Although this is debatable, there seems to be a general belief among scientists that radon levels above 16.0 pc/l pose a serious health threat. A real estate developer owns 51 residential buildings that he wants to sell. Before the buildings can be sold, the radon level in each building must be determined. If it exceeds 4.0 pc/l in any building, the buyer must be so advised. If it exceeds 16.0 pc/l in any building, some corrective measures must be taken. One question of interest in the investigation is the following.

Question 1.2.3: For what percentage of the 51 residential buildings does radon present a potential health problem? Specifically, what percentage of the 51 residential buildings has a radon level exceeding 4.0 but less than or equal to 16.0 pc/l, and what percentage of the buildings has a radon level exceeding 16.0 pc/l?

SOLUTION

1. **Data collection:** To measure the radon levels, a device called a "radon sampler" is placed in the basements of each of the 51 buildings. The measured radon concentrations in pc/l are given in Table 1.2.1.

TABLE 1.2.1 Radon Concentrations in 51 Buildings

Building number	Radon concentration (in pc/l)	Building number	Radon concentration (in pc/l)	Building number	Radon concentration (in pc/l)
1	13.6	18	2.9	35	6.5
2	2.8	19	2.0	36	11.8
3	2.9	20	2.9	37	13.2
4	3.8	21	11.2	38	2.8
5	15.9	22	1.9	39	6.9
6	1.7	23	2.0	40	0.7
7	3.4	24	6.0	41	12.9
8	13.7	25	2.9	42	3.6
9	6.1	26	7.7	43	3.6
10	16.8	27	5.1	44	8.1
11	7.9	28	13.2	45	17.0
12	3.5	29	3.8	46	8.2
13	2.2	30	13.9	47	9.8
14	4.1	31	2.4	48	13.0
15	3.2	32	7.9	49	11.3
16	2.9	33	1.4	50	4.0
17	3.7	34	5.9	51	6.0

2. **Data processing:** To help answer Question 1.2.3, we rearrange the data in Table 1.2.1 from smallest to largest radon concentrations. The rearranged data are presented in Table 1.2.2.

TABLE 1.2.2 Radon Concentrations (in pc/l) in 51 Buildings (Arranged From Smallest to Largest)

0.7	1.4	1.7	1.9	2.0	2.0	2.2	2.4	2.8	2.8	2.9
2.9	2.9	2.9	2.9	3.2	3.4	3.5	3.6	3.6	3.7	3.8
3.8	4.0	4.1	5.1	5.9	6.0	6.0	6.1	6.5	6.9	7.7
7.9	7.9	8.1	8.2	9.8	11.2	11.3	11.8	12.9	13.0	13.2
13.2	13.6	13.7	13.9	15.9	16.8	17.0				

Organization of the data in this manner helps us see that 25 of the 51 buildings (49.0%) have radon levels exceeding 4.0 but less than or equal to 16.0 pc/l, and 2 of the 51 buildings (3.9%) have radon levels exceeding 16.0 pc/l. These

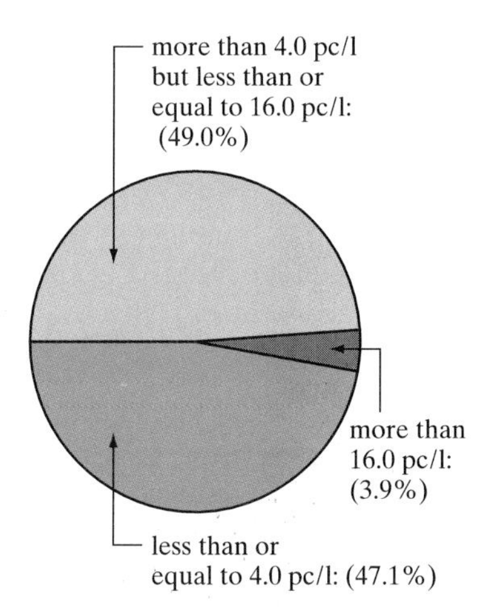

FIGURE 1.2.3
Percentage of the 51 Buildings with Various Radon Levels

percentages are shown graphically in Figure 1.2.3. This figure is called a **pie chart**. The different pieces of pie correspond to percentages of the buildings that have radon levels less than or equal to 4.0 pc/l (47.1%), more than 4.0 but less than or equal to 16.0 pc/l (49.0%), and more than 16.0 pc/l (3.9%), respectively.

3. **Data interpretation:** By examining Figure 1.2.3 or Table 1.2.2, we determine the answer to Question 1.2.3.

   ▶**Answer to Question 1.2.3:** Forty-nine percent of the 51 buildings have radon levels of more than 4.0 pc/l but less than or equal to 16.0 pc/l. Also, 3.9% of the 51 buildings have radon levels that exceed 16.0 pc/l.

If we assume that our measurements and calculations are correct, the answers to Questions 1.2.1, 1.2.2, and 1.2.3 are exact. However, situations are routinely encountered where complete data are not available and exact answers are not attainable. Two such situations are illustrated in the next two examples.

EXAMPLE 1.2.4

WAREHOUSE FIRE. A recent fire in a warehouse that contained 100,000 radios damaged an unknown number of the radios. A freight broker who purchases damaged goods offers to purchase the entire contents from the insurance company that provides coverage for the warehouse. The freight broker will eventually sort through all the radios and sell those that are not damaged. Before the broker makes an offer to the insurance company, he would like to know what proportion of the 100,000 radios are damaged and cannot be sold. Hence, the broker wants an answer to the following question:

Question 1.2.4: What proportion of the radios in the warehouse was damaged in the fire?

We illustrate how the three areas of statistical methods in Box 1.1.1 can be used to help answer the question.

SOLUTION

1. **Data collection:** To obtain an exact answer to Question 1.2.4, someone must inspect all 100,000 radios. Realizing that it is impractical to examine all 100,000 radios in a timely manner, the broker decides to use the following procedure.

A subset consisting of 50 of the 100,000 radios in the warehouse will be selected and inspected for damage. This subset is called a **sample**. On the basis of the information contained in the sample, a decision will be made as to what proportion of the entire set of 100,000 radios is damaged.

2. **Data processing:** The results of examining the 50 radios in the sample are presented in Table 1.2.3. The data in Table 1.2.3 indicate that 4 of the 50 radios selected were damaged. Hence $4/50 = 0.08 = 8\%$ of the radios in the sample were damaged. See Figure 1.2.4.

TABLE 1.2.3 Results of the Examination of 50 Radios

Radio number	Condition of radio	Radio number	Condition of radio
1	not damaged	26	not damaged
2	not damaged	27	not damaged
3	not damaged	28	not damaged
4	not damaged	29	damaged
5	not damaged	30	not damaged
6	not damaged	31	not damaged
7	not damaged	32	not damaged
8	not damaged	33	not damaged
9	not damaged	34	damaged
10	not damaged	35	not damaged
11	not damaged	36	not damaged
12	not damaged	37	not damaged
13	not damaged	38	not damaged
14	damaged	39	not damaged
15	not damaged	40	not damaged
16	not damaged	41	not damaged
17	not damaged	42	not damaged
18	not damaged	43	not damaged
19	not damaged	44	not damaged
20	not damaged	45	not damaged
21	not damaged	46	damaged
22	not damaged	47	not damaged
23	not damaged	48	not damaged
24	not damaged	49	not damaged
25	not damaged	50	not damaged

FIGURE 1.2.4
Damaged Radios

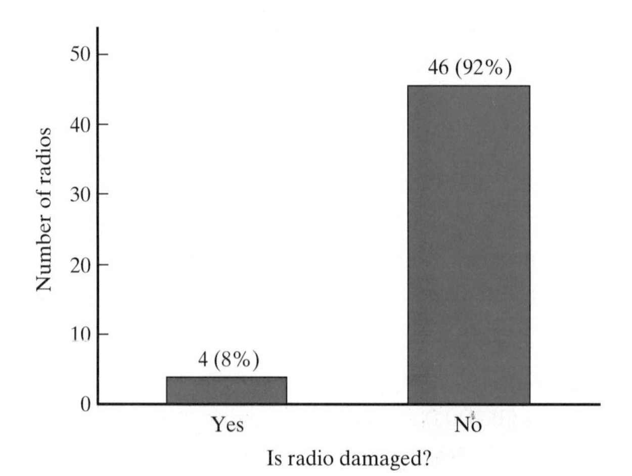

3. **Data interpretation:** The broker uses the information in Table 1.2.3 and estimates that $0.08 = 8\%$ of the 100,000 radios are damaged.

   The broker will use this information in determining how much money to pay the insurance company for the radios. The conclusions of this study are not exact and cannot be made with certainty. This is because not all of the 100,000 radios were tested, and the conclusions are based on an examination of only 50 radios. It would be very unlikely for the proportion of damaged radios in the sample of 50 radios to be exactly equal to the proportion of damaged radios in the entire warehouse of 100,000 radios. Thus, the answer to Question 1.2.4 is as follows:

▶**Answer to Question 1.2.4**: It is *estimated* that $0.08 = 8\%$ of the 100,000 radios in the warehouse are damaged. However, since only 50 radios were examined, we do not expect the estimate to be exactly equal to the proportion of the 100,000 radios in the warehouse that are damaged. We say that there is *uncertainty* in the estimate.

Notice the difference in the degree of certainty we can have in the conclusions drawn in Example 1.2.4 compared to those in Example 1.2.2. In Example 1.2.2 there is no uncertainty in the answer since all cartons of paper were examined to see how many had been damaged. In contrast, in Example 1.2.4 not all of the radios were examined and the answer is therefore uncertain. When not all items in a set are examined, answers to questions about the entire set of items are unlikely to be exactly correct. Thus, we need some way to interpret the uncertainty associated with these answers.

BOX 1.2.1: UNCERTAINTY

In most situations where statistical methods are employed, a sample of items is used to answer questions about an entire set of items. We have noted that whenever a sample is used to draw conclusions about an entire set of items, there will be uncertainty in the stated conclusions. However, it is important to understand that *uncertain conclusions are not useless conclusions*. On the contrary, the conclusions you make based on a sample are very useful when the sample has been collected properly and the resulting sample data are processed correctly.

In the remaining chapters of this book we discuss *statistical methods* where samples from a set of items are used to answer questions of interest about the entire set of items. Statistical methods allow us to quantify the amount of uncertainty associated with conclusions based on samples. We conclude this section with an additional example to illustrate how a sample can be used to answer questions.

EXAMPLE 1.2.5

TELEVISION RATINGS. The battle for weeknight insomniacs reached a fever pitch in the fall of 1993 when David Letterman left NBC for the CBS network and a \$42 million, three-year contract. In the initial months of his CBS program, *Late Show with David Letterman*, he enjoyed a ratings number that was twice as large as that of *Tonight Show with Jay Leno*, the competitor at NBC. A ratings number represents the percentage of American households with television sets that are tuned into a particular program for at least six minutes. In 1993 a single rating point translated into about 954,000 households. The ratings numbers are important because they are the primary determinant in advertising rates for commercials. A program with a large ratings number can generate millions of dollars in advertising revenue for a network. The following question may be of interest to the networks as well as to potential advertisers:

Question 1.2.5: What percentage of households with television sets was tuned to *Late Show with David Letterman* for at least six minutes last Tuesday?

SOLUTION To answer this question, a company that produces ratings numbers used computers to collect data about the viewing habits of a sample of 1,200 households. Suppose that on the Tuesday night of interest the sample data indicated that 240 (or 20%) of the surveyed homes were watching *Late Show with David Letterman*. This percentage is presented graphically with the pie chart in Figure 1.2.5.

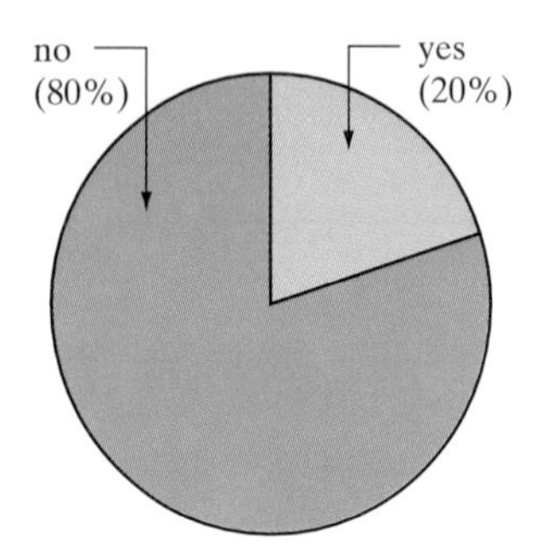

FIGURE 1.2.5
Watching *Late Show with David Letterman*

These results provide the following answer to Question 1.2.5:

▶**Answer to Question 1.2.5**: By processing the data it was determined that $240/1200 = 0.20 = 20\%$ of the households surveyed were tuned to *Late Show with David Letterman* for at least six minutes last Tuesday. Thus, it was *estimated* that 20% of all American households with television sets were tuned to *Late Show with David Letterman* last Tuesday. However, since only 1,200 households were examined, we do not expect the estimate, 20%, to be exactly correct. That is, we do not expect that exactly 20% of all American households with televisions were watching *Late Show with David Letterman* last Tuesday. There is uncertainty in the estimate.

✻ ***Summary*** Data help answer questions of interest in our everyday lives as well as in the business and scientific community. We often have to make decisions using *estimates*, and this introduces uncertainty in our decisions. Nevertheless, uncertain decisions can be very useful if they are based on data that are collected properly, processed correctly, and interpreted accurately.

THINKING STATISTICALLY: SELF-TEST PROBLEMS

1.2.1 During the past 12 months there have been several serious automobile accidents on Exit 112 of Interstate Highway 25. The Colorado Highway Patrol believes they are mainly due to speeding and has installed an electronic device to measure speeds of vehicles using the exit. The patrol wants to know what percentage of vehicles that leave Interstate 25 using Exit 112 is exceeding the speed limit of 40 miles per hour. The electronic device determines that 42% of all vehicles that used Exit 112 *yesterday* exceeded the speed limit.

a. Is there any uncertainty in using 42% as the answer to the question, "What percentage of the vehicles that left Interstate 25 *yesterday* using Exit 112 exceeded the speed limit?"
b. Suppose that the electronic measuring device was used only from 8:00 a.m. to 9:00 a.m. yesterday. During this time period 51% of the vehicles that left Interstate 25 using Exit 112 exceeded the speed limit. The percentage 51% is used as the answer to the question, "What percentage of the vehicles that left Interstate 25 yesterday using Exit 112 exceeded the speed limit?" Is this answer exact?

1.2.2 What percentage of students in this class are seniors? Is this an exact answer to the question, "What percentage of students in the entire university are seniors?"

1.2.3 Use the percentage of senior students in this class to *compute* the percentage of students in this class that are not seniors. Is there any uncertainty in using this result to answer the question, "What percentage of students in this class are not seniors?"

1.3 INTRODUCTION TO STATISTICAL INFERENCE

To continue our discussion of uncertainty when using sample data to answer questions, we need some definitions and terminology that will be used throughout this book. In particular, we explain the terms **population, variable, parameter,** and **sample**. We also introduce the concept of **statistical inference.**

Population

In Example 1.2.4 we considered a warehouse that contained 100,000 radios. The entire collection of the 100,000 radios is defined as a **population** of radios. We now formalize this definition.

DEFINITION 1.3.1: POPULATION OF ITEMS

The **population of items** is the entire set of items that one wants to investigate.

In Example 1.2.1 the population of items is the set of three restaurants. In Example 1.2.2 the population of items is the set of 5,000 cartons of paper, and in Example 1.2.3 the population of items is the set of 51 residential buildings. In Example 1.2.4 the population of items is the set of 100,000 radios, and in Example 1.2.5 the population of items is the set of all American households with a television set.

Variable

In an investigation it is generally desired to obtain information about the individual items in the population. For instance, in Example 1.2.3 the set of 51 residential buildings is the population of items. The radon concentration in each building is the information of interest. In Example 1.2.4 the set of 100,000 radios is the population of items. The condition of each individual radio, damaged or not damaged, is the information of interest. The information of interest for each population item is called a **variable** because it typically *varies* from item to item in the population.

For example, in the pizza problem the variable of interest is the price of 10 medium-sized pepperoni pizzas at a restaurant (we label this variable "Price"). As indicated by the data that were collected, the prices vary from restaurant to restaurant. In the example concerning damaged radios, we are interested in whether or not each radio is damaged. The variable of interest is labeled "Condition of radio," which varies from one radio to the next—each radio may be *damaged* or *not damaged*. We now formalize the definition of a variable.

DEFINITION 1.3.2: VARIABLE

A **variable** is information of interest about each individual item in a population. Variables are measured using numerical values (for example, radon concentration) or categories (for example, damaged or not damaged radios).

It is sometimes helpful to think of the population as being displayed in a table that records the value of the variable for each item. To demonstrate, consider two examples discussed earlier in this chapter.

1. In the radon example the population of items is the set of 51 buildings. The variable of interest is "Radon concentration" in a building. In Table 1.2.1 the column labeled "Building number" identifies the population item. The second heading, "Radon concentration (in pc/l)," is the variable of interest. The numbers in this column are the values of this variable for the items (buildings) in the population. For example, the number 13.6 is the value of this variable for item (building) 1, whereas the number 2.0 is the value of this variable for item 23.

2. In the warehouse example the population of items is the set of 100,000 radios, and the variable of interest is labeled "Condition of radio." One could make a table that represents the 100,000 items that make up the population of radios. Each radio in the population would be identified with a number, and associated with each radio would be the value (*damaged* or *not damaged*) of the variable. We don't know the value of the variable for every item (radio) in the population, but all values exist. We just don't know what they are. Table 1.2.3 reports the data for the sample of 50 radios selected from the population. The column labeled "Radio number" reports the sample item number. The column heading, "Condition of radio," is the variable of interest. Unlike the previous example, the values of the variable are not numbers. Rather, they are one of two categories—*damaged* or *not damaged*. The value of the variable for the first sample item is *not damaged*, whereas the value of the variable for the 29th sample item is *damaged*.

Population Parameter

In answering questions about populations, we generally need to know the value of a numerical quantity, called a **population parameter**, which describes a characteristic of a variable in a population.

1. In Question 1.2.1 the parameter of interest is the minimum value of the variable "Price" for 10 medium-sized pepperoni pizzas.

2. One parameter of interest in Question 1.2.3 is the percentage of the values of the variable "Radon concentration" that exceed 16.0 pc/l.

3. In Question 1.2.4 the parameter of interest is the proportion of radios with the value *damaged* for the variable "Condition of radio."

DEFINITION 1.3.3: POPULATION PARAMETER

A **population parameter** is a numerical quantity that describes a characteristic of a population. The exact value of a parameter can be obtained only if the values of a variable are known for every item in a population.

In Example 1.2.1 the population consists of three restaurants, and the variable of interest is the "Price" of 10 medium-sized pizzas at each restaurant. The parameter of interest is the minimum value of the variable "Price." In Example 1.2.1 we computed the value of the population parameter (minimum price) to be $75.00. In Example 1.2.4 the population parameter of interest is the proportion of damaged radios in the warehouse. Unlike Example 1.2.1, we cannot compute the exact value of the parameter because not all of the 100,000 radios were examined. The available data consist of values of the variable "Condition of radio" for the sample of 50 radios. Although we cannot compute the exact value of the population proportion, it does exist. It could be determined if we examined every one of the 100,000 radios in the population.

In Definition 1.2.1 we defined a **proportion**. A population proportion is a parameter that is used extensively in applied problems. Another population parameter used extensively in applied problems is a **mean**. The definition of a mean is presented in Definition 1.3.4. Throughout statistics, the word **mean** is used interchangeably with the word **average**.

DEFINITION 1.3.4: MEAN (AVERAGE)

The **mean (average)** of a set of numbers is defined by the fraction

$$\text{mean (average)} = \frac{\text{sum of all numbers in the set}}{\text{number of items in the set}}.$$

To illustrate the definition of a mean, suppose a student in a statistics class received the following grades on five one-hour tests: 75, 82, 65, 83, 70. The mean of these test grades is

$$\text{mean (average)} = \frac{75 + 82 + 65 + 83 + 70}{5} = \frac{375}{5} = 75.$$

In statistics, as in mathematics, symbols are often used as abbreviations for quantities of interest. In particular, it is common practice to use Greek letters as symbols to represent population parameters. We will follow this convention.

Convention. We will always represent a population mean with the Greek letter μ (mu). The reason we use the symbol μ for a population mean is that mu is the Greek equivalent of the lowercase English letter m, which is the first letter of the word "mean." We will always represent a population proportion with the Greek letter π (pi). The reason we use the symbol π for a population proportion is that pi is the Greek equivalent of the lowercase English letter p, which is the first letter of the word "proportion."

Samples

If every item in a population is examined and if all values of a variable are measured without error for each item, exact values of population parameters can be obtained. However, practical considerations such as cost and time often force us to examine only a subset, called a **sample**, of items from the population.

DEFINITION 1.3.5: SAMPLE

A subset of items that is selected from a population is called a **sample**.

In Example 1.2.4 it was impractical to examine all 100,000 radios, and so a subset of 50 radios was examined. These 50 radios form a sample from the population of 100,000 radios. In Example 1.2.5 it was impractical to examine all American households with a television set, and so a sample of 1,200 of these households was examined. These 1,200 households form a sample from the population of all American households with a television set. When only a subset of items (a sample) is used, we cannot determine the exact value of a parameter. In this situation we use an estimate and there is uncertainty in the result. This uncertainty can be quantified if the sample is selected using valid statistical sampling procedures. One of these sampling procedures will be discussed in Section 2.5. Additional sampling procedures are presented in Chapter 14.

Five Examples

We now present five examples that illustrate the definitions of population, variable, parameter, and sample.

EXAMPLE 1.3.1

BOTTLES OF ASPIRIN. The quality assurance manager of a company that manufactures medical products has noticed that some of the bottles of aspirin the company has distributed contained broken pills. The manager decides to examine last week's output of 120,000 bottles to see what proportion contains at least one broken pill. In this situation the population is the set of 120,000 bottles of aspirin that were produced last week. The variable of interest in this problem is labeled "Number of broken pills." There are 100 tablets in each bottle, so a numerical measurement from 0 to 100 (inclusive) is used. In your mind's eye you might think of the 120,000 bottles and the values of the variable as they are displayed in Table 1.3.1.

TABLE 1.3.1 Population of Bottles (Items) and Values of the Variable "Number of Broken Pills"

Population items	Variable
Bottle number	Number of broken pills
1	0
2	0
3	0
4	0
5	2
6	0
7	0
...	...
...	...
120,000	1

Of course, unless one observes the entire population of 120,000 bottles, not all of these values will be known. Nevertheless, these values do exist; we just don't know what they are! The parameter of interest is π, the proportion of the bottles that contains at least one broken pill. This corresponds to the proportion of bottles with values of the variable that are greater than 0. Instead of examining all 120,000 bottles, it is decided to observe a sample of 100 bottles. The results of this sample will be used to estimate the proportion of the 120,000 bottles in the population that have at least one broken pill.

EXAMPLE 1.3.2

IMMUNIZATION OF CHILDREN. A city health department wants to determine the proportion of children entering the second grade next year who have *not* received childhood immunization shots. The population of items is the set of all children in the city who are entering the second grade next year. You can envision this population as displayed in Table 1.3.2. The variable of interest is labeled "Immunization status," and the measurement consists of two categories—*immunized* or *not immunized*.

TABLE 1.3.2 Population of Children (Items) and Values of the Variable "Immunization Status"

Population items	Variable
Child number	**Immunization status**
1	immunized
2	immunized
3	not immunized
4	immunized
5	not immunized
6	not immunized
7	immunized
...	...
...	...

The parameter of interest is the proportion of children who have not been immunized, and this corresponds to the proportion of values of the variable that are categorized as *not immunized*. The parameter is represented by the Greek letter π. A sample of 375 children will be selected from the population of children in the city who will enter the second grade next year; the value of the variable will be recorded for each of the 375 children. The proportion of them who have not received the immunization shots will be determined. This is the proportion of the values of the variable in the sample that are categorized as *not immunized*. The data from this sample will be used to estimate π, the proportion of children in the population who have not been immunized.

EXAMPLE 1.3.3

ADVERTISEMENTS. The owner of a clothing store wants to determine the proportion of preferred customers who read last week's advertisement in the local newspaper. Preferred customers are customers who have the store's credit card. The collection of preferred customers is the population of items under study. The variable of interest is whether or not a preferred customer read the ad. For convenience, this variable is given the label "Read the ad." It can take one of two values—*yes* or *no*. The parameter of interest is π, the proportion of preferred customers who read the ad. This corresponds to the proportion of *yes* values in the population. Rather than contacting all of these customers, the owner selects a sample of 35 names from the list of preferred customers and telephones each one to ask if he or she read the advertisement. This set of 35 preferred customers is the sample. A partial listing of the values of the variable for the 35 customers in the sample is displayed in Table 1.3.3.

TABLE 1.3.3 Sample of Preferred Customers (Items) and Values of the Variable "Read the Ad"

Sample items	Variable
Customer number	**Read the ad**
1	yes
2	yes
3	yes
4	no
5	yes
...	...
...	...
35	no

This sample will be used to estimate π, the proportion of preferred customers in the population who read the advertisement.

EXAMPLE 1.3.4

TOMATO PLANTS. A farmer planted 3,000 tomato plants in a field. When the plants are ready for harvest, a local supermarket will purchase all of the tomatoes. To determine how much to pay the farmer, the supermarket must have an idea of how many tomatoes are in the field. This can be determined if μ, the average number of tomatoes per plant, is known (total number of tomatoes in the field = $3000 \times \mu$). The population in this problem is the set of 3,000 tomato plants. The variable of interest is number of tomatoes on a plant, which we label "Number of tomatoes." The parameter of interest is the average number of tomatoes per plant in the population. The purchasing agent for the supermarket does not want to count the tomatoes on all 3,000 plants, so she selects a sample of 100 plants and counts the number of tomatoes on each plant in the sample. This sample will be used to estimate μ.

EXAMPLE 1.3.5

CREDIT CARD USERS. The manager of a large auto repair shop wants to determine the proportion of payments that were made with a credit card during the past 12 months. The population of items is the set of all payment transactions for car repairs during the past 12 months. The variable of interest is labeled "Used a credit card," and the measurement consists of two categories—*yes* and *no*. The parameter of interest is the proportion of transactions in the population where a credit card was used (the proportion of *yes* values of the variable). This parameter is represented by the Greek letter π. A computer database program was used to print out the method of payment for each of last year's transactions; from this printout, the proportion of transactions that was paid by credit card is obtained. In this case every item in the population is observed, and so it is unnecessary to select a sample from the population.

It should be remembered that the exact value of a parameter can be determined only if the value of the variable is known for every item in a population. Thus, a sample from a population will not provide the exact value of a parameter. However, in most problems the *exact value of a parameter is not required in order to make useful decisions*. If proper statistical procedures are employed, samples can provide useful information about a parameter. This is the case in Examples 1.3.1 through 1.3.4. In Example 1.3.5 the entire population is observed, and hence the exact value of the parameter of interest (π) can be obtained.

Statistical Inference

Using samples to make decisions about populations is referred to as **statistical inference**. Statistical inference is at the very heart of the subject of statistics. A dictionary definition of inference is "the act of making a rational decision based on evidence." We define statistical inference as "the act of making a rational decision about a population based on a sample." The formal definition of statistical inference is given below.

DEFINITION 1.3.6: STATISTICAL INFERENCE

Statistical inference is the process of making decisions about a population of items based on information contained in a sample from that population.

In most practical investigations, when it is required to determine the value of a population parameter (say π or μ), it is impractical to examine every item in the population. The procedure in these cases is to collect a sample from the population, compute an estimate of the parameter, and use this estimate in place of the parameter in making decisions. Consequently, these decisions involve uncertainties, and statistical methods can be used to help understand and interpret these uncertainties. These concepts will be discussed in detail throughout this book. We begin our discussion of statistical inference with a brief explanation of one of the main topics, **estimation**.

Estimation

Estimation is the process of using sample data to calculate a number that will be used for the value of an unknown population parameter. This number is called a **point estimate** of the population parameter. For simplicity we sometimes use the word **estimate** instead of point estimate. We now present a formal definition of a point estimate of a population parameter.

DEFINITION 1.3.7: POINT ESTIMATE OF A POPULATION PARAMETER

Suppose that the objective of an investigation is to obtain the value of a population parameter. If it is impractical to measure every item in a population, the procedure is to select a sample from the population and use the sample values to compute a **point estimate** of the parameter. The symbol $\widehat{\ }$ (called hat), placed over the symbol that represents the parameter of interest, is used to indicate an estimate of the parameter. For example, if the parameter is a population mean μ, the point estimate of μ is denoted by $\widehat{\mu}$ and called "mu hat." If the parameter is a population proportion π, the point estimate of π is denoted by $\widehat{\pi}$ and called "pi hat." A point estimate of a parameter will be used in place of the parameter when making decisions.

When an estimate of an unknown population parameter is used for the value of that parameter, we do not expect the estimate to be exactly equal to the parameter. We hope, however, that the estimate will be reasonably close to the parameter. In later chapters we will describe a statistical procedure, called the **confidence interval** procedure, that can be used to help determine "how close" an estimate is to the value of the population parameter being estimated. We begin this discussion in Chapter 3. We now present some extensions of earlier examples to demonstrate the concept of statistical inference. In particular, we show how to compute estimates of the population parameters π and μ.

EXAMPLE 1.3.6

ESTIMATE OF THE PROPORTION OF DAMAGED RADIOS IN A WAREHOUSE. In Example 1.2.4 the freight broker must determine π, the proportion of damaged radios in the warehouse. Since it is not practical to examine all 100,000 radios, the exact value of π cannot be determined. A sample of 50 radios was selected, and 4 of them were damaged. Thus,

$$\widehat{\pi} = \frac{4}{50} = 0.08 = \text{the proportion of damaged radios in the sample}$$

is an *estimate* of π, where

$$\pi = \text{the proportion of damaged radios in the population.}$$

EXAMPLE 1.3.7

ESTIMATE OF THE PROPORTION OF BOTTLES OF ASPIRIN THAT CONTAIN BROKEN PILLS. In Example 1.3.1 the quality assurance manager of a company that manufactures medical products would like to know the value of the parameter π, the proportion of last week's output of 120,000 bottles of aspirin that contain broken pills. Instead of examining all 120,000 bottles, it is decided to observe an appropriately selected sample of 100 bottles. From the results in this sample an *estimate* of π will be obtained. Four bottles in the sample are observed to contain at least one broken pill. Thus,

$$\widehat{\pi} = \frac{4}{100} = 0.04 = \text{the proportion of bottles in the sample with broken pills}$$

is an *estimate* of π, where

$$\pi = \text{the proportion of bottles in the population with broken pills.}$$

The value of $\widehat{\pi}$ is displayed graphically in the pie chart in Figure 1.3.1.

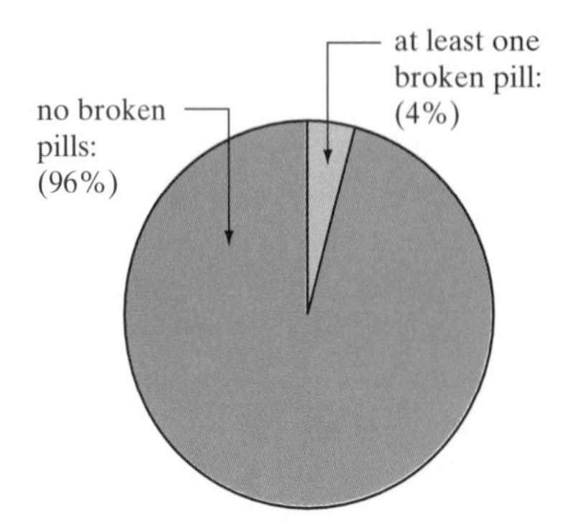

FIGURE 1.3.1
Proportion of Sampled Bottles with Broken Pills

EXAMPLE 1.3.8

ESTIMATE OF THE PROPORTION OF CHILDREN WHO HAVE NOT BEEN IMMUNIZED. In Example 1.3.2 a city health department wants to determine the value of π, the parameter that represents the proportion of children entering second grade next year who have not received the childhood immunization shots. From a sample of 375 children it was determined that 75 of them have not been immunized. Thus, $\widehat{\pi} = 75/375 = 0.20$, and

$$\widehat{\pi} = 0.20 = \text{the proportion of children in the sample who have not been immunized}$$

is an *estimate* of π, where

$$\pi = \text{the proportion of children in the population who have not been immunized.}$$

EXAMPLE 1.3.9

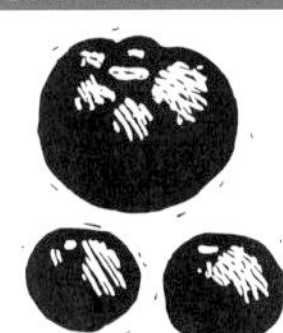

TOMATO PLANTS. In Example 1.3.4 the purchasing agent of a supermarket wants to know the value of μ, the average number of tomatoes per plant in a field of 3,000 tomato plants. The agent does not want to count the tomatoes on all 3,000 plants, so she selects a sample of 100 plants and computes $\widehat{\mu}$, the average number of tomatoes per plant for the 100 plants in the sample. She will use $\widehat{\mu}$, the average number of tomatoes per plant in the sample, to *estimate* μ, the average number of tomatoes per plant in the population. The number of tomatoes on the 100 plants is 1,403. So $\widehat{\mu} = 1403/100 = 14.03$. Thus,

$$\widehat{\mu} = 14.03 = \text{the average number of tomatoes per plant in the sample}$$

is an *estimate* of μ, where

$$\mu = \text{the average number of tomatoes per plant in the population (entire field).}$$

EXAMPLE 1.3.10

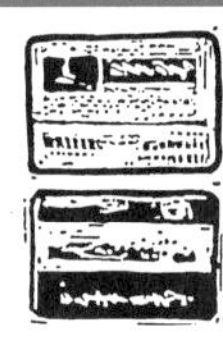

CREDIT CARD USERS. In Example 1.3.5 the manager of a large auto repair shop wants to determine π, the *parameter* that represents the proportion of payments that were made with a credit card last year (the past 12 months). The population of items is the set of all payment transactions for car repairs last year. There were 9,150 such transactions. A computer database program was used to print out the method of payment for each of last year's transactions, and from this printout, the number of transactions that were paid by credit card was observed to be 5,673. Thus $\pi = 5673/9150 = 0.62$. So

$$\pi = 0.62 = \text{the proportion of transactions in the population paid by credit card last year.}$$

In this example the entire population is observed, so there is no need to obtain a sample to estimate the parameter π. The exact value of π can be computed from the information in the database.

You should remember that there is uncertainty when using an estimate as the value of a population parameter since we do not expect the estimate to be exactly equal to the parameter. In Examples 1.3.6 through 1.3.9 there is uncertainty associated with using the point estimate as the value of the parameter. However, in Example 1.3.10 the exact value of the parameter is known since the entire population is observed. No estimate is used, and no uncertainty is involved.

If the objective of an investigation is to use sample values to compute a point estimate of a population parameter, the steps listed in Box 1.3.1 must be followed.

BOX 1.3.1: STEPS TO BE FOLLOWED FOR POINT ESTIMATION

Step 1. Define the population and identify the variable of interest.

Step 2. Define the population parameter of interest.

Step 3. Collect a sample from the population using a statistically valid sampling procedure.

Step 4. Compute a point estimate of the parameter using sample values.

In succeeding chapters we will elaborate upon these steps since they are at the heart of statistical inference. We illustrate the four steps in Box 1.3.1 with two additional examples.

EXAMPLE 1.3.11

CARPOOL. An electronics company has 3,200 employees at one of its plants. The manager wants to determine if an incentive would encourage employees to carpool if they live more than 10 miles from the plant. To help determine the impact of an incentive, the manager wants an answer to the following question:

Question 1.3.1: What proportion of the employees lives more than 10 miles from the plant?

SOLUTION The population of items is the set of 3,200 employees, and the variable of interest is "Distance from the plant" (in miles). The parameter of interest is π, the proportion of the 3,200 employees that lives more than 10 miles from the plant. Since company records do not contain the distance each employee lives from the plant, it will be necessary to obtain this information from the employees. The manager must decide whether to use the time and money necessary to ask all 3,200 employees or whether to use only a sample of employees. If a sample is used, the manager does not expect to be able to determine the exact value of π. If it is imperative for the manager to obtain the exact value of π, it would be necessary to survey all 3,200 employees. The manager decides that the expense of obtaining an exact answer is not warranted, so a sample of 100 employees is selected using a valid sampling procedure to be discussed in Chapter 2. It was observed that 82 of the employees in the sample live more than 10 miles from the plant. Thus, $\hat{\pi} = 82/100 = 0.82$, and

$\hat{\pi} = 0.82 =$ the proportion of employees in the sample that lives more than 10 miles from the plant

is an *estimate* of π, where

$\pi =$ the proportion of employees in the population that lives more than 10 miles from the plant.

EXAMPLE 1.3.12

HOUSEHOLD INCOMES. A company that owns a chain of department stores is considering opening a store in a city of about 100,000 inhabitants. The company is required to conduct a marketing research study to determine if it will be profitable to open a store in the city. The company's experience is that the stores have been successful in communities where the average annual income of households is greater than \$36,000. Thus, the company wants to know μ, the average annual income of households in the city. Since it is impractical to obtain information from all households in the city (there may be several thousand), a sample of 150 households was surveyed. The average income last year of households in the sample was \$37,500. Thus,

$$\widehat{\mu} = 37{,}500 = \text{last year's average income of households in the sample}$$

is an *estimate* of μ, where

$$\mu = \text{last year's average income of households in the population (city).}$$

In this section we have provided a brief overview of one aspect of statistical inference, point estimation. In Chapter 3 we provide a more comprehensive treatment of these concepts when the parameter of interest is π, a population proportion. Similar results are presented in Chapter 6 for μ, a population mean.

THINKING STATISTICALLY: SELF-TEST PROBLEMS

1.3.1 A medium-sized county has 10,228 automobiles registered with the automobile licensing bureau. The county commissioners want to know the proportion of these automobiles that use leaded gasoline.
a. Define the population of interest in this study.
b. Define the variable of interest in this study.
c. What is the parameter of interest?
d. It is not feasible to contact every automobile owner and ask if his or her car uses leaded gasoline, so a sample of 100 owners was selected. From this sample it was determined that 13 cars use leaded gasoline. Based on this information, what is $\widehat{\pi}$, the estimate of the proportion of cars in the county that uses leaded gasoline? Do you expect this estimate to be equal to the population parameter π?

1.3.2 In Example 1.2.2 a truck driver must determine the number of cartons of paper that was damaged as the result of a truck accident. A total of 5,000 cartons of paper was on the truck when the accident occurred.
a. Define the population of items for this study.
b. Define the variable of interest for this study.
c. Define the population parameter of interest for this study.

1.3.3 In Self-Test Problem 1.2.1 we want the answer to the question, "What proportion of the vehicles that left Interstate 25 yesterday using Exit 112 exceeded the speed limit?"
a. What is the population of items under consideration?
b. What is the variable of interest?
c. What is the parameter of interest?

1.3.4 The president of a small college asks the dean of students to determine if it would be a wise expenditure to build a movie theater on campus for the student body. Among other information, the dean wants to know μ, the average number of times students went to a movie theater during the past 30 days. She selects an appropriate sample of 20 students and asks the students how many times they went to a movie theater during the past 30 days. The results are as follows:

1, 3, 2, 1, 5, 10, 1, 2, 5, 0, 2, 4, 6, 5, 3, 0, 0, 1, 6, 8

a. What is the population of interest?
b. What is the variable of interest?
c. What is the population parameter of interest?
d. What symbol is used for the parameter?
e. What is the estimate of the population parameter?

1.4 CHAPTER SUMMARY

KEY TERMS

1. the subject of statistics
2. population of items
3. variable
4. population parameter
5. sample
6. statistical inference
7. point estimate
8. proportion
9. percentage
10. mean (average)

SYMBOLS

1. %—percentage
2. The symbol $\hat{}$ (called "hat") placed over a parameter denotes an estimate of that parameter.
3. μ—a population mean
4. $\hat{\mu}$—an estimate of μ
5. π—a population proportion
6. $\hat{\pi}$—an estimate of π

KEY CONCEPTS

1. The value of a parameter cannot be determined exactly unless every item in the population is examined.
2. A point estimate of an unknown parameter is obtained from a sample and used in place of that parameter.
3. A point estimate of a parameter will generally not be exactly equal to the parameter.
4. There is uncertainty whenever a point estimate is used in place of a population parameter.

SKILLS

1. Compute a proportion for a collection of items.
2. Convert a proportion to a percentage.
3. Convert a percentage to a proportion.
4. Compute an average of numerical values of a variable for a collection of items.
5. Estimate π using sample data.
6. Estimate μ using sample data.

1.5 CHAPTER EXERCISES

1.1 A book publisher is marketing a new statistics book and must determine a selling price. What kinds of data would the publisher use to determine an appropriate selling price for the book?

1.2 A scientific article in the *Journal of Human Behavior* reports that domestic violence is on the rise in the United States. What kinds of data do you think the researchers used to arrive at such a conclusion?

1.3 A wife and husband want to determine the amount of money they will need for groceries for their family next month. What are some data they can use to determine an appropriate amount?

1.4 The department head of a statistics department must tell the university bookstore how many books will be needed for a course next semester. What are some data the department head can use to determine this number?

1.5 I am going to drive my car 300 miles tomorrow. What are some data I can use to determine how much money I will need for gasoline?

1.6 Call three grocery stores in your community and determine which of the three stores sells a gallon of 2% milk for the lowest price today.

a. Is your answer exact?

b. Will your answer be the same if you call the same three grocery stores next week?

1.7 A computer database owned by a used-car agency includes the following information on the 2,536 cars sold during the past five years: (1) the date each car was sold; (2) the price of each car; (3) the gender of the person who bought the car; (4) whether the person paid cash or financed the purchase.

a. From the database it is determined that 304 women purchased a car from the agency during the past five years. What proportion of people who purchased a car from the agency during the past five years were women?

b. Is there any uncertainty in the answer in (a)?

c. Using the information in the database, the agency determines that 46% of the cars purchased during the past five years cost more than $9,000. Is there any uncertainty in this percentage?

d. From the database the agency determines that 1,572 of the people who purchased a car during the past five years paid cash. What proportion of people paid cash for their cars?

e. Is there any uncertainty in the answer to (d)?

f. The agency wants to know what proportion of women who bought cars during the past five years plans to purchase a car next year. Can this question be answered using the information available in the database?

g. To determine how many of the 304 women who purchased a car during the past five years plan to purchase one next year, the agency contacted 40 of the 304 women and asked them if they plan to purchase a car next year. Ten (25% of the 40) said that they planned to purchase a car next year. The sales manager stated, "Of the 304 women who purchased a car from this agency during the past five years, 25% of them plan to purchase a car next year." Is there uncertainty in this statement?

1.8 Consider the question, "What proportion of students enrolled in your university is female?"

a. What is the population of items under consideration?

b. What is the variable of interest?

c. What is the parameter of interest?

1.9 Consider Example 1.2.4, where a freight broker wants to determine the proportion of damaged radios in the warehouse.

a. What is the population of items?

b. Describe the sample.

c. What is the variable of interest?

d. What is the parameter of interest?

1.10 Why is the value of a population parameter generally unknown?

1.11 Is there ever a situation when the value of a population parameter is known? Describe such a situation.

1.12 In Example 1.3.5 the manager of the repair shop must make a decision that depends on knowing the value of the population parameter π, where π is the proportion of payments that were made by credit card during the past 12 months. Does the owner need to use statistical inference to determine the exact value of this parameter?

1.13 A state legislator made a campaign promise to reduce taxes in the state. However, it will not be possible to cut all state taxes, so she will introduce legislation to reduce either the state income tax or the state property tax. Before introducing the legislation, she would like to know the following:

1. the proportion of registered voters who prefer reducing the state income tax;

2. the proportion of registered voters who prefer reducing the state property tax;

3. the proportion of registered voters who have no preference.

She has her staff conduct a survey of voters to determine their preferences. Let π represent the proportion of registered voters whose preference is to lower the state income tax.

a. Define the population.

b. Define the variable of interest.

c. Using an appropriate sampling method, the legislator selected a sample of 489 registered voters. Of this group, 221 favored reducing the state income tax, 112 favored reducing the state property tax, and 156 had no preference. Use this information to compute an estimate of the parameter that represents the proportion of registered voters who prefer reduction of the state income tax. Is there any uncertainty in this estimate?

1.14 The following statement appeared in a newspaper: Thirty million people in the United States are left-handed.

a. Do you think this means that there were exactly 30 million left-handed people in the United States at the time the paper was published?

b. Do you think there is any uncertainty in the number 30 million?

1.15 A government agency wants to know the proportion of households in the United States that have more than one television set.

a. Define the population.

b. Define the parameter of interest.

c. An appropriately selected sample of 1,000 households was obtained. For this sample of 1,000 households it was determined that 680 have more than one television. Based on this information, what is the estimate of the population parameter π that represents the proportion of households with more than one television set? Do you expect this estimate to be exactly equal to π?

1.16 A statistics instructor has the reputation of giving exams that are considered to be unreasonably difficult. To investigate this matter, the department head wants to know the students' scores on a recent exam for a class of 48 students taught by the instructor. She also wants to know the average score for this class so she can compare it with past test results.

a. What is the population of items?

b. What is the variable of interest?

c. What is the population parameter?

d. What symbol is used to represent the population parameter?

e. It is determined that the average score for the 48 students is 88. Is this number the actual value of the parameter, or is it an estimate?

1.17 In order to qualify for a bowling tournament, a bowler must have an average of at least 200 for the most recent 30 games bowled in a sanctioned league. The tournament committee is reviewing the record of one bowler who has submitted an application to play in the tournament. The committee is interested in the bowler's average score over the most recent 30 league-sanctioned games.

a. What is the population of interest to the tournament committee when evaluating the bowler's eligibility?

b. What is the variable of interest?

c. What is the population parameter of interest?

d. What symbol is used to represent the population parameter?

1.6 SOLUTIONS FOR SELF-TEST PROBLEMS IN CHAPTER 1

1.1.2 The three areas are presented in Box 1.1.1.

1.1.3 You look to see if cars are coming and estimate their speed. If the crossing has a traffic light, you will wait for the WALK signal to appear and cross at that time.

1.1.4 You use the color of the traffic light. If it is green, you proceed. If it is yellow, you prepare to stop. If it is red, you stop.

1.1.6 Yes, I think cigarette smoking causes cancer and other diseases. This is based on newspaper stories and news reports. Also, I am somewhat influenced by the surgeon general's warning on each package of cigarettes.

1.1.7 A record of how the world population has been growing until now may help make projections about the world population in the year 2050.

1.2.1

a. If the electronic device is accurate, there is no uncertainty in the answer since the speed of *every* vehicle was measured.

b. No. There is uncertainty in using 51% as the percentage since this value was obtained by measuring the speeds of only a portion of all the vehicles that left Interstate 25 yesterday using Exit 112.

1.2.2 No. There is uncertainty in the answer since it was obtained from only a portion of all students in the university.

1.2.3 If there is no uncertainty in measuring the percentage of senior students in the class, there is no uncertainty in computing the percentage of students in the class who are not seniors. We use the relationship

$$\text{percentage of students who are not seniors} = 100\% - \text{percentage of senior students.}$$

1.3.1

a. The population is the set of 10,228 automobiles registered in the county.

b. The variable of interest is "Type of fuel" and has three possible values—*leaded gasoline*, *unleaded gasoline*, or *diesel*.

c. The parameter of interest is π, the proportion of the 10,228 automobiles that use leaded gasoline.

d. In the sample the proportion of automobiles that use leaded gasoline is 0.13. Hence, $\widehat{\pi} = 0.13$. We don't expect this estimate to be exactly equal to the population parameter π.

1.3.2

a. The population is the set of 5,000 cartons of paper.

b. The variable is "Condition of a carton." The values of this variable are *damaged* and *not damaged*.

c. The parameter is the number of damaged cartons in the population of 5,000 cartons.

1.3.3

a. The population is the set of *all* vehicles that left Interstate 25 yesterday using Exit 112.

b. The variable of interest is the speed of a vehicle as it leaves Interstate 25 using Exit 112. We label this variable "Speed."

c. The parameter of interest is π, the proportion of vehicles that left Interstate 25 yesterday using Exit 112 that exceeded the speed limit.

1.3.4

a. The population is the set of all students presently enrolled in the college.

b. The variable is the number of times a student went to a movie theater during the past 30 days. We label this variable "Number."

c. The parameter of interest is the average number of times students went to a movie theater during the past 30 days.

d. The symbol μ is used to represent the average number of times students went to a movie theater during the past 30 days.

e. The sample mean is

$$\text{mean} = \frac{65}{20} = 3.25.$$

So the estimate of the population mean is $\widehat{\mu} = 3.25$.

CHAPTER 2

POPULATIONS, VARIABLES, PARAMETERS, AND SAMPLES

2.1 INTRODUCTION

In Chapter 1 we introduced the concept of statistical inference. As you will recall, we described the process of point estimation, where sample values are used to estimate population parameters. We recommended that the steps in Box 1.3.1 be followed. We repeat these steps here.

BOX 2.1.1: STEPS TO BE FOLLOWED FOR POINT ESTIMATION

Step 1. Define the population and identify the variable of interest.

Step 2. Define the population parameter of interest.

Step 3. Collect a sample from the population using a statistically valid sampling procedure.

Step 4. Compute a point estimate of the parameter using sample values.

These steps were discussed briefly in Chapter 1, and in the following four sections we discuss them in more detail. First we revisit some earlier examples to illustrate the steps.

EXAMPLE 2.1.1

WAREHOUSE FIRE. In Example 1.2.4 a freight broker must determine π, the proportion of the 100,000 radios damaged during a recent warehouse fire.

Step 1. The population is the set of the 100,000 radios in the warehouse. The variable of interest is "Condition of radio" which can take on one of the two values—*damaged* or *not damaged*.

Step 2. The parameter of interest is the proportion of radios in the warehouse that is damaged. We represent this proportion by π.

Step 3. It is impractical to examine all 100,000 radios, so a sample of 50 radios is obtained.

Step 4. Four of the 50 radios in the sample were determined to be *damaged*. Based on this sample result, the point estimate of π, the proportion of damaged radios, is

$$\widehat{\pi} = \frac{4}{50} = 0.08.$$

EXAMPLE 2.1.2

HOUSEHOLD INCOMES. In Example 1.3.12 a company that owns a chain of department stores is considering opening a store in a city of about 100,000 inhabitants. The company wants to determine μ, the average annual income of households in the city last year.

Step 1. The population is the set of all households in the city. The variable of interest is "Last year's income."

Step 2. The parameter of interest is μ, the average household income last year.

Step 3. It is too expensive and time consuming to examine every household in the city, so a sample of 150 households is obtained.

Step 4. Since the average income last year for the 150 households was \$37,500, the estimate of μ is

$$\widehat{\mu} = 37{,}500.$$

2.2 POPULATIONS

Although we defined populations in Chapter 1, we now discuss this term in more detail. A **population of items**, which for simplicity we often call a **population**, was defined in Definition 1.3.1 as a set of items that one wants to investigate. The number of items in a population will be denoted by the uppercase letter N. In practical investigations, N may be known or unknown, finite or infinite. A population may exist in the present, in the past, in the future, or in concept only. By stating that a population exists in the present, we mean it exists when the investigation is performed, and any item in the population is available for inclusion in the sample. In the warehouse fire in Example 2.1.1, the population is the set of all radios that currently exist in the warehouse. This population currently exists, and the number of items in the population is known to be $N = 100{,}000$. In Example 2.1.2 the population of items is the set of all households that currently exist in the city. We now present an example where a population existed in the past and an example where a population will exist in the future.

EXAMPLE 2.2.1

EMPHYSEMA. The Department of Health and Human Resources wants to determine π, the proportion of people in the United States who were smokers when they were diagnosed with emphysema (a serious lung condition) last year. A smoker is defined to be a person who smoked at least 30 cigarettes the week before diagnosis. The population is the set of people in the United States who were diagnosed with emphysema last year. Technically speaking, this is a population that existed last year and may not exist at present since some people with emphysema may have died during the year.

EXAMPLE 2.2.2

TUTORS. The head of the mathematics department in a large university needs to determine whether it would be a wise expenditure to hire tutors for the first-year mathematics classes next year. Since next year's budget must be prepared now, the department head must make an immediate decision. To prepare a budget, the department head needs to know the proportion of *next year's* beginning mathematics students who would use tutors hired by the department. The population of items is the set of students who will take beginning mathematics courses *next year*. This is a future population since it will not exist until next year. Thus, it is not possible to obtain a sample from this population now. To help make the decision, the department head decides to poll the students in the beginning mathematics classes this semester and determine the proportion of them who would use tutors. In this example we have defined two populations:

population **(1)**, the population of this semester's beginning students;
population **(2)**, the population of next year's beginning students.

At the present time, a sample can be selected from population **(1)** but not from population **(2)**. The department head decides to select a sample of 50 students from population **(1)** and use the sample values to estimate π_1, the proportion of students in population **(1)** who would use tutors. The estimate is denoted by $\widehat{\pi}_1$. However, the department head wants an estimate of π_2, the proportion of students in population **(2)** who would use tutors. Since a sample is not available from population **(2)**, he will use $\widehat{\pi}_1$ as the estimate of π_2. If data are collected by a valid procedure to be discussed in Section 2.5, then $\widehat{\pi}_1$ is the estimate of π_1 using sample values. However, if the department head uses $\widehat{\pi}_1$ as the estimate of π_2, he is making an inference based on his judgment that population **(2)** is similar to population **(1)**.

Target Population and Sampled Population

Example 2.2.2 illustrates a common problem. We want to make a decision about a population from which no samples can be obtained. In this situation we must find a population that resembles the population of interest and obtain a sample from it. We are thus led to consider two types of populations—**target population** and **sampled population**.

Target population. The population to which an investigator wants to apply *final* conclusions of a study is called the **target population**. However, if any item in this population is not available for possible selection in a sample, no statistical inference can be made about this population. In many investigations the target population is a *future* population and does not exist at the time the sample is selected. Example 2.2.2 describes such a situation. Clearly, it is not possible to obtain samples from the target population of next year's beginning students at the time when the required decision must be made. Even if a target population exists in the present, it may be impractical, or inconvenient, or too expensive to obtain a sample from it. In such cases, one must select a sample from another population, a sampled population.

Sampled population. When a sample cannot be obtained from the target population, one must find another population that resembles the target population and from which a sample can be obtained. We call this the **sampled population.** Samples can be obtained from the sampled population and used to make statistical inferences about it. The sampled population is the population that is actually sampled during an investigation. It must be possible for each item in the sampled population to appear in the sample, but the sample itself consists of only those items from the sampled population that were actually selected. If samples can be obtained from the target population, the target population and the sampled population are one and the same. When samples cannot be obtained from the target population, the sampled population should be chosen to resemble the target population as closely as possible so that conclusions (using statistical inference) about the sampled population are likely to hold for the target population.

DEFINITION 2.2.1: TARGET AND SAMPLED POPULATIONS

Target population: the population to which an investigator wants to apply final conclusions of a study.

Sampled population: the population from which samples are obtained.

In Example 2.2.2 the target population is population **(2)**, the population of next year's beginning students. The sampled population is population **(1)**, the population of this semester's beginning students. In the warehouse fire example, the sampled population is also the target population. This is because the sample of 50 radios was obtained from the 100,000 radios of interest. Valid *statistical* inference can only be made about the population from which the sample was obtained (the sampled population).

DEFINITION 2.2.2: JUDGMENT INFERENCE

If the target population is not the sampled population, any inference from the sampled population to the target population is called a **judgment inference**. In addition to the sample results, a judgment inference is based on expert opinion, personal feelings, and data sources external to the sample. No statistical measure of uncertainty is available for a judgment inference. If the sample and target populations are the same, no judgment inference is necessary since statistical inference can be made about the target population.

A schematic representation of the sample, the sampled population, and the target population is shown in Figure 2.2.1 when the sampled and target populations are different.

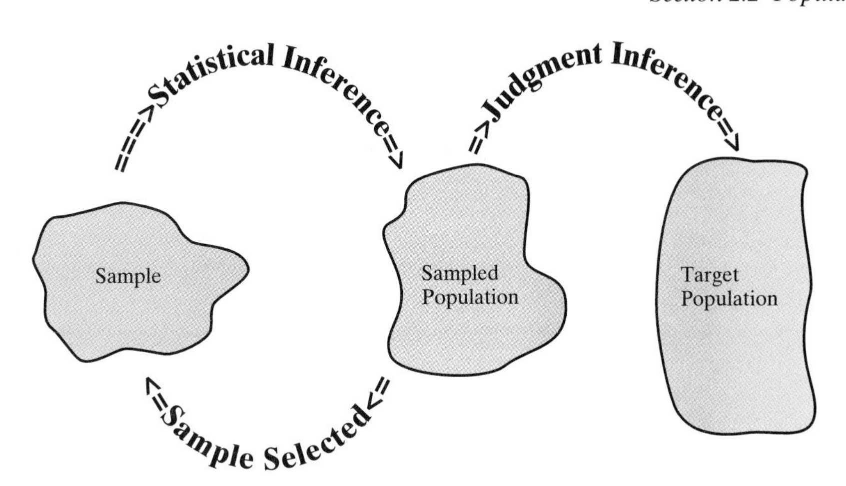

FIGURE 2.2.1
Relationships Among Sample, Sampled Population, and Target Population

---

EXAMPLE 2.2.3

OPINION SURVEY ON CASINO GAMBLING. There has been much discussion in Arizona concerning casino gambling on Indian reservations. A state congressional representative wants to determine the attitude of adults living in the Phoenix metropolitan area about this issue. The representative decides to conduct a telephone survey by having members of her staff call a sample of adults living in the metropolitan area and ask them questions about casino gambling on the reservations. For purposes of the study an adult is defined as a person of age 21 or older. A sample of 100 phone numbers will be selected from the residential section of the metropolitan phone directory. In this example there are two populations, a sampled population and a target population.

The *target population* is the set of all adults living in the Phoenix metropolitan area.

The *sampled population* is the set of all adults who are listed in the Phoenix phone directory.

The sample consists of the 100 adults whose names were selected from the phone directory. The target population and the sampled population are not the same in this problem. In particular, there are items (adults) in the target population that are not contained in the sampled population. Adults in the target population who are excluded from the sampled population include those who live in the Phoenix metropolitan area but whose names are not listed in the directory.

EXAMPLE 2.2.4

STATE'S ECONOMY. A Democratic candidate in the upcoming November gubernatorial campaign wants to determine the proportion of registered Democrats who consider the state's economy to be bad. In order to address this issue, the candidate instructed staff members to select a sample of 100 Democrats as they leave polling places during the September primary election. The voters are asked to describe the state's economy—good or bad. Voters were selected from every voting district in the state across all hours of the day.

The *target population* is the set of all registered Democrats in the state at the time of the survey.

The *sampled population* is the set of all registered Democrats who voted in the primary election.

The sample is the set of 100 Democrats who were interviewed as they left the polling places. If any registered Democrat did not vote in the election, the target and sampled populations are not the same.

Exploration

In several places throughout this book we provide the opportunity to explore concepts in more detail. These explorations are meant to ask and answer practical questions, and the answers may require calculations. The answers are shown in *italics*.

EXPLORATION 2.2.1

An investigator for a consumer group wants to inform members what the average miles per gallon (mpg) of gasoline will be for a new automobile model. The average mpg will be determined by driving automobiles of this model over a specified test route at highway speeds of 55 to 75 miles per hour. The consumer group wants to inform its members about the model's average mpg as soon as possible after it is available for public sale.

1. What is the target population? What is the variable of interest?
   The target population is the set of all cars of the new model that will be manufactured. The variable of interest is "mpg of gasoline" when a car is driven over the specified test route.
2. List five plausible values of the variable.
   Five values that seem plausible are 22, 25, 19, 26, *and* 28 *mpg.*
3. Does the target population exist in the past, the present, or the future?
   It exists in the future since it contains the set of all cars of this specific model that will be manufactured.
4. Since the target population is a future population and samples cannot be obtained from it, describe a suitable sampled population. Remember, you want to report the average mpg of all cars of this model as soon as possible after the new model is available for public sale.
   You could define the sampled population to be the first 5,000 cars of the new model that are manufactured. This sampled population would perhaps resemble the target population quite well. You could obtain a sample of 25 cars from these 5,000 cars and drive the 25 cars over the specified test route. From the average mpg of the 25 cars in the sample, you could obtain an estimate of the average mpg for the sampled population (the population of the first 5,000 cars manufactured). This represents the statistical inference from the sample to the sampled population. From this statistical inference, a judgment inference could be made about the average mpg of all cars of this model that will be manufactured. The validity of the judgment inference from the sampled population (which is the first 5,000 cars manufactured) to the target population (which is the set of all cars of this model to be manufactured) is not based entirely on sample data. Rather, its validity depends on the investigator's opinion that the sampled and target populations are similar. Figure 2.2.2 graphically illustrates the relationship between these two populations and the sample.

FIGURE 2.2.2
Relationships Among Sample, Sampled Population, and Target Population for Exploration 2.2.1

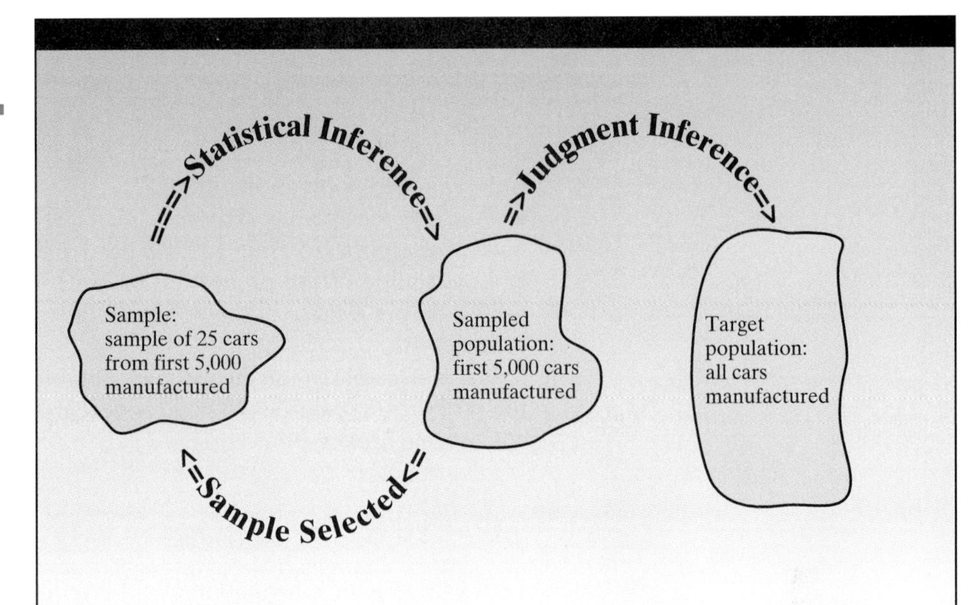

5. Can you think of any reasons why the sampled population might not resemble the target population?
The manufacturing process may change, and this may affect the average mpg of cars manufactured after the first 5,000 are produced. If this is the case, the sampled population may not resemble the target population very well.

THINKING STATISTICALLY: SELF-TEST PROBLEMS

2.2.1 In Example 1.2.4 a freight broker must determine the proportion of radios in a warehouse damaged by a recent fire.
a. What is the target population of items?
b. What is the sampled population of items?

2.2.2 Consider Example 1.2.3, where a real estate developer is trying to determine the proportion of the 51 residential dwellings that has excessive radon emissions.
a. What is the target population of items?
b. What is the sampled population of items?

2.2.3 A company that manufactures aspirin wants to determine the proportion of the 120,000 bottles of aspirin produced last week that contains at least one broken pill. To do this, a sample of 100 bottles will be selected from the 120,000 bottles.
a. What is the target population of items?
b. What is the sampled population of items?
c. Suppose that the quality assurance manager for the company wants to determine the proportion of bottles of aspirin *to be produced next month* that will contain at least one broken pill. What is the target population of items?
d. Suppose the quality assurance manager decides to select a sample of 100 bottles from *last week's* production to estimate the proportion

of *next month's* output that will contain at least one broken pill. What is the sampled population of items?
e. Suppose the quality assurance manager obtains a sample of 100 bottles of aspirin from the 120,000 bottles produced last week. What is the estimate of the proportion of bottles with broken pills in last week's output if three of the bottles in the sample contain at least one broken pill?
f. Is it reasonable to use the estimate obtained in (e) as the estimate of the proportion of bottles to be produced *next month* that will contain at least one broken pill?

2.3 VARIABLES

As stated previously, specific information about population items is of principal interest in most investigations. For instance, in Example 1.2.1 we are interested in the "Price" of pizzas at each restaurant, and in Example 1.2.3 we are interested in the "Radon concentration" of each building. This information is called a **variable** because it typically *varies* from item to item in the population. In Definition 1.3.2 we stated that variables are measured using either numerical or categorical scales. Throughout this book the names of variables will appear in quotes and the values of the variables will appear in italics.

A variable with values that are obtained by classifying an item into one of two or more categories is called a **qualitative** or **categorical variable**. For example, the values for the variable "Eye color," or "Gender," or "Watching *Late Show with David Letterman*" are all obtained by classifying items into categories. Since a person's eye color belongs to one of the categories brown, blue, hazel, gray, or green, the variable "Eye color" has five possible values—*brown, blue, hazel, gray,* and *green*. A person's gender is either male or female; so the variable "Gender" has two possible values—*male* and *female*. "Watching *Late Show with David Letterman*" is a variable with two possible values—*yes* and *no*.

DEFINITION 2.3.1: QUALITATIVE VARIABLE
A **qualitative variable** has values that are obtained by classifying items into one of two or more categories.

Quantitative variables are measured using numbers. Variables such as "Annual income," "Age," and "Height" are quantitative variables.

DEFINITION 2.3.2: QUANTITATIVE VARIABLE
A **quantitative variable** is measured using numbers.

Quantitative variables are sometimes further classified as either **discrete** or **continuous**. Variables such as "Height," "Weight," "Cholesterol concentration," and "Time" are typically classified as continuous variables. Values of continuous variables are often obtained using a measurement device such as a watch, a ruler, or a bathroom scale. Quantitative variables that are not continuous are referred to as discrete. Variables such as "Number of children in a family," "Number of patients with tuberculosis," "Number of accidents per month," and "Number of damaged radios in a warehouse" are typically classified as discrete variables. As a general rule, the values of a discrete variable are obtained by counting.

In principle, continuous variables have the property that every value between two distinct possible values is also a possible value. Suppose we are measuring the weights of people. If one person weighs 160 pounds and a second person weighs 160.5 pounds, then every value between these two numbers is a possible value. For example, it is possible for an individual to weigh 160.2479 pounds. Although any value of weight between 160 and 160.5 pounds is possible in theory, our measuring device will be accurate to only a finite number of decimal places. Also, for the sake of convenience, we may round the result to the nearest pound. However, the fact remains that if a variable is quantitative and continuous, all intermediate values between any two possible values are theoretically possible.

Discrete variables have the property that the possible values exhibit *gaps*. For instance, if one is interested in "Number of children in a family," the only possible values are the integers 0, 1, 2, 3, . . . , and so on. Values such as 1.24 and 3.5 are not possible.

To summarize, associated with each item in a population is the value of a variable; that value is either a number (for example, "Age," "Height," "Income") or a category (for example, "Condition of radio," "Gender," "Marital status," "Political affiliation," "Eye color"). Figure 2.3.1 contains a schematic representation of the types of variables discussed in this section.

FIGURE 2.3.1
Types of Variables

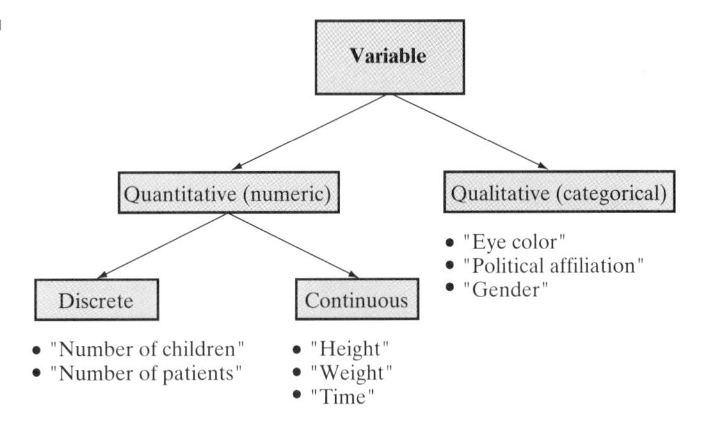

EXAMPLE 2.3.1

RADON SURVEY DATA. In Example 1.2.3 a real estate developer is interested in determining the proportion of the 51 residential dwellings with potentially harmful amounts of radon gas. The target population is the set of the 51 buildings, and the variable of interest is "Radon concentration." This is a quantitative variable that is continuous.

EXAMPLE 2.3.2

IMMUNIZATION OF CHILDREN. In Example 1.3.2 the health department wants to determine the proportion of children in the city entering the second grade next year who have *not* received childhood immunization shots. The target population is the set of all children in the city who are entering the second grade next year. The variable of interest is labeled "Immunization status." The variable can take one of two values—*immunized* or *not immunized*. This is a qualitative variable with two categories.

EXAMPLE 2.3.3

BOTTLES OF ASPIRIN. In Example 1.3.1 the manager wants to know the proportion of last week's output of 120,000 bottles of aspirin that contains at least one broken pill. In this example, the target population is the same as the sampled population—the set of 120,000 bottles of aspirin that were produced last week. The variable of interest is "Number of broken pills." This variable can take on any integer value from 0 to 100. Hence, this is a quantitative variable that is discrete.

THINKING STATISTICALLY: SELF-TEST PROBLEMS

2.3.1 A study is conducted to help determine how an individual's lifestyle might affect one's health. A variable of interest in this study is the number of cigarettes the individual smoked during the past week. This variable is labeled "Number of cigarettes."
a. Is the variable qualitative or quantitative? List some plausible values of the variable.
b. Is the variable continuous or discrete?

2.3.2 In Example 2.2.2 the head of a mathematics department must use the results of a survey of this year's beginning students to decide whether to request money to hire tutors to work with beginning students next year. The variable of interest is labeled "Use a tutor." Is this variable qualitative or quantitative?

2.3.3 A teacher wants to open a day care center in a particular neighborhood. To determine if this would be a profitable venture, she wants to know the number of children who live within a two-mile radius of the proposed site for the center. She hires a statistician to select a sample of 50 dwellings within the two-mile radius and obtain information about the number of children per dwelling. The variable of interest is labeled "Number of children" in a dwelling.
a. Is this variable qualitative or quantitative?
b. Is the variable discrete or continuous?

2.4 POPULATION PARAMETERS

When examining a population, there is generally some numerical quantity that is of primary interest to the investigator. The three numerical quantities we have discussed so far are (1) the *minimum* value of a variable, (2) the *proportion* of the values of a variable that satisfy a specific condition, and (3) the *average* of the values of a variable. As described in Definition 1.3.3, this numerical characteristic of a population is called a parameter.

In Example 1.2.1 the target population consists of three restaurants. The variable of interest is the "Price" of 10 medium-sized pepperoni pizzas, and the parameter of interest is the *minimum price*.

In Example 1.2.4 the target population is the set of 100,000 radios. The sampled and target populations are the same in this problem. The variable of interest is "Condition of radio," which takes on the value *damaged* or *not damaged*. The parameter of interest is the *proportion* of radios that are damaged.

In Example 1.3.4 the target population is the set of 3,000 tomato plants in the field. The variable of interest is "Number of tomatoes." The parameter of interest is the *average* number of tomatoes per plant.

DEFINITION 2.4.1: POPULATION PARAMETER

A population **parameter** is a numerical quantity that describes a characteristic of a population.

Defining a population parameter of interest is step 2 for point estimation described in Box 2.1.1. An investigator wants to know the value of a parameter in a target population. However, statistical inference can be made only for a parameter in a sampled population. If the sampled and target populations are not the same, no statistical inference can be made about the parameter of the target population. When we refer to a parameter, we will always mean a parameter of the *sampled* population unless we specifically state that the parameter is of a target population.

THINKING STATISTICALLY: SELF-TEST PROBLEMS

2.4.1 The owner of an orange grove consisting of 6,010 trees notices that some of the trees are diseased. It is necessary to determine the proportion of diseased trees so that a decision can be made as to whether it is necessary to spray the orchard.

a. What is the target population?

b. It is decided to select a sample of 100 trees from the orchard in order to obtain an estimate of the proportion of diseased trees. What is the sampled population?

c. What is the variable of interest?

d. Is the variable qualitative or quantitative?

e. What is the parameter of interest?

f. What symbol is used to represent the parameter?

g. Suppose that 8 of the trees in the sample of 100 are diseased. What is the value of $\widehat{\pi}$, the estimate of π?

h. Can the estimate in (g) be used in place of the parameter π in deciding whether to spray the orchard?

2.5 SAMPLES

Let's review the process of point estimation described in Box 2.1.1. In any investigation where one wants to examine a population, the investigator must define the target and sampled populations and identify the variable of interest. Whenever possible, the target and sampled populations should be the same. If they are not the same, then the sampled population should "resemble" the target population as closely as possible. The next step (step 2) is to decide what parameter is needed to answer questions of interest about the population. If every item in the sampled population is measured, the exact value of the parameter can be obtained. However, in most applied problems it is impractical to measure each item in the population, and hence the parameter cannot be known exactly. Step 3 in Box 2.1.1 is to select a sample from the sampled population using a statistically valid sampling procedure.

In this section we discuss step 3 in more detail and describe a statistically valid method for selecting a sample. In some general sense, if the sample values are *representative* of the population values, we expect the sample estimate to be a good estimate of the population parameter. Thus, it would be desirable for the sample values to be representative of the population values. But what does this mean? A sample is "representative of a population" if it possesses characteristics similar to those possessed by the population. There is no way to guarantee that a sample will necessarily be representative of a population. However, we can attempt to ensure that a sample is representative by selecting it so that no preference is given for selecting one item over any other in the population. To be somewhat facetious, suppose you want to determine π, the proportion of wage-earners in New York City whose annual salary exceeds $50,000. To estimate π, you decide to collect a sample of 100 wage-earners and ask them their annual salary. You would not use the first 100 people you meet on Wall Street as your sample! Neither would you go to a depressed area of the city and use the first 100 people you meet there as your sample! In both cases, individuals with certain incomes are shown preference over individuals with other incomes.

One procedure for obtaining a sample that shows no preference for selecting one population item over any other is called **simple random sampling**. A sample obtained using this procedure is called a **simple random sample**.

Simple Random Sampling

If a sample is obtained using the simple random sampling procedure, then these sample values can be used to make valid statistical inferences about parameters in the sampled population. In Box 2.5.1 we describe a procedure for obtaining a simple random sample from a population. Throughout we will use lowercase n to represent the size of a sample and uppercase N to represent the number of items in the sampled population.

BOX 2.5.1: A PROCEDURE FOR SELECTING A SIMPLE RANDOM SAMPLE OF SIZE n FROM A POPULATION OF SIZE N

In your mind's eye envision the following. Number the population items from 1 to N. Obtain N small plastic balls that are indistinguishable except that painted on one ball is the number 1, on another ball is painted the number 2, on another ball is painted the number 3, and so forth, until finally on the last ball the number N is painted. There is one number, from 1 to N, painted on each ball. Put the balls in a bin and shake the bin until the balls are thoroughly mixed. Then select n balls from the bin. The numbers on these n balls define the population items that belong in the sample. The values of the variable corresponding to these items are the sample values obtained by the simple random sampling procedure. We sometimes simply say that the sample was selected *at random*.

If a simple random sample of n items is selected from a population of N items by the procedure explained in Box 2.5.1, then every group of n items has the same chance of being selected. Thus, no preference is shown for selecting one group of n items over any other group of n items.

Perhaps on television you have observed how numbers are selected for the lottery in your state. Typically, in a clear plastic bin are 45 ping-pong balls with a number from 1 to 45 painted on each ball. The balls are thoroughly mixed, and then five are selected. The numbers on the selected balls are the five lottery numbers that have been selected by the simple random sampling procedure. This is the procedure described in Box 2.5.1.

The procedure for obtaining a simple random sample explained in Box 2.5.1 is impractical in most real situations since a bin and N balls may not be available. However, the procedure explained in Box 2.5.1 is important in principle since this is the idealized way to obtain a simple random sample. A more practical way to obtain the n numbers is to use a computer program, called a **random number generator**. Most statistical computer packages have a random number generator. In this book we describe the application of many statistical concepts using the package called **Minitab**. Appendix C gives a brief introduction to Minitab. Our first application of this package is to select a simple random sample of n items from a population of N items using a random number generator. If you do not have access to Minitab, you can ignore the commands and just study the output. The output is similar to the output provided in most statistical computer packages. The Minitab commands for selecting a simple random sample of size n from the integers 1 to N are as follows:

```
MTB> random n c1;
SUBC> integer 1 to N.
```

This selects a simple random sample of n numbers from the integers 1 to N and stores them in column 1. We now demonstrate the use of this command with the following example.

EXAMPLE 2.5.1

CALCULUS SCORES. The table in Appendix B reports a set of scores for 2,600 students on a standardized calculus test administered at a large state university. The score of a student is the number of correct answers on a test that contains 100 questions. The table contains three columns of numbers. Column 1 is the student identification number. These numbers range from 1 to 2,600. Column 2 contains the score of each student. Column 3 contains the number 0 or 1. The number 0 indicates that the student is not majoring in mathematics, and the number 1 indicates that the student is a mathematics major. Suppose that a professor wants to know μ, the average test score for the population of 2,600 students. Since the entire population is available, one could obtain the exact value of μ. However, for illustration we will obtain a simple random sample of size $n = 30$ from this population and estimate μ. For illustrative purposes, we consider the scores in column 2 as a population of $N = 2{,}600$ items since there are 2,600 students. The bin alluded to in Box 2.5.1 would contain 2,600 plastic balls that are indistinguishable except each has a unique number from 1 to 2,600 painted on it. We could select 30 balls using the procedure explained in Box 2.5.1, but since 2,600 balls are not readily available to us, we used the following Minitab command and subcommand and obtained the set of 30 numbers shown in Exhibit 2.5.1.

```
MTB> random 30 c1;
SUBC> integer 1 to 2600.
```

```
2446   1526    860    788   2253    706   2384   1487   1205   2385
1705   2490   1781   1604    295    415   2457    265   1888   1880
 970   1552   1591   1222   1919    651    697   1237   1104    123
```

EXHIBIT 2.5.1

These numbers are used to select the corresponding students' scores from the table in Appendix B. For instance, in this table the student numbered 2446 received a score of 67, and the student numbered 1526 received a score of 77. Table 2.5.1 contains the sample item numbers, the population item numbers, and the values of the variable "Score" for the simple random sample of $n = 30$ students.

TABLE 2.5.1 Simple Random Sample from the Population of Scores

Sample item number	Population item number	Sample value of the variable "Score"	Sample item number	Population item number	Sample value of the variable "Score"
1	2446	67	16	415	85
2	1526	77	17	2457	72
3	860	88	18	265	75
4	788	72	19	1888	56
5	2253	83	20	1880	77
6	706	62	21	970	80
7	2384	94	22	1552	95
8	1487	74	23	1591	73
9	1205	81	24	1222	58
10	2385	65	25	1919	68
11	1705	71	26	651	73
12	2490	68	27	697	42
13	1781	76	28	1237	67
14	1604	63	29	1104	69
15	295	84	30	123	70

From this sample of 30 scores, the professor can compute $\widehat{\mu}$, the estimate of μ. In particular,

$$\widehat{\mu} = \frac{67 + 77 + 88 + \cdots + 70}{30} = \frac{2185}{30} = 72.833.$$

EXAMPLE 2.5.2

STUDENTS WHO TOOK A GERMAN LANGUAGE COURSE IN HIGH SCHOOL. The head of a foreign language department in a university wants to determine π, the proportion of the 2,045 first-year students who took at least one course in German in high school. Since it is considered too costly and time consuming to consult each of the 2,045 students, the department head decides to select a simple random sample of 20 students and ask them if they had taken a high-school course in German. On the basis of this sample, the department head will estimate π. Each student in the population is assigned a unique number from 1 to 2,045 ($N = 2{,}045$). Since each of the 2,045 students can be identified and can potentially appear in the sample, the target and sampled populations are identical. The bin described in Box 2.5.1 would contain 2,045 balls that are indistinguishable except that each has a unique number from 1 to 2,045 painted on it. Instead of using balls in a bin, the 20 numbers required for the sample were selected by computer using the Minitab commands explained earlier. The following numbers were obtained:

```
1962  811  1217   538  1437  1300  1453  337   373  1478
1116  229   128  1659  2332  1583   518   94  1309  2018
```

The students corresponding to these numbers were called and asked if they took at least one course in German in high school. The results are shown in Table 2.5.2.

TABLE 2.5.2 Responses of a Simple Random Sample of 20 Students

Sample item number	Population item number	Sample value
1	1962	no
2	811	yes
3	1217	no
4	538	no
5	1437	no
6	1300	no
7	1453	no
8	337	no
9	373	no
10	1478	no
11	1116	no
12	229	no
13	128	no
14	1659	yes
15	2332	no
16	1583	no
17	518	no
18	94	no
19	1309	yes
20	2018	no

From Table 2.5.2 we observe that three of the 20 students took a course in German in high school. This provides the estimate

$$\widehat{\pi} = \frac{3}{20} = 0.150.$$

Since the data were collected from the target population using the simple random sampling procedure, $\widehat{\pi} = 0.150$ is a statistically valid estimate of π.

EXAMPLE 2.5.3

ASSEMBLY LINE OF TIRES. A company has just implemented a new process for manufacturing tires for pickup trucks. The quality control manager wants to determine if the first 30,000 tires manufactured by the new process satisfy the technical specifications established in the product design. To determine this, the manager decides to examine the first 20 of the 30,000 tires that come off the assembly line. Clearly, this is not a simple random sample of 20 tires from the population of 30,000 tires since preference is shown for selecting the *first* 20 tires made by the new process. In this situation this sample of 20 tires cannot be used to obtain a statistically valid estimate of any parameter in the population of 30,000 tires.

EXAMPLE 2.5.4

FOOTBALL GAMES. A class project requires a student to determine the proportion of university students who attended at least one of the university's football games during the past football season. A sample of 60 students is to be obtained, and from these data an estimate of the proportion who attended at least one football game is to be computed. To simplify the data collection, the student decides to visit the student recreational center and obtain information from the first 60 students encountered. This sampling procedure is not a simple random sampling procedure, because not every set of 60 students in this university has an equal chance of being selected in the sample. In particular, this procedure excludes all students in the population who do not visit the student recreational center. Clearly, the sample shows preference for selecting those students who visit the recreational center, and they may be more likely to have attended football games.

Often the population size, N, is not known. In these cases ingenious methods may be required in order to obtain a simple random sample. We illustrate with an example.

EXAMPLE 2.5.5

ADVERTISEMENT. The manager of a department store spent a large amount of money to advertise a sale to be held next Wednesday. She would like to know the effectiveness of the advertising. Of all the people who will come into the store on Wednesday, she wants to know what proportion will come because they saw the ad. If the proportion is more than 0.20, she will consider the advertising money well spent. The variable of interest is whether or not a customer came into the store because of the ad, which we abbreviate as "Saw the ad." This is a qualitative variable consisting of two categories—*yes* and *no*. The target population is the set of all customers who will come into the store on Wednesday. The manager decides to select a simple random sample of 100 customers from all customers who will come into the store on Wednesday and ask them if they came into the store because they saw the ad.

How should the 100 customers be selected? In this case, the population size N is unknown at the time the sample is being collected. This is because N represents the total number of customers on Wednesday and N won't be known until the close of business that day. By the time that N is known, the customers will be gone and no sample can be collected. One convenient way of collecting the sample is to select the first 100 people who come into the store. This procedure does not result in a simple random sample since preference is given for selecting the first 100 customers who come into the store. It may be that customers who read advertisements come to the store early before the merchandise gets picked over. If this is the case, then a sample of the first 100 customers would likely include too large a proportion of people who saw the ad.

Step 3 in Box 2.1.1 requires that a sample be obtained using a statistically valid sampling procedure. Simple random sampling is only one type of a valid sampling procedure. Some other statistically valid sampling procedures will be discussed in Chapter 14. In situations such as that discussed in Example 2.5.5, clever and ingenious methods may need to be used to obtain statistically valid samples. The advice of a professional statistician can be useful in these cases.

EXPLORATION 2.5.1

A survey was conducted among adults living in a city of about 120,000 people to determine public opinion about the death penalty. The question of interest is given below.

Question 2.5.1: What proportion of adults in the city prefers the death penalty for people convicted of murder?

A simple random sample of 60 adults living in the city was selected. Each sampled adult was asked the question, "What do you think should be the punishment for murder?" which we label "Preferred punishment." Each sampled adult was asked to select one of the following responses: (1) *death penalty*; (2) *life in prison* without parole; (3) *depends* on the circumstances; (4) *neither* death penalty, nor life in prison without parole; (5) *no opinion*. The responses of the 60 adults in the sample are presented in Table 2.5.3.

1. What are the target and sampled populations being studied?
   The target and sampled populations are the same in this study. They are the collection of all adults living in the city at the time of the survey.
2. What is the variable of interest in this study?
   The variable of interest is labeled "Preferred punishment," which is a categorical variable with five values—death penalty, life in prison, depends, neither, and no opinion.
3. Was the entire target population surveyed?
   No. This would be impractical. A simple random sample of 60 adults was obtained.
4. What proportion of adults in the sample preferred the death penalty for murder?
   Thirty-three out of 60 respondents chose the death penalty as the preferred punishment for murder. Thus, the sample proportion is 33/60 = 0.55.
5. What proportion of adults in the population prefers the death penalty for murder?
   This population proportion is unknown since not all the adults in the population were contacted.

TABLE 2.5.3 Responses to Survey About the Death Penalty

Subject number	Response	Subject number	Response
1	neither	31	death penalty
2	death penalty	32	life in prison
3	death penalty	33	life in prison
4	neither	34	no opinion
5	life in prison	35	death penalty
6	life in prison	36	death penalty
7	life in prison	37	death penalty
8	death penalty	38	death penalty
9	life in prison	39	depends
10	life in prison	40	life in prison
11	death penalty	41	no opinion
12	death penalty	42	death penalty
13	death penalty	43	death penalty
14	life in prison	44	life in prison
15	death penalty	45	life in prison
16	death penalty	46	death penalty
17	death penalty	47	death penalty
18	death penalty	48	death penalty
19	life in prison	49	death penalty
20	death penalty	50	life in prison
21	depends	51	no opinion
22	death penalty	52	death penalty
23	neither	53	life in prison
24	depends	54	death penalty
25	death penalty	55	depends
26	death penalty	56	death penalty
27	death penalty	57	death penalty
28	death penalty	58	life in prison
29	death penalty	59	life in prison
30	death penalty	60	life in prison

6. Let π represent the proportion of adults in the city who prefers the death penalty for murder. What is an estimate of π?
Since a simple random sample was used to obtain the data, the estimate of π is given by the sample proportion, $\hat{\pi} = 0.55$.

7. Suppose that 72,173 adults were living in the city at the time of the survey. How many of these adults prefer the death penalty for murder?
The exact number cannot be known without contacting every one of the 72,173 adults. However, in (6) we estimated that the proportion of adults who favors the death penalty for murder is 0.55. Hence, an estimate of the total number of adults who favor the death penalty is given by

$$\widehat{\text{total}} = 72{,}173 \times 0.55 = 39{,}695.$$

8. Based on the results in (6), should we conclude that more than 50% of the adults in the city favor the death penalty for murder?
No, not unless we know how close $\widehat{\pi} = 0.55$ is to π. This is discussed in Chapter 3.

THINKING STATISTICALLY: SELF-TEST PROBLEMS

2.5.1 Recall Example 2.5.5 in which a store manager wants to determine π, the proportion of people who will come into the store next Wednesday because they saw an advertisement.
a. Suppose the store will be open 10 hours on Wednesday. Consider a sampling procedure in which the first 10 people who come into the store at the beginning of each hour are selected. Does this method of obtaining 100 sample values result in a simple random sample?
b. Suppose the manager changes the problem slightly and decides to investigate the first 2,000 customers who will come into the store on Wednesday. She wants to determine π, the proportion who comes into the store because they saw the ad. The sampled population now is the first 2,000 customers who will enter the store on Wednesday, and the size of the population is $N = 2{,}000$. The manager uses a computer to obtain a simple random sample of 100 numbers from the set 1 to 2,000. She arranges the 100 numbers in order from smallest to largest. She will count the customers as they come into the store; if they match a number on her list, they will be selected. If at least 2,000 customers come into the store on Wednesday, does this method give a simple random sample of 100 customers from the population of the first 2,000 who enter the store?

2.5.2 Consider Self-Test Problem 2.4.1, where the owner of an orange grove wants to determine π, the proportion of the $N = 6{,}010$ trees that is diseased. He decides to select a simple random sample of $n = 50$ trees from the orchard, count the number of diseased trees, and compute $\widehat{\pi}$, the proportion of diseased trees in the sample. The sample proportion $\widehat{\pi}$ is used to estimate π. Write a paragraph explaining a procedure for obtaining a simple random sample of 50 trees.

2.5.3 A manufacturer of aspirin wants to determine the proportion of next week's production of bottles of aspirin that have at least one broken pill. The sample will be collected from next week's production of 120,000 bottles of aspirin. What are the target and sampled populations of items? Explain how to obtain a simple random sample of 250 bottles of aspirin from the sampled population.

2.5.4 A newspaper wants to know the opinion of its subscribers about removing one of the comic strips from the Sunday paper. The newspaper selected a simple random sample of 1,000 from the 42,951 subscribers and sent them a letter that included a stamped postcard addressed to the newspaper. On the postcard was the question, "Would you object if the comic strip is removed from the Sunday edition?" They were asked to answer *yes* or *no* and return the postcard within one week. Of the 1,000 subscribers in the sample, 624 returned the postcard, and 382 of them answered *no*. Can the 624 subscribers who returned the postcard be considered a simple random sample of 624 from the 42,951 subscribers?

2.6 CHAPTER SUMMARY

KEY TERMS

1. target population
2. sampled population
3. statistical inference
4. judgment inference
5. quantitative variable
6. qualitative variable
7. discrete variable
8. continuous variable
9. population parameter
10. simple random sample

SYMBOLS

1. μ—a population average
2. $\hat{\mu}$—a sample average used to estimate μ
3. π—a population proportion
4. $\hat{\pi}$—a sample proportion used to estimate π

KEY CONCEPTS

1. Statistical inference is used to draw conclusions about a sampled population based on a sample.
2. Judgment inference is used to draw conclusions about a target population based on knowledge of its relationship to a sampled population.
3. A judgment inference is not needed if the target population and the sampled population are identical.
4. Statistical inferences are valid only if the sample has been selected using a statistically valid sampling procedure.
5. Simple random sampling is a statistically valid sampling procedure.

SKILLS

1. Follow the steps in Box 2.1.1 to produce a point estimate of a population parameter.
2. Use a statistical software package to select a simple random sample of items from a population.
3. Read a newspaper article in which results of a sample are reported, and be able to determine if a judgment inference is needed to draw conclusions about the target population.

2.7 CHAPTER EXERCISES

2.1 A city planner desires to know the opinion of all registered voters in the city concerning a proposed new sales tax on hotel rooms and rental cars. A sample of voters is to be selected from a list of senior citizens living in the city who are registered to vote.

a. What is the target population, and what is the sampled population?

b. Would you have any concerns if the city planner used the results of the survey to make conclusions about the target population? In what ways might the sampled population differ from the target population?

2.2 The owner of an electronics store wants to determine if he should sell computer products in his store. The owner has never sold these items, but he is looking for a way to expand his business and increase sales. The owner has a list of all customers (called preferred customers) who have spent at least $500 in his store during the past 12 months. In order to determine the demand for computer products, the owner selects a simple random sample of the preferred customers to determine if they would be interested in purchasing these products. The owner is interested not only in the preferred customers, but also in future customers who have not previously shopped in the store.

a. Define the target population and the sampled population.

b. Does the sampled population bear a good resemblance to the target population?

2.3 The manager of a department store wants to know what proportion of customers that shopped in her store during the past 12 months listens regularly to a radio station on which she is considering placement of a commercial. The manager does not have a record of all customers who have shopped in her store during the past 12 months, but she does have a list of all customers who own a store credit card. She realizes there are many customers who do not own the store's credit card. In fact, past records indicate that less than 15% of all customers own the store's credit card. The manager selects a simple random sample of 75 customers from the list of those who own the store's credit card in order to determine the proportion of all customers who listens regularly to the radio station.

a. What is the target population, and what is the sampled population?

b. All 75 customers in the sample were contacted by telephone and asked if they listen to the radio station. It was determined that 23% of the interviewed customers listen to the radio station on a regular basis. Is 23% a valid estimate of the proportion of customers who listens to the station?

c. What factors might the manager consider in order to decide if the judgment inference from the sampled population to the target population is useful?

2.4 The police department in a large city is instructed by the mayor to study trends in violent crime rates and to compare crime rates with other cities. The department has data in its records that report the number of violent crimes committed each day for the past 12 months. The variable of interest is labeled "Number of violent crimes" (reported each day). Is this a qualitative or quantitative variable?

2.5 Although some scientists believe that the level of arsenic in drinking water is not a hazard to human health, they remain attentive to the potential risk. The federal standard for arsenic in drinking water in 1996 was 50 parts per billion (ppb). The Environmental Protection Agency (EPA) has proposed lowering the standard to 10 ppb. A study was commissioned to determine the ppb of arsenic in the Verde River which supplies municipal water to Phoenix, Arizona. The variable of interest is labeled "Concentration of arsenic." Is this a quantitative or qualitative variable?

2.6 An electric company that serves a moderate-sized city wants to determine if the customers are satisfied with the service. A letter was sent to each customer asking the question, "Are you satisfied with our service?" They were asked to mark one of the following categories: (1) very satisfied; (2) satisfied; (3) dissatisfied; (4) very dissatisfied.

a. What is the variable of interest?

b. Is the variable qualitative or quantitative?

c. The electric company also wants to know the number of children in each customer household. It telephones a sample of 35 customers and asks them how many people under age 18 live in the household. The variable is labeled "Number of people under age 18." Is this variable qualitative or quantitative?

2.7 The transportation agency of a city conducts a survey to determine how many citizens would ride the bus to work at least three times per week if the bus fees were reduced by 30%. The question asked is, "How many times per week would you ride the bus if the fees were reduced 30%?" The variable is labeled "Rides per week."

a. Is the variable qualitative or quantitative?

b. Another question asked in the survey is, "How many miles do you drive your automobiles each week?" The variable is labeled "Miles." Is this variable qualitative or quantitative?

2.8 In a weight-loss club, each new customer is asked to state how much weight he or she wants to lose. The variable is labeled "Weight." Is this variable qualitative or quantitative?

2.9 Several businesses in Mesa, Arizona, were contacted and asked if they planned to hire more people in the next 12 months. The variable is labeled "Hire more." Is this variable qualitative or quantitative?

2.10 A real estate agency wants to evaluate the rental market for apartments in the city. A survey was conducted and each occupant was asked to report the monthly rent. The variable is labeled "Monthly rent." Is this variable qualitative or quantitative?

2.11 A state university wants to know how many students graduated from the state's high schools this spring. Each high school was asked to report the number of graduates. The variable is labeled "Number of graduates." Is this variable qualitative or quantitative?

2.12 The quality control manager of a company that sells cake mix has purchased a new machine to fill boxes with 16.5 ounces of chocolate cake mix. To calibrate the machine, the manager wants to determine π, the proportion of the boxes to be filled next month that contains less than 16.4 ounces or more than 16.6 ounces of the mix. The manager wants to know as soon as possible if the machine is working properly, so he obtains a simple random sample of 120 boxes from the first 4,000 boxes filled by the new machine.

a. What are the target and sampled populations?

b. What is the variable of interest?

c. Is the variable quantitative or qualitative?

d. What is the parameter of interest?

e. Suppose that a simple random sample of 120 of the first 4,000 boxes yielded 4 boxes that contained less than 16.4 ounces of cake mix and 2 boxes that contained more than 16.6 ounces of the mix. What is the estimate of π?

2.13 A manufacturer of ballpoint pens wants to know the proportion of pens manufactured last month that was defective.

a. What is the target population of items?

b. What is the variable of interest?

c. Is the variable qualitative or quantitative?

d. What is the parameter of interest?

2.14 A periodic emissions test is required for each car registered in the state of Colorado. The State Pollution Control Division is interested in the proportion of cars that failed the test in the past two years.

a. What is the target population?

b. What is the variable of interest?

c. Is the variable qualitative or quantitative?

d. What is the parameter of interest?

2.15 Television ratings play an important role in determining advertising rates for television programs. A company that provides ratings wants to estimate the proportion of households in the United States with television sets that were tuned to the 1995 Super Bowl football game held in Miami, Florida. To obtain this estimate, a simple random sample of 1,200 households was selected from the population of households in Miami that have television sets.

a. What is the target population, and what is the sampled population for this study?

b. Do you think it is possible to make a good judgment inference about the target population based on statistical inferences for the sampled population?

c. What is the variable of interest?

d. Is the variable quantitative or qualitative?

e. What is the parameter of interest in the sampled population?

f. The sample of 1,200 households with television sets contained 480 households that were tuned to the Super Bowl. Based on this sample information, what is the estimate of π, the proportion of households in Miami with television sets that were tuned to the Super Bowl?

2.16 Recall Example 2.5.5 in which the manager of a department store wants to determine the proportion of customers who will come into the store next Wednesday because they read a store advertisement about a sale. Suppose that on the Wednesday of the sale, a drawing for a $100 gift certificate was held. To enter the contest, a customer placed an entry form consisting of the customer's name and telephone number into a bin. At closing time, the entry forms were counted; there were 2,106, with no duplicates. These 2,106 forms were thoroughly mixed and the winning entry was selected from the bin. After the winning name was selected, it was replaced in the bin and the entry forms were again thoroughly mixed. Then 100 entry forms were taken from the bin. The customers listed on the forms were telephoned and asked if they came to the store because they read the ad. This information was used to estimate the proportion of customers who came to the store Wednesday because they had read the advertisement.

a. What is the target population, and what is the sampled population?

b. Do you think the sampled population is a good representation of the target population?

c. Is the sample of 100 customers a simple random sample of customers who entered the store?

2.17 A December 1994 article in *USA Today* cited a study by the American Animal Hospital Association that asked pet owners in the United States how often they feed their pets human food. A survey of pet owners was conducted. Each pet owner in the sample was asked, "How often do you feed human food to your pet?" The responses were recorded using the following five categories: *every day; once or twice a week; once a month; only on holidays; never.*

a. What is the target population for this survey?

b. What is the variable of interest?

c. Is the variable of interest qualitative or quantitative?

2.18 A local health club is open from 6:00 a.m. to 10:00 p.m. each weekday. An investigator was hired by the club to determine what proportion of members would use the facilities between 5:00 a.m. and 6:00 a.m. if the club were to open at 5:00 a.m. each weekday. The club has 800 members, and the investigator obtained an alphabetical listing of all 800 members. Each member was assigned a number from 1 to 800 by consecutively numbering the names on the list.

a. To obtain a sample of 100 members, the investigator performed the following operation. She placed 8 slips of paper numbered from 1 to 8 in a hat. After thoroughly mixing the slips, she selected one slip at random and observed the number 3. She then selected the sample by taking the third member on the list of 800 members and every eighth member thereafter. That is, she selected the members with numbers 3, 11, 19, ..., 795. Is this a simple random sampling procedure?

b. Consider the following alternative procedure for selecting the sample in (a). Each member's name was written on a slip of paper, and the slips were indistinguishable except for the names written on them. The investigator placed these 800 slips of paper in a hat, and mixed them thoroughly. The sample was selected by pulling 100 names out of the hat. Is this a simple random sampling procedure?

2.8 SOLUTIONS FOR SELF-TEST PROBLEMS IN CHAPTER 2

2.2.1

a. The target population of items is the set of 100,000 radios in the warehouse.

b. The sampled population of items is also the set of 100,000 radios in the warehouse, since any radio can appear in the sample. Thus, the two populations are the same in this problem.

2.2.2

a. The target population is the set of 51 residential buildings.

b. The target and sampled populations are the same in this problem. Also note that the entire target population was observed.

2.2.3

a. The target population of items consists of the 120,000 bottles of aspirin produced last week.

b. The sampled population of items is the same as the target population of items.

c. The target population of items is the set of all bottles of aspirin that will be produced next month. This is a future population.

d. The sampled population of items is the 120,000 bottles of aspirin produced last week.

e. The estimate is $3/100 = 0.03 = 3\%$.

f. There are two populations here: the target population of bottles of aspirin to be produced next month, and the sampled population of bottles of aspirin produced last week. If the manager uses the proportion of bottles with broken pills in the sample as the estimate of the proportion of bottles to be produced next month that will contain broken pills, this is a judgment inference from the sampled population to the target population. Since this judgment inference is not based on sample data from the target population, we need more information about the similarities between the two populations before we can determine if the estimate is reasonable.

2.3.1

a. The variable is quantitative. Some plausible values of the variable are 100, 32, 149, and 186.

b. Discrete. If a person smoked any portion of a cigarette, it is counted as one cigarette.

2.3.2 The variable is qualitative and can take the value *yes* or *no*.

2.3.3

a. It is quantitative.

b. It is discrete.

2.4.1

a. The target population is the set of 6,010 trees.

b. The sampled population is the set of 6,010 trees. The target and sampled populations are the same in this situation.

c. The variable of interest is the disease status of each tree and is labeled "Disease status." The values of the variable are the two categories *diseased* and *healthy*.

d. The variable is qualitative.

e. The parameter of interest is the proportion of trees in the category *diseased*.

f. The symbol used for the parameter is π.

g. The value of $\widehat{\pi}$ is $8/100 = 0.08$.

h. Yes, if the sample is collected according to statistical principles to be described in Section 2.5. $\widehat{\pi} = 0.08$ is the estimate of π in both the target and the sampled populations since the two populations are the same.

2.5.1

a. No, this method does not result in a simple random sample of 100 people, since customers who come into the store after the beginning of each hour cannot appear in the sample. This is a violation of the no-preference rule.

b. Yes.

2.5.2 The target and sampled populations are the same. Specifically, they both consist of the set of 6,010 trees. It should be easy to give each tree a number from 1 to 6,010 since the orchard undoubtedly was planted in rows. Use the procedure discussed in Box 2.5.1 or a computer software package to obtain a simple random sample of 50 numbers from the set of integers 1 to 6,010. Then go into the orchard and identify the 50 trees that correspond to the 50 selected numbers. This is a simple random sample of 50 trees from the 6,010 trees in the orchard.

2.5.3 The target and sampled populations are the 120,000 bottles of aspirin that will be produced next week. In *theory*, to obtain a simple random sample of 250 bottles you should obtain a bin that contains 120,000 balls that are indistinguishable except each one has a distinct number from 1 to 120,000 painted on it. Shake the bin until the balls are thoroughly mixed, then select 250 of them. In *practice*, a computer program that generates random numbers can be used to select a simple random sample of 250 numbers from the set 1 to 120,000. After the 250 numbers are obtained, select the bottles that correspond to these numbers as they come off the production line.

2.5.4 No! This is not a simple random sample because the sample obtained gives preference to those who would respond to the survey. The problem of nonresponse is discussed later in the book.

CHAPTER 3

CONFIDENCE INTERVALS FOR A POPULATION PROPORTION

3.1 THE UNDERLYING CONCEPTS OF CONFIDENCE INTERVALS

In Chapters 1 and 2 we discussed point estimation for the parameters π and μ. We explained that when an unknown population parameter is needed to make a decision, a point estimate can be used in place of that parameter. We do not expect the point estimate to be exactly equal to the unknown population parameter. Generally, the exact value of a parameter is not needed in order to make useful decisions.

In this chapter we use the parameter π to introduce the concept of confidence intervals. The concepts apply to any population parameter, however, and in subsequent chapters we discuss confidence intervals for many other parameters. We begin our discussion with a definition.

DEFINITION 3.1.1: CONFIDENCE INTERVAL

A **confidence interval** for a population parameter is an interval with an associated confidence that the interval contains the unknown population parameter (for example, contains π). The endpoints of a confidence interval are denoted by L and U, where the number L is the **lower endpoint** of the confidence interval and the number U is the **upper endpoint** of the confidence interval. Sample data are used to compute the values of L and U.

Formulas that can be used to compute L and U for the parameter π are presented in the next section.

EXAMPLE 3.1.1

WAREHOUSE FIRE. In Example 1.2.4 a freight broker wants to estimate π, the proportion of the 100,000 radios in a warehouse that was damaged in a recent fire. Suppose that a simple random sample of 312 radios was obtained from the warehouse and that 25 of them were damaged. Thus,

$$\widehat{\pi} = \frac{25}{312} = 0.080 \text{ is the estimate of } \pi.$$

The broker does not expect this estimate to be exactly equal to the unknown parameter π, so he computes a confidence interval for π to determine "how close" $\widehat{\pi}$ is to π. Using the sample, he computes L and U, the endpoints of a confidence interval for π, such that he has 95% confidence that the interval contains π (the meaning of "95% confidence" is discussed later in this section). Using formulas to be described in the next section, he obtains $L = 0.049$ and $U = 0.111$. Thus, the broker has 95% confidence that π is in the interval 0.049 to 0.111.

Summary. Confidence intervals are used throughout statistics and are a key component of statistical inference. They can be viewed as a supplement to point estimates for determining "how close" a point estimate is to the parameter being estimated.

Confidence intervals formalize the way we typically answer a question in our everyday conversation. For example, if you are asked how long you will be gone shopping, you would not say, "I'll be gone 2 hours, 25 minutes, and 34.157 seconds." You would perhaps say something like, "I am quite confident that I will be gone between 2 and 3 hours." When a question is asked that requires a numerical answer, the exact answer is generally not necessary. It is satisfactory to state an interval with an indication of the confidence that the exact answer is contained in that interval.

Probability

In Example 3.1.1 the broker has 95% confidence that the interval 0.049 to 0.111 contains the true value of the unknown parameter π. In order to understand the meaning of the term "95% confidence" in this statement, we must first understand the following simple probability concept. In your mind's eye imagine a bin that contains 100 small plastic balls, 5 of which are white and 95 of which are black. If a *simple random sample* of one ball is to be selected from the bin, the probability that it will be black is defined to be the proportion of black balls in the bin. In this case the proportion of black balls in the bin is 0.95. In symbolic notation we write this as

$$\text{Pr(ball to be drawn will be black)} = 0.95,$$

or we write this more simply as

$$\text{Pr(ball will be black)} = 0.95.$$

The symbol Pr stands for the word "probability."

DEFINITION 3.1.2: PROBABILITY

A probability is a number between 0 and 1 inclusive.

In a group of G items, suppose that g of them have a specified characteristic and $G - g$ do not have the characteristic. If a simple random sample of one item is to be drawn from the group of G items, then

$$\begin{aligned}&\text{Pr(item will have the characteristic)}\\&\quad = \frac{g}{G} = \text{proportion of items with the characteristic.}\end{aligned}$$

Suppose a bin contains 100 balls ($G = 100$), where 95 of them are black ($g = 95$ are black) and 5 are white ($G - g = 5$ are white). If a ball is drawn at random from the bin, then

$$\text{Pr(ball will be black)} = \frac{g}{G} = \frac{95}{100} = 0.95.$$

The following physical meaning can be attached to Definition 3.1.2. Suppose a bin contains 100 balls, where 95 of them are black and 5 of them are white. In your mind's eye select a simple random sample of one ball from the bin, write down its color (W for white and B for black), and replace the ball. Repeat this process many, many times (say one billion times) and you will have a set of letters that could be represented as

$$\text{B B B W B B B B B W} \cdots \text{B.}$$

This means the first ball selected was black, the second black, the third black, the fourth white, and so on. Theory and experience tell us that 95% of the letters in this set are B and 5% are W (that is, 95% of the balls selected are black and 5% are white). These percentages are the same as the percentage of black and white balls, respectively, that are in the bin. Thus, we know that if we select a simple random sample of one ball from the bin many times and replace it each time, the proportion of balls selected that will be black is 0.95, the same as the proportion of black balls in the bin.

If a bin contains only black and white balls, the probability that a ball to be drawn at random from the bin will be black is a number between 0 and 1 inclusive and depends on the proportion of black balls in the bin. A probability equal to 0 means the ball cannot be black (that is, all balls in the bin are white). A probability equal to 1 means the ball must be black (that is, all balls in the bin are black).

From Definition 3.1.2, we note that

$$\begin{aligned}\text{Pr(ball will } \textit{not be black}) &= \text{Pr(ball will be } \textit{white})\\&= \text{proportion of white balls in the bin}\\&= \frac{G-g}{G} = 1 - \frac{g}{G} = 1 - \text{Pr(ball will be } \textit{black}).\end{aligned}$$

So

$$\text{Pr(ball } \textit{will not be black}) = 1 - \text{Pr(ball will be } \textit{black}). \qquad \textbf{Line 3.1.1}$$

EXAMPLE 3.1.2

SOME PROBABILITY COMPUTATIONS. Suppose a bin contains 300 balls, where 30 of them are white and 270 are black. One ball is to be selected at random from the bin, and we want to know the probability that the ball will be white. Since $30/300 = 0.10$, it follows that 10% of the balls in the bin are white. Hence,

$$\text{Pr(ball will be white)} = 0.10.$$

Suppose another bin contains an unknown number of balls where some are white and the remainder are black. A ball is to be drawn at random from the bin, and we want to know the probability that the ball will be white. Suppose we know that

$$\text{Pr(ball will be black)} = 0.25.$$

By the formula in Line 3.1.1, we also know that

$$\text{Pr(ball will not be black)} = 1 - \text{Pr(ball will be black)} = 1 - 0.25 = 0.75.$$

So Pr(ball will be white) = Pr(ball will not be black) = 0.75.

The Meaning of the Word "Confidence" in Confidence Interval

As we have stated, in most problems one does not need to know the exact value of a parameter such as π in order to make a decision. An interval with a high degree of confidence, say 95%, that the interval contains π is usually adequate if the interval is not too wide.

Suppose we want to know the value of π but it is impractical to observe the entire population. In this case, a simple random sample of size n is selected from this population and the sample values are used to compute $\hat{\pi}$, the estimate of π. Suppose we also compute L and U, the endpoints of a 95% confidence interval for π. We state

We have 95% confidence that π is contained in the interval L to U.

In Box 3.1.1 we explain the meaning of the term "confidence" when used in the context of confidence intervals.

BOX 3.1.1: MEANING OF CONFIDENCE

A bin contains 100 balls, of which 95 are black and 5 are white. You select a simple random sample of one ball from the bin and hold it in your hand *but never look at it.* The **confidence** (the degree of belief) that the ball hidden in your hand is black is 0.95 (equivalently, 95%). This is equal to the proportion of black balls that is in the bin.

To generalize, suppose a bin contains G balls, of which g are black and $G - g$ are white. You select a simple random sample of one ball from the bin and hold it in your hand *but never look at it.* The **confidence** that the ball hidden in your hand is black is g/G, the proportion of black balls that is in the bin.

Box 3.1.1 contains the statement, "You select a simple random sample of one ball from the bin and hold it in your hand *but never look at it.*" The phrase "never look at it" is important since if you look at the ball in your hand, you know its color and then there is no uncertainty. When we compute a confidence interval, we never know whether it contains the population parameter. We have 95% confidence that it does, but we never know for sure. Similarly, if we never look at the ball hidden in our hand, we don't know the ball's color, but we have 95% confidence that it is black since it has been drawn at random from a bin that contains 95 black balls and 5 white balls.

EXAMPLE 3.1.3

WAREHOUSE FIRE. Consider Example 3.1.1, where a broker wants to estimate π, the proportion of radios that was damaged in a warehouse fire. The sample size is $n = 312$, the point estimate of π is $\widehat{\pi} = 0.080$, and the 95% confidence interval for π is $L = 0.049$ to $U = 0.111$. Thus, we can state

We are 95% confident that π is in the interval from 0.049 to 0.111.

We don't know for sure whether π is in the interval 0.049 to 0.111, just as in Box 3.1.1 we don't know the color of the ball hidden in our hand. The confidence we have that π is in the interval 0.049 to 0.111 is 95%, the same confidence we have that a ball hidden in our hand is black if it was selected at random from a bin that contains 95 black and 5 white balls.

Another way to interpret 95% confidence is to imagine that you can obtain every possible sample of 312 radios from the 100,000 radios in the warehouse. Each sample is used to compute a 95% confidence interval for π. There would be a huge number (many billions) of different confidence intervals. We refer to this collection as **the set of all possible 95% confidence intervals for π**. Statistical theory tells us that 95% of these intervals contain π and 5% do not contain π.

Figure 3.1.1 shows a graphical representation of 95% confidence intervals calculated from 100 different random samples. Each vertical line represents a confidence interval. The lower point on the line is L and the upper point on the line is U. A horizontal line is drawn at the true value of π. If a vertical line touches the horizontal line, then the confidence interval contains the true value of the parameter π. The asterisks denote the intervals that do not contain π. We would expect about 5% of the intervals not to include π and hence to have asterisks. Exactly 5% of the confidence intervals do not contain π in Figure 3.1.1.

FIGURE 3.1.1
95% Confidence Intervals for π from 100 Random Samples

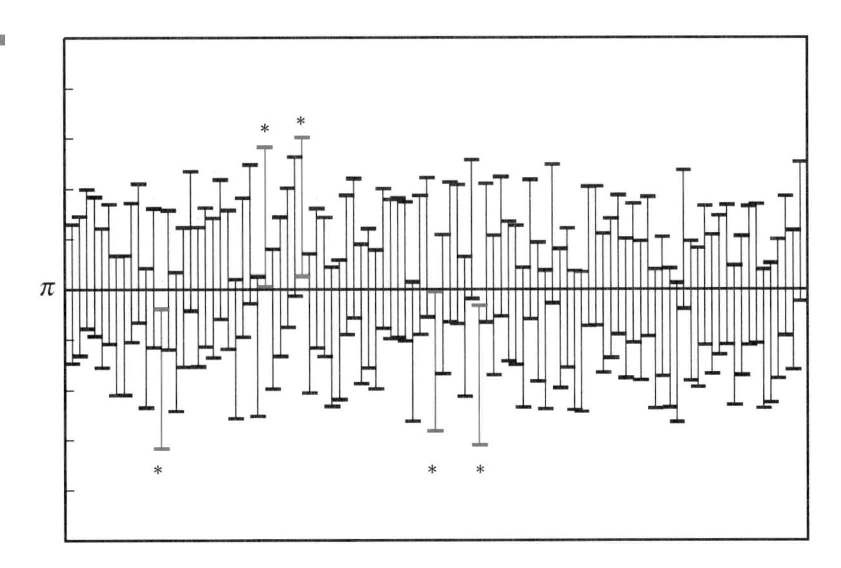

In practice, you would select only *one simple random sample* of 312 radios and compute only one 95% confidence interval for π. The computed interval is one of the intervals in the set of all possible 95% intervals. The interval actually computed can be viewed as one interval that was selected at random from the set of all possible 95% confidence intervals. Thus, the confidence is 95% that the computed interval contains π, since 95% of all possible intervals contain π.

DEFINITION 3.1.3: CONFIDENCE COEFFICIENT (CONFIDENCE LEVEL)

The proportion 0.95 (or equivalently, the percentage 95%) associated with a confidence interval is called the **confidence coefficient** or **confidence level**. We sometimes say "95% confidence interval" to mean a confidence interval with a 95% confidence coefficient. In theory, any confidence coefficient between 0% and 100% can be used. In practice, investigators generally use a confidence coefficient of 90%, 95%, or 99%.

You might wonder why investigators would not use a confidence coefficient of 100%. This would mean that they are *absolutely sure* that the computed interval contains the unknown parameter π. For a fixed sample size the only confidence interval that can contain π with absolute certainty is an interval that is so wide it is essentially useless. For instance, in Example 3.1.3, if the broker uses a 100% confidence interval, then using the formulas described in the next section, he computes $L = 0$ and $U = 1$. The broker is absolutely sure (100% confident) that π is contained in the interval 0 to 1. This says that between 0% and 100% of radios in the warehouse are damaged. This is true, but it doesn't give any useful information.

THINKING STATISTICALLY: SELF-TEST PROBLEMS

3.1.1 A bin contains 300 red balls and 700 green balls. A simple random sample of one ball is to be selected from the bin.

a. What is the probability that the ball will be green?
b. What is the probability that the ball will *not* be green?

3.1.2 Consider the following game. There are two bins labeled 1 and 2. Bin 1 contains 50 white and 50 black balls. Bin 2 contains 90 white balls and 10 black balls. I can choose either bin from which I will select a simple random sample of one ball. If the selected ball is white, I receive a valuable prize. From which bin should I select the ball? Why?

3.1.3 The owner of an orange grove obtained a simple random sample of 300 trees from the grove and used this sample to compute an estimate of π, the proportion of trees that were infected with a leaf disease. The estimate of π was $\widehat{\pi} = 0.090$, and a 95% confidence interval for π was 0.057 to 0.123. The owner stated, "I have 95% confidence that π is contained in the interval 0.057 to 0.123." Explain in detail the meaning of this statement of confidence as explained in Box 3.1.1.

3.1.4 I select a ball at random from a bin that contains 198 black balls and 2 white balls. I hold the ball in my hand but never look at it. What is my confidence that the ball hidden in my hand is white?

3.2 COMPUTING A CONFIDENCE INTERVAL FOR A POPULATION PROPORTION

In this section we show you how to compute L and U, the endpoints of a confidence interval for π. Formulas for computing confidence intervals for π are presented in Box 3.2.1.

BOX 3.2.1: CONFIDENCE INTERVAL FOR A PROPORTION π

Let π be the proportion of items in a population with a specified characteristic. In a simple random sample of size n from the population, let k be the number of items in the sample with the specified characteristic. The estimate of π is

$$\widehat{\pi} = \frac{k}{n}.$$

The endpoints L and U of a confidence interval for π are

$$L = \widehat{\pi} - (T1)\sqrt{\frac{\widehat{\pi}(1-\widehat{\pi})}{n}},$$

$$U = \widehat{\pi} + (T1)\sqrt{\frac{\widehat{\pi}(1-\widehat{\pi})}{n}},$$

where the value of $T1$ depends on the confidence coefficient specified by the investigator.

Table T-1, on the inside of the front cover of this book, contains values of $T1$ for four confidence coefficients—90%, 95%, 98%, and 99%. For convenience, Table T-1 is also shown as Table 3.2.1 here.

TABLE 3.2.1 Entries Are Values of $T1$ to Use in Formulas in Box 3.2.1

Confidence Coefficient	0.90	0.95	0.98	0.99
***T*1**	1.65	1.96	2.33	2.58

The formulas for L and U in Box 3.2.1 are approximate. The approximation is satisfactory when n is greater than 100, when k and $n - k$ are both greater than 5, and when $N > 10n$. By "approximate" we mean that the actual *confidence coefficient* may be slightly different than the one specified. If any of these conditions is not satisfied, then other, more complicated formulas available in advanced books should be used. In this case, you might want to consult a professional statistician for help. The derivations of the confidence intervals for π shown in Box 3.2.1 cannot be explained until we have developed some more concepts. These formulas are derived in Section 7.7. From Table 3.2.1 (and Table T-1), note that $T1$ increases as the confidence coefficient increases. This means that the confidence interval widens as the confidence coefficient increases.

Convention. In computing an estimate and a confidence interval for π, we will carry as many significant figures as possible in all intermediate calculations. The estimate $\widehat{\pi}$ will then be rounded to three decimals, the lower endpoint L will be rounded *down* to three decimals, and the upper endpoint will be rounded *up* to three decimals.

EXAMPLE 3.2.1

WAREHOUSE FIRE. Consider Example 3.1.3, where a fire has damaged some of the 100,000 radios stored in a warehouse. A freight broker who is considering purchasing the contents wants an estimate of π, the proportion of damaged radios. He also wants to determine a 95% confidence interval for π. Suppose that a simple random sample of 312 radios was obtained and that 25 of them were observed to be damaged. The estimate of π is $\widehat{\pi}$, where

$$\widehat{\pi} = \frac{25}{312} = 0.080.$$

Since $n > 100$, $k > 5$, $n - k > 5$, and $N > 10n$, the formulas in Box 3.2.1 can be used to compute L and U. For a 95% confidence interval, $T1 = 1.96$, and we obtain

$$L = 0.080 - 1.96\sqrt{0.080(0.920)/312} = 0.049 \text{ (rounded down)},$$
$$U = 0.080 + 1.96\sqrt{0.080(0.920)/312} = 0.111 \text{ (rounded up)}.$$

We state that

the confidence is 95% that π is contained in the interval 0.049 to 0.111.

Thus, the confidence we have that π is in the interval 0.049 to 0.111 is the same confidence we have that the ball hidden in our hand is black if it is selected at random from a bin containing 95 black balls and 5 white balls.

Margin of Error

Using the formulas in Box 3.2.1, we see that $\widehat{\pi}$ is at the center of the confidence interval for π. Hence, $\widehat{\pi}$ is $(U - L)/2$ units from L and $(U - L)/2$ units from U. The relationships among $\widehat{\pi}$, L, and U are shown in Figure 3.2.1.

FIGURE 3.2.1
Graphical Representation of the Relationships Among $\widehat{\pi}$, L, and U

L ←------ $(U-L)/2$ ------→	$\widehat{\pi}$ ←------ $(U-L)/2$ ------→	U

$\widehat{\pi}$ is halfway between L and U.

Note that $\widehat{\pi}$ is in the center of the confidence interval L to U shown in Figure 3.2.1. So if the interval contains the unknown population parameter π, then $\widehat{\pi}$ can be no more than $(U - L)/2$ units from π. The quantity $(U - L)/2$ is called the **margin of error** associated with the point estimate $\widehat{\pi}$. We use the symbol ME to denote the margin of error. Using the formulas for L and U in Box 3.2.1, we get

$$\text{ME} = \frac{U - L}{2} = (T1)\sqrt{\frac{\widehat{\pi}(1 - \widehat{\pi})}{n}}. \qquad \textbf{Line 3.2.1}$$

Hence, we can write

$$L = \hat{\pi} - \text{ME} \qquad \text{and} \qquad U = \hat{\pi} + \text{ME} \qquad \textbf{Line 3.2.2}$$

for the endpoints of a confidence interval for π.

You may have heard or read statements similar to the following on television news programs, in newspapers, or in magazines.

> A poll of 560 adults was conducted, and it was estimated that 63% of all adults favor banning cigarette advertising. The margin of error is 4%.

The phrase "margin of error" is typically used in the media to refer to a margin of error obtained from a confidence interval whose confidence coefficient is 95%. Since $\hat{\pi} = 0.63$, the formulas in Box 3.2.1 can be used to compute $L = 0.590$ and $U = 0.670$, the endpoints of a 95% confidence interval for π. Using this confidence interval, the margin of error is computed to be ME $= (0.670 - 0.590)/2 = 0.040$. Thus, $\hat{\pi} = 0.63$ and the confidence is 95% that this estimate is within 0.04 units of the unknown population parameter π. This is how a confidence interval is used to determine "how close" the estimate $\hat{\pi} = 0.63$ is to the unknown parameter π.

EXAMPLE 3.2.2

MARGIN OF ERROR FOR WAREHOUSE PROBLEM. In Example 3.2.1 the broker obtained $\hat{\pi} = 0.080$, $L = 0.049$, and $U = 0.111$. Since $(U - L)/2 = (0.111 - 0.049)/2 = 0.031$, the 95% margin of error is ME $= 0.031$. If the unknown parameter π is contained in the interval from 0.049 to 0.111 (the confidence is 95% that it is in the interval), then $\hat{\pi}$ is within ME $= 0.031$ units of π (see Figure 3.2.2). The broker has used a confidence interval to determine "how close" the estimate is to the unknown population parameter π. He states, "The estimate of π is $\hat{\pi} = 0.080$, and I have 95% confidence that this estimate is within 0.031 units of the parameter π." Or equivalently, he states, "The estimate of π is $\hat{\pi} = 0.080$, and the confidence is 95% that π is between 0.049 and 0.111. This is the same confidence I have that the ball hidden in my hand is black if it was selected at random from a bin with 95 black and 5 white balls."

FIGURE 3.2.2
A Graphical Representation of the Relationships Among $\hat{\pi}$, L, and U in Example 3.2.2

|←--------0.031--------→|←--------0.031--------→|

$L = 0.049$ $\qquad$ $\hat{\pi} = 0.080$ $\qquad$ $U = 0.111$

$\hat{\pi} = 0.080$ is halfway between $L = 0.049$ and $U = 0.111$. So if the confidence interval contains π, then $\hat{\pi}$ is within 0.031 units of π. Thus, the estimate of π is 0.080 and this estimate has a 95% margin of error of ME = 0.031.

Effect of the Confidence Coefficient and Sample Size on the Confidence Interval Width

As we have stated, when a question is asked that requires a numerical answer, the exact answer may not be necessary. It is generally satisfactory to state an interval with an associated confidence that the exact answer is contained in the interval. However, for an interval to provide a satisfactory answer, its width must be "small" and the confidence that it contains the exact answer must be "large." You can see from Box

3.2.1 that the width of a confidence interval for π is

$$\text{width} = U - L = 2(T1)\sqrt{\frac{\widehat{\pi}(1-\widehat{\pi})}{n}} = 2\ \text{ME}.$$

Thus, the larger the sample size n, the smaller the width of the interval and the smaller the margin of error. From Table 3.2.1, we note that as the confidence coefficient increases, $T1$ increases. Thus, the width of the interval and the margin of error both increase.

EXAMPLE 3.2.3

DISEASED ORANGE TREES. The owner of an orange grove wants to determine π, the proportion of diseased trees in the grove. He will use this information to help decide if it would be cost effective to spray the entire grove. The owner would like to know the *exact* value of π, but he realizes that he cannot know the exact value unless he examines every one of the 6,010 trees, which would be too expensive. The owner believes he can make the correct decision by using an estimate of π if he is fairly certain that the estimate is within 3 percentage points of the exact value of π. Thus, he wants to be 95% confident that the estimate $\widehat{\pi}$ is no more than 3 percentage points (that is, 0.03 units) from the true value π. This means that ME must be less than 0.03. He decides to obtain a simple random sample of 150 trees and examine them for the disease. He finds that 12 of the 150 trees are diseased, so the estimate of π is

$$\widehat{\pi} = \frac{12}{150} = 0.080.$$

Using the formulas in Box 3.2.1, we find that the endpoints of a 95% confidence interval for π are

$$\begin{aligned} L &= 0.080 - 1.96\sqrt{0.080(0.920)/150} \\ &= 0.036 \quad \text{(rounded down according to the convention)}, \\ U &= 0.080 + 1.96\sqrt{0.080(0.920)/150} \\ &= 0.124 \quad \text{(rounded up according to the convention)}. \end{aligned}$$

From this confidence interval we obtain $\text{ME} = (0.124 - 0.036)/2 = 0.044$. Since ME is greater than 0.03, the owner's specification that $\widehat{\pi}$ be no more than 0.03 units from π with 95% confidence is *not* satisfied. To satisfy the specification, a larger sample size is needed.

Instead of 150 trees, suppose that a simple random sample of 500 trees had been obtained and that 45 of them were observed to be diseased. From these data the estimate of π is

$$\widehat{\pi} = \frac{45}{500} = 0.090.$$

Using the formulas in Box 3.2.1, we find that the endpoints of a 95% confidence interval for π are

$$\begin{aligned} L &= 0.090 - 1.96\sqrt{0.090(0.910)/500} = 0.064, \\ U &= 0.090 + 1.96\sqrt{0.090(0.910)/500} = 0.116. \end{aligned}$$

From this we get ME = $(0.116 - 0.064)/2 = 0.026$; since this is less than 0.03, the owner's specification is now satisfied. The owner's specification was not satisfied when a sample size of 150 was used but was satisfied when a sample size of 500 was used. This demonstrates the importance of using an appropriate sample size. This will be discussed further in Section 3.3.

EXPLORATION 3.2.1

Here we describe a procedure that shows how a confidence interval for π can be used to determine the number of items in a population.

Fish biologists, interested in estimating the total number of fish in a lake, often use a technique called the "capture–mark–recapture" method. To use this method, the biologist catches q fish, marks them in some manner, and then releases them back into the lake. Later n fish are caught, and k of them are observed to have the mark that was placed on them by the biologist. From this information, the biologist can estimate N, the total number of fish in the lake.

1. Suppose that the biologist caught 80 fish, marked them, and threw them back into the lake. Subsequently, the biologist caught 1,000 fish and 10 of them had the mark she had placed on them. What are the values of q, n, and k?
The values of q, n, and k are $q = 80$, $n = 1{,}000$, and $k = 10$.

2. Using the values in (1), estimate π, the proportion of fish in the lake that is marked.
If a simple random sample of $n = 1,000$ fish was caught, of which $k = 10$ contained the biologist's mark, the estimate of π is $\hat{\pi} = k/n = 10/1000 = 0.010$.

3. What is the estimate of the total number of fish in the lake?
Let N represent the total number of fish in the lake. Since π is the proportion of the fish in the lake that were marked by the biologist, it follows that

$$q = \text{total number of fish in the lake that are marked} = N \times \pi.$$

Solve for N to obtain

$$N = \frac{q}{\pi}.$$

Thus, an estimate of N is

$$\hat{N} = \frac{q}{\hat{\pi}}.$$

Since $q = 80$ and $\hat{\pi} = 0.01$, we obtain $\hat{N} = 80/0.01 = 8{,}000$. So we estimate that there are 8,000 fish in the lake.

4. Obtain a 95% confidence interval for N.
To obtain a 95% confidence interval for N, first find a 95% confidence interval for π. Using the formulas in Box 3.2.1, we find that the endpoints L and U of the 95% confidence interval for π are

$$L = 0.01 - 1.96\sqrt{0.01(0.99)/1000} = 0.003,$$
$$U = 0.01 + 1.96\sqrt{0.01(0.99)/1000} = 0.017.$$

The confidence is 95% that π is contained in the interval 0.003 *to* 0.017, *and we write this interval as*

$$0.003 \leq \pi \leq 0.017.$$

Now substitute q/N for π, and set $q = 80$ to obtain

$$0.003 \leq \frac{80}{N} \leq 0.017.$$

Solve this for N and obtain

$$4{,}705 \leq N \leq 26{,}667.$$

The confidence is 95% that there are between 4,705 *and* 26,667 *fish in the lake.*

THINKING STATISTICALLY: SELF-TEST PROBLEMS

3.2.1 A machine that puts aspirin tablets into bottles was found to be breaking some of the pills. The company's quality assurance manager wants to estimate π, the proportion of the 120,000 bottles produced during the past week that contains at least one broken aspirin. She wants to be 95% confident that the estimate, $\widehat{\pi}$, will be no more than 1 percentage point from the true value of π. From a simple random sample of 200 bottles it was found that 8 bottles had at least one broken aspirin.

a. Find a point estimate of π, the proportion of the 120,000 bottles that has at least one broken aspirin.
b. Find a 95% confidence interval for π using the formulas in Box 3.2.1.
c. Give an interpretation for the 95% confidence interval.
d. What is the 95% margin of error?
e. What confidence do you have that π *is not* in the stated confidence interval?
f. Do the results of the investigation satisfy the manager's specification of being 95% confident that $\widehat{\pi}$ is no more than 1 percentage point from π?

3.3 SAMPLE SIZE TO USE FOR ESTIMATING A POPULATION PROPORTION

In Example 3.2.3 the owner of an orange grove wants to be 95% confident that the estimate of π is within 3 percentage points of the true value, where π is the proportion of trees in the grove that is diseased. With a sample of 150 trees, the estimate of π did not satisfy the owner's specification. However, the estimate of π using a sample of 500 trees did satisfy the specification. This leads to the question, "Before the sample is obtained, how can one determine the sample size needed to meet an investigator's specification?" In this section we answer this question.

Suppose an investigator needs an estimate of π and wants to be 95% confident that this estimate will be within a distance of d units from the true value. That is, the investigator wants to be 95% confident that the margin of error ME will be less than or equal to d, where d is specified by the investigator. We have learned that ME will decrease as the sample size n increases. Thus, the investigator wants to determine how large n must be so that ME will be at most d with a confidence of 95%. Instead of 95%, the investigator can use 99% or any other confidence coefficient.

Let L and U be the endpoints of a confidence interval computed for π using the formulas in Box 3.2.1. The confidence interval can be written as

$$\widehat{\pi} - \text{ME} \quad \text{to} \quad \widehat{\pi} + \text{ME},$$

where $\text{ME} = (U - L)/2$. So if $\widehat{\pi}$ is to be within d units of π, it follows that ME must be less than or equal to d. That is, the inequality $(U - L)/2 \leq d$ must be satisfied. If we substitute for L and U from Box 3.2.1, we obtain

$$\frac{(U - L)}{2} = (T1)\sqrt{\frac{\widehat{\pi}(1 - \widehat{\pi})}{n}} \leq d,$$

where $T1$ is obtained from Table T-1. Solving this inequality for n gives

$$n \geq \frac{\widehat{\pi}(1 - \widehat{\pi})(T1)^2}{d^2}.$$

In practice, we want to use the smallest value of n that satisfies this inequality. This value of n is

$$n = \frac{\widehat{\pi}(1 - \widehat{\pi})(T1)^2}{d^2}. \qquad \textbf{Line 3.3.1}$$

To use the formula in Line 3.3.1, we must have a value to use for $\widehat{\pi}$. Since we have not yet selected a sample (we cannot select a sample until we decide on n), we cannot compute $\widehat{\pi}$. However, from previous studies, an investigator may know that π must be between two numbers, a and b. In this case the value of $\widehat{\pi}$ to use in Line 3.3.1 is the value between a and b that is closest to 0.50. If no previous information about π is available, this means we know only that π is between 0 and 1. In this case $a = 0$ and $b = 1$, and so the value 0.50 is the value closest to 0.50. Thus, 0.50 is used for $\widehat{\pi}$ in Line 3.3.1. If we substitute 0.50 for $\widehat{\pi}$ into Line 3.3.1, we get

$$n = \left(\frac{T1}{2d}\right)^2. \qquad \textbf{Line 3.3.2}$$

If the solution for n in Line 3.3.1 or 3.3.2 is not an integer, round up to the next integer and use it as the value of n.

BOX 3.3.1: GUIDELINES FOR DETERMINING THE SAMPLE SIZE REQUIRED SO THAT $\widehat{\pi}$ WILL BE WITHIN d UNITS OF π WITH A SPECIFIED CONFIDENCE

Suppose an investigator wants to determine the sample size n to use so that with a specified confidence $\widehat{\pi}$ will be within d units of the population parameter π. If it is known that π is between a and b inclusive, the value of n to use is

$$n = \frac{\hat{\pi}(1-\hat{\pi})(T1)^2}{d^2}, \qquad \textbf{Line 3.3.1}$$

where $\hat{\pi}$ is the value between a and b that is closest to 0.50. If no information about the value of π is available, use $\hat{\pi} = 0.50$ and obtain

$$n = \left(\frac{T1}{2d}\right)^2. \qquad \textbf{Line 3.3.2}$$

EXAMPLE 3.3.1

WAREHOUSE FIRE. In Example 3.2.1 the freight broker wants to estimate π, the proportion of the 100,000 radios in a warehouse that is damaged. The broker wants to select a large enough sample so that he can be 95% confident that the estimate $\hat{\pi}$ will be within $d = 0.030$ units of the unknown value of π. The value of d was selected based on the amount of money the broker was willing to risk on his decision to purchase fire-damaged goods. Since no previous information about π is available, he computed n using the formula in Line 3.3.2. For a 95% confidence coefficient with $d = 0.030$, he obtained

$$n = \left(\frac{1.96}{2(0.030)}\right)^2 = 1{,}068 \quad \textit{(rounded up)}.$$

If a simple random sample of 1,068 radios is selected, the freight broker has 95% confidence that $\hat{\pi}$ will be no more than 0.030 units from the population parameter π. For example, suppose that a simple random sample of 1,068 radios was selected and $\hat{\pi} = 0.130$ for this sample. Using the formulas in Box 3.2.1, a 95% confidence interval for π is $L = 0.109$ to $U = 0.151$. The 95% margin of error is

$$\text{ME} = \frac{U-L}{2} = \frac{0.151 - 0.109}{2} = 0.021.$$

Thus, ME $= 0.021$, which is less than $d = 0.030$.

If no information about the value of π is available, Table 3.3.1 provides the values of n needed for 95% and 99% confidence intervals for π for several values of d. These values were computed using the formula in Line 3.3.2.

TABLE 3.3.1 Maximum Sample Sizes Needed so That $\hat{\pi}$ Is Within d Units of π

d	n for 95%	n for 99%
0.010	9,604	16,641
0.020	2,401	4,161
0.030	1,068	1,849
0.040	601	1,041
0.050	385	666
0.100	97	167

THINKING STATISTICALLY: SELF-TEST PROBLEMS

3.3.1 A company that manufactures a new type of lawn mower wants to determine π, the proportion of mowers that require minor adjustments before they can be used. The company's vice president wants to obtain an estimate of π and be 99% confident that the estimate is within 5 percentage points (0.050 units) of the true value of π. Using this information, he will decide whether to have all lawn mowers checked before sending them to retail outlets. The target population is the set of all lawn mowers that will be produced this year. The sampled population is the first 10,000 mowers produced. A simple random sample of size n will be selected from the first 10,000 mowers produced.

a. If no previous information about π is available, find n so that the vice president is 99% confident that $\widehat{\pi}$ differs from π by less than 5 percentage points.

b. If no previous information about π is available, find n so that the vice president is 95% confident that $\widehat{\pi}$ differs from π by less than 5 percentage points.

c. What sample size should be used if the vice president wants to be 99% confident that $\widehat{\pi}$ is within 0.100 units of π and no previous information about π is available?

d. Based on previous experience with newly manufactured lawn mowers, the vice president knows that π is between 0.03 and 0.20. What sample size should be used if the vice president wants to be 99% confident that $\widehat{\pi}$ is within 0.050 units of π?

3.4 CHAPTER SUMMARY

KEY TERMS

1. confidence interval
2. confidence coefficient
3. margin of error

SYMBOLS

1. L—lower bound of a confidence interval
2. U—upper bound of a confidence interval
3. Pr—probability
4. ME—margin of error
5. d—the maximum distance that an investigator wants $\widehat{\pi}$ to be from π with a specified confidence
6. $T1$—a table value used to compute a confidence interval for π

KEY CONCEPTS

1. If a point estimate is used in place of a parameter to help make a decision, the investigator should determine "how close" the point estimate is to the parameter.
2. The confidence that a 95% confidence interval contains the true value of an unknown parameter is the same as the confidence that a ball hidden in your hand is black if it was selected at random from a bin that contains 95 black balls and 5 white balls.

3. A confidence interval will not always contain the true value of the parameter.
4. For a given level of confidence, the width of a confidence interval for π (and the margin of error) will decrease as the sample size increases.
5. For a given sample size, the width of a confidence interval for π (and the margin of error) will increase as the confidence coefficient increases.

SKILLS

1. Compute a confidence interval for π using the formulas in Box 3.2.1.
2. Determine the sample size needed to estimate π within d units for a given level of confidence using the rules in Box 3.3.1.

3.5 CHAPTER EXERCISES

3.1 A bin contains 1,000 balls, where 150 are white, 350 are black, 400 are green, and the remaining balls are red. A simple random sample of one ball is to be selected from the bin.

a. What is the probability that the ball will be green?

b. What is the probability that the ball will be either green or red?

3.2 A bin contains 1,070 balls, where 450 are blue and 620 are yellow.

a. If a simple random sample of one ball is to be selected, what is the probability that it will be yellow?

b. Suppose you repeat the following process many times (say one billion times): Select one ball at random, record its color, and replace it in the bin. What proportion of the balls selected would you expect to be blue?

3.3 A bin contains 1,000 balls that are either black or white. One ball is selected at random from the bin and its color is recorded. The ball is then replaced in the bin. This process is repeated many times (say one billion times), and 92% of the balls selected are black. How many of the balls in the bin would you expect to be black?

3.4 A quality assurance manager wants to determine the value of π, the proportion of a shipment of 13,000 computer printers that needs minor adjustments at initial startup. Based on a simple random sample of 100 printers, she computes $\hat{\pi} = 0.10$ and makes the following statement: "I have 95% confidence that π is within 5.9 percentage points of this estimate."

a. Interpret this statement.

b. Compute the 95% confidence interval for π using the formula in Line 3.2.2.

c. What is the confidence that π is not contained in the interval in (b)?

3.5 Suppose an electronics store receives a shipment of 1,000 memory chips, 10 of which are defective. You plan to buy one chip from this batch. Assuming that the chip you buy is selected at random from the entire shipment of 1,000, what is the probability that the chip you will purchase is *not* defective?

3.6 A newspaper article reported that a 99% confidence interval for π, the proportion of high-school seniors who has used marijuana, is from 0.22 to 0.28. Interpret the meaning of this confidence interval.

3.7 In a study of a large high school it was reported that 26% of this year's students plan to attend a university. If a student is selected at random from this high school, what is the probability that this student plans to attend a university?

3.8 In Self-Test Problem 3.2.1 a manager wants to estimate π within 1 percentage point with 95% confidence, where π is the proportion of the 120,000 bottles of aspirin produced last week that contains at least one broken pill. Suppose the manager decides to obtain a simple random sample of 1,000 bottles. In this sample 30 bottles are observed to contain broken pills.

a. Compute a point estimate of π and the 95% margin of error for the estimate.

b. Find a 95% confidence interval for π using the formulas in Line 3.2.2.

c. Do the results in (b) satisfy the manager's specification that $\hat{\pi}$ is to be no more than 1 percentage point from π with 95% confidence?

3.9 Suppose a simple random sample of 300 radios was obtained from the warehouse described in Example 3.2.2 and 45 of them were damaged. Find a 99% confidence interval for $1 - \pi$, the proportion of radios that were *not* damaged.

3.10 A survey was conducted in Mesa, Arizona, to determine π, the proportion of registered voters who favor a measure that would ban smoking in all public places without a separate ventilation system. A newspaper article that reported the results of the survey stated that 72% of the registered voters in the sample support the measure. A 95% confidence interval for π was reported to be $L = 0.670$ and $U = 0.770$.

a. Explain the meaning of the confidence interval.

b. Is it reasonable to conclude that a majority (greater than 50%) of the registered voters in Mesa favors the measure to ban smoking in all public places without a separate ventilation system?

c. Can we say with absolute certainty that a majority favors selected from the measure?

3.11 A simple random sample was selected from registered voters in Mesa, Arizona, to determine π, the proportion of registered voters who favors a measure to establish a midnight curfew for all youths under the age of 17. The reported results of the survey stated that 51.0% of the sampled voters favor the curfew. The reported 95% confidence interval is $L = 0.490$ to $U = 0.530$.

a. What sample size was used?

b. Based on the results of the 95% confidence interval, is it reasonable to conclude with 95% confidence that a majority (greater than 50%) of the registered voters in Mesa favors the measure to establish a curfew?

3.12 According to a travel reduction survey at a large state university, 13% of employees rode bicycles to campus for work at least three days a week during the past year. The survey was based on a simple random sample of 100 employees. Use this information and compute a 95% confidence interval for π, the proportion in the population who rode bicycles to work at least three days a week during the past year.

3.13 An investigator for a large state university is studying the relationship between drop-out rate of first-year students and the size of their high-school graduating class. One parameter of interest is π, the proportion of the 25,000 students currently enrolled in the university whose high-school graduating class contained more than 200 students. A simple random sample of 300 students was selected; 150 of them reported that they graduated in a high-school class of more than 200 students.

a. Estimate π, the proportion of students in the university whose high-school graduating class contained more than 200 students.

b. Find a 95% confidence interval for π.

c. Suppose that the person conducting the study states, "I want to be 99% confident that the estimate of π will be within 6.0 percentage points (0.060) of π." If no previous information about π is available, what sample size should be used?

3.14 A manufacturer of personal computers is considering several marketing strategies for selling a new computer model. Based on marketing research, the company has decided to target sales to consumers between the ages of 21 and 35 with annual incomes exceeding $35,000. In order to make this target group aware of the new product, the company plans to use advertisements on national television that appeal to this target group. Two television commercials have been produced by a marketing firm hired by the company. The company wants to determine which of the commercials is more effective in eliciting interest from the target group of consumers. One of the commercials stresses the business and educational benefits of owning a computer, and the other commercial stresses the potential entertainment value. A simple random sample of size $n = 100$ consumers was selected from a database of consumers who satisfy the age and income requirements. Each of the 100 consumers was shown both commercials and asked which one they thought was more effective in generating interest for the computer. A total of 70 consumers preferred the commercial that stressed the business and educational applications, and the remaining 30 preferred the commercial that stressed the entertainment value.

a. Estimate π, the proportion of consumers in the sampled population who would prefer the commercial that stresses business and educational applications. Find a 95% margin of error for the estimate.

b. If no previous information about π is available, how many consumers must be surveyed if the company wants to be 95% confident that the estimate of π will be no more than 0.020 units from the true value of π?

c. Suppose a simple random sample of 2,500 consumers is selected and 1,750 of them prefer the commercial that emphasizes the business and educational applications. Compute a 95% confidence interval for the proportion of consumers in the sampled population who prefer the commercial that emphasizes the business and educational applications.

d. Suppose that the company allocates only $2,000 to collect the sample and that each sample response costs $5. If the company wants to be 95% confident that the estimate of π will be within 0.05 units of π, has enough money been allocated? Assume that no previous information about π is available.

3.15 The 1994 Greater Phoenix Manufacturers Survey reported that 81% of the surveyed respondents believe economic conditions are better this year than last year. The sampled population consisted of all manufacturers in the metropolitan Phoenix area. The sample consisted of 200 of these manufacturers.

a. Determine the number of respondents in the sample who believe economic conditions are better this year than they were last year.

b. Construct a 99% confidence interval for the proportion of manufacturers in the sampled population who believe economic conditions are better this year than they were last year.

c. How many manufacturers must be interviewed if $\hat{\pi}$ is to be no more than 0.040 units from the true value with a confidence coefficient of 99% and no previous information about π is available?

3.16 The superintendent of a large school district wants to estimate π, the proportion of first-graders who has not had the childhood immunization shots. She plans to use a simple random sample of n first-graders to obtain the estimate, and she wants to be 95% confident that $\hat{\pi}$ will be no more than 0.05 units from the true value of π.

a. What is the value of n to use if no previous information about π is available?

b. What is the value of n to use if π is known to be less than 0.10?

3.6 SOLUTIONS FOR SELF-TEST PROBLEMS IN CHAPTER 3

3.1.1 a. Pr(ball will be green) = 700/1000 = 0.70, the same as the proportion of the 1,000 balls in the bin that are green.

b. Using Line 3.1.1 we get

$$\text{Pr(ball will not be green)} = 1 - \text{Pr(ball will be green)}$$
$$= 1 - 0.70 = 0.30.$$

The answer can also be obtained by noticing that 300 of the balls are not green. Thus, Pr(ball will not be green) = 300/1000 = 0.30 = proportion of balls in the bin that are *not* green.

3.1.2 I should select the ball from bin 2. If I select a ball from bin 2, the probability that it will be white is 0.90, but if I select a ball from bin 1, the probability that it will be white is 0.50. Thus, I have a larger probability of receiving the prize if I select a ball from bin 2.

3.1.3 The confidence the owner has that the interval from 0.057 to 0.123 contains π is the same confidence the owner has that the ball hidden in his hand is black if selected at random from a bin that contains 95 black and 5 white balls.

3.1.4 The confidence is 2/200 = 0.010 (1.0%), the same as the proportion of white balls in the bin.

3.2.1 a. The estimate of π is

$$\widehat{\pi} = \frac{8}{200} = 0.040.$$

b. Using the procedure in Box 3.2.1 we obtain

$$L = 0.04 - 1.96\sqrt{0.04(0.96)/200} = 0.012,$$
$$U = 0.04 + 1.96\sqrt{0.04(0.96)/200} = 0.068.$$

The 95% confidence interval is 0.012 to 0.068. The manager has 95% confidence that between 1.2% and 6.8% of the 120,000 bottles contain broken aspirins.

c. Your confidence that the interval from 0.012 to 0.068 contains π is the same as your confidence that a ball hidden in your hand is black if selected at random from a bin that contains 95 black and 5 white balls.

d. The margin of error is

$$(T1)\sqrt{\frac{\widehat{\pi}(1-\widehat{\pi})}{n}}.$$

For a 95% margin of error, $T1 = 1.96$ and the 95% margin of error is 0.028. This can also be obtained by computing $(U - L)/2 = (0.068 - 0.012)/2 = 0.028$.

e. Your confidence that the interval from 0.012 to 0.068 *does not* contain π is the same confidence you have that a ball hidden in your hand is *not* black (is white) if selected at random from a bin that contains 95 black and 5 white balls. This confidence is 5%, the same as the percentage of white balls in the bin. Hence, your confidence that the interval from 0.012 to 0.068 *does not* contain π is $1 - 0.95 = 0.05 = 5\%$.

f. No. Since ME = 0.028 is greater than 0.010, the specification is not satisfied. For example, one has 95% confidence that π is a value in the interval 0.012 to 0.068, so it could be 0.067. In this case, π would be 2.7 percentage units from the point estimate of π, which is $\widehat{\pi} = 0.04$.

3.3.1 a. Using the formula in Line 3.3.2 with $d = 0.050$ and $T1 = 2.58$, we obtain $n = 666$.

b. Using the formula in Line 3.3.2 with $d = 0.050$ and $T1 = 1.96$, we obtain $n = 385$.

c. Using the formula in Line 3.3.2 with $d = 0.100$ and $T1 = 2.58$, we obtain $n = 167$.

d. Of all the values between 0.03 and 0.20, the value 0.20 is the closest to 0.50. So we substitute 0.20 for $\widehat{\pi}$ in the formula in Line 3.3.1 and get $n = 427$.

CHAPTER 4

PROCESSING DATA

4.1 INTRODUCTION

As stated in Chapter 1, statistical methods can be viewed as the science of collecting, processing, and interpreting data. In Chapter 2 we discussed simple random sampling as one method of collecting data. In Chapter 3 we described how confidence intervals can be used to help interpret data. In this chapter we describe ways to process data. Processing data includes organizing, summarizing, analyzing, and displaying data in tabular and graphical form so they are more easily understood. When a data set is collected, it may not be known exactly how it will be used. In many situations it will be used several ways by many people for different purposes. When organizing a data set, first decide how you intend to use it. After making this decision, you must put it into a form that best helps answer questions of interest.

In the previous three chapters we discussed data sets that were either populations or samples. In this chapter the data under consideration might be either sample data or population data, so we use the terms **data set** or **set of data** to describe either type of data.

4.2 ORGANIZING AND SUMMARIZING DATA SETS

When a data set is collected, it is not always in a form that can be easily interpreted. Hence, it may be difficult to obtain the desired information until it is organized and summarized properly. Often a simple reorganization of the data set is all that is needed to glean important information from it. We illustrate with an example.

EXAMPLE 4.2.1

RADON GAS EMISSIONS. In Example 1.2.3 it is of interest to determine the proportion of the 51 residential buildings owned by a real estate developer in which radon presents a health hazard. The data set as it was collected is given in Table 1.2.1 and for convenience is repeated in Table 4.2.1. In addition to determining the proportion of buildings in which radon emissions present health problems, the developer wants to identify the specific buildings that have radon emission levels less than or equal to 4.0 pc/l, the specific buildings that have radon emission levels greater than 4.0 pc/l but less than or equal to 16.0 pc/l, and the specific buildings that have radon emission levels exceeding 16.0 pc/l.

TABLE 4.2.1 Radon Concentrations in 51 Buildings

Building number	Radon concentration (in pc/l)	Building number	Radon concentration (in pc/l)	Building number	Radon concentration (in pc/l)
1	13.6	18	2.9	35	6.5
2	2.8	19	2.0	36	11.8
3	2.9	20	2.9	37	13.2
4	3.8	21	11.2	38	2.8
5	15.9	22	1.9	39	6.9
6	1.7	23	2.0	40	0.7
7	3.4	24	6.0	41	12.9
8	13.7	25	2.9	42	3.6
9	6.1	26	7.7	43	3.6
10	16.8	27	5.1	44	8.1
11	7.9	28	13.2	45	17.0
12	3.5	29	3.8	46	8.2
13	2.2	30	13.9	47	9.8
14	4.1	31	2.4	48	13.0
15	3.2	32	7.9	49	11.3
16	2.9	33	1.4	50	4.0
17	3.7	34	5.9	51	6.0

The buildings with levels of radon emissions in specific ranges are easier to identify if the data are reorganized by rearranging the radon concentrations from smallest to largest. The results are presented in Table 4.2.2. From Table 4.2.2 you can easily identify the buildings with various radon emission levels. For example, buildings 40, 33, 6, 22, . . . , 50 have levels of radon less than or equal to 4.0 pc/l.

Statistical software and spreadsheet programs have many commands that can be used to organize data. To illustrate, we will show how to use Minitab to construct Table 4.2.2 from Table 4.2.1. The data in Table 4.2.1 are stored in the file **radon.mtp** on the data disk. Column 1 in the file contains "Building number" and column 2 contains "Radon concentration." To obtain the rearranged data shown in Table 4.2.2, retrieve the data and enter the following command in the session window of Minitab.

```
MTB > sort c2 c1 c4 c3
```

TABLE 4.2.2 Radon Concentrations Arranged from Smallest to Largest

Building number	Radon concentration (in pc/l)	Building number	Radon concentration (in pc/l)	Building number	Radon concentration (in pc/l)
40	0.7	12	3.5	32	7.9
33	1.4	42	3.6	44	8.1
6	1.7	43	3.6	46	8.2
22	1.9	17	3.7	47	9.8
19	2.0	4	3.8	21	11.2
23	2.0	29	3.8	49	11.3
13	2.2	50	4.0	36	11.8
31	2.4	14	4.1	41	12.9
2	2.8	27	5.1	48	13.0
38	2.8	34	5.9	28	13.2
3	2.9	24	6.0	37	13.2
16	2.9	51	6.0	1	13.6
18	2.9	9	6.1	8	13.7
20	2.9	35	6.5	30	13.9
25	2.9	39	6.9	5	15.9
15	3.2	26	7.7	10	16.8
7	3.4	11	7.9	45	17.0

The rearranged values of "Radon concentration" are found in column 4 with the associated value of "Building number" in column 3. Thus, columns 3 and 4 in the output from the previous Minitab command are columns 1 and 2, respectively, in Table 4.2.2. Other software packages also have commands that allow you to sort a data set.

The order in which data are recorded usually corresponds to the order in which they are collected. In fact, when collecting data, it is best to use a format that facilitates data recording and minimizes keyboarding errors when entering the data into a computer. However, this format does not always facilitate easy interpretation of the data. In reorganizing a data set, you should give considerable thought as to how it should be organized so that the required information can be obtained easily and quickly. In most studies you may need to organize the data set in several ways so that different questions can be answered with each arrangement.

As Example 4.2.1 demonstrated, sometimes a mere reorganization of the data is all that is needed to obtain desired information. More commonly, to help answer questions about a data set, one must also summarize the data in some manner. The next example demonstrates such a situation.

EXAMPLE 4.2.2

DEATH PENALTY SURVEY. This is a continuation of Exploration 2.5.1, where a survey was conducted among adults in a city of about 120,000 people to determine public opinion about the death penalty. A simple random sample of 60 adults living in the city was selected. Each sampled adult was asked, "What do you think should be the punishment for murder?" Each person in the sample was requested to choose one of the following responses: (1) *death penalty*; (2) *life in prison* without parole; (3) *depends* on the circumstances; (4) *neither* death penalty, nor life in prison without parole; (5) *no opinion*. The respondents were also asked to state their gender—*male* or *female*. The data are presented in Table 4.2.3 in the order collected during the survey.

TABLE 4.2.3 Preferred Punishment for Murder (in Order Collected)

Subject number	Gender of subject	Response	Subject number	Gender of subject	Response
1	M	neither	31	F	death penalty
2	M	death penalty	32	F	life in prison
3	M	death penalty	33	F	life in prison
4	F	neither	34	M	no opinion
5	F	life in prison	35	M	death penalty
6	M	life in prison	36	M	death penalty
7	F	life in prison	37	M	death penalty
8	M	death penalty	38	F	death penalty
9	F	life in prison	39	F	depends
10	F	life in prison	40	F	life in prison
11	M	death penalty	41	F	no opinion
12	M	death penalty	42	M	death penalty
13	M	death penalty	43	M	death penalty
14	M	life in prison	44	F	life in prison
15	M	death penalty	45	M	life in prison
16	M	death penalty	46	M	death penalty
17	M	death penalty	47	F	death penalty
18	M	death penalty	48	M	death penalty
19	F	life in prison	49	M	death penalty
20	F	death penalty	50	F	life in prison
21	M	depends	51	F	no opinion
22	F	death penalty	52	F	death penalty
23	M	neither	53	F	life in prison
24	M	depends	54	F	death penalty
25	F	death penalty	55	M	depends
26	F	death penalty	56	M	death penalty
27	F	death penalty	57	M	death penalty
28	M	death penalty	58	F	life in prison
29	M	death penalty	59	F	life in prison
30	F	death penalty	60	F	life in prison

One question of interest is, "Do the attitudes of males and females differ with regard to the preferred punishment for murder?" This question can be conveniently answered if the data are summarized as shown in Table 4.2.4.

TABLE 4.2.4 Summary of Responses in Table 4.2.3

Preferred punishment	Number of males	Proportion of males	Number of females	Proportion of females	Total number	Proportion of total
Death penalty	22	0.710	11	0.380	33	0.550
Life in prison	3	0.097	14	0.483	17	0.283
Depends	3	0.097	1	0.034	4	0.067
Neither	2	0.064	1	0.034	3	0.050
No opinion	1	0.032	2	0.069	3	0.050
Total	31	1.000	29	1.000	60	1.000

The proportion 0.710 shown in the first row of the column labeled "Proportion of males" is obtained by dividing the number located in the first row of the column labeled "Number of males" (22) by the total number of males reported in the last row of this column (31). That is, $22/31 = 0.710$. The other proportions in the table are computed in a similar manner.
The simple summarization presented in Table 4.2.4 allows an investigator to easily determine the number and proportion of males and females, as well as the total number and proportion of adults, who prefer each form of punishment. For example, Table 4.2.4 can be used to answer the question, "What is the difference between the proportion of males who favor the death penalty and the proportion of females who favor the death penalty for this sample of 60 adults?" The answer is $0.710 - 0.380 = 0.330$.

In Example 4.2.2 we summarized a *qualitative* variable by determining the number (frequency) of items in each category of the qualitative variable. A similar type of summarization is often needed for quantitative variables. This is discussed next.

Using Tables to Show How Quantitative Variables Are Distributed

In many investigations the data set being studied is very large. For example, suppose a nutritionist is studying the cholesterol of children who eat certain types of prepared food that contain large amounts of fat. This data set may consist of several thousand children. In studies where the number of items in the data set is very large, the investigator may want to summarize these data so they can be more easily understood. An investigator often wants to determine how the values of a quantitative variable are "distributed." By "distributed" we mean the investigator wants to determine the proportion of values that lie in specified intervals. Among other things, this information tells us where the values are centered and how they are spread out around this center.

Frequency tables and relative frequency tables are useful for determining how a variable is distributed. First we divide the range of the data values into an appropriate number of intervals, called **class intervals**. We then determine how the data values are distributed across these intervals. Class intervals are defined so they do not overlap. Every value in the data set appears in exactly one interval. It is generally recommended that 5 to 15 class intervals be used depending on the number of items in the data set. Computer programs that construct frequency tables apply certain rules to determine the most appropriate number of class intervals. After the class intervals are defined, each item in the data set is placed into the class interval that corresponds to the value of the quantitative variable of interest.

DEFINITION 4.2.1: FREQUENCY AND RELATIVE FREQUENCY TABLES

A **frequency table** lists the frequency (number) of values in a data set that are in each class interval.

A **relative frequency table** lists the relative frequency (proportion) of values in a data set that are in each class interval.

We illustrate by using the calculus scores in the table in Appendix B. Suppose that an instructor is interested in summarizing the calculus scores using the 12 class intervals shown in Table 4.2.5. The frequency and relative frequency of the scores in each class interval are displayed in Table 4.2.6.

TABLE 4.2.5 Class Intervals of Scores

Class intervals		Class intervals	
(1)	$40 \le scores < 45$	(7)	$70 \le scores < 75$
(2)	$45 \le scores < 50$	(8)	$75 \le scores < 80$
(3)	$50 \le scores < 55$	(9)	$80 \le scores < 85$
(4)	$55 \le scores < 60$	(10)	$85 \le scores < 90$
(5)	$60 \le scores < 65$	(11)	$90 \le scores < 95$
(6)	$65 \le scores < 70$	(12)	$95 \le scores \le 100$

TABLE 4.2.6 Frequency and Relative Frequency of Scores

Column 1	Column 2	Column 3
Class interval	**Frequency**	**Relative frequency**
40 to 45	10	0.004
45 to 50	22	0.008
50 to 55	58	0.023
55 to 60	150	0.058
60 to 65	246	0.095
65 to 70	341	0.131
70 to 75	464	0.178
75 to 80	449	0.173
80 to 85	370	0.142
85 to 90	260	0.100
90 to 95	151	0.058
95 to 100	79	0.030
Total	2,600	1.000

DEFINITION 4.2.2: CLASS INTERVAL

The notation *a to b* (or *between a and b*) is used to define a **class interval**. This means that the class interval from a to b contains all of the items with values in the range $a \le value < b$. The interval that contains the item with the maximum value contains both boundary points (that is, contains all items with values in the range $a \le value \le b$). Endpoints of class intervals are called **cutpoints**. The average of the cutpoints of any class interval is called the **midpoint** of the interval. In this book all class intervals have equal width.

For example, in Table 4.2.6 the score 45 is included in the interval 45 to 50, but the score 50 is included in the interval 50 to 55. The scores 95 and 100 are both included in the interval 95 to 100 since this is the interval that contains the maximum score. In Table 4.2.6 the numbers 40, 45, 50, 55, 60, 65, 70, 75, 80, 85, 90, 95, and 100 are cutpoints. The width of each interval is five. The midpoint for the class interval 40 to 45 is $(40 + 45)/2 = 42.5$.

The values in the frequency column of Table 4.2.6 are obtained by counting the frequency (number) of scores in each class interval. For example, in Appendix B there are 10 scores greater than or equal to 40 and less than 45, so the frequency of scores in the interval 40 to 45 is 10. The values in the relative frequency column of Table 4.2.6 are obtained by dividing each number in the frequency column by the total number of items in the data set, which is 2,600. For example, in the interval 40 to 45, the frequency is 10, so the relative frequency is $10/2{,}600 = 0.003846$, which is rounded to 0.004.

Relative frequencies can be used to compute the *proportion* of observations in the interval a to b, where a and b are any two cutpoints in a relative frequency table. To demonstrate, suppose we want to know the proportion of scores in the class interval 55 to 60. Using the relative frequency in column 3 of Table 4.2.6, we obtain

$$\text{the proportion of scores from 55 to 60} = \frac{150}{2{,}600} = 0.058 \quad \text{(rounded)}.$$

Suppose we want to know the proportion of scores from 65 to 80. Using the relative frequencies in column 3 of Table 4.2.6, we obtain

$$\text{the proportion of scores from 65 to 80} = 0.131 + 0.178 + 0.173 = 0.482.$$

Assume we have a set of items where the value of a quantitative variable is recorded for each item. If an item is to be chosen at random from the set of items, the probability that the value of the item will be in the interval from a to b is the proportion of items with values that are contained in this interval (see Definition 3.1.2). Hence, this probability is the sum of the relative frequencies for all class intervals between a and b.

Suppose a simple random sample of one student is to be selected from the set of students in Appendix B. The probability that this student received a score between 60 and 80 is the sum of the relative frequencies for the class intervals between 60 and 80 in Table 4.2.6. We obtain

$$0.095 + 0.131 + 0.178 + 0.173 = 0.577.$$

So the probability is 0.577 that a student chosen at random from the set of students will have a score between 60 and 80.

Since both a frequency table and a relative frequency table are *summaries* of data sets, not all questions about a data set can be answered using these tables. For example, suppose one student is to be selected at random from the set of students in Appendix B, and one wants to compute the probability that this student received a score between 68 and 73. Since 68 and 73 are not cutpoints in Table 4.2.6, this question cannot be answered using Table 4.2.6. To answer the question, one must consult the entire data set in Appendix B.

In this section we have illustrated how a simple reorganization or summarization of a data set makes it easier to interpret and obtain needed information. In the next two sections we demonstrate how graphical representations of these summaries can be used to further aid the process of data interpretation.

THINKING STATISTICALLY: SELF-TEST PROBLEMS

4.2.1 Use Table 4.2.6 to find the number of students who received scores in the interval 75 to 85.

4.2.2 Use Table 4.2.6 to find the number of students who received scores of 80 or more.

4.2.3 Use Table 4.2.6 to find the proportion of students who received a score less than 60.

4.2.4 Using Table 4.2.6, can you determine the number of students who received a score less than 72?

4.2.5 One student is to be selected at random from the 2,600 students summarized in Table 4.2.6. What is the probability that the student who will be selected received a score of 90 or higher?

4.3 USING HISTOGRAMS TO VISUALIZE HOW DATA ARE DISTRIBUTED

In addition to reporting tables, pictorial representations of data are useful in helping to answer questions. You may have heard the saying, "A picture is worth a thousand words." In statistics we present "pictures" in the form of charts and plots to give the reader a quick overview of values of all items in a data set.

Histograms are graphical displays of frequency and relative frequency tables of a quantitative variable. The histogram is very helpful in understanding the concepts of statistical inference related to a population mean that are discussed in Chapters 6 and 7. In this section we discuss histograms, and in Section 4.4 we present other graphical displays of data.

Figures 4.3.1 and 4.3.2 show histograms that are used to display the information in Table 4.2.6. Figure 4.3.1 shows a frequency histogram, and Figure 4.3.2 shows a relative frequency histogram. Histograms are graphs with vertical bars (rectangles) over each class interval. In a frequency histogram, the height of a rectangle over a class interval corresponds to the frequency (number) of observations in that interval. In a relative frequency histogram, the height of a rectangle corresponds to the relative frequency of observations in that interval. Thus, the shapes of a frequency histogram and a relative frequency histogram with the same class intervals are the same for any given set of data. The height of each rectangle is listed at the top of the rectangle in Figures 4.3.1 and 4.3.2. The heights of the rectangles in Figure 4.3.1 are the frequencies listed in column 2 of Table 4.2.6. The heights of the rectangles in Figure 4.3.2 are the relative frequencies listed in column 3 of Table 4.2.6. The cutpoints are also displayed in each figure and represent boundaries of the class intervals shown in column 1 of Table 4.2.6. Thus, from the histograms in Figures 4.3.1 and 4.3.2, one can reconstruct Table 4.2.6, and vice versa.

FIGURE 4.3.1
Frequency Histogram

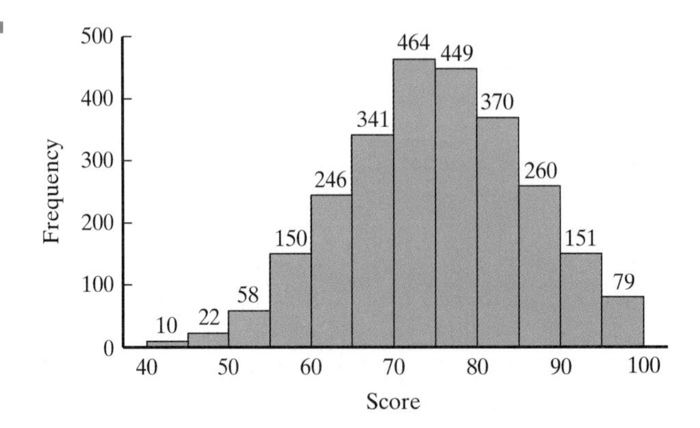

FIGURE 4.3.2
Relative Frequency Histogram

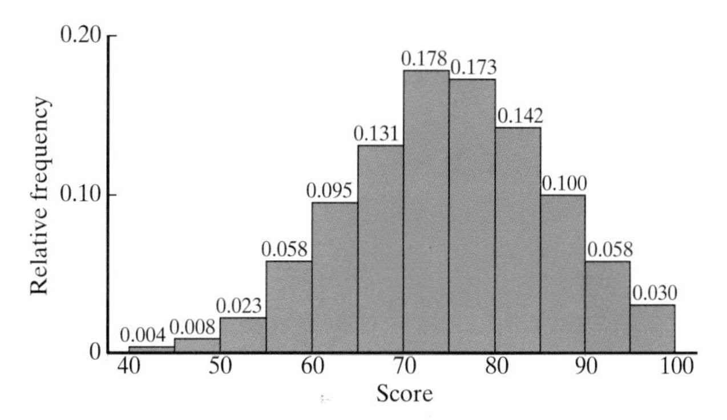

Shapes of Histograms

As we previously stated, histograms can be used to visualize how values of a quantitative variable are distributed. The distribution of a quantitative variable is often described by the shape of the histogram. We now present three relative frequency histograms that have very different shapes. Figure 4.3.3 is a relative frequency histogram that shows the number of home runs hit by the 455 major league baseball players who hit at least one home run in 1995. Notice that the histogram has a long right tail. This is because only a few superstars hit a relatively large number of home runs. A histogram with a long right tail such as the one shown in Figure 4.3.3 is described as **skewed to the right**.

FIGURE 4.3.3
Relative Frequency Histogram for Home Runs Hit in Major League Baseball for 1995 (skewed to the right)

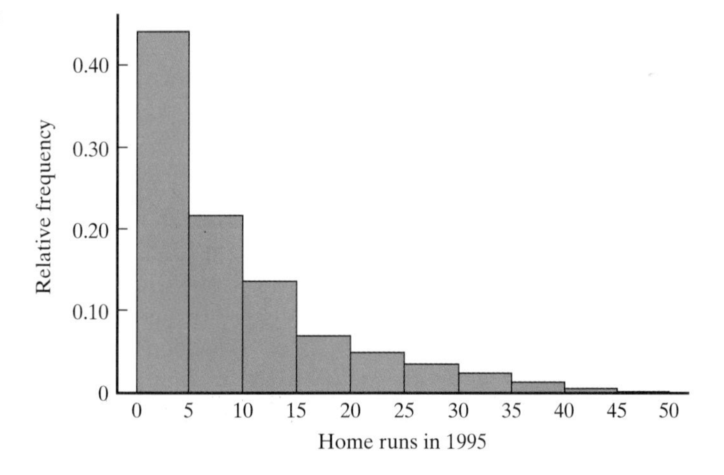

In Figure 4.3.4 the relative frequency histogram has a long left tail and is described as **skewed to the left**. The data represented in this figure are the scores on a midterm exam in an elementary statistics course. In this case the majority of the class did well on the exam; the rectangles with the greatest frequencies appear with the higher test scores. Unfortunately, a few students did not perform well, and they appear in the tail at the left side of the histogram.

The relative frequency histogram in Figure 4.3.5 represents the relative frequencies of weights (in pounds) of the 1,500 watermelons in a patch. The number above a rectangle corresponds to the relative frequency of watermelons that have

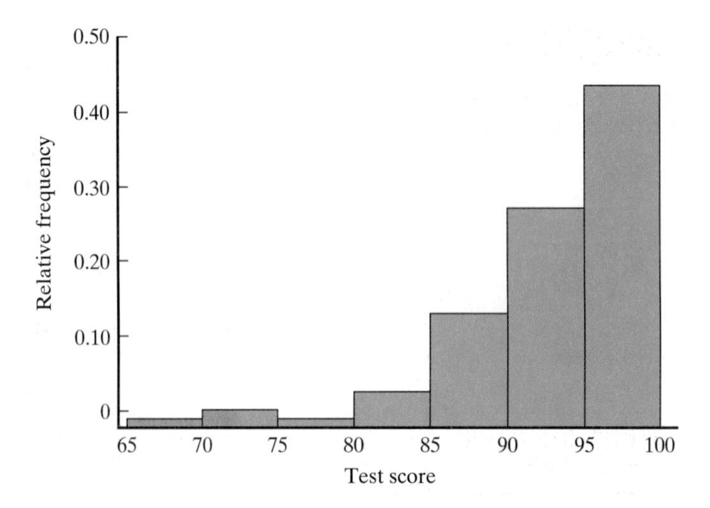

FIGURE 4.3.4
Relative Frequency Histogram of Elementary Statistics Test Scores (skewed to the left)

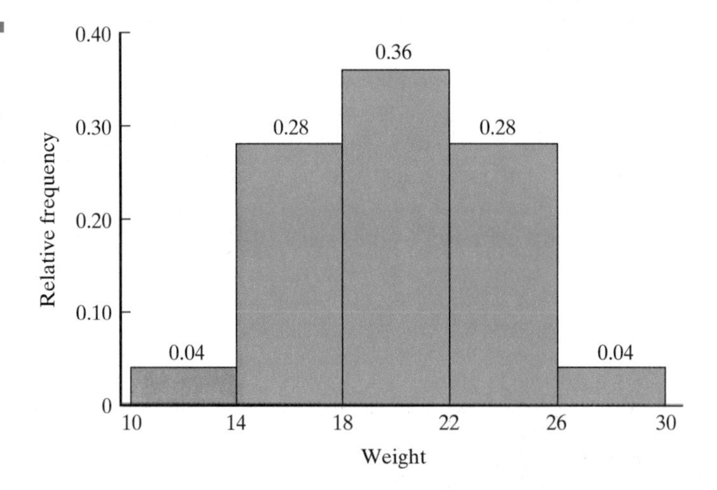

FIGURE 4.3.5
Relative Frequency Histogram of Watermelon Weights (in pounds)

weights in that class interval. Notice that the histogram in Figure 4.3.5 is not skewed in either direction. Such a histogram is described as **symmetric**. A histogram with c class intervals is defined as symmetric if the relative frequency in the ith class interval is equal to the relative frequency in the $(c + 1 - i)$th class interval for $i = 1, 2, \ldots, c$. For example, in Figure 4.3.5 there are 5 class intervals, so $c = 5$. The relative frequency in the second class interval $(i = 2)$ is 0.28. Because the histogram is symmetric, this is equal to the relative frequency in class interval $(c+1-i) = (5+1-2) = 4$.

Using Minitab to Construct a Histogram

Computer programs are generally used to construct histograms. We will describe how to use Minitab to obtain the frequency histogram shown in Figure 4.3.1. Retrieve the file **grades.mtp** and enter the following Minitab command and subcommands in the session window.

```
MTB > histogram c2;
SUBC> cutpoint 40:100/5;
SUBC> bar;
SUBC> type 1;
```

```
SUBC> color 6;
SUBC> symbol;
SUBC> type 0;
SUBC> label.
```

The subcommand `cutpoint 40:100/5` instructs Minitab to use class intervals from 40 to 100 with widths equal to 5. If you want Minitab to determine the class intervals when constructing a histogram, you can use the single command

```
MTB > histogram ci
```

where the data are in column `i`. If this command is used, the midpoints of the class intervals are labeled rather than the cutpoints.

Stem-and-Leaf Plot

A **stem-and-leaf plot** is another useful tool for summarizing values of a quantitative variable. It is an alternative to a frequency table or a frequency histogram. To demonstrate, the following set of numbers represents the scores of 33 students on a midterm exam in an elementary statistics course:

92, 93, 68, 92, 100, 74, 78, 94, 84, 84, 96, 100, 90, 96, 86, 96

94, 90, 90, 92, 76, 82, 80, 82, 94, 88, 80, 96, 94, 94, 92, 86, 96

The scores are taken from an alphabetical listing of the students, and the instructor wants to quickly summarize the data to display the level of student performance on the exam. A stem-and-leaf plot for these data is shown in Figure 4.3.6.

FIGURE 4.3.6
Stem-and-Leaf Plot of Exam Scores

Stem	Leaf
6	8
7	4 6 8
8	0 0 2 2 4 4 6 6 8
9	0 0 0 2 2 2 2 3 4 4 4 4 6 6 6 6 6
10	0 0

In this data set, the rightmost digit in each number is called a *leaf*. The remaining digit(s) form a *stem*. For example, for the score 68, the stem is 6 and the leaf is 8. For the score 100, the stem is 10 and the leaf is 0. The stems are written in a column, and the leaves for each stem are ordered from smallest to largest and listed in a row to the right of the stem. For example, the second row in Figure 4.3.6 has a stem of "7" and a leaf of "4," of "6," and of "8." This represents the scores 74, 76, and 78. From Figure 4.3.6 we see that nine students scored between 80 and 88 (inclusive), five students scored 96, and two students scored 100. If Figure 4.3.6 is rotated 90 degrees, it has the appearance of a frequency histogram. You can see that the stem 9 has the largest frequency of items. Unlike a frequency table, the individual values for all items in the data set can be recovered from a stem-and-leaf plot. Stem-and-leaf plots are especially useful if the data set is small and if the item numbers have two or three digits. However, some investigators find it useful to construct a stem-and-leaf plot for large data sets. Most computer software packages have programs to produce these plots.

4.4 MORE DISPLAYS

In the previous section we showed how histograms are used as pictorial representations of frequency and relative frequency tables for a quantitative variable. We now present several other charts that can be used to present data for quantitative and qualitative variables. These include

1. pie charts (used to display one qualitative variable),
2. bar charts (used to display the relationship between two variables, where one is quantitative and one is qualitative),
3. scatter plots (used to display the relationship between two quantitative variables),
4. line plots and area charts (used to display the relationship between two quantitative variables, where one of the variables is a measure of time).

Each of the preceding displays is informative, and in any particular problem it may be appropriate to present several displays. In other problems only one display may be appropriate. After reading this section you will be able to determine which display is best suited to your needs in any particular situation. We now discuss each of these displays and show Minitab commands that can be used to produce them. If you don't have the Minitab software package, you can ignore the commands and just observe the displays. Many software packages can be used to produce these displays.

Pie Charts

A **pie chart** is a popular pictorial representation often seen in newspapers and magazines. It is used to display a qualitative variable that has two or more values. The pie chart is a graphical representation of the frequency (number) or the proportion (equivalently, the percentage) of items in each category. To demonstrate, Figure 4.4.1 presents a pie chart for Table 4.4.1. This table is the subset of Table 4.2.4 that summarizes the male responses in the survey to determine public opinion about the death penalty.

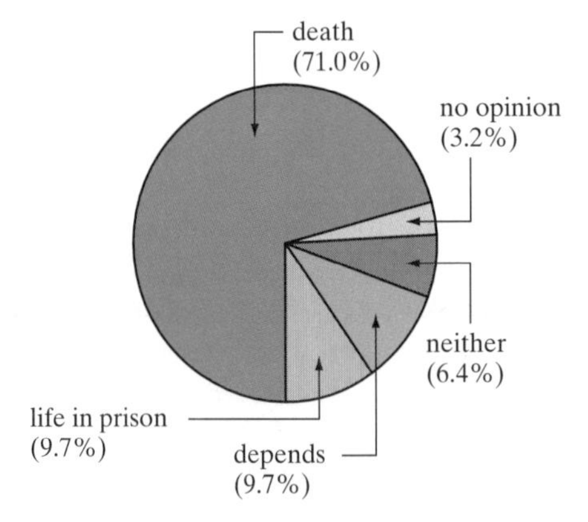

FIGURE 4.4.1
Pie Chart of Percentage of Male Responses of Preferred Penalty for Murder

From Table 4.4.1 we see that 71.0% of the men in the sample favor the death penalty as the preferred punishment for murder. Thus, this category represents 71.0% of the total area of the pie in Figure 4.4.1. In a similar manner, the other pieces of the pie represent the percentage of males in the other four categories of "Preferred punishment." In the next example we demonstrate how Minitab can be used to construct a pie chart.

TABLE 4.4.1 Summary of Male Responses of Preferred Penalty for Murder

Preferred punishment	Number of males (frequency)	Proportion of males (relative frequency)
Death penalty	22	0.710
Life in prison	3	0.097
Depends	3	0.097
Neither	2	0.064
No opinion	1	0.032

EXAMPLE 4.4.1

UNIVERSITY GRADUATES. Table 4.4.2 reports the frequency and relative frequency of graduates from a university in May 1996 categorized by the type of degree. "Type of degree" is a qualitative variable with the values *bachelor's*, *master's*, and *doctoral*. These data are illustrated with the pie chart shown in Figure 4.4.2.

TABLE 4.4.2 Summary of Type of Degree for May 1996 Graduates

Type of degree	Frequency	Relative frequency
Bachelor's	2,543	0.676
Master's	897	0.238
Doctoral	323	0.086

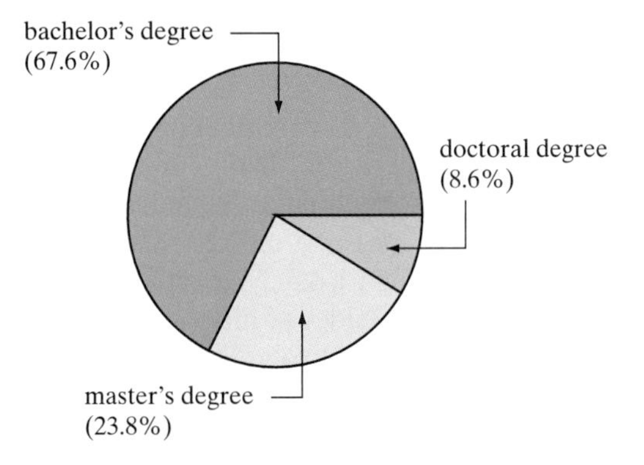

FIGURE 4.4.2
Types of Degree for May 1996 Graduates

To obtain Figure 4.4.2 using Minitab, place the values of the variable "Type of degree" (*bachelor's degree*, *master's degree*, and *doctoral degree*) in the first three rows of column 1 of the worksheet. Enter the frequencies shown in Table 4.4.2 in column 2 of the worksheet. Enter the following command and subcommand in the session window:

```
MTB > %pie c1;
SUBC> counts c2;
SUBC> label 3.
```

When a qualitative variable has many categories, where each one makes up a small percentage of the total, a pie chart may not be a very informative display. There would be many very small pieces of pie, and it would be difficult to distinguish individual slices. In this situation it may be possible to combine some of the categories and obtain a smaller set of meaningful groups.

Bar Charts

A bar chart was used in Figure 1.2.1 to display the price of pizza at three different restaurants. **Bar charts** are useful for viewing the relationship of two variables, where one variable is quantitative and the other is qualitative. We illustrate with an example.

EXAMPLE 4.4.2

EMISSIONS STANDARD FOR MANUFACTURING PLANTS. A large manufacturing plant wants to locate in Arizona and has asked the County Pollution Control Division to outline the emission limits for several chemical compounds. Table 4.4.3 reports the daily limits for the compounds.

TABLE 4.4.3 Daily Pollution Limits

Item number	Variable 1 (qualitative) Chemical compound or gas	Variable 2 (quantitative) Daily limit in pounds
1	Carbon monoxide	550
2	Oxides of nitrogen	1,600
3	Oxides of sulfur	430
4	Total suspended particulates	145
5	Total volatile organic compounds	300
6	Nonprecursor organic compounds	130

Variable 1 in Table 4.4.3, "Chemical compound or gas," is a qualitative variable, and Variable 2, "Daily limit in pounds," is quantitative. A bar chart is useful to display this type of information. Figure 4.4.3 presents a **vertical bar chart** of the daily pollution limits in Table 4.4.3. The numbers on the vertical axis correspond to values of the quantitative variable, "Daily limit in pounds." Each bar in the chart starts at zero and has a height equal to the daily pollution limit for the category represented by the bar. Notice that unlike a histogram, the bars in a bar chart do not touch. The bars in Figure 4.4.3 are represented by the following code: 1 = Carbon monoxide, 2 = Oxides of nitrogen, 3 = Oxides of sulfur, 4 = Total suspended particulates, 5 = Total volatile organic compounds, and 6 = Nonprecursor organic compounds.

FIGURE 4.4.3
Daily Pollution Limits (in pounds)

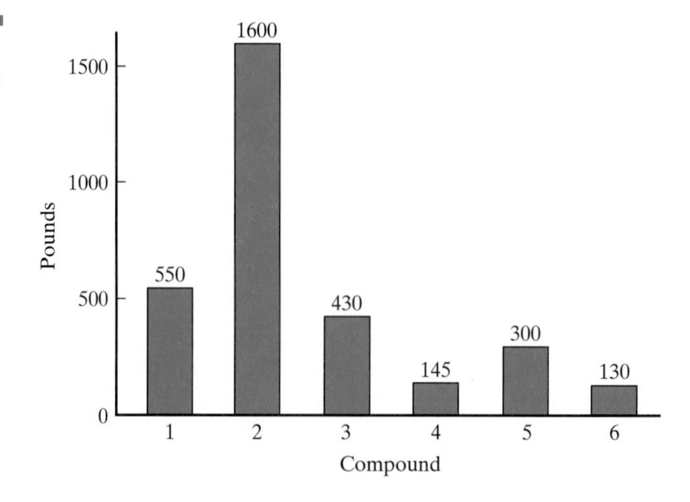

To construct Figure 4.4.3 using Minitab, enter the values 1, 2, 3, 4, 5, and 6 into column 1 of a worksheet. Enter the values 550, 1600, 430, 145, 300, and 130 into column 2. Then enter the following command and subcommands in the session window:

```
MTB > chart c2*c1;
SUBC> bar;
SUBC> type 1;
SUBC> color 1;
SUBC> symbol;
SUBC> type 0;
SUBC> label.
```

Scatter Plots

Scatter plots are useful when you want to examine the relationship between two quantitative variables, Y and X, measured on each item in a data set. A plot of "Y versus X" with Y on the vertical axis and X on the horizontal axis can reveal the association and relationship between Y and X. We illustrate with an example.

EXAMPLE 4.4.3

WEIGHT AND LENGTH OF MAMMALS. A wildlife biologist who is interested in the relationship of the length and weight of a certain species of mammal at age 12 months collected the data shown in Table 4.4.4 on 25 offspring of this species.

TABLE 4.4.4 Weight and Length of Mammals

Animal number	Weight (in pounds)	Length (in inches)
1	16.4	23.1
2	16.3	22.8
3	18.2	24.0
4	14.8	22.1
5	14.7	21.5
6	16.4	23.2
7	17.3	22.7
8	13.0	20.2
9	18.8	24.2
10	16.3	23.5
11	15.0	22.5
12	16.8	22.6
13	16.0	23.2
14	18.4	23.4
15	13.5	21.2
16	12.4	20.6
17	15.6	22.2
18	16.6	23.4
19	16.4	23.1
20	14.1	21.0
21	17.9	23.1
22	17.7	22.8
23	16.3	23.5
24	16.6	22.3
25	15.2	22.5

Figure 4.4.4 contains a scatter plot of the mammal data with the quantitative variable "Length" plotted on the x- (horizontal) axis and the quantitative variable "Weight" plotted on the y- (vertical) axis. From a scatter plot you can observe patterns that might exist. Notice that "Weight" tends to increase as "Length" increases in Figure 4.4.4. One can use this information to estimate how much "Weight" tends to increase per unit increase in "Length."

FIGURE 4.4.4
Scatter Plot of Weight and Length

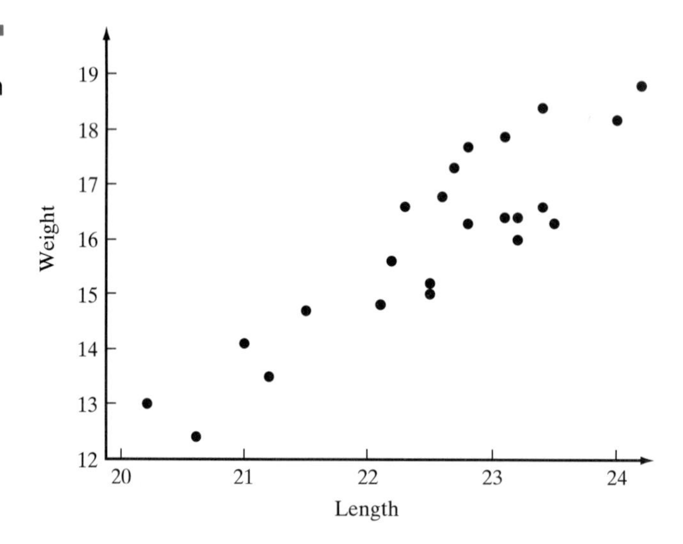

The data shown in Table 4.4.4 are stored in the file **mammalwt.mtp** on the data disk. Column 1 contains the animal number. Column 2 contains the value of "Weight" (in pounds), and column 3 contains the value of "Length" (in inches) of each offspring at age 12 months. To obtain the scatter plot in Figure 4.4.4 using Minitab, retrieve the data and enter the following command in the session window:

```
MTB > plot c2*c3
```

Line Plots and Area Charts

Sometimes two quantitative variables Y and X are of interest where X is a measure of time. The X variable may appear in units of seconds, minutes, hours, days, quarters, years, or any other unit of time. These data are often called **time-series** data. We illustrate with an example.

EXAMPLE 4.4.4

MILK PRICES. A consumer group has received several complaints from the residents of a large retirement community about the price of milk in a neighborhood market that is the most convenient store for the residents to shop for groceries. The complaint is that the price of 2% milk seems to increase around the 20th of each month, the date that residents receive their retirement checks. The consumer group decides to investigate. The group wants to answer the question, "For the past 30 days, how did the daily price of a quart of 2% milk at this neighborhood market compare with the price at a large supermarket three miles away?"
To investigate, the price of 2% milk was obtained from both stores for each of the past 30 days. These daily prices were compared and recorded as shown in Table 4.4.5. The number recorded each day is the daily price of a quart of 2% milk at the

neighborhood market minus the price at the supermarket. If the number recorded is negative, the price at the supermarket is higher. If the number recorded is positive, the price at the neighborhood market is higher.

TABLE 4.4.5 Comparison of Daily Milk Prices at Two Stores for the Past 30 Days

Day	Price difference (in cents)	Day	Price difference (in cents)
1	0	16	+1
2	0	17	+1
3	−1	18	+2
4	−1	19	+3
5	−3	20	+4
6	−3	21	+12
7	−3	22	+11
8	0	23	+10
9	+2	24	+9
10	+2	25	+6
11	+2	26	+5
12	+2	27	+4
13	+2	28	+3
14	+2	29	+1
15	+2	30	0

From Table 4.4.5 you can readily see how the daily prices compare. For example, on day 21 the price of a quart of milk is 12 cents higher at the neighborhood market than at the supermarket. Figure 4.4.5 contains a line plot of the data given in Table 4.4.5 with the time variable "Day" on the x- (horizontal) axis and "Price difference" on the y- (vertical) axis. There appears to be a general upward trend in price differences over the first 21 days, with an unusually large increase on day 21. The price differences then decrease every day from day 21 to day 30.

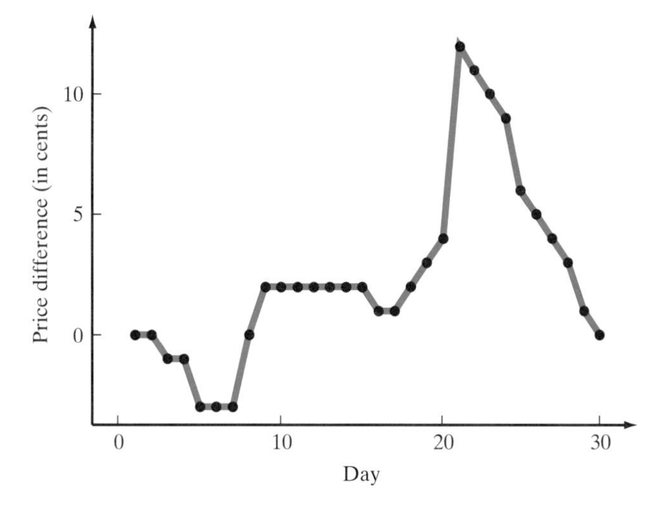

FIGURE 4.4.5
Line Plot of the Comparison of the Daily Price of One Quart of Milk

The data shown in Table 4.4.5 are in the file **milkpric.mtp**. Column 1 contains the values for "Day" and column 2 contains the values for "Price difference." To construct Figure 4.4.5 using Minitab, retrieve the data and enter the following command and subcommands in the session window:

```
MTB > plot c2*c1;
SUBC> symbol;
SUBC> size 0.5;
SUBC> connect.
```

Figure 4.4.6 contains the same plot shown in Figure 4.4.5 except that the area between the horizontal axis and the line is shaded. Some investigators find that shading makes the chart easier to read. This is called an **area chart**. The following Minitab command and subcommand can be used to obtain the area chart in Figure 4.4.6:

```
MTB > plot c2*c1;
SUBC> area.
```

FIGURE 4.4.6
Area Chart of the Comparison of the Daily Price of One Quart of Milk

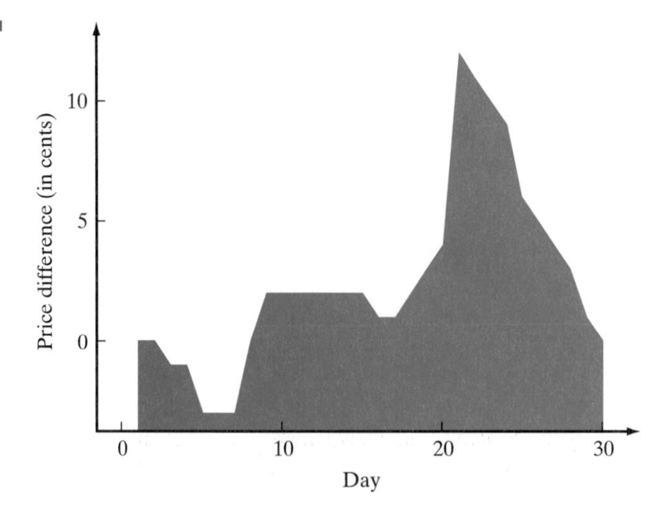

Recall that the consumer group in Example 4.4.4 seeks an answer to the question, "For the past 30 days, how did the daily price of a quart of 2% milk at this neighborhood market compare with the price at a large supermarket three miles away?" Which of the following formats do you find most useful for answering the question?

1. Table 4.4.5
2. Figure 4.4.5
3. Figure 4.4.6

If you are presenting data to your employer, think how that person can get the most information quickly. Figures 4.4.5 and 4.4.6 give a quick overview of the data; but if detailed numbers are needed, it may be necessary to refer to Table 4.4.5. As this problem illustrates, sometimes *both* a picture and a table are needed to adequately answer a question. For the milk-price data set, Table 4.4.5 and Figure 4.4.6 together provide a pleasing and informative way to present the data.

THINKING STATISTICALLY: SELF-TEST PROBLEM

4.4.1 Students conducted an experiment to determine how certain factors affect the amount of popcorn produced using a microwave oven. The results of the experiment are shown in Table 4.4.6. The volume of popcorn is measured in (fluid) ounces by pouring the popcorn into a measuring cup after it has been popped.

TABLE 4.4.6 Popcorn Data

Trial	Type of popcorn	Popping time (minutes)	Volume (ounces)
1	plain	2	26
2	plain	2	20
3	plain	2	32
4	plain	3	52
5	plain	3	54
6	plain	3	42
7	plain	4	96
8	plain	4	82
9	plain	4	83
10	buttered	2	62
11	buttered	2	56
12	buttered	2	54
13	buttered	3	68
14	buttered	3	38
15	buttered	3	64
16	buttered	4	80
17	buttered	4	85
18	buttered	4	83

a. Three variables are measured for each trial. Name each variable and indicate whether it is quantitative or qualitative.
b. What display would be the most appropriate for presenting the average volume for each type of popcorn?
c. What display would be the most appropriate for examining the relationship between "Popping time" and "Volume"?

4.5 CHAPTER SUMMARY

KEY TERMS

1. frequency table
2. relative frequency table
3. class interval
4. cutpoint
5. midpoint
6. histogram
7. skewness
8. symmetry
9. stem-and-leaf plot
10. pie chart
11. bar chart
12. scatter plot
13. line plot
14. area chart

SYMBOL

1. *a* to *b*—defines the class interval with values in the range $a \leq value < b$. The interval that contains the maximum data value contains both boundary points *a* and *b*.

KEY CONCEPTS

1. Data must be organized and summarized to facilitate presentation and interpretation.
2. Graphical displays are invaluable in helping to make decisions based on data.
3. If an item is to be chosen at random from a set of items, the probability that an associated value of a quantitative variable will be in the interval from *a* to *b* is the proportion of items with values that are contained in this interval.
4. A histogram is used to display a frequency or relative frequency table based on the values of a quantitative variable.
5. A pie chart is used to display one qualitative variable.
6. A bar chart is used to display the relationship between two variables, where one is quantitative and one is qualitative.
7. A scatter plot is used to display the relationship between two quantitative variables.
8. A line plot is used to display the relationship between two quantitative variables, where one of the variables is a measure of time.

SKILLS

1. Construct a frequency table and a histogram for the values of a quantitative variable.
2. Describe the shape of a histogram.
3. Use a statistical software package to create a pie chart, a bar chart, a scatter plot, a line plot, and an area plot.
4. Determine the most appropriate visual aid(s) to display a data set.

4.6 CHAPTER EXERCISES

4.1 Each pumpkin in a patch was weighed when it was harvested. The frequencies of weights (in pounds) are exhibited in Table 4.6.1.

TABLE 4.6.1 Frequency of Pumpkin Weights

Class interval (in pounds)	Frequency
10 to 12	87
12 to 14	322
14 to 16	808
16 to 18	1473
18 to 20	1811
20 to 22	1525
22 to 24	938
24 to 26	349
26 to 28	110

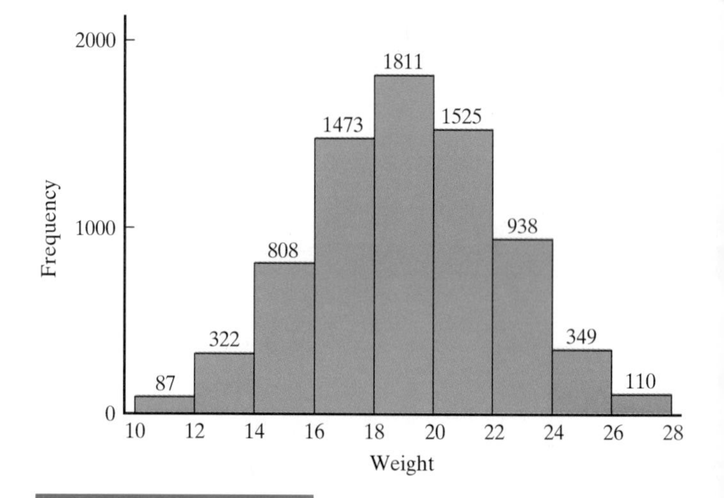

FIGURE 4.6.1
Frequency Histogram of Pumpkin Data

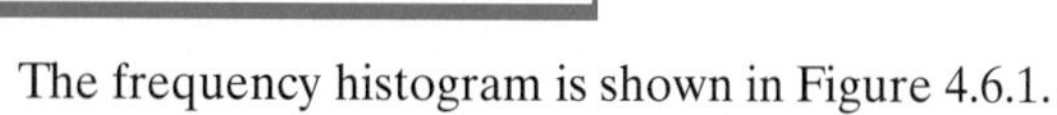
The frequency histogram is shown in Figure 4.6.1.

a. How many pumpkins were in the patch?

b. Compute the relative frequencies of weights in Table 4.6.1.

c. What is the probability that a pumpkin chosen at random from all pumpkins in the patch will weigh less than 18 pounds?

d. What proportion of the pumpkins weighs between 16 and 26 pounds?

e. What proportion of pumpkins weighs at least 24 pounds or less than 14 pounds?

f. What proportion of pumpkins weighs between 10 and 30 pounds?

g. If a pumpkin is chosen at random from all pumpkins in the patch, what is the probability that the weight of the chosen pumpkin will be in the interval 16 to 18?

4.2 A study was conducted to determine the amount of annual rainfall in regions of Arizona. To conduct the study, water gauges were placed at various locations in the state to measure the amount of rainfall in the 12-month period from January 1993 through December 1993. In order to determine the location of the gauges, a map was used to separate the state into 600 geographic regions of roughly the same size. A simple random sample of 104 of the regions was selected from the set of 600 regions, and gauges were placed in a central location for each of the 104 regions in the sample. At the end of the 12-month period, the rainfall measured by the gauges was recorded to the nearest tenth of an inch (0.1). Table 4.6.2 uses 10 class intervals to report the frequencies and relative frequencies for the 104 regions.

TABLE 4.6.2 Annual Rainfall (in Inches)

Class interval (in inches)	Frequency	Relative frequency
2.5 to 5.5	8	0.08
5.5 to 8.5	21	0.20
8.5 to 11.5	24	0.23
11.5 to 14.5	18	0.17
14.5 to 17.5	17	0.16
17.5 to 20.5	5	0.05
20.5 to 23.5	5	0.05
23.5 to 26.5	5	0.05
26.5 to 29.5	0	0.00
29.5 to 32.5	1	0.01

The frequency histogram of the annual rainfall is shown in Figure 4.6.2.

a. How many of the 104 regions reported an annual rainfall of 2.5 to 11.5 inches?

b. How many of the 104 regions reported an annual rainfall of less than 17.5 inches?

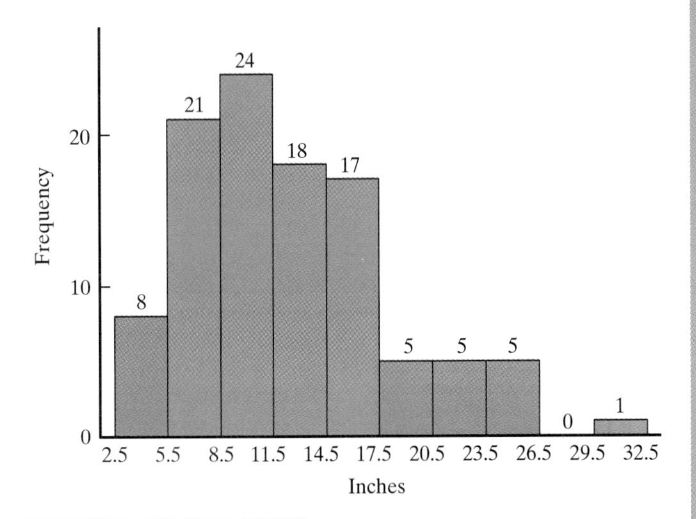

FIGURE 4.6.2
Frequency Histogram of Annual Rainfall

c. What proportion of the regions had an annual rainfall from 20.5 to 32.5 inches?

d. Is the histogram shown in Figure 4.6.2 symmetric, skewed to the right, or skewed to the left?

4.3 In addition to recording the rainfall at each gauge site in Exercise 4.2, the elevation was recorded. The elevations (in feet) for the 104 randomly selected sites are shown in Table 4.6.3.

TABLE 4.6.3 Elevation (in Feet)

Class interval (in feet)	Frequency	Relative frequency
0 to 1,000	5	0.048
1,000 to 2,000	5	0.048
2,000 to 3,000	14	0.135
3,000 to 4,000	14	0.135
4,000 to 5,000	22	0.211
5,000 to 6,000	12	0.115
6,000 to 7,000	16	0.154
7,000 to 8,000	11	0.106
8,000 to 9,000	3	0.029
9,000 to 10,000	2	0.019
Total	104	1.000

a. How many regions have an elevation less than 7,000 feet?

b. How many regions have an elevation greater than or equal to 5,000 feet?

c. What proportion of the regions has an elevation in the interval 2,000 to 6,000 feet?

4.4 The owner of an ice cream shop, The Creamery Club, wishes to determine the impact of the store's advertising. For the past 12 months, the store has advertised in a movie theater located in the same shopping mall. The owner of the ice cream shop is interested in determining if people who visit the movie theater remember the store's advertisement. The owner decided to survey people as they leave the theater. People who responded to the survey were given a card for a free ice cream cone. All respondents were asked the following question: "Do you remember the advertisement for The Creamery Club?" The interviewer also recorded the age of the respondent. Table 4.6.4 reports responses for 30 respondents. The data are stored in the file **icecream.mtp** on the data disk.

TABLE 4.6.4 Responses for Awareness of Advertisement

Sample item number	Do you remember the advertisement?	Age
1	no	42
2	no	18
3	no	41
4	no	41
5	no	44
6	no	35
7	no	57
8	yes	35
9	no	50
10	no	26
11	no	43
12	no	43
13	no	22
14	no	68
15	no	35
16	no	21
17	no	36
18	no	33
19	yes	19
20	no	56
21	no	36
22	no	47
23	no	21
24	yes	36
25	no	36
26	yes	53
27	yes	21
28	no	25
29	no	66
30	yes	19

a. Define the two variables in this study. Denote whether each variable is quantitative or qualitative.

b. Complete Table 4.6.5 to display the number of respondents who remembered the advertisement.

TABLE 4.6.5 Number of Respondents Who Remembered the Advertisement

Response	Frequency
no	
yes	

c. Use Minitab or any other computer software package to construct a pie chart to display the data in Table 4.6.5. If you do not have a software package, construct the pie chart by hand.

d. Construct a stem-and-leaf plot for the variable "Age." Use the integers 1 through 6 as the stems in your plot.

4.5 The data in Table 4.6.6 appeared in the January 28, 1994, issue of the *Mesa Tribune*. It reports net profits and losses for a major airline for the 12 quarters (12 three-month periods) for 1991 to 1993. All amounts are reported in millions of dollars. Losses are reported as negative values. Construct a line chart of these data and use it to describe the general trend of profits. If you want to use Minitab, the data are contained in the file **airline.mtp** on the data disk.

TABLE 4.6.6 Net Profits and Losses (in Millions of Dollars)

Year	Quarter	Net profit or loss
1991	1	−49.9
1991	2	−30.9
1991	3	−85.0
1991	4	−56.0
1992	5	−10.0
1992	6	−33.4
1992	7	−70.8
1992	8	−17.6
1993	9	2.1
1993	10	10.1
1993	11	14.4
1993	12	10.4

4.6 Super Bowl XXX was played in January 1996 in Tempe, Arizona, between the Dallas Cowboys and the Pittsburgh Steelers. Tickets to the game were distributed as follows: Arizona Cardinals (host team)—10%, league office—25%, Dallas Cowboys—17.5%,

Pittsburgh Steelers—17.5%, and the other National Football League teams—30%. Construct a pie chart that represents this information. The data are contained in the file **football.mtp** on the data disk.

4.7 Two variables were measured for recent graduates from a college of business. The variable "Gradgpa" is the grade point average at graduation. The variable "Appldgpa" is the grade point average when the student applied for admission to the college. The college uses "Appldgpa" to help decide whether a student should be admitted to thc college since this variable is a good measure of the student's potential in the program. The data are in the file **busgrade.mtp** on the data disk. Column 1 contains values for "Gradgpa" and column 2 contains values for "Appldgpa." Construct a scatter plot with "Gradgpa" on the vertical axis and "Appldgpa" on the horizontal axis. Does there appear to be a relationship between the two variables?

4.8 Table 4.6.7 reports information collected from 20 students who recently completed a course in intermediate statistics. Student performance is measured by total points. It is of interest to determine whether factors can be identified that affect performance. Four variables were measured for each student. The four variables are shown in columns 2 to 5 in Table 4.6.7.

a. Name each of the four variables and indicate whether each is quantitative or qualitative.

b. What type of display would be the most appropriate for examining the relationship between "GPA" and "Total points"?

c. What proportion is represented in each slice of a pie chart that displays the proportion of students in each of the four categories of "Major"?

d. Complete Table 4.6.8 by filling in the Frequency column.

TABLE 4.6.8 Frequency Table of Total Points

Total points	Frequency
280 to 295	
295 to 310	
310 to 325	
325 to 340	
340 to 355	
355 to 370	

e. How many students scored less than 325 total points?

f. Sketch a frequency histogram of "Total points" using the class intervals defined in Table 4.6.8.

g. How many students had a value for "Total points" less than 340 but at least 295?

TABLE 4.6.7 Intermediate Statistics Data

Student number	Major	Grade point average (GPA)	Grade in elementary statistics	Total points (400 possible)
1	Accounting	3.30	A	361
2	Economics	2.00	B	281
3	Economics	3.10	B	295
4	Finance	3.75	A	342
5	Accounting	3.94	A	351
6	Marketing	2.89	B	317
7	Marketing	2.50	C	287
8	Finance	3.50	A	358
9	Finance	3.25	B	330
10	Economics	2.67	B	321
11	Accounting	3.10	B	280
12	Accounting	3.00	B	297
13	Economics	3.30	B	293
14	Finance	3.65	B	324
15	Finance	3.64	B	311
16	Finance	2.69	B	297
17	Marketing	2.60	C	287
18	Marketing	3.30	A	358
19	Accounting	3.45	A	350
20	Accounting	2.78	B	331

4.7 SOLUTIONS TO SELF-TEST PROBLEMS IN CHAPTER 4

4.2.1 The answer is $449 + 370 = 819$.

4.2.2 The answer is $370 + 260 + 151 + 79 = 860$.

4.2.3 Using the relative frequencies in Table 4.2.6, we obtain

$$0.004 + 0.008 + 0.023 + 0.058 = 0.093 = 9.3\%.$$

4.2.4 The answer to this question cannot be obtained from Table 4.2.6 since 72 is not a cutpoint. You can obtain the answer by using the table in Appendix B and counting the number of scores that are less than 72. Table 4.2.6 is only a summary of the data in Appendix B.

4.2.5 From Table 4.2.6 we see that $0.058 + 0.030 = 0.088 = 8.8\%$ of the students received a score of 90 or higher. The probability that a student to be selected at random obtained a score of 90 or higher is therefore 0.088, the same as the proportion of students in the data set with scores of 90 or higher.

4.4.1

a. The three variables are "Type of popcorn" (qualitative), "Popping time" (quantitative), and "Volume" (quantitative).

b. Since "Type of popcorn" is qualitative and "Volume" is quantitative, a bar chart would be the most appropriate display.

c. Since "Popping time" and "Volume" are both quantitative, a scatter plot would be the most appropriate display.

CHAPTER 5

MEDIAN AND INTERQUARTILE RANGE; MEAN AND STANDARD DEVIATION

5.1 INTRODUCTION

In addition to various tables and displays discussed in Chapter 4, numerical summaries are also useful for understanding data sets. Numerical summaries are especially useful if the data set contains a large number of items. For example, suppose you have a population consisting of the ages, heights, and weights of $N = 5{,}974$ people who were admitted to a large hospital during the past few years suffering from a heart attack. Since N is so large, it may be necessary to use just a few numbers that summarize the 5,974 values of age, height, and weight in order to make decisions about the population. The particular summary numbers to use depend on the information needed to make a decision. In this chapter we consider some numerical measures that are used to summarize the values of a population. We begin the presentation with an example.

EXAMPLE 5.1.1

COST OF HOUSING. A couple who will be retiring next year is considering moving to a large retirement community and wants to determine the cost of two-bedroom houses in the community. They wrote to a real estate agent who sells property in the community and asked her for this information. The agent could send the couple a computer printout of the prices of the 2,623 two-bedroom houses that were sold in the community during the past two years. However, the couple is not interested in searching through the entire set of 2,623 prices. Rather, they want a few numbers that summarize or represent the prices of two-bedroom houses in the community. What summary numbers might be useful to them? The cost of the lowest-priced house and the cost of the highest-priced house would provide some information. Perhaps a more useful number would be the price that is in the "middle" or "center" of all 2,623 prices.

Center of a Population

When making a decision about a population, an investigator often wants to use a *single number* to represent *all* the values of a population. The question that comes to mind is this: "What single number *best* represents the entire population?" Certainly it must be a number that is somewhere in the "center" of the population.

EXAMPLE 5.1.2

INCOMES OF RESIDENTS. The couple in Example 5.1.1 also wants to know the annual income of residents in the retirement community. They feel that if their annual income is significantly larger or smaller than most of the residents, they may be uncomfortable living in the community. They don't want to know the annual incomes of all 14,826 households in the community. Rather, they want a single number that represents a "middle" income to compare with their income. A "middle" income could be defined as an income such that one-half of the households have annual incomes less than this value and one-half have annual incomes greater than this value. Such a number would help the couple determine if their income is such that they would feel comfortable living in this community.

EXAMPLE 5.1.3

YOUR GRADE. In any course you have taken in college or high school you received a final grade. Although you received scores on quizzes, one-hour tests, homework, and the final exam, only one grade was entered on your permanent record. This grade was based on a single number (perhaps the average score) that summarized all the scores you received in the course. The score on which your final grade was based was neither the smallest nor the largest score you received on tests, quizzes, or homework. It was a number somewhere in the "middle" or "center" of these scores.

Sometimes more than one "center" value can be defined for a population, and an investigator must decide which one should be used for the problem under study. This is illustrated in the next example.

EXAMPLE 5.1.4

CENTER OF A CITY. A city council voted to build a city recreation hall to be located in the geographical center of the city, as shown in Figure 5.1.1. However, 95% of the city's inhabitants live in the shaded area shown in Figure 5.1.2, and a citizen's committee argues that the recreation hall should be located in the center of this populated area rather than in the geographical center of the city.
Clearly, one can argue that both "centers" are reasonable places to build the recreation hall. As this example demonstrates, the appropriate way to define "center" depends on how it will be used. The situation described in this example is similar to problems encountered in statistics when a decision depends on a

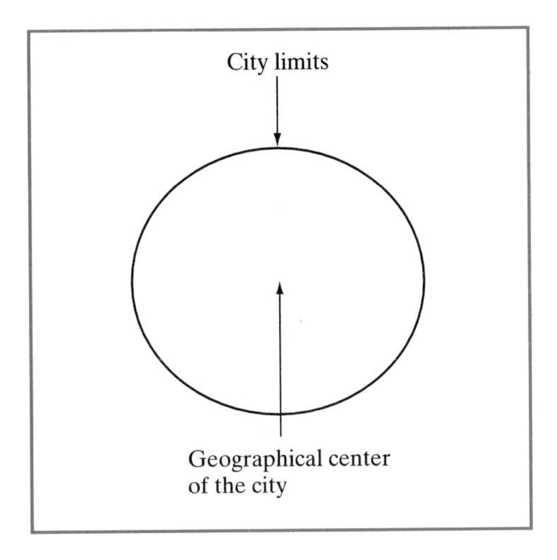

FIGURE 5.1.1
Geographical Center of the City

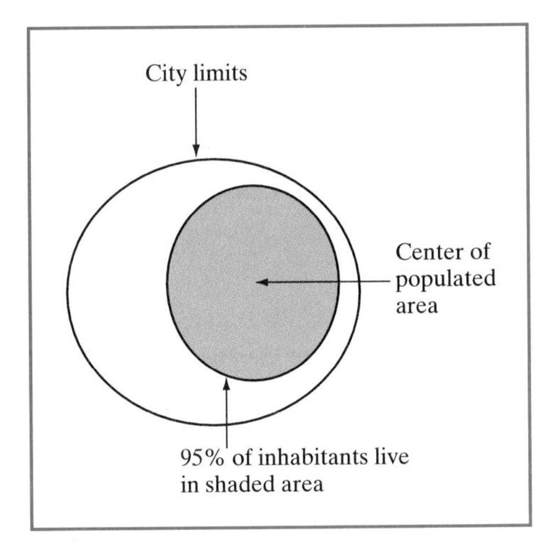

FIGURE 5.1.2
Center of the Populated Area of the City

"central" number that best describes the population. That is, the appropriate way to define the "center" of a population depends on how it is to be used.

In statistical applications, two "centers" of a population that are used extensively are the **mean** and the **median.** These two centers will be discussed in the next three sections.

THINKING STATISTICALLY: SELF-TEST PROBLEMS

5.1.1 Find a newspaper or magazine article in which a few numbers are used to summarize a large population of numbers. Do the summary numbers give you some general understanding about the population?

5.1.2 Determine how your final grade will be decided in this statistics course. Does it depend on a numerical score? Is this numerical score the "center" of all the numerical scores you will receive on homework, quizzes, one-hour tests, and the final examination?

5.1.3 Manufacturing specifications for a particular type of bolt require that it be between 2.50 inches and 2.60 inches in length. A population of numbers consists of the lengths of 1,000 bolts. The population is summarized with two numbers, the minimum length and the maximum length. The minimum length is 2.55 inches and the maximum length is 2.59 inches. Can you draw any conclusions about the population of 1,000 bolts based on these two summary numbers?

5.2 POPULATION MEDIAN AND INTERQUARTILE RANGE

Suppose a professor is interested in making a detailed examination of the calculus scores in Appendix B. Since the population of scores is so large, it is virtually impossible to obtain useful information from it in its present form. The frequency histogram in Figure 4.3.1 (which is repeated here in Figure 5.2.1) can be used to visually examine how the scores are distributed.

Suppose the professor wants to use a "central" number that best describes the entire population. Specifically, she wants to use a "central" number that is the middle of the population of calculus scores when they are arranged from smallest to largest. Such a central value is called the **median** of the population.

Convention. The symbol M will be used for a population median. You recognize the symbol M as the uppercase "em" in the English alphabet. This symbol is also the uppercase Greek letter *mu*, and so we continue our convention of using Greek letters to represent population parameters.

The median of a population of N numbers is the middle number when the numbers are arranged in order from smallest to largest. At least 50% of the population values are less than or equal to the median, and at least 50% of the population values are greater than or equal to the median. The median of any set of numbers is computed using the instructions in Box 5.2.1.

BOX 5.2.1: INSTRUCTIONS FOR COMPUTING A MEDIAN

The following steps can be used to compute the median of a population of N numbers:

1. Arrange the N population numbers in order from smallest to largest, and denote them by $Y_1, Y_2, Y_3, \ldots, Y_N$, where Y_i is the value of the population item in position i when counting up from the smallest value.
2. If N is an odd number, the median is the number that appears in position $(N + 1)/2$ when counting up from the smallest number. This is the middle number in the ordered set of N numbers.
3. If N is an even number, we define the median as the average of the two numbers that appear in positions $N/2$ and $(N/2) + 1$ when counting up from the smallest number. This is the average of the two middle numbers in the ordered set of N numbers.

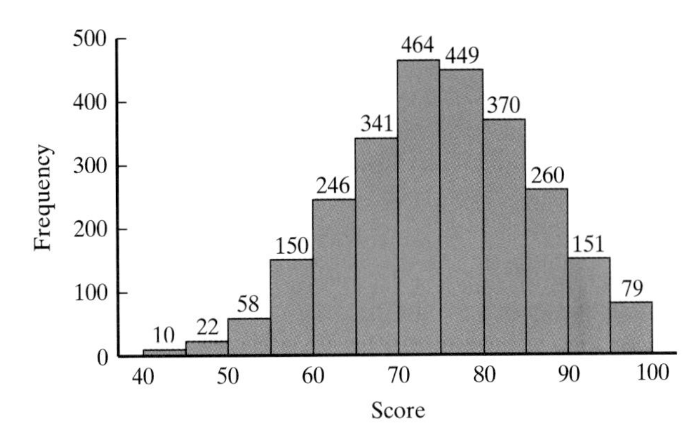

FIGURE 5.2.1
Frequency Histogram of Grades Data

EXAMPLE 5.2.1

TEMPERATURES IN PHOENIX. The numbers in Line 5.2.1 are the daily high temperatures (in degrees Fahrenheit) in Phoenix, Arizona, on nine consecutive days in March.

83 79 86 87 90 86 88 92 78 **Line 5.2.1**

Suppose we want to compute the median of this population of nine numbers. First arrange the temperatures from smallest to largest and obtain

Position number:	1	2	3	4	5	6	7	8	9
Temperature:	78	79	83	86	86	87	88	90	92

Since $N = 9$ is an odd number, the median is the temperature in position $(N + 1)/2 = (9 + 1)/2 = 5$. The temperature in this position is 86, which is the middle of this set of nine temperatures, so $M = 86$. Notice that five of the nine temperatures are less than or equal to M and six of the nine numbers are greater than or equal to M. Thus, *at least* 50% of the temperatures are less than or equal to M and *at least* 50% of the temperatures are greater than or equal to M.
If 79 is deleted from the set of numbers in Line 5.2.1, the modified set arranged from smallest to largest is

Position number:	1	2	3	4	5	6	7	8
Temperature:	78	83	86	86	87	88	90	92

In this case $N = 8$ is an even number. To find the median, select the temperatures located in the two positions $N/2 = 8/2 = 4$ and $(N/2) + 1 = 5$. These values are 86 and 87, so

$$M = \text{average of 86 and 87} = \frac{86 + 87}{2} = 86.5.$$

In this case exactly 50% of the temperatures are less than or equal to M, and exactly 50% of the temperatures are greater than or equal to M.

Quartiles

Sometimes an investigator wants to know more about a population than just the median. She may want some information about how the population values are distributed (spread out) around the median. Population quartiles are often used for this purpose.

Quartiles are three numbers that divide a population of numbers into four groups such that about 25% of the population values are in each group. The symbol $Q1$ is used to denote the first quartile, and at least 25% of the population values are less than or equal to $Q1$. The symbol $Q2$ is used to denote the second quartile, and at least 50% of the population values are less than or equal to $Q2$. The symbol $Q3$ is used to denote the third quartile, and at least 75% of the population values are less than or equal to $Q3$. The second quartile, $Q2$, is also the median, M. If $(N+1)/4$ is an integer (whole number), the quartiles of a population of N numbers are computed using the instructions in Box 5.2.2. If $(N+1)/4$ is not an integer, the formula for computing quartiles requires interpolation, and a computer package is generally used to perform the computations.

BOX 5.2.2: HOW TO COMPUTE QUARTILES WHEN $(N+1)/4$ IS AN INTEGER

The following four steps can be used to compute the quartiles of a population of N numbers when k is an integer, where $k = (N+1)/4$:

1. Arrange the N population numbers in order from smallest to largest, and denote them by $Y_1, Y_2, Y_3, \ldots, Y_N$, where Y_i is the value of the population item in position i when counting up from the smallest.
2. The first quartile $Q1$ is Y_k, the population number that appears in position k from the smallest.
3. The second quartile $Q2$ is Y_{2k}, the population number that appears in position $2k$ from the smallest. $Q2$ is also the median, M.
4. The third quartile $Q3$ is Y_{3k}, the population number that appears in position $3k$ from the smallest.

EXAMPLE 5.2.2

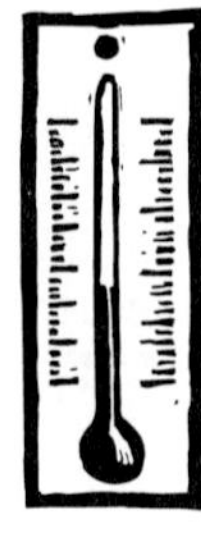

MORE TEMPERATURES. Consider the following 11 ordered temperatures.

Position number:	1	2	3	4	5	6	7	8	9	10	11
Temperature:	78	79	83	86	86	87	88	90	92	98	99

Since $N = 11$, it follows that $(N+1)/4 = (11+1)/4 = 3$ is an integer, so $k = 3$ and the instructions in Box 5.2.2 can be used to compute the quartiles. Thus,

$$Q1 = Y_3 = 83, \ Q2 = Y_6 = 87, \ Q3 = Y_9 = 92.$$

The **interquartile range (IQR)** of a population is defined as $\text{IQR} = Q3 - Q1$. At least 50% of the values in a population are between $Q1$ and $Q3$, inclusive. For the population of 11 temperatures in Example 5.2.2, $\text{IQR} = 92 - 83 = 9$. In this case, $7/11 = 0.636 = 63.6\%$ of the temperatures are between $Q1$ and $Q3$ inclusive.

Five-Number Summary

For any population of numbers, the five numbers

minimum, $Q1$, M, $Q3$, maximum

are referred to as a **five-number summary** of the data set. These five numbers provide a great deal of information about the distribution of population values. A five-number summary can be obtained by using Minitab. To illustrate, consider the population of calculus scores in Appendix B. The scores are stored in column 2 in the file **grades.mtp** on the data disk. Retrieve the data and enter the following command in the session window:

```
MTB > describe c2
```

The output is shown in Exhibit 5.2.1.

```
Descriptive Statistics

Variable        N      Mean    Median    TrMean     StDev    SEMean
Score        2600    74.520    75.000    74.622    10.923     0.214

Variable      Min       Max        Q1        Q3
Score      41.000   100.000    67.000    82.000
```

EXHIBIT 5.2.1
Minitab Output for Calculus Scores

From this exhibit we see that

$$\text{minimum} = 41, \quad Q1 = 67, \quad M = 75, \quad Q3 = 82, \quad \text{maximum} = 100.$$

We note that

$$\text{IQR} = 82 - 67 = 15.$$

These five numbers give us a great deal of information about how the population values are distributed.

Boxplots

Several summary numbers of a population are often presented graphically in a display called a **boxplot**. Figure 5.2.2 is the boxplot display for the calculus score data in Appendix B.

The box in the middle of the diagram has the upper line located at the third quartile ($Q3$), the lower line located at the first quartile ($Q1$), and the middle line located at the median (M). The height of the box ($Q3 - Q1$) is the interquartile range. For the calculus data $Q1 = 67$ and $Q3 = 82$, and so the interquartile range is $Q3 - Q1 = 82 - 67 = 15$. The vertical lines above and below the box are called **whiskers**. The lower whisker extends from the bottom of the box to the smallest observation that is greater than $Q1 - 1.5(\text{IQR})$. The upper whisker extends from the

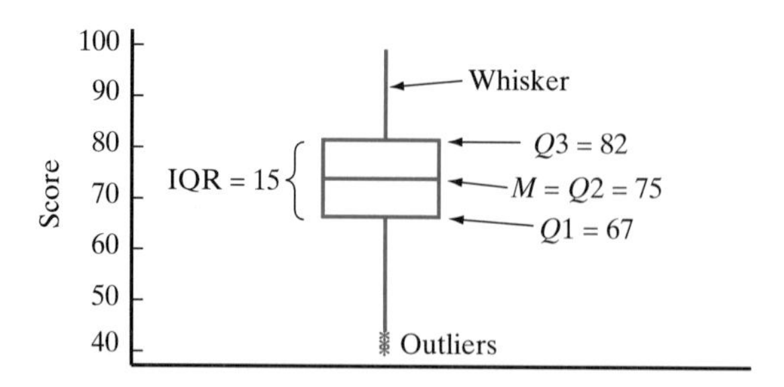

FIGURE 5.2.2
Boxplot of the Calculus Scores in Appendix B

top of the box to the largest observation that is less than $Q3 + 1.5(\text{IQR})$. Any observation that lies beyond either whisker is identified on the graph with an asterisk ($*$). These observations are called **outliers** and identify observations that are far removed from the rest of the population values. Outliers may have an unduly large influence on parameters computed from the population. It is prudent to examine these observations to determine if something may have gone wrong in collecting or measuring them. If these values have been incorrectly measured and/or recorded, they should be replaced with the correct values. The outliers in Figure 5.2.2 correspond to the scores 41, 42, and 44.

A boxplot can be obtained using Minitab by entering the following command in the session window where the data are in column `i`:

```
MTB > boxplot ci
```

Since a boxplot is a display that can be used to determine how data are distributed, it can be used as an alternative or as a supplement to a histogram. Boxplots are particularly useful for comparing two or more populations. Suppose we want to compare the scores of nonmath majors with the scores of math majors in the data set in Appendix B. Figure 5.2.3 shows the calculus test scores in two separate boxplots. The boxplot on the left represents the scores for nonmath majors (these are denoted by 0 in column 3), and the boxplot on the right represents the scores of math majors (these are denoted by 1 in column 3). From this figure we can see that the median score of math majors is higher than the median score of nonmath majors.

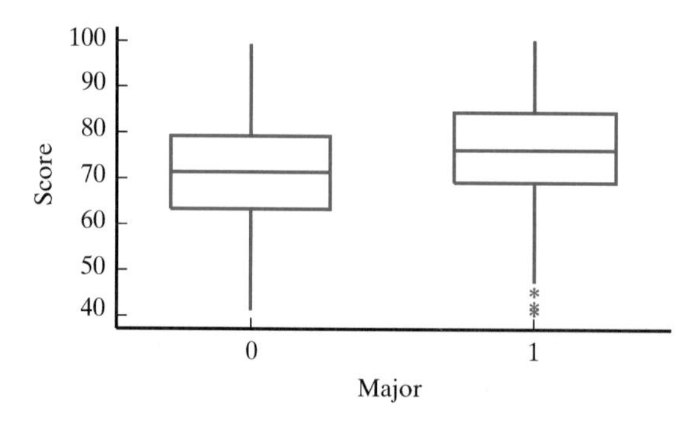

FIGURE 5.2.3
Boxplots of Calculus Scores by Major

The Minitab command for obtaining the boxplots in Figure 5.2.3 using the file **grades.mtp** is

```
MTB > boxplot c2*c3
```

where the calculus scores are in column 2 and the numbers in column 3 identify whether a student is a math major (1) or a nonmath major (0).

THINKING STATISTICALLY: SELF-TEST PROBLEMS

5.2.1 The following population represents the amounts of money (in thousands of dollars) that 23 patients spent for surgery and follow-up care in a private hospital.

23 12 55 63 23 12 10 9 13 25 86 93
23 24 31 29 36 30 29 19 32 21 18

a. Find the median, the quartiles, and the interquartile range (IQR).
b. Find the five-number summary for this population.
c. Repeat (a) if the number 93 is replaced with 206. Do you think that 206 is an outlier?

5.2.2 The following population represents the amounts of money (in thousands of dollars) that 27 patients spent for surgery and follow-up care in a nonprofit hospital.

55 21 22 41 32 13 10 18 6 12 23 52 33 23
25 11 28 32 37 39 9 22 18 16 14 23 53

a. Find the median, the quartiles, and the interquartile range (IQR).
b. Are there any outliers in this population?
c. Find the five-number summary for this population.

5.3 POPULATION MEAN AND STANDARD DEVIATION

In this section we discuss a population **mean**, which is a "central" value of a population that is used extensively in statistical applications. The mean is sometimes called the **average**, so we will use the terms *mean* and *average* interchangeably. Box 5.3.1 contains the definition of a population mean, which is denoted by the symbol μ as explained in Section 1.3.

BOX 5.3.1: POPULATION MEAN (AVERAGE)

Let $Y_1, Y_2, Y_3, \ldots, Y_N$ be a population of N numbers. The mean of the population is denoted by μ and defined as

$$\mu = \frac{\text{total of the } N \text{ population numbers}}{N} = \frac{\text{SUMY}}{N} = \frac{Y_1 + Y_2 + \cdots + Y_N}{N}.$$

Note. From the definition of a population mean given in Box 5.3.1, it follows that

$$\text{total of the } N \text{ population numbers} = N\mu.$$

The formula in Box 5.3.1 uses the abbreviation SUMY. This abbreviation stands for the words **SUM** of the **Y** values. To illustrate the use of this formula, consider the following population of scores on 10 homework assignments for a student in a statistics class:

76 82 58 90 72 85 75 68 79 86 **Line 5.3.1**

We denote the individual values of this population of $N = 10$ numbers by

$$Y_1, Y_2, \ldots, Y_{10},$$

where Y_i is the value of the ith population item. Thus, $N = 10$ and

$$\begin{array}{lllll} Y_1 = 76, & Y_2 = 82, & Y_3 = 58, & Y_4 = 90, & Y_5 = 72, \\ Y_6 = 85, & Y_7 = 75, & Y_8 = 68, & Y_9 = 79, & Y_{10} = 86. \end{array}$$

The population mean (average) is

$$\mu = \frac{\text{SUMY}}{N} = \frac{Y_1 + Y_2 + Y_3 + Y_4 + Y_5 + Y_6 + Y_7 + Y_8 + Y_9 + Y_{10}}{10},$$

which is

$$\mu = \frac{76 + 82 + 58 + 90 + 72 + 85 + 75 + 68 + 79 + 86}{10} = \frac{771}{10} = 77.1.$$

The total of the population in Line 5.3.1 is $N\mu = 10(77.1) = 771$.

EXAMPLE 5.3.1

HOSPITAL STUDY. A hospital administrator wants to study characteristics of patients who were admitted to the hospital last year. This information will be used to help determine if the facilities should be enlarged. Last year's hospital records contain many items of information on several thousand patients, and so they represent a very large population of numbers. One variable of interest is the number of days that a patient remained in the hospital. The administrator wants to know the total of this variable for all patients admitted last year. Since N, the total number of patients who entered the hospital, is known, the average number of days (μ) that patients remained in the hospital can be used to determine the total number of days that patients used this hospital. This total is

$$\text{total} = N\mu.$$

Thus, in this problem, the mean μ is the "central" population parameter of interest.

When a mean is used to represent an entire population of values, we would like to have a measure that tells us how well it represents the population. The mean is a good number to represent all values in a population if all the population values are close to the mean. We now define a quantity called the **standard deviation** that helps us determine how close the values in a population are to the mean. In Example 5.3.1 suppose that the average length of hospital stay for the population of last year's patients is $\mu = 6$ days. This is important information, but the administrator also needs to know whether most patients stay between 5 and 7 days (with an average of 6 days) or whether they are evenly spread out between 1 day and 11 days (with an average of 6 days). Thus, in understanding how well the *average* represents a population, it is important to know how the numbers are spread out around the average. The standard deviation is a numerical measure that provides this information.

Standard Deviation

The **standard deviation** is a population parameter used to measure the spread of a population around the *mean*. Whenever a mean is used to represent a population, the standard deviation should be examined to determine how well the mean represents the population. The symbol σ, which is the lowercase Greek letter *sigma*, will be used to denote a population standard deviation. The reason we use the symbol σ is that it is the Greek equivalent of the English letter "s," which is the first letter in the words *standard deviation*. Box 5.3.2 provides a method for computing the standard deviation of a population.

BOX 5.3.2: HOW TO COMPUTE A POPULATION STANDARD DEVIATION

Let $Y_1, Y_2, \ldots, Y_N$ be a population of N numbers. The following four steps can be used to compute σ, the population standard deviation:

1. Compute μ, the population mean:

$$\mu = \frac{\text{SUMY}}{N} = \frac{Y_1 + Y_2 + Y_3 + \cdots + Y_N}{N}.$$

2. Subtract μ from each number in the population and square each result. That is, compute

$$(Y_1 - \mu)^2, \quad (Y_2 - \mu)^2, \quad (Y_3 - \mu)^2, \ldots, \quad (Y_N - \mu)^2.$$

3. Compute SSDY, the sum of the N numbers in (2). That is, compute

$$\text{SSDY} = (Y_1 - \mu)^2 + (Y_2 - \mu)^2 + (Y_3 - \mu)^2 + \cdots + (Y_N - \mu)^2.$$

4. The population standard deviation is defined as $\sigma = \sqrt{\text{SSDY}/N}$.

The abbreviation SSDY in Box 5.3.2 stands for **S**um of **S**quared **D**eviations of **Y** values from the mean. We illustrate the formulas in Box 5.3.2 by computing the standard deviation of the population of six numbers:

$$1 \quad 18 \quad 13 \quad 5 \quad 6 \quad 11.$$

The mean of these six numbers is $\mu = 9$. Using (2) and (3) in Box 5.3.2, we obtain

$$\begin{aligned}\text{SSDY} &= (1-9)^2 + (18-9)^2 + (13-9)^2 + (5-9)^2 + (6-9)^2 + (11-9)^2 \\ &= 64 + 81 + 16 + 16 + 9 + 4 = 190.\end{aligned}$$

Thus, $\sigma = \sqrt{190/6} = 5.6273$.

To see how the standard deviation measures the spread of a set of numbers, notice that in step (2) of Box 5.3.2 each number is subtracted from the mean and the result is squared. If a number is much larger (or smaller) than the mean, the square of the deviation is also large, and this contributes to a large standard deviation. On the other hand, if all the numbers are close to the mean, and hence close to each other, the standard deviation is not so large. For example, consider the set of $N = 14$ numbers displayed in Line 5.3.2.

23 23 23 23 23 23 23 23 23 23 23 23 23 23 **Line 5.3.2**

The mean of these 14 numbers is 23; using the computing instructions in Box 5.3.2, we get SSDY $= 0^2 + 0^2 + \cdots + 0^2 = 0$, and hence $\sigma = 0$. Thus, if all values are equal, they are equal to the mean, and $\sigma = 0$. Now change the first and last numbers in Line 5.3.2 to 46 and 0, respectively. This yields the set of numbers in Line 5.3.3.

46 23 23 23 23 23 23 23 23 23 23 23 23 0 **Line 5.3.3**

The mean of the numbers in Line 5.3.3 is still 23, but the standard deviation is 8.693. This tells us that the numbers in Line 5.3.3 are spread out more than those in Line 5.3.2. As you can see, although $\mu = 23$ in both data sets, μ represents the data in Line 5.3.2 better than it represents the data in Line 5.3.3.

Z-scores

The difference between the value of a population item and the mean of the population is sometimes expressed in standard deviation units. For example, suppose a population average is $\mu = 80$ and the population standard deviation is $\sigma = 4$. If an item in that population has value 88, we say, "The value of this item is 2 standard deviation units larger than the mean." The difference in standard deviation units between the *value* of a population item and the *mean* of that population is called the **Z-score** of that population item. If the Z-score of an item is negative, it means the value of the population item is smaller than the mean. Notice that a Z-score has no unit of measurement since it simply expresses how many standard deviations a population value is from the mean. Box 5.3.3 explains how to compute a Z-score.

BOX 5.3.3: HOW TO COMPUTE A Z-SCORE

The following instructions can be used to compute the Z-score for a specified population value Y.

1. Compute the population mean μ using the instructions in Box 5.3.1.
2. Compute the population standard deviation σ using the instructions in Box 5.3.2.
3. Compute the Z-score for the Y value using the formula

$$Z = \frac{Y - \mu}{\sigma}. \quad \textbf{Line 5.3.4}$$

For the population with $\mu = 80$ and $\sigma = 4$, the Z-score for a population value 76 is

$$Z = \frac{76 - 80}{4} = -1.$$

The Z-score for a population value 82 is

$$Z = \frac{82 - 80}{4} = 0.5.$$

By using the Z-score of any population value, we are able to determine how many standard deviations that value is from the population mean.

In the previous section we described how boxplots can be used to locate outliers (unusual observations) in a data set. Z-scores can be used for this same purpose. Many investigators consider an observation with a Z-score that exceeds the value 3 in magnitude to be an outlier. That is, Z-scores greater than 3 or less than -3 are considered to be outliers. Other investigators may consider an observation to be an outlier if its Z-score is less than -2 or greater than 2.

Chebyshev's Result

There is a remarkable result discovered by the Russian mathematician Chebyshev that uses the standard deviation to determine the proportion of values in a population that is within a specified distance from the mean.

BOX 5.3.4: CHEBYSHEV'S RESULT

1. *At least* 75% of the values in any population of numbers are within 2 standard deviations of the mean (that is, at least 75% of the population values have Z-scores between -2 and 2 inclusive).
2. *At least* 88% of the values in any population of numbers are within 3 standard deviations of the mean (that is, at least 88% of the population values have Z-scores between -3 and 3 inclusive).
3. For any positive number k, *at least* $(1 - 1/k^2) \times 100\%$ of the values in any population of numbers are within k standard deviations of the mean (that is, at least $(1 - 1/k^2) \times 100\%$ of the population values have Z-scores between $-k$ and k inclusive).

To demonstrate the results in Box 5.3.4, consider the population of calculus scores in Appendix B. Several population parameters, reported in Exhibit 5.2.1, were computed using Minitab. This exhibit is repeated in Exhibit 5.3.1.

```
Descriptive Statistics

Variable        N     Mean    Median   TrMean    StDev   SEMean
score        2600   74.520    75.000   74.622   10.923    0.214

Variable      Min      Max        Q1       Q3
score      41.000  100.000    67.000   82.000
```

EXHIBIT 5.3.1
Minitab Output for Calculus Scores

Exhibit 5.3.1 was produced using the Minitab command `describe`. Using this command, Minitab computes `StDev`, the standard deviation, using the formula

$$\sqrt{\text{SSDY}/(N-1)} \quad \text{instead of the formula} \quad \sqrt{\text{SSDY}/N} \quad \text{shown in Box 5.3.2.}$$

Thus, you must multiply `StDev` in Exhibit 5.3.1 by $\sqrt{(N-1)/N}$ to get the correct answer. Using this, we get $\sigma = 10.923\sqrt{(N-1)/N} = 10.921$. The reason that

Minitab uses the formula $\sqrt{SSDY/(N-1)}$ will be explained in Section 6.3. For this population, $\mu = 74.5$ and $\sigma = 10.9$ (each number is rounded to one decimal). Using Chebyshev's result in Box 5.3.4, we can state at least 75% of the calculus scores are within $2\sigma = 2(10.9) = 21.8$ units of the mean $\mu = 74.5$. That is, at least 75% of the students' calculus scores are between the numbers $\mu - 2\sigma = 74.5 - 21.8 = 52.7$ and $\mu + 2\sigma = 74.5 + 21.8 = 96.3$. An equivalent way to state this is as follows: At least 75% of the students' Z-scores are between -2 and 2 inclusive.

Chebyshev's result is valid for **every** population of numbers. For any particular population, the percentage of values within k standard deviations of the mean may actually be much greater than the percentage stated in Box 5.3.4, but it can *never* be less. For instance, for the population of calculus scores in Appendix B, 96% of the scores are within two standard deviations of the mean instead of the minimum percentage of 75% given by Chebyshev's result.

Variance

A parameter that is closely related to the standard deviation is the **variance**. The variance of a population of numbers is defined as the square of the population standard deviation. It is denoted by σ^2 (called "sigma squared"). The variance is a useful parameter in theoretical statistics. For practical applications, the standard deviation is more useful than the variance because the standard deviation has the same units of measurement as the population mean. For example, a population of numbers for the quantitative variable "Weight" might be measured in pounds. The population mean and the population standard deviation are in units of pounds. The population variance is in units of "pounds squared," a term that has no meaningful interpretation.

THINKING STATISTICALLY: SELF-TEST PROBLEMS

5.3.1 The following population of nine numbers reports the amounts of money (in dollars) that the first nine customers in a convenience store spent for gasoline last Saturday.

6 13 2 51 32 16 8 19 24

a. Compute the population mean.
b. Compute the population standard deviation.
c. Compute the Z-score for the population value 6. Compute the Z-score for the population value 24.
d. Use Chebyshev's result in Box 5.3.4 to find two numbers a and b such that at least 75% of the population is between a and b. Count the number of values in the population that are between a and b, and verify that it is at least 75% of the population.

5.3.2 **a.** Write down any set of five numbers and consider it to be a population of five numbers, $Y_1, Y_2, \ldots, Y_5$.
b. What is the value of Y_4 in this population? What is the value of Y_3 for this population?
c. What is the value of the population mean, μ?
d. Show that for this population $(Y_1-\mu)+(Y_2-\mu)+\cdots+(Y_5-\mu) = 0$.
e. Is the following true for *any* population of N numbers $Y_1, Y_2, \ldots, Y_N$?

$$(Y_1 - \mu) + (Y_2 - \mu) + (Y_3 - \mu) + \cdots + (Y_N - \mu) = 0.$$

Verify your answer.

5.3.3 An alternative formula to the one in Box 5.3.2 is sometimes used to compute the population standard deviation. This formula is algebraically equivalent to the one in Box 5.3.2 (it will give the same result), but it may be easier to use if N is large. To compute the population standard deviation for a population of N numbers given by $Y_1, Y_2, \ldots, Y_N$ using this formula, follow steps (1) to (4) below.

(1) Square each number and add them together. This is denoted by

$$\text{SUMSQY} = Y_1^2 + Y_2^2 + Y_3^2 + \cdots + Y_N^2.$$

The abbreviation SUMSQY stands for **SUM** of **SQ**uares of **Y** values.

(2) Compute the sum of all the numbers. This is denoted by

$$\text{SUMY} = Y_1 + Y_2 + Y_3 + \cdots + Y_N.$$

(3) Compute $\text{SSDY} = \text{SUMSQY} - (\text{SUMY})^2/N$.

(4) The population standard deviation is $\sigma = \sqrt{\text{SSDY}/N}$.

Use this formula to compute σ for the population in Self-Test Problem 5.3.1.

5.4 SHALL I USE THE MEAN OR SHALL I USE THE MEDIAN?

In this chapter we presented two parameters—the median and the mean—that can be used to describe the center of a population of numbers. Perhaps you have read articles in the popular press or heard reports on television or the radio in which either the mean or median was used to describe a population. For example, we have all read statements such as, "Married couples in this city have an average of 1.8 children," or "The median salary of high-school teachers in this state is $28,000." Because both the mean and median are used so frequently, it is important to have a good understanding of these two "central" measures in order to determine when each should be used. Of course, if the mean and median are equal, then it makes no difference which is used to describe the center of a population. Thus, in the remainder of this section, we assume that the mean and median are *not* equal. That is, we assume that the mean and median differ by an amount that is of practical importance in the problem under study.

The histograms that were presented in Figures 4.3.3 to 4.3.5 are reproduced in Figures 5.4.1 to 5.4.3, with the values of the mean and median shown on each figure. In Figure 5.4.1 the mean is greater than the median because just a few players hit an unusually large number of home runs. In Figure 5.4.2 very few students scored unusually low on the test, and this resulted in the mean being less than the median. These two figures illustrate that the mean is more affected than the median by extreme population values. The histogram in Figure 5.4.3 is symmetric, and the mean and the median are equal.

FIGURE 5.4.1
Relative Frequency Histogram for Home Runs Hit in Major League Baseball for 1995 (skewed to the right)

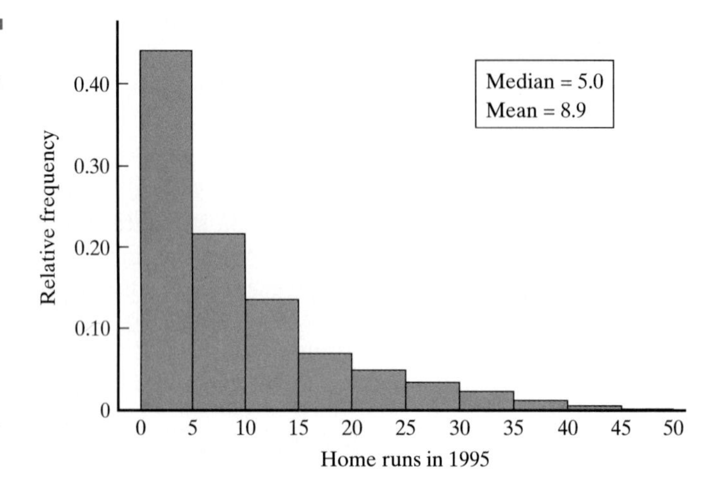

FIGURE 5.4.2
Relative Frequency Histogram of Elementary Statistics Test Scores (skewed to the left)

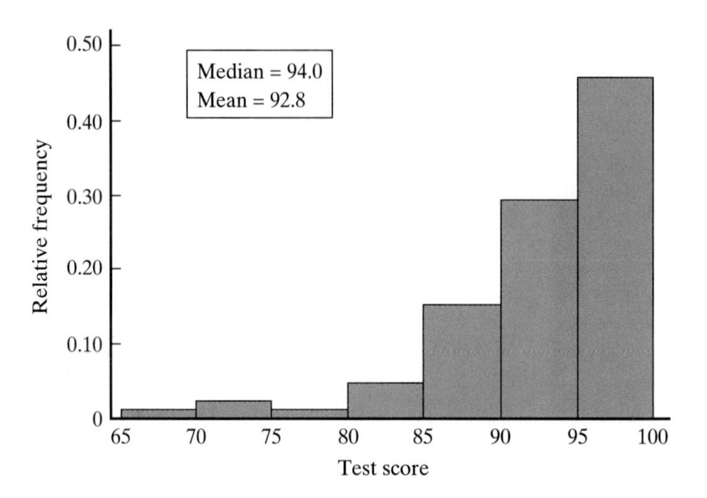

FIGURE 5.4.3
Relative Frequency of Weights of Watermelons

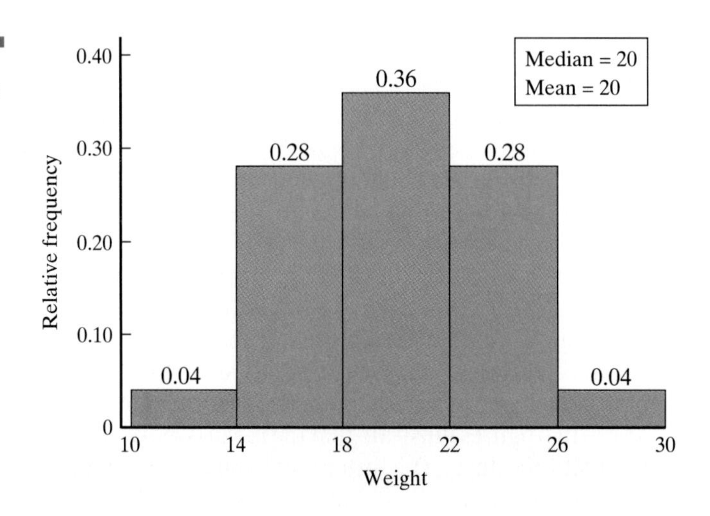

Here are two general rules to use in deciding when to use a mean and when to use a median for a population "center":

1. If it is necessary to determine a population total, the "center" to use is the mean.
2. If it is necessary to determine a population "center" such that at least 50% of the population items have values less than or equal to this center and at least 50% of the population items have values greater than or equal to the center, then the "center" to use is the median.

The following three examples typify situations where interest focuses on a population total and the mean is the "center" that should be used.

EXAMPLE 5.4.1

CAFETERIA SPACE. A high-school administrator in charge of planning cafeteria space for a school district must determine the *total* number of students who will be enrolled in this district next year. Since N, the total number of dwellings in the district, is known, the total number of students is

$$\text{total number of students} = N\mu,$$

where μ is the *average* number of children per dwelling in the district who are of high-school age. Since the administrator wants to determine a *total*, the mean is the "center" of interest. The total number of children cannot be obtained by using the median.

EXAMPLE 5.4.2

TOTAL CORN YIELD. The U.S. Department of Agriculture is interested in determining the *total* bushels of corn produced in the state of Iowa last year. The department knows N, the total number of acres of corn planted in the state last year. If μ, the *average* bushels per acre of corn planted, is known, then the *total* bushels can be obtained using the formula

$$\text{total bushels of corn} = N\mu.$$

In this example the mean is of interest since the total bushels of corn cannot be obtained from the median bushels of corn per acre.

EXAMPLE 5.4.3

PARKING GARAGES. A city manager is interested in estimating the *total* number of passengers who rode downtown in the $N = 30{,}000$ vehicles that parked in the six city garages during the past five-day workweek. The total number of passengers is

$$\text{total} = 30{,}000\mu,$$

where μ is the average number of passengers per vehicle that parked in the six garages. From a simple random sample of 100 vehicles, an estimate of μ was obtained and denoted by $\widehat{\mu}$. The estimate of the total number of passengers is

$$\widehat{\text{total}} = 30{,}000\widehat{\mu}.$$

In this problem an estimate of the *total* number of passengers is of interest. If $\widehat{\mu}$, an estimate of the mean, is known, an estimate of the total number of passengers can be obtained. An estimate of the total number of passengers cannot be obtained using an estimate of the median.

If a population total is not of interest but a measure of center is desired such that at least 50% of the population items have values less than or equal to the center and at least 50% have values greater than or equal to this center, then the median is the center that must be used.

EXAMPLE 5.4.4

ANNUAL BONUS. The president of a small company informs his 11 employees that the average annual bonus this year is \$1,000 per employee. The employees think that \$1,000 *average* per employee seems to be a generous bonus, but they are more interested in the median bonus. When they ask for the median value, the president informs them that the median bonus is \$5. Now the employees are not happy since at least one-half of them will receive \$5 or less. The money to be distributed to the 11 employees is as follows:

\$2 \$5 \$5 \$5 \$5 \$5 \$5 \$5 \$5 \$8 \$10,950.

This example is extreme, but it shows that the mean and median can sometimes be very different, and it is important to understand when each should be used to describe a population "center" of interest.

EXAMPLE 5.4.5

AMERICAN BILLIONAIRES. *Fortune* magazine reported the wealth in billions of dollars of seven Americans in 1992: H. R. Walton, \$24.0; William Gates, \$5.9; J. P. Getty, \$4.5; Estée Lauder, \$4.3; W. R. Hearst, Jr., \$3.4; H. R. Perot, \$3.3; and S. M. Redstone, \$3.3. The median for this population of seven numbers is $M = \$4.3$ (billion) and the mean is $\mu = \$6.96$ (billion). If you wanted to compute the total wealth of these seven billionaires, then the mean must be used since the total wealth $= N\mu = 7(\$6.96) = \$48.7$ billion. However, the median may be more meaningful in describing the center of these numbers since the population mean is greater than six of the seven values in the population.

EXAMPLE 5.4.6

BASEBALL STRIKE. In a hearing conducted by the National Labor Relations Board, players from the Major League Baseball Players Association (MLBPA) claimed that the *median* annual salary for a baseball player in 1994 was \$450,000, which they argued was far too low for baseball players with such outstanding abilities. However, using the same population of salary numbers, owners of the baseball teams reported that the *average* annual salary for a baseball player was \$1,168,263! The reason for the large difference between the mean and median is that there are a few unusually gifted baseball players who command very high salaries relative to the rest of the players; hence, the mean is (considerably) larger than the median. In this problem the mean was more meaningful to the owners because they were interested in the *total* of the salaries. In contrast, the median was more meaningful to the MLBPA because it believed a middle salary value provides a better description of the financial position of the players. Thus, one group was interested in the mean and the other group was interested in the median, even though they were both using the same population of numbers.

Although we have focused on the mean and the median, there are some situations where neither parameter is the most informative for making a decision.

EXAMPLE 5.4.7

DEPTH OF RIVER. It is reported that the *average* depth of a river channel between two cities is 15 feet. This is not very useful information for the captain of a barge who wants to go from one city to the other and needs a depth of at least 10 feet everywhere. The river could be 4 feet deep for one-half the distance and 26 feet deep for one-half the distance and have an average depth of 15 feet. What is of interest to the captain is the *minimum* depth of the channel.

In this chapter we have described numerical measures for a population of numbers. As we discussed in previous chapters, it will often not be possible to determine the values of a quantitative variable for all items in a population. Often we will have only a sample from the population. In these situations we will compute estimates from the sample to use in place of the unknown parameter values. In Chapter 6 we describe methods for obtaining point estimates and confidence intervals for the population mean based on a simple random sample, and in Chapter 7 we consider the same topics for the population median.

THINKING STATISTICALLY: SELF-TEST PROBLEMS

5.4.1 A recent college graduate is considering employment with a pharmaceutical company. Before applying for the job, the graduate wants to examine the company's salary structure. What population measures would be most meaningful for this graduate?

5.4.2 An agricultural research scientist has developed a new diet supplement for chickens. He states that if chickens are fed this supplement for 10 weeks, they will gain an *average* of 0.15 pounds more per week than if fed a standard diet. Which is a more meaningful center in this problem, the mean or the median?

5.4.3 A consumer group has monitored the prices of unleaded regular gasoline at a service station for the past year. A brochure was published for its members that contained the following statement: "The price of unleaded gasoline tends to vary from day to day. During the past year the *average* daily price on weekdays was $1.21 per gallon and on weekends it was $1.25." Is this information useful to customers of this station?

5.4.4 A company that manufacturers light bulbs advertises that in laboratory tests their bulbs lasted an average of 1,000 hours.

a. Interpret the meaning of this statement.

b. Is the average important information to a company that owns a building that uses hundreds of light bulbs?

5.4.5 A sports magazine reports that a running back on a football team averaged 3.4 yards per carry last season. Is the *average* yards gained per carry important information to a team that wants to trade for a running back next season?

5.4.6 Consider the following population of numbers:

50 36 80 72 90 1,000 72.

a. Compute the population mean.
b. Compute the population median.
c. Remove 1,000 from the population and recompute the mean and median.
d. Describe how the median and mean changed when 1,000 was removed from the population. What can you conclude about the sensitivity of the mean and median to extreme values in a population?

5.5 CHAPTER SUMMARY

KEY TERMS

1. population median
2. population quartiles
3. interquartile range
4. boxplot
5. outlier
6. population mean
7. population total
8. population standard deviation
9. Z-score
10. Chebyshev's result
11. variance

SYMBOLS

1. Q_1—first quartile of a population of numbers
2. Q_3—third quartile of a population of numbers
3. M—median of a population of numbers
4. μ—mean of a population of numbers
5. σ—standard deviation of a population of numbers
6. σ^2—variance of a population of numbers

KEY CONCEPTS

1. The mean and the median are both used to describe the center of a population of numbers.
2. If a population total is desired, then the mean is the center to use.
3. If a population center is desired such that at least 50% of the values are less than or equal to the center and at least 50% of the values are greater than or equal to the center, then the median is the center to use.
4. Outliers should be investigated to determine why their values are unusual.
5. The population standard deviation is used to describe how close values are to the population mean.

SKILLS

1. Compute a mean and a standard deviation for a population of numbers.
2. Compute the Z-score for an individual value.
3. Compute the median, the first quartile, and the third quartile for a population of numbers. Use the quartiles to determine the interquartile range.
4. Use a boxplot to identify the first and third quartiles, the median, and any outliers that might exist in a population of numbers.

5. Use a statistical software package to compute the numerical measures discussed in this chapter.

5.6 CHAPTER EXERCISES

5.1 The following data represent a population $Y_1, Y_2, \ldots, Y_N$, which is the set of annual incomes (in thousands of dollars) of some government workers the year they retired.

31 45 62 46 31 34 78 98 34 52 39 52
56 58 94 43 32 45 63 61 67 85 37 83
56 71 68 67 59 68

a. What are the values of Y_{10}, Y_{21}, and Y_{42}?
b. Is any other population value equal to Y_{15}?

5.2 The following data represent a population $Y_1, Y_2, \ldots, Y_{21}$, which is the set of ages of 21 people who applied for work at a fast-food restaurant.

18 23 19 28 45 32 19 20 18 24 23
51 22 18 18 29 28 17 19 18 23

a. What are the values of Y_{13}, Y_{19}, and Y_5?
b. What is the value of $Y_2 + Y_{16}$?
c. What is the maximum of the Y values?
d. What is the minimum of the Y values?

5.3 A professional football team has five running backs on its roster. Shown below are the weights in pounds of the five players.

179 209 204 235 214

Compute the median weight for this population of five running backs.

5.4 The following data represent the weights of 15 people who were admitted to a small hospital last week with heart problems.

179 194 198 245 249 251 255 260
264 264 266 278 281 281 285

Find the median weight of these 15 people.

5.5 The median, M, of a population of numbers is a number such that at least 50% of the observations in the population are less than or equal to M and at least 50% of the observations are greater than or equal to M. In Exercise 5.4 what is the set of numbers that is less than or equal to the median? What proportion of the numbers in the data set is less than or equal to M?

5.6 The following numbers are the annual salaries (in thousands of dollars) of 11 high-school principals in a school district:

52 67 59 55 55 72 63 55 57 52 53.

a. Compute the median.
b. List the values that are greater than or equal to M.
c. What proportion of the values is greater than or equal to M?
d. Suppose the value 53 is deleted. There are now $N = 10$ values, given by

52 67 59 55 55 72 63 55 57 52.

What is M for this set of numbers?

5.7 The following population of numbers represents the money (in thousands of dollars) that each of seven patients spent for surgery and follow-up hospital care.

23 12 55 63 23 12 8

a. Find the mean.
b. Find σ.
c. Find the median.

5.8 Consider the following three populations of numbers.

population (1)	1 2 4 7 2 4 9 10 3 18
population (2)	101 102 104 107 102 104 109 110 103 118
population (3)	100 200 400 700 200 400 900 1,000 300 1,800

Notice that 100 is added to each number in population (1) to get population (2), and each number in population (1) is multiplied by 100 to get population (3).
a. Guess which population has the largest mean.
b. Guess whether population (1) or (2) has the larger standard deviation.
c. Compute the standard deviations of populations (1) and (2) to check your guess. Use the formula in Self-Test Problem 5.3.3.
d. Guess whether population (1) or (3) has the larger standard deviation.
e. Compute the standard deviations for populations (1) and (3).

5.9 An instructor in a statistics course announced that the average score for the population of 99 students who took the first exam was 72. After returning to her office, the instructor discovered an exam that had been misplaced and not included in the information presented to the class. The score on the misplaced exam was 90. What is the average score for the complete set of 100 exams?

5.10 An instructor in a statistics course announced that the median score for the population of 99 students who took the first exam was 70. After returning to her office, the instructor discovered an exam that had been misplaced and not included in the information presented to the class. The score on the misplaced exam was 90. Can the instructor compute the median score for the complete set of 100 exams using only the misplaced exam score and the information given to the students?

5.11 A statistics instructor announced that a course grade of C would be given to all students with a total score within two standard deviations of the class mean. Based on Chebyshev's result in Box 5.3.4, determine the *minimum* proportion of C's that will be assigned.

5.12 A distant relative has died and left you a collection of nine old baseball cards. The values of these cards in dollars are shown below:

15 20 30 45 15 10 50 20 20.

a. Compute the mean of these nine values.

b. Compute the standard deviation of these nine values.

c. Compute the median of these nine values.

d. Yesterday you received a letter that included a 1952 Mickey Mantle rookie card that is valued at $3,000. Recompute the average card value and the median card value of your collection of 10 cards. Is the mean or the median more affected by the value of the Mickey Mantle card?

e. Compute the standard deviation for the set of 10 cards. Compare this value to the standard deviation for the original set of nine cards.

f. Verify that instruction (1) of Chebyshev's result in Box 5.3.4 holds for the set of 10 cards.

5.13 The government reports that the *average* life expectancy of a person with a certain type of cancer is 3.45 years after diagnosis.

a. How is this number computed?

b. Is the mean or the median more meaningful to a person with this type of cancer? Explain.

5.14 A newspaper reports, "In Carson, Iowa, the average family has 3.2 children."

a. Interpret the meaning of this statement.

b. Is it possible that a family in Carson has 8 children?

5.15 A newspaper reported, "The average citizen in this city rode the bus 13.7 times last year." Interpret the meaning of this statement.

5.16 The first workday of each month a person invests $75 in a mutual fund. Because the price per share varies, the shares purchased each month vary. For example, if the price per share one month is $7.50, the person can purchase 10 shares. However, if the price per share is $15, the person can purchase only 5 shares. On December 31 of each year, the total number of shares purchased during the year must be determined for tax purposes.

a. Will knowledge of the mean shares per month or the median shares per month provide more useful information?

b. Suppose the average number of shares purchased per month was 5.3 during 1994. What is the total number of shares purchased during 1994?

c. What is the total value of the shares on December 31 if the price per share on that day was $16?

5.17 Explain why $\sigma \geq 0$ for any set of numbers.

5.7 SOLUTIONS FOR SELF-TEST PROBLEMS IN CHAPTER 5

5.1.3 Yes. One conclusion is that all of the 1,000 bolts meet the manufacturing specifications.

5.2.1

a. Order the numbers from smallest to largest to obtain

9 10 12 12 13 18 19 21 23 23 23 24
25 29 29 30 31 32 36 55 63 86 93.

From this ordered set we obtain $M = 24$, which is the number in position $(N+1)/2 = 24/2 = 12$, counting up from the smallest number. Since $k = (N+1)/4 = 24/4 = 6$ is an integer, we can use the procedure in Box 5.2.2 to compute $Q1$ and $Q3$. Here $k = 6$, so $Q1 = 18$ and $Q3 = 32$. Thus, IQR = 14.

b. minimum = 9, $Q1 = 18$, $M = 24$, $Q3 = 32$, maximum = 93

c. The answers are the same as in (a). The value 206 is an outlier because it exceeds $Q3 + 1.5(\text{IQR}) = 32 + 1.5(14) = 53$.

5.2.2

a. The 27 numbers arranged from smallest to largest are

6 9 10 11 12 13 14 16 18
18 21 22 22 23 23 23 25 28
32 32 33 37 39 41 52 53 55.

From this we get $M = 23$, $Q1 = 14$, $Q3 = 33$, and IQR = 19.

b. No. $Q1 - 1.5\text{IQR} = -14.5$, and $Q3 + 1.5\text{IQR} = 61.5$. All 27 numbers are within these limits, and so there are no outliers.

c. minimum = 6, $Q1 = 14$, $M = 23$, $Q3 = 33$, maximum = 55

5.3.1

a. $\mu = 19.0$.

b. $\sigma = \sqrt{1{,}842/9} = 14.31$.

c. The Z-score for $Y = 6$ is Z-score $= (6-19)/14.31 = -0.908$. The Z-score for $Y = 24$ is $(24 - 19)/14.31 = 0.349$.

d. From Chebyshev's result we know that at least 75% of the population is between $\mu - 2\sigma$ and $\mu + 2\sigma$. So at least 75% of the population is between 19–2(14.31) and 19+2(14.31); that is, between

$$-9.62 \quad \text{and} \quad 47.62.$$

By looking at the nine observations, we see that all but one of the population values are between -9.62 and 47.62 (that is, $8/9 = 0.889 = 88.9\%$ of the population values are between -9.62 and 47.62). Thus, at least 75% of the population values are between these numbers.

5.3.2

a. We selected the five numbers 21, 6, 18, 3, and 7.

b. $Y_4 = 3$. $Y_3 = 18$.

c. $\mu = 11$.

d. The mean of these five numbers is $\mu = 11$, and

$$(21 - 11) + (6 - 11) + (18 - 11) + (3 - 11) + (7 - 11)$$
$$= 10 - 5 + 7 - 8 - 4 = 17 - 17 = 0.$$

e. Yes. It is true for any set of N numbers as shown below:

$$(Y_1 - \mu) + (Y_2 - \mu) + (Y_3 - \mu) + \cdots + (Y_N - \mu)$$
$$= (Y_1 + Y_2 + Y_3 + \cdots + Y_N) - (\mu + \mu + \mu + \cdots + \mu)$$
$$= N\mu - N\mu = 0.$$

5.3.3 SUMSQY = 5,091, SUMY = 171, SSDY = 5,091 – $(171)^2/9 = 1{,}842$. Thus, $\sigma = \sqrt{1{,}842/9} = 14.31$.

5.4.1 Perhaps the singlemost meaningful number is the minimum salary, since the starting salary for the graduate will undoubtedly be close to this value. If the graduate is concerned with longterm growth, then the median would be useful. If the median annual salary is obtained, the graduate will know that at least half of the employees get a larger salary than the median. If she has a good work record, she would hope to get at least the median salary.

5.4.2 The mean is more important since the total gain in weight is of interest. A total can be obtained by using the mean, but it cannot be obtained by using the median.

5.4.3 Yes. Since automobile owners are interested in the *total* amount of money they will pay for gasoline for the year, they are interested in the *average* daily price. In the absence of other information, customers may want to buy gasoline at this service station only on weekdays.

5.4.4

a. Many light bulbs manufactured by the company were tested. The average burn time for bulbs tested was 1,000 hours.

b. Yes. The total burnout time for all bulbs purchased is important.

5.4.5 Yes. Since the *total* yards that a running back gains is important, the *average* yards per carry is an important parameter.

5.4.6

a. The mean is $\mu = 200$.

b. The median is $M = 72$.

c. After removing the number 1,000 from the population, the new population is

$$50, \quad 36, \quad 80, \quad 72, \quad 90, \quad 72.$$

The median is $M = 72$. The mean is $\mu = 66.67$.

d. The median did not change, but the mean decreased from 200 to 66.67. The mean is more sensitive to extreme values than the median.

CHAPTER 6

THE *NORMAL* POPULATION

6.1 INTRODUCTION

Let us review the process of statistical inference we have discussed up to now. In any investigation where one wants to examine a population, the variable of interest must be identified and the target and sampled populations defined. Whenever possible, these two populations should be the same. If they are not, the sampled population should "resemble" the target population as closely as possible. The next step is to decide what parameter is needed to help answer questions about the population. If every item in the sampled population is measured, the exact value of the parameter can be obtained. However, in most applied problems it is impractical to measure every item in the population, and the value of the parameter will not be known exactly. In this case one must obtain a sample from the sampled population and use the sample values to obtain a *point estimate* of the parameter. This estimate will be used in place of the parameter when making decisions. One wants to know how close the estimate is to the unknown parameter or determine a range of values that is likely to include the parameter. Confidence intervals can be used for these purposes.

In Chapter 3 we discussed confidence intervals for π, the proportion of a population that possesses a specified characteristic. These confidence intervals for π are valid for *any population* for which the characteristic has been defined and from which a simple random sample has been selected. In this chapter we describe formulas that can be used to obtain confidence intervals for μ, a population mean, and for σ, a population standard deviation. Strictly speaking, these formulas are valid only if the population is of a special type called a **normal population**. When used to describe a population, the word "normal" does not have the dictionary definition of "standard" or "regular." In statistics the word "normal" is a technical term meaning a specific type of population. To emphasize this meaning, we will write the word *normal* in italics. The *normal* population is by far the most frequently encountered type of population in both theoretical and applied statistics. It has dominated statistical theory and practice for the past 60 years and continues to do so today. For this reason, this chapter is devoted entirely to the *normal* population.

In Section 6.2 we define the *normal* population, and in Section 6.3 we discuss confidence intervals for μ, the mean of a *normal* population. Section 6.4 provides a formula for determining the sample size necessary to obtain a specified margin of error for the estimate of μ. Section 6.5 presents the confidence interval for σ, the standard deviation of a *normal* population. In Section 6.6 we explain how an investigator can use a simple random sample to help decide whether a population is a *normal* population. In two appendices at the end of this chapter (Sections 6.10 and 6.11), we provide a more detailed discussion of confidence intervals for μ and σ, respectively.

6.2 THE *NORMAL* POPULATION

In this section we discuss the *normal* population in some detail. The *normal* population is a conceptual population that contains an infinite number of values. It is often used as a model to describe populations used in applications. In this book we consider a *normal* population as having N values where N is finite but extremely large.

In Chapter 4 we discussed how a population of numbers is described by the shape of its relative frequency histogram. Because we consider a *normal* population to have an extremely large number of values, the class intervals of the relative frequency histogram can be chosen to be so small that the relative frequency histogram appears as a smooth curve, as shown in Figure 6.2.1. Due to its shape, the curve that represents the relative frequency histogram of a *normal* population is sometimes referred to as a **bell curve**.

FIGURE 6.2.1
Relative Frequency Histogram of a *Normal* Population

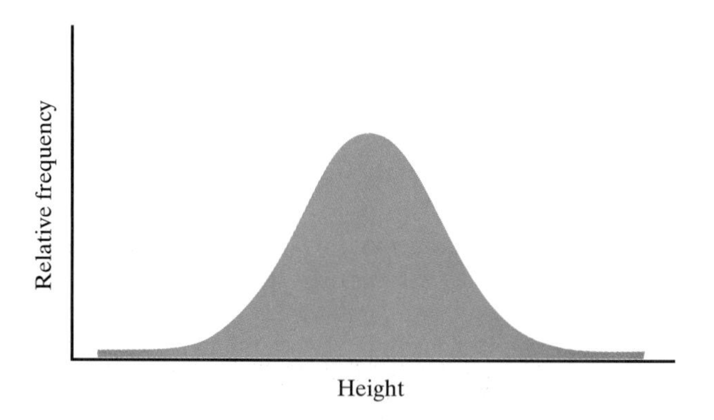

To better understand why the relative frequency histogram of a *normal* population appears as a curve, we will demonstrate how a relative frequency histogram changes as the widths of the class intervals decrease for a population of size N when N is large. Consider the heights of a population of $N = 100{,}000$ men who recently completed physical exams as part of a national health survey (artificial data were used to construct the relative frequency histograms in Figures 6.2.2, 6.2.3, and 6.2.4). A relative frequency histogram of this population with class intervals that are each three inches wide is shown in Figure 6.2.2.

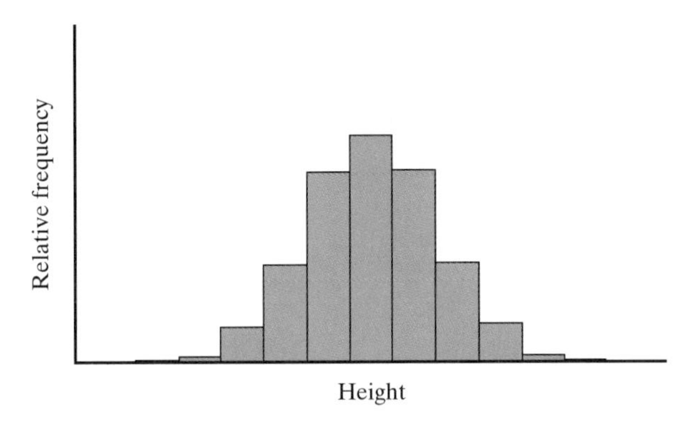

FIGURE 6.2.2
Relative Frequency Histogram for Population of Heights of Men Using Three-Inch Class Intervals

Figure 6.2.3 shows a relative frequency histogram for the same population of men, but with class intervals of width one inch.

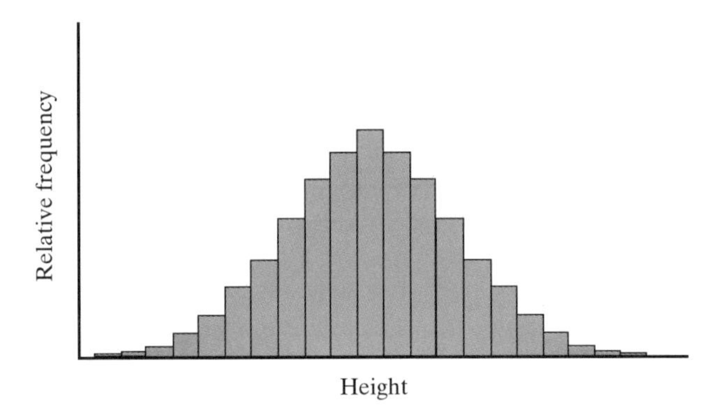

FIGURE 6.2.3
Relative Frequency Histogram for Population of Heights of Men Using One-Inch Class Intervals

Figure 6.2.4 shows a relative frequency histogram for the same population of men, but with class intervals of width one-half inch.

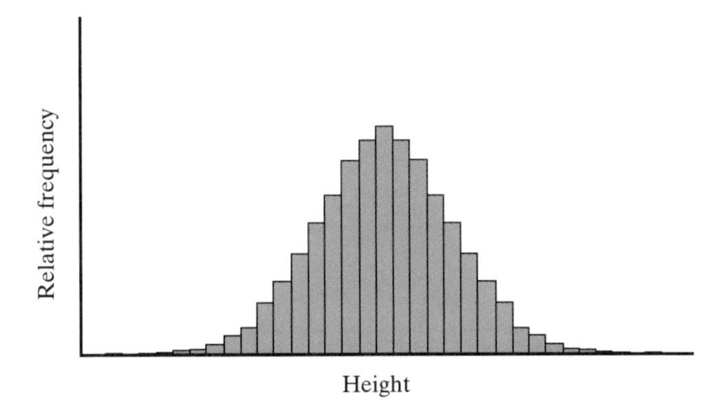

FIGURE 6.2.4
Relative Frequency Histogram for Population of Heights of Men Using One-Half-Inch Class Intervals

Notice that as the widths of the class intervals decrease, the relative frequency histogram of this population of heights begins to smooth out and appears more like a curve. For this particular population of $N = 100{,}000$ men, as the widths of the class intervals decrease, the relative frequency histogram is beginning to look like the relative frequency histogram of a *normal* population (the bell curve) shown in Figure 6.2.1.

Using the *Normal* Population as a Model for Real Populations

The population of $N = 100{,}000$ men in the previous discussion has a relative frequency histogram that looks similar to the bell curve shown in Figure 6.2.1. In practice, we would never have a population that would provide a relative frequency histogram that looks *exactly* like the bell curve shown in Figure 6.2.1, but a bell curve is a good representation of the relative frequency histogram of many populations encountered in practice. Thus, the *normal* population serves as a **model** for these populations. Throughout the rest of this book we will present many results that are based on the assumption that a simple random sample has been selected from a *normal* population. In practice, the results obtained using this assumption will be approximately correct if a simple random sample has been selected from any population of size N, where N is large and where the population relative frequency histogram is well represented by the bell curve shown in Figure 6.2.1.

Convention. When we write the statement "a population is *normal*," or when we write the statement "a population $Y_1, Y_2, \ldots, Y_N$ is *normal*," we mean that the population has a relative frequency histogram that is well represented by the bell curve shown in Figure 6.2.1.

Although the meaning of the term "well represented" is subjective, we provide a procedure in Section 6.6 that will help you decide if a population can be considered *normal*. It is a common practice to use mathematical models such as the *normal* population to approximate real-life situations. For instance, a theoretical rectangle is a geometric figure that does not exist in practice. However, as the next example illustrates, a theoretical rectangle can be a useful model in applications.

EXAMPLE 6.2.1

NEW CARPET. Suppose you want to put a new carpet on your living room floor and you need to know how many square yards of carpet to buy. The floor resembles a theoretical rectangle, yet it is not exactly a theoretical rectangle since the angles formed by intersecting sides of the room are not *exactly* 90 degrees (for example, when measured to several decimal places of accuracy). However, the room is approximately a theoretical rectangle, so you use the formula *area* = *width* × *length* to find the approximate area of the floor. Although the room is not exactly a theoretical rectangle, you use this idealized geometric figure as a model to compute the approximate amount of carpet needed. You examine the room and decide it is approximately a theoretical rectangle, just as an investigator might examine a population and decide that it is approximately a *normal* population.

Some Attributes of a *Normal* Population

Let μ and σ denote the mean and the standard deviation, respectively, of a *normal* population. A *normal* population has the following attributes:

1. A *normal* population is *symmetric*; among other things, this implies (a), (b), and (c) below, where z is any number greater than zero.

(a) The proportion of the values of a *normal* population that is less than $\mu - z\sigma$ is equal to the proportion of the population values that is greater than $\mu + z\sigma$. See Figure 6.2.5.

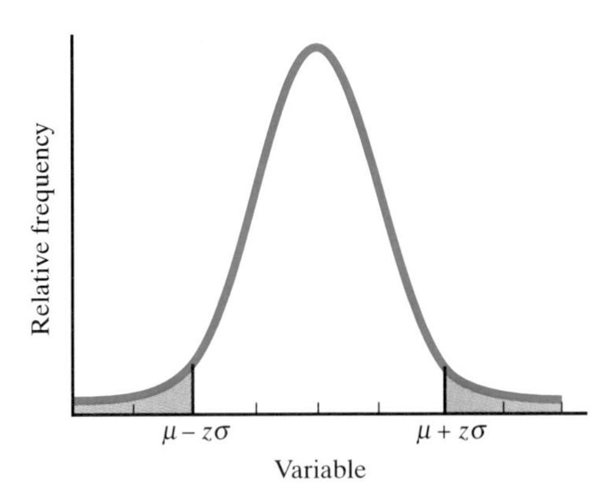

FIGURE 6.2.5
Proportion of Values $< \mu - z\sigma$ Is Equal to Proportion of Values $> \mu + z\sigma$

(b) The proportion of the values of a *normal* population that is less than $\mu + z\sigma$ is equal to the proportion of the population values that is greater than $\mu - z\sigma$. See Figure 6.2.6.
(c) In a *normal* population the mean is equal to the median.

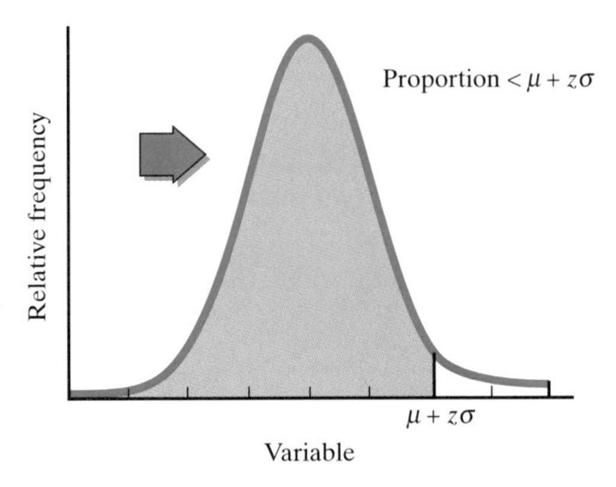

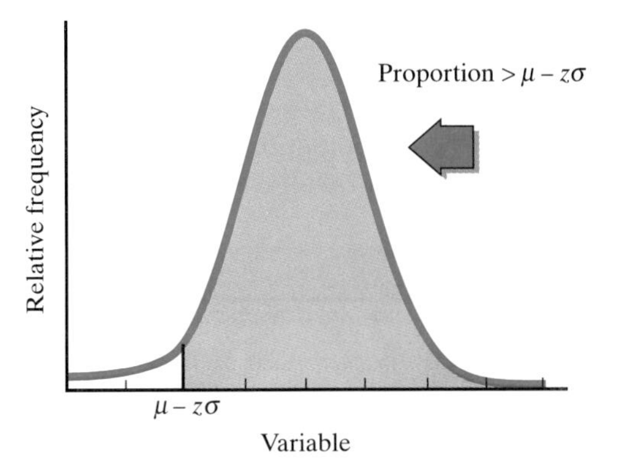

FIGURE 6.2.6
Proportion of Values $< \mu + z\sigma$ Is Equal to Proportion of Values $> \mu - z\sigma$

2. A *normal* population depends only on two parameters, μ and σ. This means that for a *normal* population where μ and σ are known, the proportion of population values in any interval a to b ($a < b$) can be evaluated.
3. If σ remains fixed but μ changes, the relative frequency histogram of a *normal* population has the same *shape* but its location changes. See Figure 6.2.7.

FIGURE 6.2.7
Two *Normal* Populations with the Same Standard Deviations but Different Means

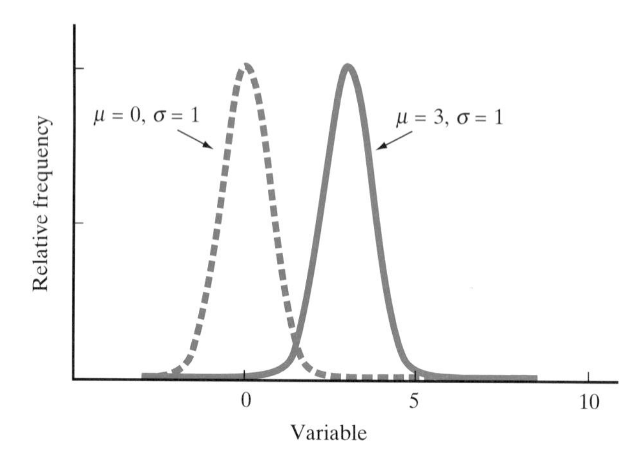

4. If μ remains fixed but σ changes, the relative frequency histogram of a *normal* population has the same *location* but its shape changes. See Figure 6.2.8.

FIGURE 6.2.8
Two *Normal* Populations with the Same Means but Different Standard Deviations

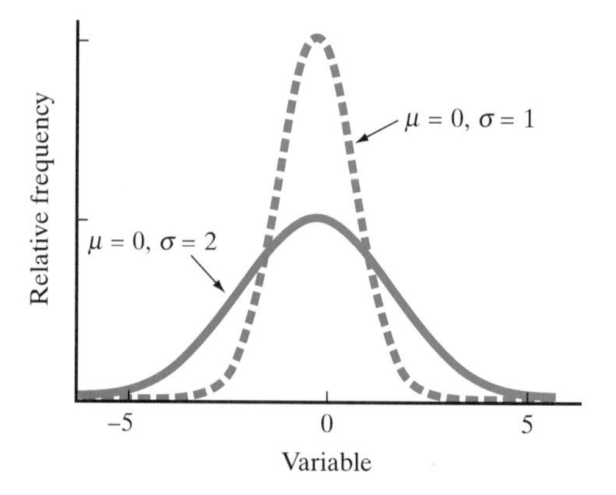

Proportions of *Normal* Population Values That Are in Specified Intervals

In Box 5.3.4 we stated Chebyshev's result, which uses the standard deviation to tell how "close" the values of a population are to its mean. The results in Box 5.3.4 hold for *any* population. A similar result is given in Box 6.2.1 for a *normal* population.

BOX 6.2.1: USING σ TO DETERMINE HOW CLOSE THE VALUES IN A *NORMAL* POPULATION ARE TO THE MEAN

Consider a *normal* population with mean μ and standard deviation σ.

1. The interval $\mu - \sigma$ to $\mu + \sigma$ contains 68.26% of the population values.
2. The interval $\mu - 2\sigma$ to $\mu + 2\sigma$ contains 95.45% of the population values.
3. The interval $\mu - 3\sigma$ to $\mu + 3\sigma$ contains 99.73% of the population values.

As stated in Box 6.2.1, if a population is *normal*, approximately 95% of the population values are within two standard deviations of the mean. By Chebyshev's results in Box 5.3.4, for *any* population at least 75% of the population values are within two standard deviations of the mean. If a population is *normal*, the results in Box 6.2.1 are more precise than Chebyshev's results for stating how "close" the values are to the mean. Figure 6.2.9 illustrates the results of Box 6.2.1.

FIGURE 6.2.9
Proportions of *Normal* Population Values Within 1, 2, and 3 Standard Deviations of the Mean

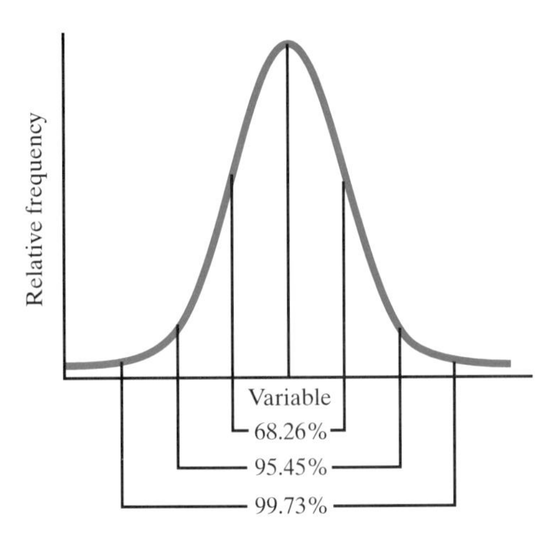

We present an example to illustrate how the results in Box 6.2.1 can be used to determine the proportion of values in a *normal* population that are within 2σ units of μ.

EXAMPLE 6.2.2

HEIGHTS OF MEN. Suppose the heights of a population of men between the ages of 40 and 60 years can be represented by a *normal* population with mean $\mu = 67$ inches and standard deviation $\sigma = 3$ inches. Suppose we want to know the proportion of men in this population who have heights between 61 and 73 inches. Since $\sigma = 3$, it follows that $\mu - 2\sigma = 67 - 2(3) = 61$ and $\mu + 2\sigma = 67 + 2(3) = 73$. So the proportion of population values between 61 and 73 is the proportion of population values between $\mu - 2\sigma$ and $\mu + 2\sigma$. By the results in Box 6.2.1, about 95% (actually 95.45%) of the population values are within 2σ units of the mean in a *normal* population. Hence, about 95% of the men in the population have heights between 61 and 73 inches.

Proportion of Values Between a and b in a *Normal* Population When μ and σ Are Known

Here we consider the problem where an investigator wants to determine the proportion of values in a *normal* population that are between a and b (where $a < b$) and where the parameters μ and σ are both known. Figure 6.2.10 shows the relative frequency histogram of a *normal* population where the "rectangles" over the class intervals from a to b have been shaded.

By Definition 3.1.2 the proportion of population items that are in an interval a to b can be expressed as a probability. Specifically, the probability that an item to be chosen at random from a *normal* population will be in the interval from a to b is equal to the proportion of population values in the interval. If Y represents a simple random sample of one value to be selected at random from a *normal* population, the

FIGURE 6.2.10
Proportion of *Normal* Population Values Between *a* and *b*

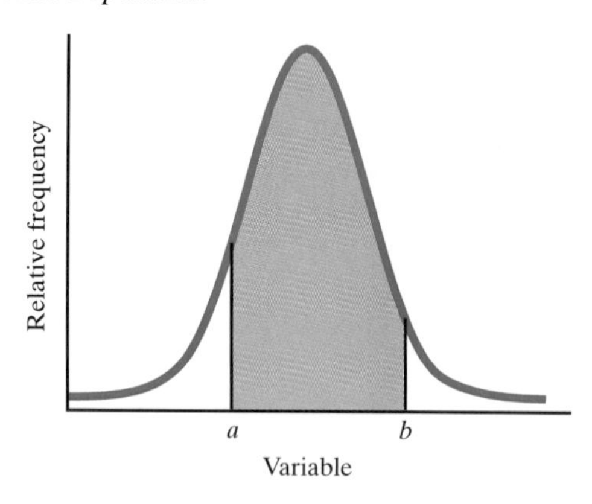

probability that Y will be between a and b is denoted by

$$\Pr(a \le Y < b) = \text{proportion of } Y \text{ values in the population that are between } a \text{ and } b.$$

Note that $a \le Y < b$ is the same notation used in Section 4.2 to define the class interval a to b. However, for the *normal* population, the proportion $\Pr(a \le Y < b)$ is unchanged if the boundary points a and b are included or excluded. That is, for the *normal* population,

$$\Pr(a \le Y < b) = \Pr(a < Y < b) = \Pr(a \le Y \le b) = \Pr(a < Y \le b). \quad \textbf{Line 6.2.1}$$

These proportions are equal since the *normal* population consists of an infinite number of values, and the inclusion or exclusion of a single value does not alter these proportions. Similarly, in applied problems where N is very large, the inclusion or exclusion of a single value will generally not alter these proportions significantly. To determine the proportion of values that are between a and b in a *normal* population requires either a computer or a table of *normal* proportions. Such a table is provided in Table 6.2.1, and later we describe Minitab commands that can be used to compute these proportions. For a *normal* population with mean μ and standard deviation σ, Table 6.2.1 lists the proportion of population values that are less than specified numbers which are expressed in terms of standard deviation units from the mean. All values are rounded to three decimals. To illustrate the use of this table, suppose $Y_1, Y_2, \ldots, Y_N$ is a *normal* population with mean μ and standard deviation σ. By line (7) of the table, for $z = 1.282$ the proportion of values in this population that are less than $\mu + z\sigma$ (that is, less than $\mu + 1.282\sigma$) is 0.900. For your convenience, Table 6.2.1 is also on the inside front cover of this book.

Table 6.2.1 can be used to obtain the proportion of values in a *normal* population that are less than a specified value b. Let $Y_1, Y_2, \ldots, Y_N$ be a *normal* population with mean μ and standard deviation σ. Let

$$\Pr(Y \le b)$$

denote the proportion of values in this population that are less than or equal to a specified value b. Equivalently, $\Pr(Y \le b)$ is the probability that a value selected at random from this population is less than or equal to b. From Line 6.2.1, it follows that

$$\Pr(Y \le b) = \Pr(Y < b) \quad \textbf{Line 6.2.2}$$

for any number b, so Table 6.2.1 can be used to evaluate both $\Pr(Y \le b)$ and $\Pr(Y < b)$. Box 6.2.2 describes a procedure that can be used to evaluate $\Pr(Y < b)$.

TABLE 6.2.1 Proportion of *Normal* Population Values Less Than $\mu + z\sigma$ for Several Values of z [This Can Be Written as $\Pr(Y < \mu + z\sigma)$.]

Line	z	$\Pr(Y < \mu + z\sigma)$
(1)	3.000	0.999
(2)	2.576	0.995
(3)	2.326	0.990
(4)	2.000	0.977
(5)	1.960	0.975
(6)	1.645	0.950
(7)	1.282	0.900
(8)	1.000	0.841
(9)	0.842	0.800
(10)	0.524	0.700
(11)	0.253	0.600
(12)	0.000	0.500
(13)	−0.253	0.400
(14)	−0.524	0.300
(15)	−0.842	0.200
(16)	−1.000	0.159
(17)	−1.282	0.100
(18)	−1.645	0.050
(19)	−1.960	0.025
(20)	−2.000	0.023
(21)	−2.326	0.010
(22)	−2.576	0.005
(23)	−3.000	0.001

BOX 6.2.2: PROCEDURE FOR EVALUATING $\Pr(Y < b)$

The following steps can be used for evaluating

$$\Pr(Y < b),$$

the proportion of values in a *normal* population that are less than a specified value b. The population mean and standard deviation are denoted by μ and σ, respectively, and the quantities b, μ, and σ are known.

1. Set $b = \mu + z\sigma$, and substitute the known values of b, μ, and σ.
2. Solve the equation $b = \mu + z\sigma$ for z. This yields $z = (b - \mu)/\sigma$. The value of z is the number of standard deviation units that b is from μ.
3. Look up $\Pr(Y < \mu + z\sigma)$ in Table 6.2.1 for the value of z computed in step 2.

We illustrate in the following example.

EXAMPLE 6.2.3

USING TABLE 6.2.1. A simple random sample of one value, denoted by Y, is to be selected from a *normal* population $Y_1, Y_2, \ldots, Y_N$, with $\mu = 20$ and $\sigma = 10$. Suppose we want to evaluate

$$\Pr(Y < 43.26).$$

That is, in a *normal* population with mean equal to 20 and standard deviation equal to 10, we want to evaluate the proportion of population values that are less than b, where $b = 43.26$. We will use steps (1) to (3) in Box 6.2.2 to perform this computation.

1. Set $b = \mu + z\sigma$ and substitute for b, μ, and σ to obtain

$$43.26 = 20 + 10z.$$

2. Solve the equation $43.26 = 20 + 10z$ for z to obtain $z = (43.26 - 20)/10 = 23.26/10 = 2.326$.
3. So $\Pr(Y < 43.26) = \Pr(Y < \mu + 2.326\sigma)$, and from line (3) in Table 6.2.1 we get

$$\Pr(Y < \mu + 2.326\sigma) = 0.990.$$

So, the proportion of population values less than 43.26 is equal to the proportion of population values less than $\mu + 2.326\sigma$, which is 0.990. We write this as

$$\Pr(Y < 43.26) = \Pr(Y < \mu + 2.326\sigma) = 0.990.$$

The shaded area in Figure 6.2.11 represents the proportion 0.99 of the population values that are less than $\mu + 2.326\sigma$.

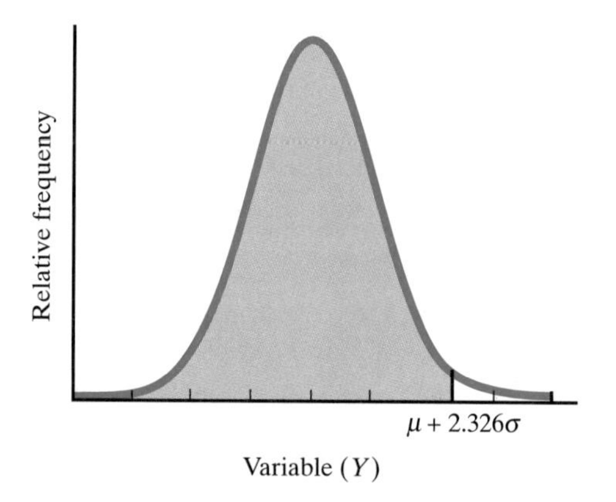

FIGURE 6.2.11
Proportion of Population Values $< \mu + 2.326\sigma$

EXAMPLE 6.2.4

BOXES OF GELATIN MIX. A company packages several thousand boxes of raspberry gelatin mix each month. Each box is labeled "8.4 grams net weight." To satisfy a legal requirement, no more than 1.0% of the boxes can contain less than 8.4 grams of gelatin. If the set of weights for the monthly output of boxes of gelatin is a *normal* population with mean 8.5 grams and standard deviation 0.05 grams, is the legal requirement satisfied?

To answer this question, let $Y_1, Y_2, \ldots, Y_N$ denote the population of weights of gelatin mix in the N boxes packaged in a month, where $\mu = 8.5$ and $\sigma = 0.05$. We must determine the proportion of Y values in the population that are less than 8.4 grams. That is, we must evaluate the proportion $\Pr(Y < 8.4)$. If this proportion is less than or equal to 1.0%, the legal requirement is satisfied. If this proportion is greater than 1.0%, the legal requirement is not satisfied. To evaluate

$$\Pr(Y < b), \quad \text{where } b = 8.4,$$

we use steps (1) to (3) in Box 6.2.2.

1. Set $b = \mu + z\sigma$, and substitute $b = 8.4$, $\mu = 8.5$, and $\sigma = 0.05$ to obtain

$$8.4 = 8.5 + 0.05z.$$

2. Solving the equation $8.4 = 8.5 + 0.05z$ for z, we get $z = (8.40 - 8.50)/0.05 = -2.000$.
3. Look up $\Pr(Y < \mu - 2.000\sigma)$ in Table 6.2.1. From line (20) in Table 6.2.1 we see that $\Pr(Y < \mu - 2.000\sigma) = 0.023$. So $\Pr(Y < 8.4) = \Pr(Y < \mu - 2.000\sigma) = 0.023$, and hence 2.3% of the boxes packaged in a month contain less than 8.4 grams of gelatin. The legal specification is *not* satisfied.

Since the legal specification is not satisfied, suppose the operator of the packaging machine wants to know what the value of μ must be so that the legal specification is satisfied. That is, the operator wants to determine the value of μ so that $\Pr(Y < 8.4) = 0.01$, where $\sigma = 0.05$. From line (21) in Table 6.2.1 we see that if $z = -2.326$, then $\Pr(Y < \mu + z\sigma) = 0.01 = 1.0\%$. Substitute $b = 8.4$, $z = -2.326$, and $\sigma = 0.05$ into the equation $b = \mu + z\sigma$ and solve for μ. Substituting, we get $8.4 = \mu - 2.326(0.05)$, and the solution for μ is $\mu = 8.4 + 0.116 = 8.516$. So if $\mu = 8.516$, we get

$$0.01 = \Pr(Y < \mu - 2.326\sigma) = \Pr(Y < 8.516 - 2.326 \times 0.05) = \Pr(Y < 8.4),$$

and the legal specification is satisfied.

Two relationships that are helpful when using Table 6.2.1 are given in Lines 6.2.3 and 6.2.4:

$$\Pr(a < Y < b) = \Pr(Y < b) - \Pr(Y \le a), \qquad \text{Line 6.2.3}$$

$$\Pr(Y > b) = 1 - \Pr(Y \le b). \qquad \text{Line 6.2.4}$$

The relationship in Line 6.2.3 states that the proportion of population values between a and b is equal to the proportion of population values less than b minus the proportion of population values less than or equal to a. The relationship in Line 6.2.4 states that the proportion of population values greater than b is 1 minus the proportion of population values less than or equal to b.

EXAMPLE 6.2.5

CUSTOMER PURCHASES. To encourage customers to make larger purchases, the manager of a small shop plans to offer a small discount on any purchase over \$100.00. The manager will examine last year's purchases to determine the amount of discount to offer. The manager wants to determine the proportion of last year's purchases that were less than \$100.00. The relative frequency histogram of last year's purchases is similar to the relative frequency histogram of a *normal* population (the bell curve in Figure 6.2.1), so the manager assumes that the population of last year's purchases is a *normal* population. Using last year's purchases, it was found that $\mu = \$90.20$ and $\sigma = \$5.00$. We need to find $\Pr(Y < 100)$, the proportion of values in this population that are less than \$100.00. To do this, we must determine how many standard deviation units 100.00 is from μ. We use steps (1) to (3) in Box 6.2.2 and set $b = \mu + z\sigma$. Substituting the values of b, μ, and σ, we get $100.00 = 90.20 + 5.00z$. Solving this equation for z, we get $z = 1.96$. Thus,

$$\Pr(Y < 100.00) = \Pr(Y < \mu + 1.96\sigma).$$

From line (5) of Table 6.2.1 the answer is $\Pr(Y < \mu + 1.96\sigma) = 0.975$. So 97.5% of the purchases were less than \$100.00 last year.

Suppose the manager also wants to know the proportion of purchases that were between \$86.00 and \$100.00 last year. We write this proportion as

$$\Pr(86.00 < Y < 100.00).$$

By Line 6.2.3 it follows that

$$\Pr(86.00 < Y < 100.00) = \Pr(Y < 100.00) - \Pr(Y \le 86.00).$$

We have computed $\Pr(Y < 100.00)$, so we must now compute $\Pr(Y \le 86.00)$. Set $86.00 = \mu + z\sigma$, substitute for μ and σ, and get $86.00 = 90.20 + 5.00z$. Solving this for z gives $z = -0.84$. From line (15) in Table 6.2.1 we see that

$$\Pr(Y \le 86.00) = \Pr(Y \le \mu - 0.84\sigma) = 0.200 \quad \text{(approximately)}.$$

Substitute $\Pr(Y < 100.00) = 0.975$ and $\Pr(Y \le 86.00) = 0.200$ into Line 6.2.3 to obtain

$$\Pr(86.00 < Y < 100.00) = 0.975 - 0.200 = 0.775.$$

Thus, 77.5% of last year's purchases were between \$86.00 and \$100.00.

Minitab Instructions for Computing $\Pr(Y < b)$

To compute

$$\Pr(Y < b),$$

which is the proportion of population values less than b for a *normal* population with mean μ and standard deviation σ, the following Minitab command can be used:

```
MTB > cdf b;
SUBC> normal mu sigma.
```

The expressions `cdf` and `normal` are instructions to Minitab. The quantities `b`, `mu` and `sigma` are numerical values supplied by the user.

To illustrate, we refer to Example 6.2.3. We want to determine

$$\Pr(Y < 43.26),$$

which is the proportion of population values less than 43.26 where the population is a *normal* population with mean 20 and standard deviation 10. To use Minitab, enter the following command and subcommand in the session window:

```
MTB > cdf 43.26;
SUBC> normal 20 10.
```

A portion of the output is

```
         x        P( X <= x)
     43.26           0.990
```

Notice that the Minitab output uses X where we use Y, and Minitab uses x where we use b. The result, $\Pr(Y < 43.26) = 0.990$, is what we obtained in Example 6.2.3.

To evaluate $\Pr(86 < Y < 100)$ in Example 6.2.5, use Minitab to evaluate $\Pr(Y < 100)$ and to evaluate $\Pr(Y \leq 86)$. Then use Line 6.2.3 to obtain

$$\Pr(86 < Y < 100) = \Pr(Y < 100) - \Pr(Y \leq 86).$$

To compute $\Pr(Y < 100)$, use the following Minitab command and subcommand:

```
MTB > cdf 100;
SUBC> normal 90.2 5.
```

A portion of the output is

```
         x        P( X <= x)
  100.0000          0.9750
```

So $\Pr(Y < 100) = 0.9750$. To compute $\Pr(Y < 86)$, use the following Minitab command and subcommand:

```
MTB > cdf 86;
SUBC> normal 90.2 5.
```

A portion of the output is

```
         x        P( X <= x)
   86.0000          0.2005
```

So $\Pr(Y < 86) = 0.2005$. Thus, using Minitab, we get

$$\Pr(86 < Y < 100) = 0.9750 - 0.2005 = 0.7745,$$

which is the same value (within rounding error) we obtained using Table 6.2.1.

THINKING STATISTICALLY: SELF-TEST PROBLEMS

6.2.1 A company that assembles automobiles plans to purchase 1,000,000 bolts from a company that manufactures machine bolts. If the lengths of the bolts form a *normal* population with mean 6.00 centimeters (cm) and standard deviation 0.05 cm, use Table 6.2.1 to answer the following questions.

a. A bolt is usable only if it is between 5.90 and 6.15 (cm) long. What proportion of the 1,000,000 bolts is usable?

b. If a bolt is selected at random from the 1,000,000 bolts, what is the probability that the bolt is between 5.90 and 6.15 cm long?

c. If a bolt is selected at random from the 1,000,000 bolts, what is the probability that the bolt is less than 6.00 cm long?

d. What proportion of the bolts is between 5.95 cm and 6.10 cm in length?

6.2.2 Use Table 6.2.1 to verify the results in Box 6.2.1.

6.2.3 Consider a *normal* population with $\mu = 100$ and $\sigma = 20$, and let Y denote a value to be selected at random from this population.

a. Find $\Pr(Y > 140)$.
b. Find $\Pr(Y < 80)$.
c. Find $\Pr(60 < Y < 80)$.

6.3 CONFIDENCE INTERVAL FOR THE MEAN OF A *NORMAL* POPULATION

In Section 6.2 we showed you how to find the proportion of values between a and b in a *normal* population when μ and σ are known. While these parameters are known in some problems, in most problems they are not known and must be estimated. In this section we show how to use sample values to compute point estimates for μ and σ for *any* population and how to compute confidence intervals for μ when the population is *normal*. Recall from Box 5.3.1 and Box 5.3.2 that the mean and the standard deviation of a population of N numbers $Y_1, \ldots, Y_N$ are defined as

$$\mu = \frac{Y_1 + Y_2 + \cdots + Y_N}{N} = \frac{\text{SUMY}}{N}$$

and

$$\sigma = \sqrt{\frac{(Y_1 - \mu)^2 + (Y_2 - \mu)^2 + \ldots + (Y_N - \mu)^2}{N}} = \sqrt{\frac{\text{SSDY}}{N}}.$$

The estimate of the population mean is the sample mean, which is defined in Definition 6.3.1.

DEFINITION 6.3.1: SAMPLE MEAN $\bar{y}$

Let the lowercase letters $y_1, y_2, y_3, \ldots, y_n$ denote a simple random sample of size n from a population whose mean is μ.

The sample mean is denoted by $\bar{y}$ (called "y bar") and defined as

$$\bar{y} = \text{sample mean} = \frac{y_1 + y_2 + y_3 + \cdots + y_n}{n}.$$

The sample mean is the estimate of the population mean μ. Hence,

$$\hat{\mu} = \bar{y}.$$

The estimate of the population standard deviation is the sample standard deviation, which is defined in Definition 6.3.2.

DEFINITION 6.3.2: SAMPLE STANDARD DEVIATION *s*

Let the lowercase letters $y_1, y_2, y_3, \ldots, y_n$ denote a simple random sample of size n from a population whose mean and standard deviation are denoted by μ and σ, respectively.

The sample standard deviation is denoted by s and defined as

$$s = \sqrt{\frac{(y_1 - \overline{y})^2 + (y_2 - \overline{y})^2 + \cdots + (y_n - \overline{y})^2}{n-1}} = \sqrt{\frac{\text{SSD}y}{n-1}}.$$

The sample standard deviation s is the estimate of the population standard deviation σ. Hence,

$$\widehat{\sigma} = s.$$

The denominator $n - 1$ in the calculation of the sample standard deviation is called the **degrees of freedom** and abbreviated as df or DF. This terminology and these abbreviations are used extensively throughout statistics. It may seem natural to divide SSDy by n instead of $n - 1$ to obtain s in Definition 6.3.2. However, it can be shown that dividing by $n - 1$ (rather than n) results in a slightly "better" estimate of σ.

As discussed in Chapter 3, point estimates are particularly useful if we know how close the point estimate might be to the true population parameter. Confidence intervals provide this information. A confidence interval for the population mean of a *normal* population is presented in Box 6.3.1.

BOX 6.3.1: CONFIDENCE INTERVAL FOR THE MEAN OF A *NORMAL* POPULATION

Let $Y_1, Y_2, \ldots, Y_N$ represent a *normal* population with mean μ and standard deviation σ, both unknown. From a simple random sample of size n from this population, $\widehat{\mu}$ and $\widehat{\sigma}$ are computed. A confidence interval for μ is L to U, where

$$L = \widehat{\mu} - (T2)(\widehat{\sigma}/\sqrt{n}) \quad \text{and} \quad U = \widehat{\mu} + (T2)(\widehat{\sigma}/\sqrt{n}).$$

The quantity $T2$ is a *table value*, which is found in Table T-2 on the inside front cover of this book. The quantity $T2$ depends on the specified confidence coefficient and on the degrees of freedom $n - 1$ used to compute s.

To find the value of $T2$, enter the column of Table T-2 that corresponds to the specified confidence coefficient (0.90, 0.95, 0.98, or 0.99). Then go down the column labeled DF (degrees of freedom) until you come to the value $n - 1$. The intersection of this row with the confidence coefficient column is the value of $T2$ to use. Table T-2 is called a ***t* table**, and the confidence interval in Box 6.3.1 is often called a ***t* interval**. Appendix 1 (Section 6.10) contains a more detailed discussion of this confidence interval.

EXAMPLE 6.3.1

WEIGHTS OF MEN. An investigator for an insurance company wants to determine μ, the average weight of men 40 to 60 years old who were admitted to a weight-loss clinic during the past year with symptoms of heart problems. A simple random sample of 40 men was obtained, and their weights recorded. The sample weights are presented in Table 6.3.1.

TABLE 6.3.1 40 Sample Weights

Patient	Weight	Patient	Weight
1	220	21	214
2	214	22	211
3	209	23	213
4	208	24	223
5	231	25	229
6	215	26	222
7	213	27	217
8	210	28	226
9	215	29	208
10	219	30	217
11	214	31	209
12	223	32	219
13	225	33	207
14	210	34	215
15	210	35	217
16	217	36	223
17	213	37	217
18	227	38	226
19	226	39	215
20	220	40	214

From Table 6.3.1 we obtain $\overline{y} = 217.02$ and $s = 6.36$, so $\widehat{\mu} = 217.02$ pounds and $\widehat{\sigma} = 6.36$ pounds. From previous experience with populations similar to this sampled population, the investigator believes that the population of weights is *normal*. To determine "how close" $\widehat{\mu} = 217.02$ pounds is to μ, the investigator will obtain a 95% confidence interval for μ using the formulas for L and U in Box 6.3.1. To obtain $T2$, enter Table T-2 in the column labeled 0.95. Since DF (degrees of freedom) $= n - 1 = 40 - 1 = 39$, go down the column labeled DF until you come to row 39. The number 2.02 is in the 0.95 column of this row, and so $T2 = 2.02$. Using this value for $T2$ in the formulas for L and U, we obtain

$$L = 217.02 - 2.02(6.36/\sqrt{40}) = 214.98$$
$$U = 217.02 + 2.02(6.36/\sqrt{40}) = 219.06.$$

We have 95% confidence that μ is in the interval 214.98 to 219.06 pounds.

Margin of Error

From Box 6.3.1 we see that $\widehat{\mu}$ is at the center of the confidence interval for μ. Thus, as defined in Chapter 3, the margin of error for the estimate $\widehat{\mu}$ is

$$ME = \frac{U - L}{2} = (T2)(\widehat{\sigma}/\sqrt{n}). \qquad \textbf{Line 6.3.1}$$

This margin of error can be computed for any confidence coefficient by selecting the appropriate value of $T2$. To emphasize the level of confidence associated with a margin of error, it is customary to use phrases like "the 95% margin of error" or "the 99% margin of error." An alternative way to describe a confidence interval for μ (say, a 95% confidence interval) is to state

the estimate of μ is $\widehat{\mu}$ with a 95% margin of error of $(T2)(\widehat{\sigma}/\sqrt{n})$.

The unit of measurement for the margin of error is the same as the unit of measurement for the sample mean, $\bar{y}$. For instance, in Example 6.3.1 $ME = (2.02) \times (6.36/\sqrt{40}) = 2.04$ pounds. We therefore state

the estimate of μ is 217.02 pounds with a 95% margin of error of 2.04 pounds.

This means that the estimate of μ is 217.02 pounds, and we are 95% confident that this estimate is within 2.04 pounds of μ.

Using Minitab to Compute a Confidence Interval for μ

Minitab can be used to compute the confidence interval in Box 6.3.1. Enter the following command in the session window:

```
MTB > tinterval confidence ci
```

where `confidence` is the specified confidence coefficient and the data are in column `i`. We illustrate using the weight data in Example 6.3.1. These data are stored in column 2 in the file **weight.mtp** on the data disk. Retrieve the data, and enter the following command in the session window:

```
tinterval 0.95 c2
```

This command tells Minitab to compute the confidence interval in Box 6.3.1 where the value 0.95 is the desired confidence coefficient, and `c2` indicates that column 2 contains the data. The output is shown in Exhibit 6.3.1.

```
------------------------------------------------------------------------------
Confidence Intervals

Variable      N       Mean     StDev  SE Mean        95.0 % C.I.
Weight       40     217.02      6.36     1.01   (  214.99,  219.06)
------------------------------------------------------------------------------
```

EXHIBIT 6.3.1

The number 6.36 under the label `StDev` in Exhibit 6.3.1 is s, so $\widehat{\sigma} = 6.36$. The column labeled `95.0 % C.I.` contains the 95% confidence interval for μ, which is 214.99 to 219.06 pounds. Apart from round-off error, this is the same interval computed in Example 6.3.1 using the formulas in Box 6.3.1.

THINKING STATISTICALLY: SELF-TEST PROBLEMS

6.3.1 Consider the population of 2,600 calculus scores in Appendix B. We denote this population of scores by $Y_1, Y_2, Y_3, \ldots, Y_{2600}$. The population mean and the standard deviation are denoted by μ and σ, respectively.

Since the entire population is known, the values of μ and σ can be computed. However, for illustration, a simple random sample of size 50 is selected from the population for the purpose of obtaining estimates of μ and σ. This sample is denoted by $y_1, y_2, y_3, \ldots, y_{50}$. The sample values are shown in Table 6.3.2 and stored in column 2 in the file **50grades.mtp** on the data disk.

TABLE 6.3.2 Sample of 50 Calculus Scores

Sample number	Score	Sample number	Score
1	90	26	83
2	66	27	85
3	80	28	75
4	90	29	82
5	59	30	47
6	98	31	80
7	74	32	59
8	75	33	67
9	79	34	72
10	85	35	66
11	92	36	82
12	56	37	88
13	73	38	83
14	79	39	84
15	81	40	73
16	80	41	82
17	75	42	84
18	73	43	76
19	92	44	68
20	82	45	73
21	68	46	59
22	76	47	85
23	73	48	68
24	82	49	71
25	63	50	70

The output that resulted from the Minitab command `describe` is shown in Exhibit 6.3.2.

```
-----------------------------------------------------------

Descriptive Statistics

Variable        N    Mean   Median   TrMean   StDev   SEMean
Score          50   76.06    76.00    76.34   10.25     1.45

Variable      Min     Max       Q1       Q3
Score       47.00   98.00    69.50    83.00
-----------------------------------------------------------
```

EXHIBIT 6.3.2

a. Use Exhibit 6.3.2 to find $\widehat{\mu}$ and $\widehat{\sigma}$.
b. Use Table T-2 to find $T2$ for computing a 90% confidence interval for μ.
c. Compute a 90% confidence interval for μ, assuming the population of scores is *normal* with mean μ and standard deviation σ.

6.4 SAMPLE SIZE TO USE FOR ESTIMATING A POPULATION MEAN

In this section we describe a procedure for determining the sample size required so that the estimate $\widehat{\mu}$ will be within d units of μ with a specified confidence. The value of d is specified in advance by the investigator. We begin with an example.

EXAMPLE 6.4.1

AUTOMOBILE MAINTENANCE. A used-car dealership that sells several thousand cars each year wants to offer a warranty policy for each used car that is sold. This policy will pay for all repairs on a car for one year if the repairs are made in a company garage. A surcharge will be added to the price of each car to pay for the warranty. A company accountant is asked to determine the required amount of the surcharge. The accountant reasons that if N cars are sold with the warranty, the total repair cost of the N cars will be N multiplied by μ, the average repair cost per car. So the parameter of interest is μ, the average repair cost per car sold with the warranty. However, this parameter is not known because the target population (cars sold with a warranty) is a future population of cars. Thus, it is necessary to obtain a sample from a sampled population that is similar to the target population. The accountant decides to use the set of cars sold by the dealership during the past 24 months as the sampled population. The variable of interest is "Repair cost during the first year," and the parameter of interest is μ, the average first-year repair cost for the cars in the sampled population. From past experience, the accountant assumes that the population is *normal* with mean μ and standard deviation σ. The company has a record of all customers who bought cars during the past 24 months, and a simple random sample of these customers is to be selected. This sample will be used to compute the estimate $\widehat{\mu}$. The company president wants to be quite confident that $\widehat{\mu}$ will be within \$5.00 of μ. Specifically, he wants to be 95% confident that the estimate of μ (namely $\widehat{\mu}$) is within $d = \$5.00$ of μ. The accountant must determine n, the sample size required so that the president's specification will be satisfied. That is, the accountant must determine the required sample size n so that the confidence is 95% that $\widehat{\mu}$ will be within $d = \$5.00$ of μ.

Using the formulas for L and U in Box 6.3.1, we notice that $\widehat{\mu}$ is at the center of the confidence interval. So if μ is contained in the confidence interval, then $\widehat{\mu}$ will be within d units of μ if

$$\frac{U - L}{2} \leq d. \qquad \text{Line 6.4.1}$$

Substituting L and U from Box 6.3.1 into Line 6.4.1, we get

$$\frac{U - L}{2} = (T2)(\widehat{\sigma}/\sqrt{n}) \leq d. \qquad \text{Line 6.4.2}$$

Solving the inequality in Line 6.4.2 for n, we obtain

$$n \geq \left[\frac{(T2)\widehat{\sigma}}{d}\right]^2.$$

An investigator would use the smallest value of n that satisfies Line 6.4.2, which is

$$n = \left[\frac{(T2)\widehat{\sigma}}{d}\right]^2. \qquad \textbf{Line 6.4.3}$$

In general, n in Line 6.4.3 will not be an integer, so round n up to the nearest integer and use this value for the sample size.

There are a couple of problems in using Line 6.4.3 to compute n. First, $T2$ depends on the unknown value of n since DF is $n - 1$ in Table T-2. To solve this problem, we assume that n will be at least 31, so we set $T2$ equal to values in Table T-2 with $DF = 30$. Thus, we use $T2 = 2.04$ for a 95% confidence interval and $T2 = 2.75$ for a 99% confidence interval. The second problem encountered in using the formula in Line 6.4.3 is that an estimate of σ is required, and this estimate must be known before the sample is collected. This estimate of σ might be obtained in one of the three ways described in Box 6.4.1.

BOX 6.4.1: METHODS TO ESTIMATE σ FOR PLANNING SAMPLE SIZE

1. An investigator might be able to obtain an estimate of σ based on information from previous studies that are similar to the present study.
2. A small sample might be collected from the population with the sole purpose of obtaining an estimate $\widehat{\sigma}$ to be used in the formula in Line 6.4.3.
3. An investigator familiar with thc population might know that most of the population values are between the numbers a and b. If the population is *normal*, we know that about 95% of the values are between $\mu - 2\sigma$ and $\mu + 2\sigma$. So let $a = \mu - 2\sigma$ and $b = \mu + 2\sigma$. Subtract the first equation from the second and obtain $b - a = 4\sigma$, or $\sigma = (b - a)/4$. Thus, we can substitute

$$\widehat{\sigma} = \frac{b - a}{4} \qquad \textbf{Line 6.4.4}$$

into the formula for n in Line 6.4.3.

The value of n obtained using Line 6.4.3 will not guarantee that the margin of error of $\widehat{\mu}$ will be less than d (that is, it will not guarantee that $\widehat{\mu}$ will be within d units of μ with the *specified* confidence), but the margin of error should be close to the specified value d.

EXAMPLE 6.4.2

AUTOMOBILE MAINTENANCE. We continue with Example 6.4.1, where a used-car dealership wants to add a surcharge on each car it sells to pay for a one-year warranty. The amount to be added to the price of each car is μ, the average 12-month repair cost of the population of cars that were sold during the past 24 months. But since μ is unknown, it must be estimated; the company president wants to be 95% confident that the estimate is within $d = \$5.00$ of the population mean μ. From the service records in the company garage for the past five years, it was determined that a reasonable estimate of σ is \$21.63. Using this value for $\widehat{\sigma}$ in Line 6.4.3, we compute

$$n = \left[\frac{(T2)\widehat{\sigma}}{d}\right]^2 = \left[\frac{(2.04)(21.63)}{5.00}\right]^2 = 77.88,$$

which we round up to 78. So if a simple random sample of size 78 is obtained and $\widehat{\mu}$ is computed, it will be within about \$5.00 of μ with confidence 95%. Suppose that the service records used to estimate σ in Line 6.4.3 are not readily available, but from experience the accountant believes that the annual repair cost for most cars is between \$120 and \$200. In this case Line 6.4.4 is used to obtain $\widehat{\sigma}$, where $a = 120$ and $b = 200$. This provides $\widehat{\sigma} = (200 - 120)/4 = \$20$. If this value is used in Line 6.4.3, we obtain $n = 66.58$, which is rounded up to 67.

6.4.1 The department head wants to obtain an estimate of μ, the mean of the 2,600 calculus scores in Appendix B, based on a simple random sample of size n from the population. The department head wants the estimate to be within 2 points of μ with 95% confidence. From previous experience, the department head knows that most of the scores will be between 55 and 100. What sample size is needed to obtain the desired result?

6.5 CONFIDENCE INTERVAL FOR σ IN A *NORMAL* POPULATION

In this section we present the formulas for computing L and U, the endpoints of a confidence interval for the standard deviation of a *normal* population. Information about how these formulas are derived is provided in Appendix 2 of this chapter (Section 6.11).

BOX 6.5.1: CONFIDENCE INTERVAL FOR σ IN A *NORMAL* POPULATION

Let $y_1, y_2, y_3, \ldots, y_n$ represent a simple random sample of size n from a *normal* population with mean μ and standard deviation σ, where both are unknown. Let $\widehat{\sigma}$ be the estimate of σ computed from these sample values. A confidence interval for σ is L to U, where

$$L = (T3L)\widehat{\sigma} \qquad \text{and} \qquad U = (T3U)\widehat{\sigma}.$$

The quantities $T3L$ and $T3U$ are *table values* that are found in Table T-3 on the page facing the inside of the front cover of this book.

The symbol $T3L$ stands for the table value that is used to compute L, the lower endpoint of a confidence interval for σ. $T3$ in the symbol $T3L$ tells us that the table value is found in Table T-3, and the L in $T3L$ tells us that this is the table value used to compute L. The symbol $T3U$ stands for the table value used to compute U, the upper endpoint of a confidence interval for σ. $T3$ in $T3U$ tells us that the table value is found in Table T-3, and the U in $T3U$ tells us that this is the table value used to compute U.

The table values $T3L$ and $T3U$ are different than the table value for the confidence interval for the mean reported in Table T-2. The reason for the difference is explained more fully in Appendix 2. You are encouraged to read the material in Appendix 2 if you want a more in-depth understanding of the table values $T3L$ and $T3U$. If not, it is only important that you understand that all confidence intervals are interpreted in the same manner that we described in Chapter 3. In particular, if you select a simple random sample of size n from a *normal* population and use Box 6.5.1 to compute a 95% confidence interval for σ, your confidence that the computed interval contains σ is the same as your confidence that a ball hidden in your hand is black if it was selected at random from a bin with 95 black balls and 5 white balls.

EXAMPLE 6.5.1

WEIGHTS OF MEN. In Example 6.3.1 an investigator for an insurance company wants to examine the weights of men 40 to 60 years old who were admitted to a weight-loss clinic with heart problems. The investigator assumes that the population of weights is *normal*; among other things, he wants to obtain a point estimate and a 95% confidence interval for σ, the population standard deviation. From Exhibit 6.3.1 we obtain $n = 40$ and $\widehat{\sigma} = s = 6.36$ pounds. From Table T-3 for a confidence coefficient equal to 95% and with $DF = 40 - 1 = 39$, we obtain $T3L = 0.819$ and $T3U = 1.28$. Substituting these values along with $\widehat{\sigma} = 6.36$ into the formulas in Box 6.5.1, we get $L = 5.20$ and $U = 8.15$. We have 95% confidence that σ is between 5.20 and 8.15 pounds. Notice that unlike the previous confidence intervals we have studied, the point estimate, $\widehat{\sigma}$, is not in the center of the confidence interval for σ.

6.5.1 The department head wants a point estimate and a confidence interval for σ, the standard deviation for the 2,600 calculus scores in Appendix B. Since the entire population is available, the exact value of σ can be determined. However, for illustration we will use a simple random sample of size 50 from the population and obtain an estimate of σ. The sample is shown in Table 6.3.2 and stored in column 2 in the file **50grades.mtp** on the data disk. From the data we obtain $n = 50$ and $s = 10.25$. Find $\widehat{\sigma}$ and a 99% confidence interval for σ.

6.5.2 Find $T3L$ and $T3U$ for computing a 99% confidence interval for σ for $n = 12$, 21, and 34.

6.5.3 Find $T3L$ and $T3U$ for computing a 95% confidence interval for σ for $n = 12$, 21, and 34.

6.6 USING SAMPLE VALUES TO HELP DECIDE IF A POPULATION IS *NORMAL*

The confidence interval formulas for μ, presented in Box 6.3.1 (and for σ in Box 6.5.1), are strictly valid only if the population is *normal.* In a practical problem, all values in a population are generally not known, in which case one would not know whether the population under study is *normal.*

When making statistical inferences about μ (and σ) in applied problems, investigators customarily proceed as if the sampled population is a *normal* population. Unless investigators have considerable experience with similar populations, *they should* **always** *use sample values to help decide if the sampled population is indeed normal.* If an investigator concludes that the population is *normal*, then the formulas in Box 6.3.1 and Box 6.5.1 can be used to obtain confidence intervals for μ and σ, respectively. In Chapter 7 we discuss procedures for obtaining a confidence interval for μ if the population is *not normal.*

There are several statistical procedures that investigators can use to help decide if a simple random sample appears to have been selected from a *normal* population. Most of them rely on the availability of a computer for carrying out the calculations. One of the procedures that uses a graph is discussed next.

Normal Rankit-plot and Nscores

A graphical procedure that can be used to help decide whether or not a simple random sample appears to have been selected from a *normal* population is called a **normal rankit-plot**. Suppose a simple random sample of size n is available from the population under study. These sample values are compared with n "specially constructed" values, called **normal scores** and abbreviated as **nscores**. Suppose a simple random sample of size n is selected from a *normal* population (with mean 0 and standard deviation 1), and the sample values are ordered from smallest to largest. Repeat this for every possible sample of size n. Obtain the average of the smallest ordered sample values (this is the smallest nscore); obtain the average of the second smallest ordered sample values (this is the second smallest nscore), and so on. These n averages are the nscores. The calculation of nscores is somewhat involved and requires a computer.

If a scatter plot of the data values against the nscores displays a very nearly straight-line relationship between them, then the investigator concludes that the sample was selected from a *normal* population. If the points exhibit a departure from a straight line, then the investigator concludes that the sample was not selected from a *normal* population. In Box 6.6.1 we describe the procedure to construct a normal rankit-plot.

BOX 6.6.1: HOW TO CONSTRUCT A NORMAL RANKIT-PLOT FOR DECIDING IF A SIMPLE RANDOM SAMPLE CAME FROM A *NORMAL* POPULATION

1. Obtain a simple random sample of size n from an *unknown* population, and order the sample values from smallest to largest. Denote this ordered sample by

$$y_1, y_2, \ldots, y_n.$$

2. Determine the values of the nscores and order them from smallest to largest value. Denote the ordered nscores by

$$x_1, x_2, \ldots, x_n.$$

3. Construct a scatter plot with the sample values on the vertical axis and the nscores on the horizontal axis by plotting y_1 against x_1, y_2 against x_2, and so on. This plot is called the normal rankit-plot.

If a scatter plot of the simple random sample values $(y_1, y_2, \ldots, y_n)$ versus the computed nscores $(x_1, x_2, \ldots, x_n)$ appears to follow a straight line pattern, then the investigator proceeds as if the sample were selected from a *normal* population. If the scatter plot exhibits a departure from a straight line, then the investigator concludes that the sample was not selected from a *normal* population. As with all graphical procedures, the decision whether the plotted points exhibit a nearly linear relationship is subjective. It may happen that one investigator examines the plot and decides that the sample came from a *normal* population, whereas another investigator looks at the same plot and reaches the opposite conclusion. Even though this procedure is subjective, it is useful in identifying extreme cases in which a population is clearly not *normal*. We will present several examples that will help you better understand how to interpret the normal rankit-plot.

To construct a normal rankit-plot using Minitab, suppose the simple random sample of size n is in column `i` of a Minitab worksheet. To compute nscores, enter the following command in the session window:

```
MTB > nscores ci cj
```

This command tells Minitab to compute the nscores corresponding to the n data values in column `i` and put them in column `j`, where n is the number of sample observations. Minitab automatically matches the smallest data value with the smallest nscore, the second smallest data value with the second smallest nscore, and so on. Therefore, no ordering of the data is necessary when using Minitab.

To plot the sample values against the nscores, use the Minitab command

```
MTB > %fitline ci cj
```

This command shows the normal rankit-plot and in addition places a reference line on the scatter plot. This line aids in deciding if the points hover around a straight line.

EXAMPLE 6.6.1

A SAMPLE FROM A POPULATION THAT IS SKEWED TO THE RIGHT. A simple random sample of size 35 was obtained from the population whose relative frequency histogram is shown in Figure 6.6.1. This population is skewed to the right and is clearly not a *normal* population. A normal rankit-plot for this sample is shown in Figure 6.6.2.

FIGURE 6.6.1
Histogram of a Population That Is Skewed to the Right

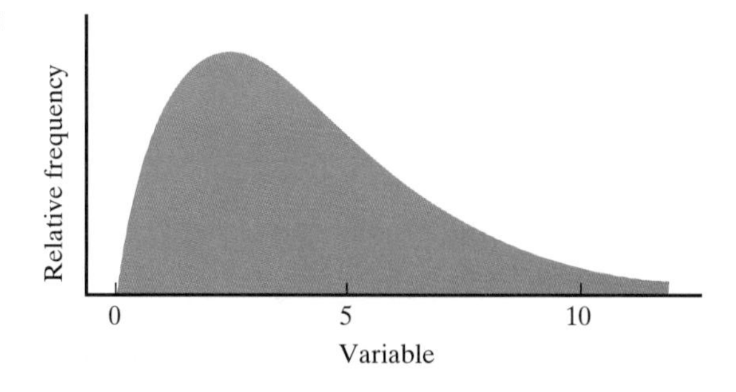

Notice that the points in Figure 6.6.2 show a definite deviation from a straight-line pattern. Thus, we conclude that the simple random sample did *not* come from a *normal* population.

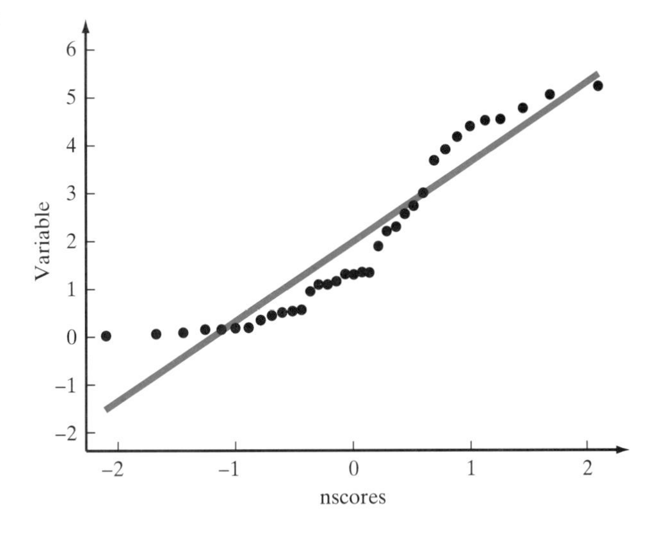

FIGURE 6.6.2
A Normal Rankit-Plot for a Sample from the Population Whose Histogram Is Shown in Figure 6.6.1 (skewed to the right)

EXAMPLE 6.6.2

SAMPLE FROM A *NORMAL* POPULATION. A simple random sample of size 35 was obtained from a *normal* population. A normal rankit-plot of the data is shown in Figure 6.6.3.

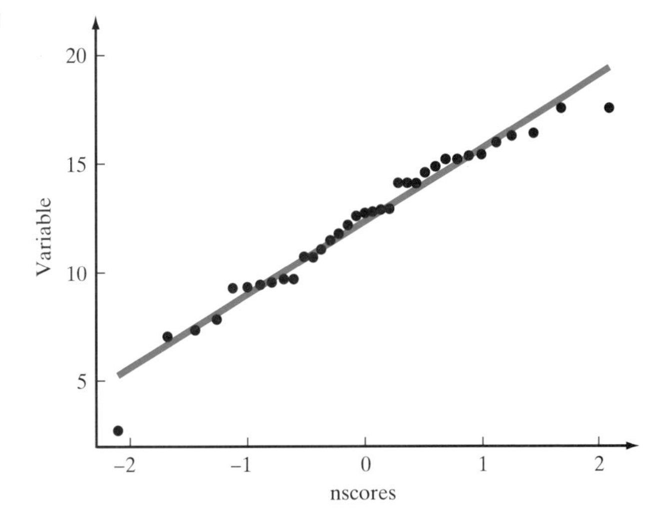

FIGURE 6.6.3
Normal Rankit-Plot for a Simple Random Sample from a *Normal* Population

The points in the plot of Figure 6.6.3 hover around the reference line in a tight pattern. Thus, there is no compelling reason to doubt that the simple random sample came from a *normal* population.

In this section we have demonstrated how the normal rankit-plot can be used to decide if a simple random sample was selected from a *normal* population. When the sample size is very large, histograms can also be used for this purpose. Many software programs overlay the theoretical *normal* curve on histograms to aid in determining how well the *normal* population model represents the histogram. One problem with this procedure is that the shape of a histogram is somewhat influenced by the selected width of the class intervals.

THINKING STATISTICALLY: SELF-TEST PROBLEM

6.6.1 A simple random sample of 85 students was selected from last year's graduating class at a small college. Each student's grade point average is recorded in column 2 in the file **studtgpa.mtp** on the data disk. To obtain the normal rankit-plot using Minitab, retrieve the data and enter the following commands in the session window:

```
MTB > nscore c2 c3
MTB > %fitline c2 c3
```

The normal rankit-plot is shown in Figure 6.6.4.

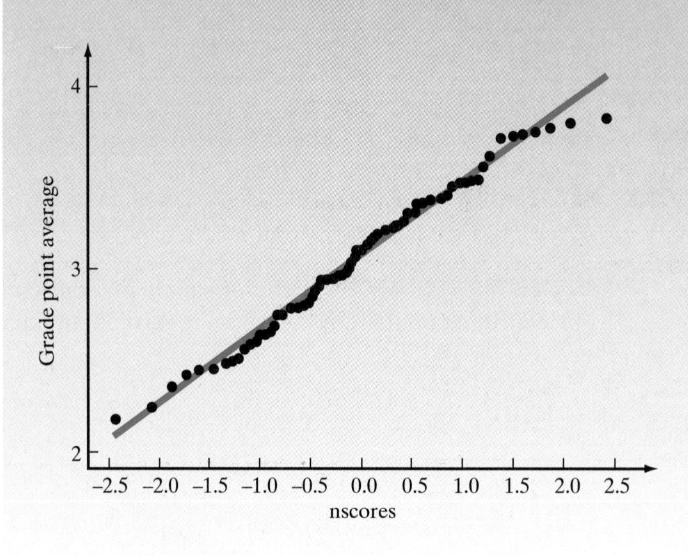

FIGURE 6.6.4
Normal Rankit-Plot of GPA Data

Does it seem appropriate to conclude that the population of grade point averages is a *normal* population?

6.7 CHAPTER SUMMARY

KEY TERMS

1. *normal* population
2. bell curve
3. margin of error
4. nscores
5. *normal* rankit-plot

SYMBOLS

1. $\bar{y}$—the sample mean
2. s—the sample standard deviation
3. $DF = n - 1$ = degrees of freedom associated with the sample standard deviation
4. $T2$—table value used in the confidence interval for the mean of a *normal* population
5. $T3L$ and $T3U$—table values used in the confidence interval for the standard deviation of a *normal* population

KEY CONCEPTS

1. The sample mean is the estimate of the population mean (that is, $\bar{y} = \hat{\mu}$).
2. The sample standard deviation is the estimate of the population standard deviation (that is, $s = \hat{\sigma}$).
3. The *normal* population is a conceptual population that contains an infinite number of values.
4. The *normal* population serves as a model for many real populations.
5. The *normal* population is symmetric.
6. The mean and the median are equal in a *normal* population.
7. The probability that an item selected at random from a *normal* population will have a value in a specified interval a to b is equal to the proportion of items in the population with values in that interval.
8. In fact, 68.26% of the values in a *normal* population are within one standard deviation of the mean; 95.45% are within two standard deviations of the mean; and 99.73% are within three standard deviations of the mean.
9. The confidence intervals for the population mean and the population standard deviation presented in this chapter are strictly valid only if the population is *normal*.
10. If a simple random sample is selected from a *normal* population, the scatter plot of the ordered sample values versus the nscores should appear to follow a straight line pattern.
11. When a simple random sample is selected for obtaining a confidence interval for a population mean μ, the sample should first be used to construct a normal rankit-plot to help decide if the sampled population is *normal*.

SKILLS

1. Use Table 6.2.1 to determine the proportion of values in a specified interval a to b in a *normal* population, where μ and σ are known.
2. Compute a confidence interval for the population mean of a *normal* population using a simple random sample of size n.
3. Use Line 6.4.3 to determine the sample size needed to estimate a population mean within d units of the true value for a specified level of confidence.
4. Compute a confidence interval for the population standard deviation of a *normal* population using a simple random sample of size n.
5. Construct a *normal* rankit-plot to determine if a sample has been selected from a *normal* population.

6.8 CHAPTER EXERCISES

6.1 A company manufactures a breakfast cereal and each month it packages several thousand boxes that are labeled "453 grams net weight." The set of weights for the monthly output of boxes is a *normal* population with mean 461 grams and standard deviation 3 grams. To satisfy a legal requirement, no more than 1% of the boxes can contain less than 453 grams of cereal.

a. Is the legal requirement satisfied?

b. Suppose a new machine is purchased to fill boxes of cereal, and the standard deviation of box weights for this machine is $\sigma = 1.29$ grams. To save money, the company wants μ to be as small as possible. What is the smallest value of μ that satisfies the legal requirement (that is, so that no more than 1% of the boxes have net weights less than 453 grams)?

6.2 A food vendor at a football game must determine how many hot dogs to purchase for an upcoming game. If more hot dogs are purchased than can be sold, the unsold hot dogs must be discarded. Alternatively, if too few hot dogs are ordered, money is lost by not making potential sales. To help determine how many hot dogs to purchase, the vendor collected data for the past 10 football seasons. Over the past 10 seasons, the average attendance at a football game has been 40,000 people and the standard deviation 3,000 people. A histogram of the data suggests that the attendance figures can be considered to be a *normal* population.

a. The vendor wants to order enough hot dogs to meet the demand for 90% of the football games. Find the number b such that for 90% of the football games the attendance will be less than b.

b. In the past, the vendor sold one hot dog for every two people who attended a game. How many hot dogs should the vendor order for each game?

6.3 A report from the American Automobile Manufacturers Association stated that the average number of miles driven last year by all cars in the United States was 12,500 miles. Let Y represent the variable "Miles driven last year," and assume the population of Y values is a *normal* population with a standard deviation of 3,000 miles.

a. Describe the target population referred to in this statement.

b. What is the probability that a car selected at random from the population was driven less than 18,000 miles last year?

c. What proportion of cars in the United States was driven between 9,500 miles and 18,500 miles last year?

6.4 As supervisor of a sales staff of 400 employees, you are responsible for recommending salary adjustments to the company vice president. You have been given a sum of money that is to be distributed among the top 20% of employees based on their amount of total sales during the past year. You access a database that contains the total sales for each employee for the past year and produce a relative frequency histogram for the population of total sales for each employee. This histogram looks very similar to the relative frequency histogram of a *normal* population. Using the database, you also determine that the average sales for the population is $400,000 and the population standard deviation is $30,000. Find a number b such that 20% of the employees have total sales larger than b and hence qualify for a bonus.

6.5 A simple random sample of 40 employees selected from a population of employees at a large state university has an average commuting distance of 8.95 miles (that is, $\bar{y} = 8.95$ miles).

a. What is the point estimate for μ, the average commuting distance for the sampled population of all university employees?

b. Assume the population of commuting distance values is *normal*. Construct a 95% confidence interval for μ if the sample standard deviation is $s = 2.2$ miles.

6.6 The manufacturer of a new model of automobile claims that it will get an average of 30 miles per gallon (mpg) of gasoline for highway driving. A company that makes a gasoline additive believes that the 30 mpg can be increased by at least 4 miles per gallon if one pint of the additive is used for every 40 gallons of gasoline. The company wants to make this claim in its advertising. To help verify the claim, an investigator for the company purchased 36 new cars of this model, used the additive, and drove them over a specified test route. The sampled population is the set of automobiles of this model produced during the month the sample was collected. About 80,000 cars were produced during this period. The mean of this population is μ, and the standard deviation is σ. Theoretically, the investigator could have driven each of the 80,000 cars over the test route, using the additive; and from these 80,000 cars, the investigator could have selected a simple random sample of 36 cars and observed the mpg of each of them. This is not practical, so the 36 cars were selected first, and only those 36 were driven over the test route and used the additive. So the sampled population is conceptual. The sample data are stored in column 2 in the file **additive.mtp** on the data disk. The sample mean and the sample standard deviation are $\bar{y} = 35.6$ and $s = 3.9$, respectively (each rounded to one decimal).

a. What is the estimate of μ and σ?

b. Find a 95% confidence interval for μ assuming the population of mpg values is *normal*.

6.7 A consumer's group is interested in determining the price of regular unleaded gasoline in a large city on the past July 4. The sampled (and target) population of items is the set of all service stations in the city on July 4. A statistician in the consumer's group selected a simple random sample of 20 service stations and asked them to report the price of gasoline for the past July 4. The sample mean and the sample standard deviation are $\bar{y} = \$1.19$ per gallon and $s = \$0.024$ per gallon, respectively.

a. What are the estimates of μ and σ?

b. Suppose the investigator has experience analyzing gasoline prices and can safely say that this population of gasoline prices is a *normal* population. Find a 99% confidence interval for μ, the average price of gasoline at all service stations in the city this past July 4.

6.8 Two months ago the County Air Pollution Control Division informed a large manufacturing company that the company's emission of carbon monoxide had exceeded the monthly limit set by the state. The company must convince the Control Division that it was in compliance for the past 30 days. To determine if the company was in compliance during the past 30 days, a simple random sample of emissions was obtained for 100 of the 8,640 five-minute time periods of the past 30 days. Let μ denote the average carbon monoxide emission for the population of $N = 8{,}640$ five-minute periods, and assume this population of emission values is a *normal* population. The amount of carbon monoxide in the 100 sampled periods was measured; the sample mean and the sample standard deviation were computed to be $\bar{y} = 1.920$ pounds per five minutes and $s = 0.230$ pounds per five minutes, respectively.

a. What are the estimates of μ and σ?

b. For the company to be in compliance, the total carbon monoxide emitted for the past 30 days must have been less than 15,000 pounds. What is the estimate of total carbon monoxide emitted for the 30 days under consideration?

c. Find a 99% confidence interval for μ.

d. Find a 99% confidence interval for the total amount of carbon monoxide emitted during the 30 days under study. Based on this interval, does it appear that the company was in compliance for the past 30 days (that is, were fewer than 15,000 pounds of carbon monoxide emitted)?

6.9 A company that manufactures chicken feed has developed a new product. The company claims that at the end of 12 weeks after hatching, the average weight of chickens using this product will be at least 3.0 pounds. The owner of a large chicken farm decided to examine this new product, so he fed the new ration to all 12,000 of his newly hatched chickens. At the end of 12 weeks he selected a simple random sample of 40 chickens and weighed them. The sample mean and the sample standard deviation are $\bar{y} = 3.72$ pounds and $s = 0.63$ pounds, respectively.

a. What is the estimate of μ, the average weight of the 12,000 chickens at the end of 12 weeks?

b. Find a 99% confidence interval for μ assuming the population of weights is a *normal* population.

6.10 To evaluate the effectiveness of a new type of plant food that was developed for tomatoes, an experiment was conducted in which a simple random sample of 25 seedlings was obtained from a large greenhouse that has thousands of such seedlings. Each of the 25 plants received 10 grams of this new type of plant food each week for 14 weeks. The number of tomatoes produced by each plant in the sample is shown in Table 6.8.1 and stored in column 2 in the file **tomatoes.mtp** on the data disk.

a. What is the sampled population of items?

b. Compute a 95% confidence interval for μ, the average number of tomatoes that would have been produced by all seedlings in the greenhouse if they had received 10 grams of the new type of plant food each week for 14 weeks. Assume that the population of values from the seedlings is a *normal* population. The sample mean and the sample standard deviation for the sample of 25 tomatoes in **tomatoes.mtp** is $\bar{y} = 35.280$ tomatoes and $s = 4.774$ tomatoes, respectively.

c. What is the 95% margin of error?

d. A farmer plans to grow 2,000 tomato seedlings of this variety and feed them 10 grams of the new type of plant food each week for 14 weeks. Estimate the total number of tomatoes the farmer can expect to harvest, and find a 95% confidence interval for this total.

TABLE 6.8.1 Sample of 25 Tomatoes

Sample number	Number of tomatoes	Sample number	Number of tomatoes
1	28	14	30
2	42	15	41
3	34	16	40
4	35	17	29
5	26	18	30
6	37	19	32
7	41	20	29
8	37	21	37
9	38	22	35
10	31	23	43
11	38	24	40
12	36	25	37
13	36		

6.11 A new cream has been developed to shrink the fat in a person's thigh. A study was conducted to determine the effect of the cream. In particular, a simple random sample of 40 subjects was selected from the 1,200 members of a local health club. Each person in the study treated one of their thighs with the cream for a period of five weeks. Half of the subjects treated the right thigh and half treated the left thigh. The exercise and eating habits of all 40 subjects were comparable. At the end of the experiment, the difference in the circumference of each person's thighs was determined by subtracting the circumference of the untreated thigh from the circumference of the treated thigh. Thus, a positive difference means the treated thigh is greater in circumference than the untreated thigh, and a negative difference means the treated thigh is smaller in circumference than the untreated thigh. For example, if the difference is -1.50 inches, this means the treated thigh is 1.50 inches less in circumference than the untreated thigh. Assume that if the entire 1,200 members of the health club had participated in the study, the population of differences would be a *normal* population. The sample mean for the 40 subjects is -0.50 inches, and the sample standard deviation is 2.0 inches.

a. Determine the 95% confidence interval for μ, the average difference in thigh circumference that would result if everyone in the sampled population used the cream.

b. What conclusion can you make regarding the effectiveness of the cream?

6.12 A research company is conducting a study to determine the average hours of television watched each week by children under the age of 10. The sampled population will be all children under the age of 10 who attend any of the seven elementary schools in a school district. This population is assumed to be *normal*. The sample will be obtained from the list of children who attend these schools. The parents of each sampled child will be contacted to provide information about their child's television viewing habits. Based on past studies of a similar nature, the research company expects individual responses to vary between 15 hours a week and 80 hours a week. Determine the required sample size so the company will be 95% confident that the estimated average is within 2 hours of the true sampled population average.

6.13 A problem that confronts high-school teachers is the lack of self-esteem among high-school students. Students who do not have high self-esteem find it difficult to achieve their full academic potential. Research has shown that one indicator of self-esteem is the number of days a student is absent from school. Students with comparatively high self-esteem generally have fewer absences from school than students with lower self-esteem. The principal of a large high school wants to determine the average number of student absences during the past academic year in the high school. The number of absences for each student is not readily available in a form that can be easily processed on a computer. Rather, this information must be collected manually from student report cards. For this reason, the principal decides to select a simple random sample of students from the high school. Although the high school has students in grades 9 to 12, the sampled population will consist only of students in grades 10 to 12, because students presently in grade 9 were not at the high school during the previous academic year. The principal recently read a newspaper article that reported a sample of students selected nationwide had an average of 20 days absent a year with a standard deviation of 6 days. Use this information to determine the required number of students to sample if the principal desires an estimate of the population average that is within one day of the true value with 95% confidence. Assume the population is *normal*.

6.14 A study was conducted for the state of Arizona to determine factors that can be used for predicting water runoff. One purpose of the study was to provide information that can be used for determining the water level in the dams north of Phoenix. If water runoff is expected to be high, then water must be released from the dams in order to avoid flooding. To collect the data, gauges were placed at various locations in the state to measure water runoff over a 12-month period. In order to determine the location of the gauges, a map was used to divide the state into 800 geographic regions of roughly the same size. A simple random sample of 40 regions was selected from the set of 800 regions, and gauges were placed in a central location for each of the 40 regions in the sample. At the end of the 12-month period, the water runoff was measured in cubic feet at each of the sites. The values of runoff for the 40 sites are stored in column 2 in the file **runoff.mtp** on the data disk. A normal rankit-plot of the 40 values is shown in Figure 6.8.1. Do you think the population is *normal*?

6.15 The manufacturing of circuit boards involves a process in which gold plating is placed on each board. During this process, measurements of the gold thickness are obtained for a sample of circuit boards in order to determine if the boards have the required level of gold plating. A simple random sample of 40 boards is selected from the population of boards manufactured on a given day. The thickness of gold plating is measured in microinches for each sampled board using a method called X-ray fluorescence. A microinch is one millionth of an inch. The sample of 40 observations is in column 2 of the file **circuit.mtp** on the data disk. A normal rankit-plot for these data is shown in Figure 6.8.2.

a. Does it seem appropriate to assume the population of gold-plating thickness values is *normal*?

b. In any manufacturing process it is important to know the amount of variability in the process. Compute a 95% confidence interval for σ, the standard deviation of the population of gold thickness values. The sample

FIGURE 6.8.1
Normal Rankit-Plot for Water Runoff

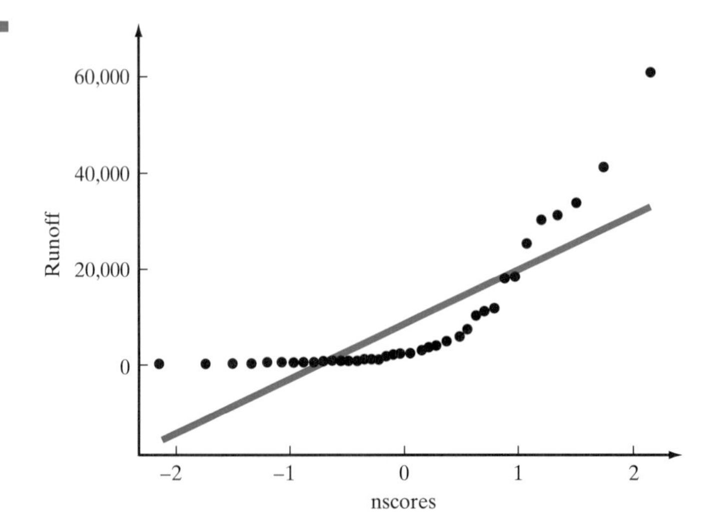

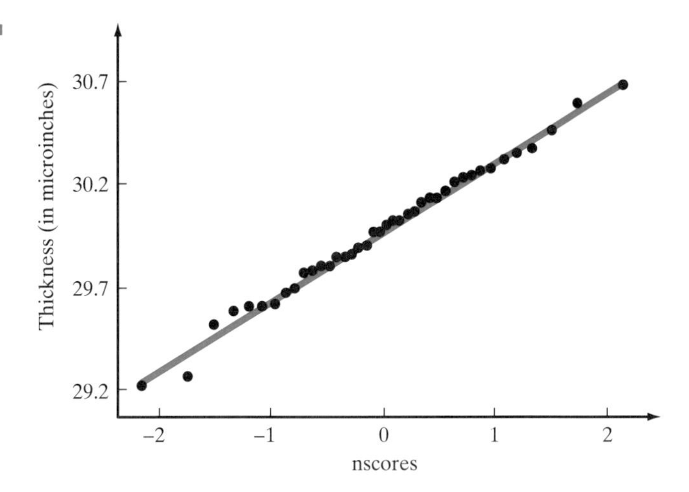

FIGURE 6.8.2
Normal Rankit-Plot of Gold Thickness

standard deviation for the sample in **circuit.mtp** is 0.33 microinches (rounded to two decimals).

c. It is desired to estimate μ, the average thickness of the gold plating for the population of circuit boards that will be manufactured during the next day of production. Determine the sample size that should be selected in order to estimate μ within 0.1 microinch with 95% confidence. Use the estimate of σ in part (b) to help obtain the answer to this question.

6.16 A liquid chemical used in the manufacturing of circuit boards is stored in a large tank. In order to ensure that the chemical provides the desired effect, the acid content of the liquid must be continually monitored. To do this, simple random samples of the liquid are collected over time and the acid concentration of the sampled liquid is measured. The unit of measurement is the number of acid equivalents per liter. The data file **liquid.mtp** contains a simple random sample of 40 observations taken from the tank within a 30-minute period. A normal rankit-plot for these data is shown in Figure 6.8.3.

a. Does it appear that the population of acid-concentration values is a *normal* population?

b. Compute a 95% confidence interval for σ, the standard deviation of the population of all acid-concentration measurements in the tank. The sample standard deviation for the sample in **liquid.mtp** is 0.00582 equivalents per liter.

c. It is desired to estimate μ, the average equivalents per liter of the storage tank within 0.001 equivalents per liter of the true value with 95% confidence. Use the estimate of σ in part (b) to help obtain the sample size required to meet this specification.

6.17 Four simple random samples of size 40 were obtained from four different populations. The samples are stored in columns 1, 2, 3, and 4 in the file **samples.mtp** on the data disk. Only one of the samples came from a *normal* population. Use *normal* rankit-plots to determine which sample was obtained from a *normal* population.

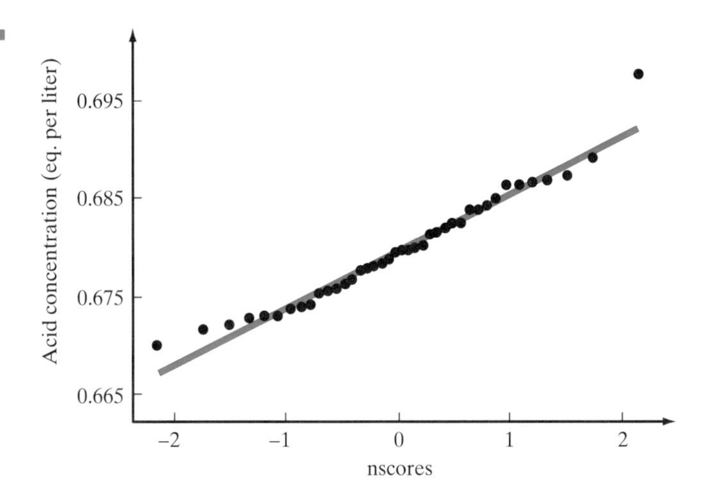

FIGURE 6.8.3
Normal Rankit-Plot of Acid Concentration

6.9 SOLUTIONS FOR SELF-TEST PROBLEMS IN CHAPTER 6

6.2.1

a. We need to compute $\Pr(5.90 < Y < 6.15)$, where Y is a simple random sample of size 1 from a *normal* population with $\mu = 6.00$ and $\sigma = 0.05$. From Line 6.2.3 we get

$$\Pr(5.90 < Y < 6.15) = \Pr(Y < 6.15) - \Pr(Y \leq 5.90).$$

We use Box 6.2.2 to compute $\Pr(Y < 6.15)$ and $\Pr(Y \leq 5.90)$. To compute $\Pr(Y < 6.15)$, set $b = \mu + z\sigma$; substitute $b = 6.15$, $\mu = 6.00$, and $\sigma = 0.05$; and solve for z. We get $6.15 = 6.00 + 0.05z$, and the solution is $z = 3.0$. From line (1) of Table 6.2.1 we see that $\Pr(Y < \mu + 3.0\sigma) = 0.999$. Thus, $\Pr(Y < 6.15) = 0.999$. In a similar manner we see from line (20) in Table 6.2.1 that $\Pr(Y \leq 5.90) = 0.023$. Thus, $\Pr(5.90 < Y < 6.15) = \Pr(Y < 6.15) - \Pr(Y \leq 5.90) = 0.999 - 0.023 = 0.976$. So 97.6% of the bolts are usable.

b. Since 97.6% of the bolt lengths are between 5.90 cm and 6.15 cm, by Definition 3.1.2 the probability is 0.976.

c. We must compute $\Pr(Y < 6.00)$, which is equal to the proportion of bolts that are less than 6.00 cm in length. Using the procedure in Box 6.2.2, set $b = \mu + z\sigma$; substitute $b = 6.00$, $\mu = 6.00$, $\sigma = 0.05$; and solve for z. We get $6.00 = 6.00 + 0.05z$, and the solution for z is $z = 0.00$. We see from line (12) in Table 6.2.1 that $\Pr(Y < \mu + 0 \times \sigma) = 0.50$. So the probability is 0.50 that a bolt selected at random is less than 6.00 cm long.

d. We must compute $\Pr(5.95 < Y < 6.10) = \Pr(Y < 6.10) - \Pr(Y \leq 5.95)$. To compute $\Pr(Y < 6.10)$, set $b = \mu + z\sigma$; substitute $b = 6.10$, $\mu = 6.00$, $\sigma = 0.05$; and solve for z. We get $6.10 = 6.00 + 0.05z$, and the solution for z is $z = (6.10 - 6.00)/0.05 = 2.0$. By line (4) of Table 6.2.1 we get $\Pr(Y < \mu + 2.0\sigma) = 0.977$. By a similar procedure we get $\Pr(Y \leq 5.95) = 0.159$. So $\Pr(Y < 6.10) - \Pr(Y \leq 5.95) = 0.977 - 0.159 = 0.818$. So 81.8% of the bolts are between 5.95 cm and 6.10 cm in length.

6.2.2 Let Y represent a simple random sample of one value to be selected from a *normal* population with mean μ and standard deviation σ. Statement (1) in Box 6.2.1 means $\Pr(\mu - \sigma < Y < \mu + \sigma) = 0.6826$. By Line 6.2.3, $\Pr(\mu - \sigma < Y < \mu + \sigma) = \Pr(Y < \mu + \sigma) - \Pr(Y \leq \mu - \sigma)$. From lines (8) and (16) of Table 6.2.1 we see that $\Pr(Y < \mu + \sigma) = 0.841$ and $\Pr(Y \leq \mu - \sigma) = 0.159$. Thus, $\Pr(\mu - \sigma < Y < \mu + \sigma) = 0.841 - 0.159 = 0.682 =$ 68.2%. This is equal to result (1) to the degree of accuracy afforded by Table 6.2.1. Statements (2) and (3) are verified in the same manner.

6.2.3

a. From Line 6.2.4, $\Pr(Y > 140) = 1 - \Pr(Y \leq 140)$. Here $b = 140$, $\mu = 100$, and $\sigma = 20$, so that $z = (140 - 100)/20 = 2$. From line (4) of Table 6.2.1, $\Pr(Y \leq \mu + 2\sigma) = 0.977$, and so $\Pr(Y \leq 140) = 0.977$. Thus, $\Pr(Y > 140) = 1 - 0.977 = 0.023$.

b. Here $b = 80$, $\mu = 100$, and $\sigma = 20$, so that $z = (80 - 100)/20 = -1$. From line (16) of Table 6.2.1, $\Pr(Y < \mu - 1\sigma) = 0.159$, and so $\Pr(Y < 80) = 0.159$.

c. From Line 6.2.3, $\Pr(60 < Y < 80) = \Pr(Y < 80) - \Pr(Y \leq 60)$. Since $\Pr(Y < 80)$ was computed in the previous part of this question, you only need to compute $\Pr(Y \leq 60)$. To do this, set $b = 60$, $\mu = 100$, $\sigma = 20$ and obtain $z = (60 - 100)/20 = -2.0$. From line (20) of Table 6.2.1, $\Pr(Y \leq \mu - 2\sigma) = 0.023$, and so $\Pr(Y \leq 60) = 0.023$. Thus, $\Pr(60 < Y < 80) = \Pr(Y < 80) - \Pr(Y \leq 60) = 0.159 - 0.023 = 0.136$.

6.3.1

a. From Exhibit 6.3.2 we observe that $\bar{y} = 76.06$ and $s = 10.25$. Hence, $\hat{\mu} = 76.06$ and $\hat{\sigma} = 10.25$.

b. Since the sample size is $n = 50$, $DF = n - 1 = 49$. In Table T-2 use the column 90% and row $DF = 49$ to get $T2 = 1.68$.

c. Using the formulas in Box 6.3.1, $L = 76.06 - (1.68)(10.25)/\sqrt{50} = 73.62$ and $U = 76.06 + (1.68)(10.25)/\sqrt{50} = 78.50$.

6.4.1 The value for $\hat{\sigma}$ in Line 6.4.3 is obtained by using Line 6.4.4 to get $\hat{\sigma} = (100 - 55)/4 = 11.25$. Using this value in Line 6.4.3 with $T2 = 2.04$ gives

$$n = \left[\frac{(2.04)(11.25)}{2}\right]^2 = 131.68.$$

So a simple random sample of 132 scores is needed to meet the desired specification. This will result in an estimate of μ that will be within approximately 2 points of μ with 95% confidence.

6.5.1 Using $s = 10.25$, we get $\hat{\sigma} = 10.25$. To use the procedure in Box 6.5.1 to obtain a 99% confidence interval for σ, we need the table values $T3L$ and $T3U$. The confidence coefficient is 99%, and $DF = n-1 = 49$; so from Table T-3 we obtain $T3L = 0.791$ and $T3U = 1.34$. Using the procedure in Box 6.5.1, we get

$$L = 10.25(0.791) = 8.10, \quad \text{and}$$
$$U = 10.25(1.34) = 13.74.$$

So we have 99% confidence that σ is between 8.11 and 13.74.

6.5.2 For $n = 12$, $T3L = 0.641$ and $T3U = 2.06$.
For $n = 21$, $T3L = 0.707$ and $T3U = 1.64$.
For $n = 34$, $T3L = 0.757$ and $T3U = 1.44$.

6.5.3 For $n = 12$, $T3L = 0.708$ and $T3U = 1.70$.
For $n = 21$, $T3L = 0.765$ and $T3U = 1.44$.
For $n = 34$, $T3L = 0.807$ and $T3U = 1.32$.

6.6.1 The points in this plot hover around a straight line. This is what we would expect for a sample from a *normal* population. Thus, it appears that the simple random sample of 85 grade point averages came from a *normal* population.

6.10 APPENDIX 1: DERIVATION OF A CONFIDENCE INTERVAL FOR μ

In this appendix we discuss the formulas for L and U, the endpoints of a confidence interval for μ reported in Box 6.3.1. A confidence interval for the mean of a *normal* population uses a special population called a t **population** that is derived from the *normal* population. In Box 6.10.1 we explain how a t population is obtained.

BOX 6.10.1: THE *T* POPULATION

In your mind's eye, envision *every* possible simple random sample of size n ($n \geq 2$) that can be obtained from a *normal* population with mean μ and standard deviation σ. For each of these samples envision that $\widehat{\mu}$, $\widehat{\sigma}$, and t are computed where

$$t = \frac{\sqrt{n}(\widehat{\mu} - \mu)}{\widehat{\sigma}}. \qquad \textbf{Line 6.10.1}$$

The collection of all possible values of t in Line 6.10.1 forms a population called a t **population with** $n - 1$ **degrees of freedom.**

t Population

The t population was discovered by the British scientist W. S. Gossett, who published his result under the pseudonym of *Student.* In his honor, the t population is often referred to as *Student's t* population. The relative frequency histogram of a t population is symmetric about zero, and it has a slightly different shape for each different value of the degrees of freedom $n-1$. Figure 6.10.1 shows the curves representing the relative frequency histograms of two t populations, one with 30 degrees of freedom and the other with 3 degrees of freedom.

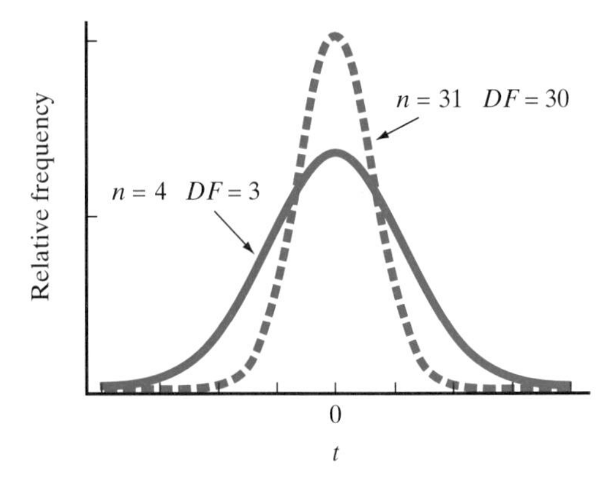

FIGURE 6.10.1
Two t Populations with Different Degrees of Freedom

When the degrees of freedom DF is large (say, greater than 30), the relative frequency histogram of the t population is quite similar in appearance to the relative frequency histogram of a *normal* population (a bell curve) with mean 0 and standard deviation 1.

The table values in Table T-2, which are denoted by $T2$, are used in Box 6.3.1 to compute confidence intervals for μ. These table values are derived from a t population. In particular, for any specified value for DF, the confidence coefficient in Table T-2 represents the proportion of values in the t population in the interval $-T2$ to $T2$. Because the t population is symmetric, the proportion of population values greater than $T2$ is equal to the proportion of values less than $-T2$ (see Figure 6.10.2).

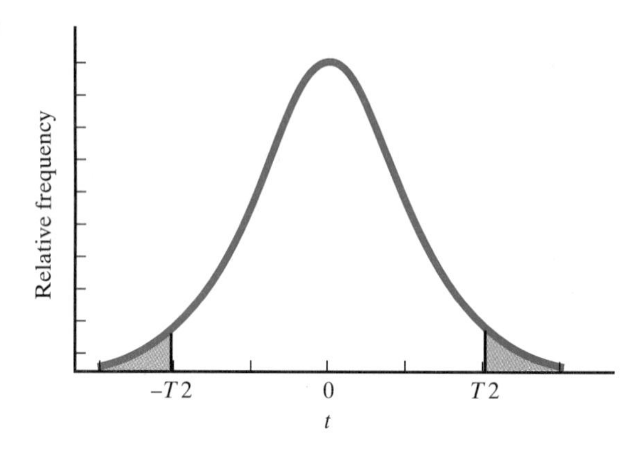

FIGURE 6.10.2
Proportion of t Population Values Less Than $-T2$ Is Equal to Proportion of t Population Values Greater Than $T2$

Suppose an investigator wants to obtain a 95% confidence interval for μ (any confidence coefficient can be used). From Table T-2, choose a value $T2$ that corresponds to 95% confidence. In this case $T2$ is a value such that a proportion 0.95 of values in a t population with $n-1$ degress of freedom are between $-T2$ and $T2$ (this is the unshaded portion of the relative frequency histogram of the t population shown in Figure 6.10.2). One value is selected at random from the t population with $n-1$ degrees of freedom and is denoted by t. If this value is hidden in our hand (and not observed), then we have 95% confidence that the hidden value of t is in the interval $-T2$ to $T2$, and we write this interval as

$$-T2 \le t \le T2.$$

Thus, we have 95% confidence that t satisfies the inequalities $-T2 \le t \le T2$. If we substitute $\sqrt{n}(\widehat{\mu} - \mu)/\widehat{\sigma}$ from Line 6.10.1 for t, we have 95% confidence that the following inequalities are satisfied:

$$-T2 \le \frac{\sqrt{n}(\widehat{\mu} - \mu)}{\widehat{\sigma}} \le T2. \qquad \textbf{Line 6.10.2}$$

Using algebra to reorganize the quantities in Line 6.10.2, we obtain

$$\widehat{\mu} - (T2)\frac{\widehat{\sigma}}{\sqrt{n}} \le \mu \le \widehat{\mu} + (T2)\frac{\widehat{\sigma}}{\sqrt{n}}.$$

So we have 95% confidence that the interval

$$\widehat{\mu} - (T2)\frac{\widehat{\sigma}}{\sqrt{n}} \quad \text{to} \quad \widehat{\mu} + (T2)\frac{\widehat{\sigma}}{\sqrt{n}} \qquad \textbf{Line 6.10.3}$$

contains μ. A comparison of Line 6.10.3 with the formulas in Box 6.3.1 shows that the left side of Line 6.10.3 is L, the lower endpoint of the confidence interval for μ shown in Box 6.3.1. Similarly, the right side of Line 6.10.3 is U, the upper endpoint of the confidence interval for μ shown in Box 6.3.1.

6.11 APPENDIX 2: DERIVATION OF A CONFIDENCE INTERVAL FOR σ

In this appendix we discuss the formulas for L and U, the endpoints of a confidence interval for the standard deviation of a *normal* population. These formulas are reported in Box 6.5.1. A confidence interval for the standard deviation of a *normal* population is based on a specific population called a **chi-squared population**. The Greek symbol χ^2 is sometimes used for the words "chi-squared."

BOX 6.11.1: THE CHI-SQUARED POPULATION

In your mind's eye, imagine *every* possible simple random sample of size n ($n \geq 2$) that could be selected from a *normal* population with mean μ and standard deviation σ. For each of these samples compute $\widehat{\sigma}$ and χ^2, where

$$\chi^2 = \frac{(n-1)\widehat{\sigma}^2}{\sigma^2}. \qquad \textbf{Line 6.11.1}$$

The collection of all possible values of χ^2 in Line 6.11.1 forms a population called a **chi-squared population with $n - 1$ degrees of freedom.**

Chi-squared Population

The relative frequency histogram of a chi-squared population is slightly different for each different value for the degrees of freedom $n - 1$, but they are all skewed to the right. Figure 6.11.1 shows the curves representing the histograms of two chi-squared populations, one with 4 degrees of freedom and the other with 15 degrees of freedom.

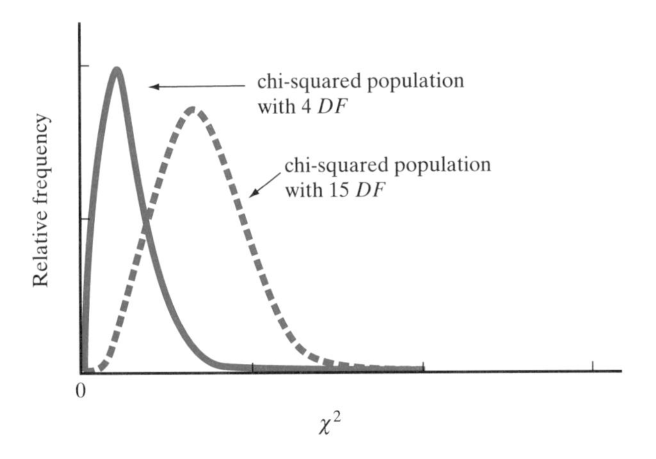

FIGURE 6.11.1
Relative Frequency Histogram of Two Chi-squared Populations

Table T-4 is located on the last page (just before the inside back cover) of your book. Using this table, one can find the proportion of values in a chi-squared population that are between χ_1^2 and χ_2^2. In particular, for a specified value for DF, the two numbers under the label 95% are χ_1^2 and χ_2^2, such that 95% of the values of a chi-squared population are between them. The proportion of values less than χ_1^2 is equal to the proportion of values greater than χ_2^2 (2.5% in this case). This is shown in Figure 6.11.2.

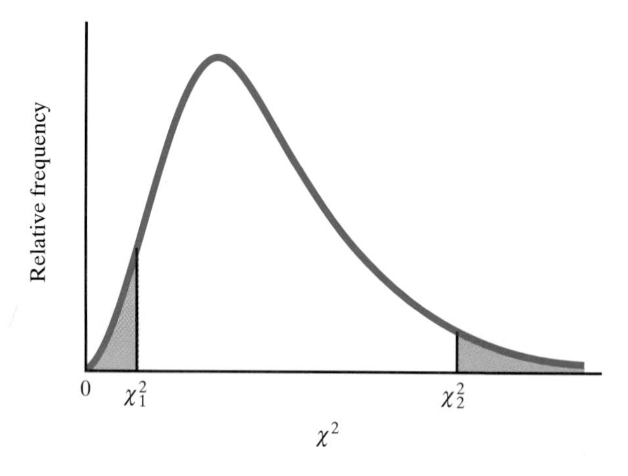

FIGURE 6.11.2
Proportion of Values Less Than χ_1^2 Is Equal to the Proportion of Values Greater Than χ_2^2 in a χ^2 Population

Suppose an investigator wants to obtain a 95% confidence interval for σ (any confidence coefficient can be used). From Table T-4 the two quantities χ_1^2 and χ_2^2 are obtained such that 95% of the values of a chi-squared population with $n - 1$ degrees of freedom are between them (this is the nonshaded portion of the relative frequency histogram of the chi-squared population shown in Figure 6.11.2). One value is selected at random from this chi-squared population and is denoted by χ^2. If this value is hidden in our hand (and not observed), we have 95% confidence that it is in the interval χ_1^2 to χ_2^2. We write this as

$$\chi_1^2 \le \chi^2 \le \chi_2^2. \qquad \textbf{Line 6.11.2}$$

Substituting the quantity $(n - 1)\widehat{\sigma}^2/\sigma^2$ from Line 6.11.1 for χ^2 in Line 6.11.2, we get

$$\chi_1^2 \le (n-1)\frac{\widehat{\sigma}^2}{\sigma^2} \le \chi_2^2. \qquad \textbf{Line 6.11.3}$$

Reorganizing the quantities in Line 6.11.3, we obtain

$$\widehat{\sigma}\sqrt{\frac{(n-1)}{\chi_2^2}} \le \sigma \le \widehat{\sigma}\sqrt{\frac{(n-1)}{\chi_1^2}}. \qquad \textbf{Line 6.11.4}$$

Thus, we have 95% confidence that σ is contained in the interval displayed in Line 6.11.4. Substitute $T3L$ for

$$\sqrt{\frac{(n-1)}{\chi_2^2}},$$

and substitute $T3U$ for

$$\sqrt{\frac{(n-1)}{\chi_1^2}}$$

in Line 6.11.4. Then Line 6.11.4 can be written as

$$(T3L)\widehat{\sigma} < \sigma < (T3U)\widehat{\sigma}. \qquad \textbf{Line 6.11.5}$$

You will notice that $(T3L)\widehat{\sigma}$ is L, the lower endpoint of the confidence interval for σ shown in Box 6.5.1. Also, $(T3U)\widehat{\sigma}$ is U, the upper endpoint of the confidence interval for σ shown in Box 6.5.1. You can use either Table T-3 and Line 6.11.5 or Table T-4 and Line 6.11.4 to compute a confidence interval for σ. The use of $T3L$ and $T3U$ in Table T-3 is slightly simpler, so we have used these in defining the confidence interval for σ in Box 6.5.1.

CHAPTER 7

INFERENCE FOR A POPULATION MEAN AND MEDIAN FOR ANY POPULATION

7.1 INTRODUCTION

In many practical problems the relative frequency histogram of the population under study is well represented by the relative frequency histogram of a *normal* population (by the bell curve in Figure 6.2.1). In this case the formulas in Box 6.3.1 are used to compute a confidence interval for a population mean μ and for a population median M since in a *normal* population $M = \mu$. However, in some practical problems a confidence interval for μ is required and the population is *not normal*. Under certain circumstances, the formulas in Box 6.3.1 can also be used to compute a confidence interval for a population mean even if the population is *nonnormal*. This result, which is based on a fundamental theorem called the **central limit theorem**, is discussed in Section 7.2. In Section 7.3 we provide a formula for computing a confidence interval for a population median. This formula can be used for any population, either *normal* or *nonnormal*, when the variable is quantitative and continuous. In Appendix 1 of this chapter (Section 7.7) we discuss the central limit theorem and explain how it is used to obtain confidence intervals for the mean of a *nonnormal* population. In that section we also show how the central limit theorem is used to obtain the formulas that were presented in Chapter 3 to compute confidence intervals for a population proportion π.

7.2 CONFIDENCE INTERVALS FOR THE MEAN OF A *NONNORMAL* POPULATION

In Box 6.3.1 we presented formulas for constructing a confidence interval for the mean of a *normal* population. If the population under study is *nonnormal*, the formulas in Box 6.3.1 can still be used to construct a confidence interval for a population mean if the sample size is "sufficiently large." In this case the confidence coefficient of the resulting interval may not be exactly equal to the specified confidence coefficient, but it will be close. This result is based on the **central limit theorem**, which is discussed in Appendix 1 (Section 7.7). It is not imperative that you read this appendix in order to apply the methods presented in this section, but it will enhance your understanding of why the procedures in Box 6.3.1 can be used even if the population is *nonnormal*.

As stated above, the formulas in Box 6.3.1 can be used to compute a confidence interval for the mean of a *nonnormal* population if the sample size is sufficiently large. The term "sufficiently large" is subjective, but in this book we will take it to mean "at least 30." Thus, if $n \geq 30$, the formulas in Box 6.3.1 can be used to compute a confidence interval for the mean of a *nonnormal* population. In this case the actual confidence coefficient should be close enough to the specified confidence coefficient for the result to be useful in making decisions. However, you should be aware that if the population being studied is considerably different from a *normal* population, a sample of size 50, 100, or even larger may be required before the actual confidence coefficient is close enough to the stated confidence to be useful. If the population being studied is *nonnormal*, then the larger the sample size n, the closer the actual confidence coefficient will be to the specified confidence coefficient. Also, as the sample size increases, one can expect the margin of error and the width of the confidence interval to decrease.

Box 7.2.1 contains guidelines to use when constructing a confidence interval for a population mean. Before constructing a confidence interval for a population mean, you should always examine a normal rankit-plot to help decide if the sample has been selected from a *normal* population.

BOX 7.2.1: GUIDELINES TO USE WHEN COMPUTING A CONFIDENCE INTERVAL FOR μ

1. The formulas for L and U in Box 6.3.1 can be used to compute a confidence interval for μ for any sample size $n \geq 2$ when the population is *normal*.
2. When the population is *nonnormal*, the formulas in Box 6.3.1 can be used if (a) the sample size n is at least 30 and (b) the population size N is much larger than the sample size n; as a guideline, we use $N > 10n$. However, if the population is considerably different from a *normal* population, a sample size of 50, 100, or even larger may be required for the actual confidence coefficient to be close enough to the specified confidence coefficient. In most situations, experience suggests that the formulas in Box 6.3.1 can be used when (a) and (b) are satisfied; we will use these guidelines in this book. However, if you have concerns that the population differs so much from a *normal* population that these guidelines may not apply, you should consult a professional statistician for advice.

3. Recall from Chapter 5 that if the population relative frequency histogram is symmetric, then the population mean is equal to the population median. In Section 7.3 we provide formulas for computing a confidence interval for the median for any population, *normal* or *nonnormal*. So if $n < 30$, this procedure can be used to obtain a confidence interval for the mean of a *nonnormal* population if the population is symmetric.
4. If the population under study is *nonnormal*, not symmetric, and $n < 30$, you should consult a professional statistician for advice. There are some advanced methods available for this situation, but they are not discussed in this book. These methods include transforming the population values so that the resulting population is *normal*.

EXAMPLE 7.2.1

ARSENIC DATA. A factory that discharges waste water into the sewage system is required to monitor the arsenic levels in its waste water and report the results to the Environmental Protection Agency (EPA) at regular intervals. Sixty beakers of waste water from the discharge are obtained at randomly chosen times during a certain month. The measurement of arsenic is in nanograms per liter for each beaker of water obtained. These data are shown in Table 7.2.1 and stored in column 2 in the file **arsenic.mtp** on the data disk.

TABLE 7.2.1 Arsenic Concentrations (nanograms/liter)

37.6	152.7	9.5	36.2	21.6	3.7	17.4	40.6	24.1	87.7
20.7	21.5	38.8	33.6	182.2	15.4	12.0	59.2	14.5	37.6
23.2	24.1	33.8	5.1	46.9	21.1	16.5	6.6	15.7	26.1
19.4	25.6	10.7	37.5	45.2	8.6	9.9	15.3	8.3	23.5
24.0	56.7	63.5	7.9	48.9	15.3	3.5	17.4	35.0	33.2
4.8	81.3	3.2	20.9	12.2	17.6	10.6	128.5	6.2	30.0

Since the total amount of arsenic in the waste water is of interest to the EPA, the population mean μ is the parameter of interest, and a 95% confidence interval for μ is required. A normal rankit-plot of these data is shown in Figure 7.2.1. This plot provides evidence that the sample was *not* selected from a *normal* population. However, since the sample size 60 exceeds 30, this seems large enough to conclude that the formulas in Box 6.3.1 will provide a confidence interval for μ with a confidence coefficient close to the specified 95%.
The sample mean is $\bar{y} = 31.84$, and the sample standard deviation is $s = 34.02$. Hence, $\hat{\mu} = 31.84$ and $\hat{\sigma} = 34.02$. Since $n = 60$, the degrees of freedom, DF, is $n - 1 = 59$. For a 95% confidence interval we obtain $T2 = 2.00$ from Table T-2. Hence, we get

$$L = 31.84 - (2.00)\frac{34.02}{\sqrt{60}} = 23.05$$

FIGURE 7.2.1
A Normal Rankit-Plot of the Arsenic Data

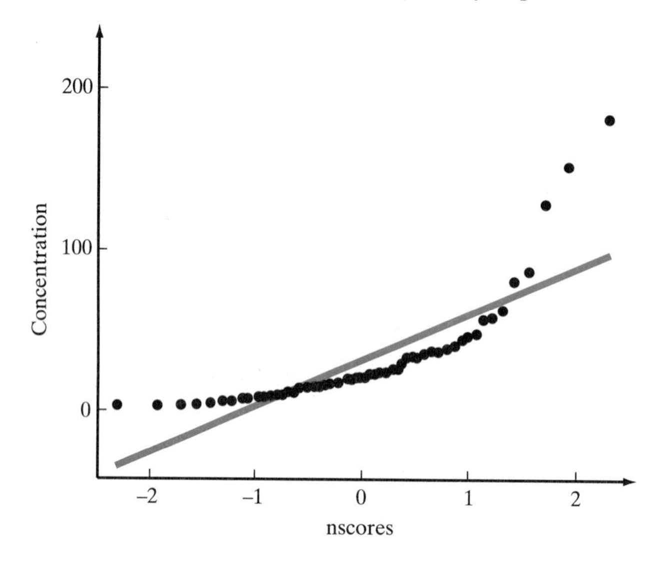

and

$$U = 31.84 + (2.00)\frac{34.02}{\sqrt{60}} = 40.63.$$

Therefore, we are (approximately) 95% confident that μ is between 23.05 and 40.63 nanograms/liter.

THINKING STATISTICALLY: SELF-TEST PROBLEMS

7.2.1 Consider Self-Test Problem 6.2.1, where a company that assembles automobiles plans to purchase 1,000,000 bolts. A simple random sample of 45 bolts is obtained from the population of 1,000,000 bolts. The sample mean length of the bolts is 6.02 cm, and the sample standard deviation is 0.048 cm.

a. What are the estimates of μ and σ if the population is not *normal*?
b. A bolt is usable if it is between 5.90 and 6.15 cm in length. Is either the mean or the median useful in determining if the population of bolts should be purchased?

7.2.2 A study was conducted for the state of Arizona to determine factors that could be used for predicting water runoff. To collect the data, gauges were placed in a central location for each of 40 regions selected at random from a population of 800 regions. At the end of the 12-month period, the water runoff was measured in cubic feet at each of the sites. A normal rankit-plot of the 40 values is shown in Figure 7.2.2.

a. Do you think the population is *normal*?
b. Which measure of center—the mean or the median—seems more appropriate in this problem?
c. Does it seem appropriate to use the formulas in Box 6.3.1 to construct a confidence interval for the population mean in this problem?

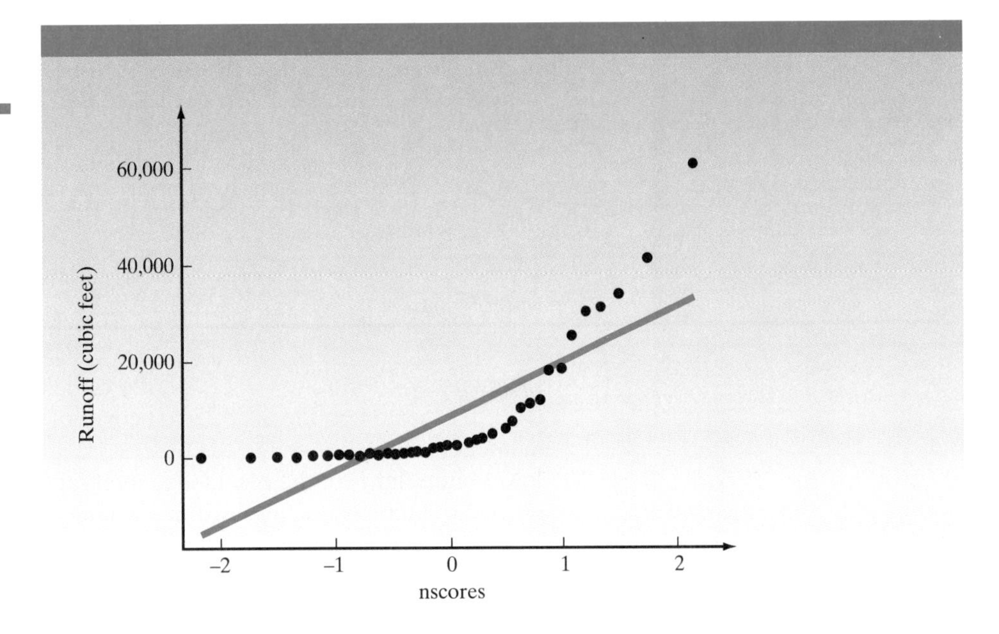

FIGURE 7.2.2
Normal Rankit-Plot for Water Runoff

7.3 POINT ESTIMATE AND CONFIDENCE INTERVAL FOR A POPULATION MEDIAN

In this section we show you how to obtain a point estimate and a confidence interval for a population median. Recall from Chapter 5 that at least half of the values in a population are less than or equal to the median and at least half of the values are greater than or equal to the median.

BOX 7.3.1: POINT ESTIMATE OF A POPULATION MEDIAN

Consider a population of size N whose median is denoted by M. A simple random sample of size n is obtained from this population, and the sample values ordered from smallest to largest are denoted by

$$y_1, y_2, y_3, \ldots, y_n.$$

Using these sample values, the instructions in Box 5.2.1 can be used to compute $\widehat{M}$, the sample median. The sample median $\widehat{M}$ is the estimate of the population median M.

We don't expect the estimate $\widehat{M}$ to be exactly equal to M; thus, we would like to know "how close" the estimate is to M. A confidence interval for M can be used for this purpose. In Box 7.3.2 we describe how to compute L and U, the endpoints of a confidence interval for a population median M.

BOX 7.3.2: CONFIDENCE INTERVAL FOR A POPULATION MEDIAN

Suppose the variable of interest in a population is quantitative, and M denotes the median of this population variable. Use steps (1) to (4) below to compute a confidence interval for M.

1. Specify a confidence coefficient.
2. Select a simple random sample of size n and order the sample values from smallest to largest. Denote the ordered sample values by

$$y_1, y_2, y_3, \ldots, y_n. \qquad \textbf{Line 7.3.1}$$

3. Calculate q, where

$$q = \frac{(n+1) - (T1)\sqrt{n}}{2}. \qquad \textbf{Line 7.3.2}$$

The quantity $T1$ in Line 7.3.2 is a table value that can be found in Table T-1 located on the inside front cover of this book. If q is not an integer, round it *down* to the nearest integer.

4. The lower endpoint L of the confidence interval for M is denoted by y_q and is the sample value in position q when counting *up* from the smallest of the sample values shown in Line 7.3.1. The upper endpoint U of a confidence interval for M is the sample value in position q when counting *down* from the largest of the sample values shown in Line 7.3.1. Equivalently, U is the value in position $n+1-q$ when counting *up* from the smallest sample value (this value is denoted by y_{n+1-q}). Thus, a confidence interval for M with a specified confidence coefficient is the interval from L to U where

$$L = y_q \quad \text{and} \quad U = y_{n+1-q}. \qquad \textbf{Line 7.3.3}$$

The confidence that M is contained in the interval L to U in Line 7.3.3 is (at least) equal to the confidence coefficient specified in step (1).

The procedure in Box 7.3.2 can be used when $n \geq 10$. We provide an illustration in the following example.

EXAMPLE 7.3.1

CALCULUS SCORES. In Self-Test Problem 6.3.1 a 90% confidence interval was computed for the mean of the population of 2,600 calculus scores shown in the table in Appendix B. The confidence interval was based on the simple random sample of size $n = 50$ scores stored in column 2 of the file **50grades.mtp** on the data disk. We will now use this sample to compute a 90% confidence interval for the population median. The sample data (arranged from smallest to largest) are shown in Table 7.3.1.

Using the sample values, we compute the sample median in the same manner as the population median described in Box 5.2.1. Since the sample size ($n = 50$) is an even number, the sample median is the average of the sample values in positions $n/2 = 25$ and $(n/2) + 1 = 26$. Thus, the estimate of M is

$$\widehat{M} = \frac{76 + 76}{2} = 76.$$

We estimate that at least half of the values in the population are less than or equal to 76 and at least half are greater than or equal to 76. The boxplot of the sample is

TABLE 7.3.1 Simple Random Sample of Size 50 from the Calculus Score Population Arranged from Smallest to Largest

Sample position number	Sample value "Score"	Sample position number	Sample value "Score"	Sample position number	Sample value "Score"	Sample position number	Sample value "Score"
1	47	14	71	27	79	40	84
2	56	15	72	28	79	41	84
3	59	16	73	29	80	42	85
4	59	17	73	30	80	43	85
5	59	18	73	31	80	44	85
6	63	19	73	32	81	45	88
7	66	20	73	33	82	46	90
8	66	21	74	34	82	47	90
9	67	22	75	35	82	48	92
10	68	23	75	36	82	49	92
11	68	24	75	37	82	50	98
12	68	25	76	38	83		
13	70	26	76	39	83		

presented in Figure 7.3.1. Notice that the minimum score of 47 is an outlier. It is very low relative to the other scores in the sample. In an actual application, always examine any outlier so you can determine if a mistake has been made in observing or recording this value.

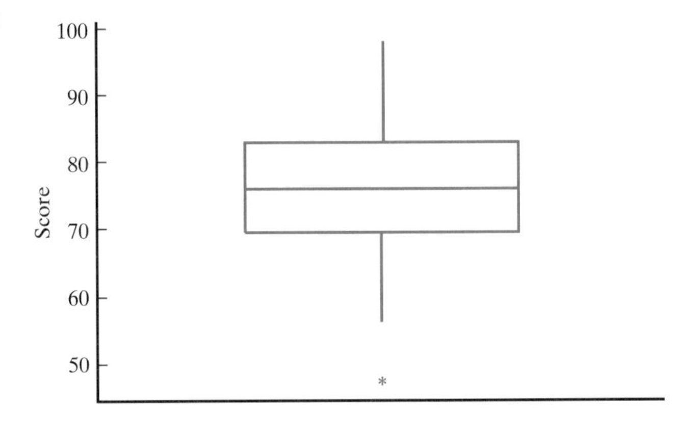

FIGURE 7.3.1
Boxplot of a Simple Random Sample of 50 Calculus Scores

We determine q by using the formula in Line 7.3.2 in Box 7.3.2. Since the confidence coefficient is 90%, we use Table T-1 to obtain $T1 = 1.65$. Thus,

$$q = \frac{(50 + 1) - 1.65\sqrt{50}}{2} = 19.67,$$

which we round down to 19. The lower endpoint of the confidence interval is $L = 73$, which is the sample value in position 19 in Table 7.3.1 when counting up from the smallest sample value. The upper endpoint of the confidence interval is $U = 81$, which is the sample value in position 19 in Table 7.3.1 when counting down from the largest sample value. Equivalently, this is the sample value in position $n + 1 - q = 50 + 1 - 19 = 32$ when counting up from the smallest value. We conclude that the confidence is (at least) 90% that M is in the interval 73 to 81.

As noted earlier, unlike in real applications the entire population is known in this problem. We computed the population median M and obtained $M = 75$. Hence, the estimate $\widehat{M} = 76$ is quite close. Notice also that the population median $M = 75$ is contained in the confidence interval 73 to 81.

The confidence interval for a median described in Box 7.3.2 is valid for any population, either *normal* or *nonnormal*, if the variable is quantitative. A statistical procedure that is valid for *any* population is referred to as a **nonparametric** procedure or a **distribution-free** procedure.

EXPLORATION 7.3.1

Researchers have long been interested in the connection, if any, between diet and longevity. An investigator conjectures that a low-fat diet she has developed will increase longevity in a certain species of rat. To study this, she wants to determine M, the median lifetime in days of this species of rat if they had all been fed the low-fat diet. She cannot feed every rat in the species her low-fat diet, so she selected a simple random sample of 120 rats from a rat breeder who has a colony of 200,000 rats. The investigator fed these 120 rats the low-fat diet until they all died. Table 7.3.2 shows the lifetime in days (arranged from smallest to largest) for each rat in the sample. These data are stored in column 2 in the file **lowfat.mtp** on the data disk.

1. What is the target population of items?
   The target population of items is the collection of all rats of this particular species if each had been fed the low-fat diet. In this problem, the target population is conceptual.
2. What is the sampled population of items?
   For this investigation, the sampled population of items is the colony of 200,000 rats owned by the breeder if each rat had been fed the low-fat diet. The sampled population is also conceptual. The sampled population would have actually existed if the investigator had fed all 200,000 rats the low-fat diet and then selected a simple random sample of 120 of the rats. Equivalently, the investigator selected a simple random sample of 120 rats first and then fed only these rats the low-fat diet.
3. What is the population parameter of interest?
   The parameter of interest is M, the median lifetime of the 200,000 rats in the sampled population if they had all been fed the low-fat diet.
4. Find a point estimate of M.
   From the data in Table 7.3.2 we compute the average of the values in positions $120/2 = 60$ *and* $(120/2) + 1 = 61$ *to get*

$$\widehat{M} = \frac{588 + 592}{2} = 590 \text{ days.}$$

5. From past experience the investigator knows that the median number of days that this species of rat lives when fed an ordinary diet is about 580 days. From this investigation, can we conclude that the median survival time of the sampled population of the 200,000 rats would be more than 580 days if all rats had been fed the low-fat diet?

TABLE 7.3.2 Longevity (in days arranged from smallest to largest) of a Simple Random Sample of 120 Rats Fed a Low-fat Diet

Rat number	Days lived	Rat number	Days lived	Rat number	Days lived	Rat number	Days lived
1	143	31	485	61	592	91	700
2	182	32	485	62	592	92	702
3	191	33	491	63	598	93	718
4	217	34	497	64	601	94	723
5	279	35	500	65	606	95	726
6	307	36	503	66	614	96	727
7	309	37	503	67	619	97	732
8	320	38	508	68	622	98	736
9	339	39	509	69	629	99	745
10	345	40	510	70	633	100	746
11	348	41	517	71	634	101	753
12	365	42	520	72	638	102	754
13	370	43	525	73	643	103	757
14	371	44	529	74	653	104	764
15	377	45	540	75	662	105	768
16	390	46	544	76	663	106	784
17	398	47	546	77	663	107	786
18	402	48	553	78	669	108	787
19	408	49	555	79	670	109	788
20	415	50	563	80	682	110	791
21	426	51	569	81	684	111	795
22	434	52	571	82	686	112	795
23	455	53	575	83	687	113	812
24	458	54	576	84	691	114	833
25	458	55	576	85	691	115	856
26	463	56	578	86	692	116	880
27	477	57	586	87	692	117	913
28	477	58	587	88	693	118	913
29	477	59	588	89	695	119	961
30	481	60	588	90	699	120	1,029

The point estimate certainly suggests that this is true. However, a confidence interval should be examined to help determine "how close" the estimate $\widehat{M}$ is to M.

6. Find a 95% confidence interval for M.
Since n is greater than 10, we find q by using the formula in Line 7.3.2:

$$q = \frac{(120 + 1) - 1.96\sqrt{120}}{2} = 49.76,$$

which we round down to 49. From Table 7.3.2 we obtain $L = 555$, which is the sample value in position $q = 49$ when counting up from the smallest sample value. We also obtain $U = 638$, which is the value in position $n + 1 - q = 120 + 1 - 49 = 72$ when counting up from smallest sample value. We are (at least) 95% confident that the median lifetime of the 200,000 rats would be *between 555 and 638 days if all rats had been fed the low-fat diet. Since this interval contains values less than 580 as well as values greater than 580, the data do not clearly indicate whether or not the median lifetime of the 200,000 rats in the sampled population would exceed 580 days if all rats were fed the low-fat diet. In particular one cannot rule out the fact that the median lifetime of the 200,000 rats in the sampled population might be less than 580 days.*

7. A boxplot of the sample data is shown in Figure 7.3.2. Notice there are two asterisks that denote outliers. What should the investigator do about these two sample values?
The investigator should thoroughly examine the records and equipment to make sure that no mistake was made in measuring or recording the longevity of these two rats. If a mistake was made, the two values should be removed from the sample and the data reanalyzed. If no mistake in measuring or recording was made, these two data points should not be removed since they belong to the population being studied.

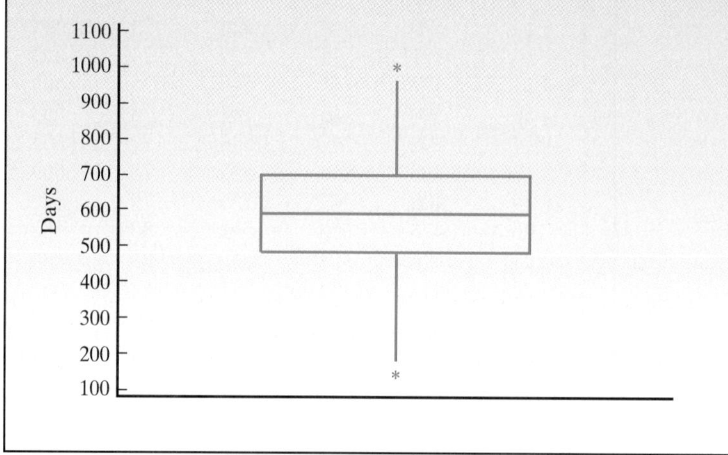

FIGURE 7.3.2
Boxplot of a Sample of Rats Fed a Low-fat Diet

THINKING STATISTICALLY: SELF-TEST PROBLEMS

7.3.1 A simple random sample of 52 customers was selected from all customers of an electric company and asked to participate in a program to reduce the demand for electricity. Before the study begins, each customer in the sample is asked to report last year's annual income for their household. These data are shown below (in thousands of dollars) and ordered from left to right in terms of magnitude. The data are also stored in the file **customer.mtp** on the data disk.

```
23  29  30  31  31  31  32  33  34  35  38  41  42
43  47  48  49  50  50  52  54  55  57  58  60  61
62  63  63  64  64  65  66  68  70  70  72  72  72
72  74  76  76  76  78  79  79  80  81  82  84  86
```

a. What are the target and sampled populations?
b. What is the estimate of the median annual income of all customers in the sampled population?

c. What is the value of q to use to obtain a 95% confidence interval for M?
d. What is the value of q for a 99% confidence interval for M?
e. Find a 95% confidence interval for M, the median annual income of all customers.

7.4 CHAPTER SUMMARY

KEY TERM

1. central limit theorem

SYMBOLS

1. M—a population median
2. $\widehat{M}$—the estimate of a population median
3. q—the position of a sample value used to form a confidence interval for a population median counting up from the smallest value or counting down from the largest value

KEY CONCEPTS

1. The procedure shown in Box 6.3.1 for computing a confidence interval for the mean of a *normal* population can be used to compute a confidence interval for the mean of *any* population if the sample size is "sufficiently large."
2. The central limit theorem provides the justification for using the procedure described in Box 6.3.1 for computing a confidence interval for a population mean when the population is *nonnormal*, but the sample size is "sufficiently large."
3. The confidence interval for the population median in Box 7.3.2 can be used for any population, either *normal* or *nonnormal*, if the variable is quantitative and $n \geq 10$.

SKILLS

1. Compute a confidence interval for the population mean of a *nonnormal* population.
2. Compute $\widehat{M}$, the estimate of a population median.
3. Compute a confidence interval for a population median.

7.5 CHAPTER EXERCISES

7.1 Suppose you want to construct a confidence interval for the mean of a *nonnormal* population. If you had a choice of selecting a simple random sample of size 35 or a simple random sample of size 45 from this population, which would you prefer? Explain your reasoning.

7.2 The vice president of a large company that manufactures bathroom accessories is interested in leasing a fleet of 100 automobiles for the sales staff. She will lease 100 cars of a specific model if she is convinced that the average miles per gallon (mpg) of gasoline for this model is at least 30 when driven over a specified test route. The vice president is interested in the average miles per gallon because the total gallons used over the term of the lease will determine the total fuel expense. The sampled population is the set of cars of this model that are available at the time the sample is selected.

The target population is the set of cars of this model that will be available at the time the cars are leased. A simple random sample of 36 cars was selected from the sampled population and driven over the test route. The mpg values of these 36 cars are shown in Table 7.5.1 and stored in column 2 of the file **mpgcar.mtp** on the data disk.

TABLE 7.5.1 Simple Random Sample of 36 Cars

Car number	Miles per gallon	Car number	Miles per gallon
1	27	19	32
2	25	20	30
3	25	21	32
4	26	22	26
5	26	23	29
6	29	24	26
7	26	25	25
8	27	26	32
9	26	27	28
10	32	28	27
11	31	29	32
12	31	30	30
13	28	31	29
14	26	32	34
15	30	33	29
16	29	34	32
17	32	35	30
18	30	36	28

The vice president examined a normal rankit-plot that suggested the sample was selected from a *normal* population. However, this result surprised her because for similar studies in the past, the populations always appeared to be *nonnormal.* Exhibit 7.5.1 shows some of the summary measures the vice president calculated for the sample.

a. What are the estimates of μ and σ?

b. Compute a 99% confidence interval for μ using Box 6.3.1.

c. How would you address the vice president's concern that the sample may have come from a *nonnormal* population even though the normal rankit-plot suggests otherwise?

7.3 A consumer group that monitors prices of various commodities wants to determine how the price of gasoline in Denver, Colorado, compares with the price in smaller communities. A simple random sample of 30 gasoline stations was obtained on March 30 in the Denver metropolitan area, and the price per gallon of 85 octane unleaded gasoline was recorded. The data are shown in Table 7.5.2 and stored in column 2 in the file **station.mtp** on the data disk. The data are rounded to the nearest cent.

TABLE 7.5.2 Price of Unleaded Gasoline in a Sample of 30 Stations (arranged from smallest to largest)

Sample number	Price per gallon	Sample number	Price per gallon
1	1.12	16	1.20
2	1.12	17	1.21
3	1.12	18	1.25
4	1.12	19	1.25
5	1.13	20	1.27
6	1.14	21	1.27
7	1.15	22	1.28
8	1.15	23	1.28
9	1.16	24	1.32
10	1.16	25	1.33
11	1.18	26	1.34
12	1.19	27	1.35
13	1.19	28	1.36
14	1.19	29	1.37
15	1.20	30	1.38

a. What measure of center—the mean or the median—is more meaningful for the consumer group to report?

b. Use the formulas in Box 6.3.1 to construct a 90% confidence interval for the average price per gallon of 85 octane unleaded gasoline for the population of gasoline stations in Denver on

```
---------------------------------------------------------------
Descriptive Statistics

Variable          N      Mean    Median    TrMean     StDev    SEMean
mpg              36    28.806    29.000    28.781     2.539     0.423

Variable        Min       Max        Q1        Q3
mpg          25.000    34.000    26.000    31.000
---------------------------------------------------------------
```

EXHIBIT 7.5.1

March 30. The sample mean for the data in Table 7.5.2 is 1.226, and the sample standard deviation is 0.0852.

7.4 A study was conducted in a small African community to help determine the amount of food that will be needed in the next generation. One of the variables of interest was the number of children a woman has before she is 35 years old. A simple random sample of 20 women aged 35 years or more was selected from the community, and the number of children each woman bore before the age of 35 was recorded. This resulted in the following data:

2, 1, 0, 2, 2, 4, 2, 3, 2, 6, 1, 3, 1, 1, 6, 4, 9, 1, 1, 9.

Use the sample of size 20 to obtain an estimate and a 95% confidence interval for the population median M.

7.5 A company that is considering opening a new store in a city wants to estimate the median annual income of all households in the city. A simple random sample of 75 households was selected, and last year's income was obtained for each household in the sample. The sample is shown in Table 7.5.3 (arranged from smallest to largest) and stored in column 2 in the file **income.mtp** on the data disk. Find the point estimate and a 95% confidence interval for M, the median of last year's income of all households in the city.

7.6 A simple random sample was obtained of 22 employees in an insurance company in order to see how the company's salaries compare with salaries in a government agency. The annual salaries in thousands of dollars are shown in Table 7.5.4 and stored in column 2 in the file **insalary.mtp** on the data disk.

TABLE 7.5.4 Salaries (in thousands of dollars) in a Sample of 22 Employees (ordered from smallest to largest)

Employee	Salary	Employee	Salary
1	27.0	12	42.7
2	33.0	13	45.3
3	38.9	14	47.7
4	39.0	15	51.2
5	40.1	16	53.6
6	40.7	17	64.9
7	40.8	18	69.1
8	40.8	19	83.6
9	41.1	20	94.5
10	42.0	21	94.7
11	42.3	22	94.8

a. Find the estimate for M, the median annual salary of all employees in the company.

b. Compute a 99% confidence interval for M.

TABLE 7.5.3 Last Year's Income for a Sample of Households (ordered from smallest to largest)

Household	Income	Household	Income	Household	Income
1	$25,000	26	$46,000	51	$64,000
2	25,000	27	47,000	52	65,000
3	28,000	28	49,000	53	65,000
4	28,000	29	49,000	54	65,000
5	29,000	30	49,000	55	65,000
6	30,000	31	50,000	56	66,000
7	31,000	32	51,000	57	66,000
8	31,000	33	51,000	58	68,000
9	33,000	34	53,000	59	68,000
10	33,000	35	53,000	60	69,000
11	33,000	36	54,000	61	70,000
12	34,000	37	55,000	62	71,000
13	34,000	38	57,000	63	71,000
14	36,000	39	57,000	64	73,000
15	36,000	40	57,000	65	74,000
16	40,000	41	57,000	66	74,000
17	42,000	42	58,000	67	74,000
18	43,000	43	58,000	68	74,000
19	43,000	44	58,000	69	74,000
20	44,000	45	59,000	70	76,000
21	44,000	46	60,000	71	77,000
22	45,000	47	61,000	72	79,000
23	45,000	48	62,000	73	80,000
24	46,000	49	63,000	74	80,000
25	46,000	50	63,000	75	80,000

7.7 The purchasing department of a government agency wants to purchase a shipment of 6,000 light bulbs to be used in one of its buildings. The purchasing officer will buy the bulbs from a small company if she is convinced that at least one-half of the 6,000 bulbs will last over 2,000 hours. The purchasing officer collected a simple random sample of 20 bulbs from the same lot as the 6,000 bulbs and tested them. The lifetimes of the 20 bulbs were ordered from smallest to largest and are shown below. The sample is stored in column 2 in the file **bulbs.mtp** on the data disk.

```
1562 1672 1740 1820 1895 1974 2010 2063 2140 2211
2258 2266 2312 2329 2342 2345 2381 2401 2414 2496
```

a. Find an estimate for M, the median lifetime for the population of 6,000 light bulbs.

b. Compute a 95% confidence interval for M. Should the purchasing officer buy the bulbs from the small company?

7.8 Faced with a sharp rise in traffic accidents, the police department in a small city is considering the use of photo radar to cite speeding motorists. The city council requested more information before it would approve implementation of the program. Among other things, the council members wanted information about the speeds that motorists were presently driving on city streets. In order to collect this information, electronic devices were placed at various locations in the city to record the speeds of passing vehicles. A simple random sample of 25 vehicles was collected on a residential street in one 24-hour period where the posted speed limit was 30 miles per hour. The speeds of the 25 vehicles, ordered from smallest to largest, are as follows:

```
28 29 29 30 32 33 33 35 36
36 37 38 38 38 39 39 40 42
43 44 44 44 45 47 48
```

These values are stored in column 2 of the data set **speed.mtp** on the data disk.

a. Define the target and sampled populations.

b. Compute $\widehat{M}$, the estimate of the median speed for the population of all cars that traveled on this street during the 24-hour period.

c. Compute a 95% confidence interval for M.

d. Does it appear that a majority of vehicles in the sampled population were exceeding the posted speed limit of 30 miles per hour?

7.6 SOLUTIONS FOR SELF-TEST PROBLEMS IN CHAPTER 7

7.2.1

a. By Definitions 6.3.1 and 6.3.2, $\widehat{\mu} = 6.02$ and $\widehat{\sigma} = 0.048$. These estimates are valid whether or not the population is *normal.*

b. No. The parameter of interest is π, the proportion of bolts between 5.90 and 6.15 cm.

7.2.2

a. No. The plotted points do not seem to have a straight-line pattern.

b. There is interest in the total runoff across the state for the 12-month period, and so the mean is the appropriate measure of center.

c. Yes, the confidence interval presented in Box 6.3.1 is (approximately) correct in this case since the sample size is greater than 30.

7.3.1

a. The target and sampled populations are the same: the set of all customers of the electric company at the time the sample was obtained.

b. Since $n = 52$ is an even number, the sample median is the average of the values in positions $n/2 = 26$ and $(n/2) + 1 = 27$. This gives $\widehat{M} = (61 + 62)/2 = 61.5$.

c. From Line 7.3.2 the value of q to use for the procedure in Box 7.3.2 is $q = [(52 + 1) - (1.96)\sqrt{52}]/2 = 19.43$, which we round down to 19.

d. From Line 7.3.2, the value of q to use for the procedure in Box 7.3.2 is $q = [(52 + 1) - (2.58)\sqrt{52}]/2 = 17.20$, which we round down to 17.

e. Based on part (c), the lower bound is the value in position 19 when counting up from the smallest sample value. This gives $L = 50$. The value of U is found in position $n+1-q = 52+1-19 = 34$ when counting up from the smallest sample value. This value is $U = 68$. The confidence interval is from 50 to 68. Thus, we are at least 95% confident that the median annual income of all customers is between $50,000 and $68,000.

7.7 APPENDIX 1: THE CENTRAL LIMIT THEOREM

In this appendix we discuss the central limit theorem, which we introduced in Section 7.2. First we define a **derived population of sample means**.

The collection of all possible sample means that can be obtained from all possible samples from a population under study plays an important role in the derivation of confidence intervals for a population mean. This population of all possible sample means is described in Box 7.7.1.

BOX 7.7.1: THE DERIVED POPULATION OF SAMPLE MEANS

Let $Y_1, Y_2, \ldots, Y_N$ denote *any* population of N numbers with mean μ and standard deviation σ. In this section we will label this as the sampled population under study. In your mind's eye, envision the following:

Obtain *every* possible simple random sample of size n from the sampled population. Let k denote the number of possible samples of size n. For each of these k samples compute the sample mean, and denote these k sample means by

$$\bar{y}_1, \bar{y}_2, \bar{y}_3, \ldots, \bar{y}_k.$$

This collection of all possible sample means (where each mean has been computed from a sample of size n from the sampled population) is called a **derived population of sample means** (that is, this population of sample means has been *derived* from the sampled population).

For a simple illustration, in Example 7.7.1 we will examine a sampled population of size $N = 6$. From this population we will construct the derived population of sample means for samples of size $n = 2$.

EXAMPLE 7.7.1

A SIMPLE DERIVED POPULATION. Consider a sampled population consisting of the following six numbers:

$$2 \quad 4 \quad 8 \quad 5 \quad 7 \quad 4.$$

The mean of this sampled population is

$$\mu = \frac{2+4+8+5+7+4}{6} = 5,$$

and the standard deviation is

$$\sigma = \sqrt{\frac{(2-5)^2 + (4-5)^2 + (8-5)^2 + (5-5)^2 + (7-5)^2 + (4-5)^2}{6}} = 2.0.$$

There are 15 different possible samples of size $n = 2$ from this sampled population; they are listed in the following table along with the sample mean $\bar{y}$ for each of the possible 15 samples.

Sample number	Sample values	Sample mean
1	2, 4	3.0
2	2, 8	5.0
3	2, 5	3.5
4	2, 7	4.5
5	2, 4	3.0
6	4, 8	6.0
7	4, 5	4.5
8	4, 7	5.5
9	4, 4	4.0
10	8, 5	6.5
11	8, 7	7.5
12	8, 4	6.0
13	5, 7	6.0
14	5, 4	4.5
15	7, 4	5.5

Thus, the derived population of all possible sample means for samples of size 2 from the sampled population consists of the following means:

3.0 5.0 3.5 4.5 3.0 6.0 4.5 5.5 4.0 6.5 7.5 6.0 6.0 4.5 5.5.

We denote the mean and the standard deviation of this derived population of sample means by μ_{SM} and σ_{SM}, respectively, where SM stands for "population of sample means." We obtain

$$\mu_{\text{SM}} = \frac{3.0 + 5.0 + 3.5 + \cdots + 4.5 + 5.5}{15} = 5$$

and

$$\sigma_{\text{SM}} = \sqrt{\frac{(3.0 - 5.0)^2 + (5.0 - 5.0)^2 + \cdots + (5.5 - 5.0)^2}{15}}$$

$$= \sqrt{\frac{24}{15}} = 1.265 \quad \text{(rounded to three decimals).}$$

So

$$\mu_{\text{SM}} = 5 \quad \text{and} \quad \sigma_{\text{SM}} = 1.265. \qquad \textbf{Line 7.7.1}$$

In most practical problems, the size N of the sampled population is large and there would be an extremely large number of possible samples of size n. It would be impractical to list them all and compute all of the possible sample means. Fortunately, there are formulas we can use to obtain μ_{SM} and σ_{SM} without having to list all the values in the derived population. These formulas are presented in Box 7.7.2.

BOX 7.7.2: MEAN AND STANDARD DEVIATION OF THE DERIVED POPULATION OF SAMPLE MEANS

Consider *any* sampled population of N numbers with mean equal to μ and standard deviation equal to σ. In your mind's eye, envision a derived population consisting of all possible sample means obtained from all possible samples of size n from this sampled population. We have two populations—the sampled population and the derived population—and each has a mean and a standard deviation. These are represented as follows:

Parameter	Sampled population	Derived population
Mean	μ	μ_{SM}
Standard deviation	σ	σ_{SM}

There is a relationship between the means (and between the standard deviations) of the sampled and the derived populations. These relationships are shown in (1) and (2) below.

1. The mean of the derived population of sample means is equal to the mean of the sampled population. That is,

$$\mu_{SM} = \mu. \qquad \textbf{Line 7.7.2}$$

2. The relationship between σ_{SM} (the standard deviation of the derived population) and σ (the standard deviation of the sampled population) is

$$\sigma_{SM} = \frac{f\sigma}{\sqrt{n}}, \qquad \text{where } f = \sqrt{\frac{N-n}{N-1}}. \qquad \textbf{Line 7.7.3}$$

When N is very large compared to n (say, $N > 10n$), the value of f is very close to 1 and σ_{SM} is very nearly equal to $\sigma/\sqrt{n}$. So for practical problems we use

$$\sigma_{SM} = \frac{\sigma}{\sqrt{n}}.$$

In Example 7.7.2 we demonstrate how the formulas in Box 7.7.2 can be used to compute the mean and the standard deviation of the derived population of sample means using the mean and the standard deviation of the sampled population.

EXAMPLE 7.7.2

A SIMPLE DERIVED POPULATION (CONTINUED). In Example 7.7.1 we used the sampled population data and computed μ and σ to obtain $\mu = 5$ and $\sigma = 2.0$. We used the derived population data and computed μ_{SM} and σ_{SM} to obtain $\mu_{SM} = 5$ and $\sigma_{SM} = 1.265$. Instead of using the derived population data, we could use the formulas in Box 7.7.2 to compute μ_{SM} and σ_{SM}. We get

$$\mu_{SM} = \mu = 5$$

and

$$\sigma_{SM} = \sqrt{\frac{N-n}{N-1}}\left(\frac{\sigma}{\sqrt{n}}\right) = \sqrt{\frac{6-2}{6-1}}\left(\frac{2.0}{\sqrt{2}}\right)$$

$$= \sqrt{\frac{16}{10}} = 1.265 \quad \text{(rounded to three decimals).}$$

These are the same values we obtained in Example 7.7.1, where we listed all 15 possible samples, computed the mean of each of the 15 samples, and calculated the mean and the standard deviation of these 15 means.

If the sampled population is *normal*, then the derived population of sample means is also *normal*. Additionally, the central limit theorem tells us that under certain circumstances the relative frequency histogram of the derived population of sample means is well represented by a bell curve, regardless of the shape of the relative frequency histogram of the sampled population. We state this result formally in Box 7.7.3.

BOX 7.7.3: CENTRAL LIMIT THEOREM

Consider *any* sampled population of N numbers with mean μ and standard deviation σ, where N is large. In your mind's eye, construct the derived population consisting of the means of all samples of size n from the sampled population. The mean and the standard deviation of the derived population are μ_{SM} and σ_{SM}, respectively, where

$$\mu_{SM} = \mu \qquad \text{and} \qquad \sigma_{SM} = \frac{\sigma}{\sqrt{n}}.$$

If n is "sufficiently large" and $N > 10n$, the derived population of sample means is a *normal* population with mean μ_{SM} and standard deviation σ_{SM}.

From a practical perspective, the central limit theorem is important because when we have a simple random sample that is "sufficiently large," the formulas for computing confidence intervals for the mean of a *normal* population can also be used to compute confidence intervals for the mean of *any* (sampled) population. This result is demonstrated next.

Using the Central Limit Theorem to Obtain Confidence Intervals for the Mean of a *Nonnormal* Population

Consider a *nonnormal* (sampled) population with mean μ and standard deviation σ, both unknown. The problem is to obtain a confidence interval for μ. We select a simple random sample of size n (where n is "sufficiently large") from the *nonnormal* (sampled) population and compute $\bar{y}$ and s. Thus, the estimates of μ and σ are

$$\hat{\mu} = \bar{y} \qquad \text{and} \qquad \hat{\sigma} = s.$$

Since $\bar{y}$ is one of the items in the derived population of sample means, this value of $\bar{y}$ can be considered to be a simple random sample of size 1 from the derived population. We will use this simple random sample of size 1 from the derived population to construct a confidence interval for μ_{SM}, the mean of the derived population. By the central limit theorem, the derived population is *normal*, so we can use the formulas in Box 6.3.1 to obtain a confidence interval L to U for μ_{SM}, which can be written as

$$L \leq \mu_{SM} \leq U, \qquad \textbf{Line 7.7.4}$$

where

$$L = \widehat{\mu}_{SM} - (T2)\frac{\widehat{\sigma}_{SM}}{\sqrt{1}} \quad \text{and} \quad U = \widehat{\mu}_{SM} + (T2)\frac{\widehat{\sigma}_{SM}}{\sqrt{1}}. \qquad \textbf{Line 7.7.5}$$

By Line 7.7.2, $\mu_{SM} = \mu$, so L and U in Line 7.7.5 are also the endpoints of a confidence interval for μ, the mean of the sampled population. This is the confidence interval we desire. To compute L and U in Line 7.7.5, we need $\widehat{\mu}_{SM}$. But $\overline{y}$ is an estimate of μ, and since $\mu_{SM} = \mu$, it follows that $\overline{y}$ is also an estimate of μ_{SM}. Hence, $\widehat{\mu}_{SM} = \widehat{\mu} = \overline{y}$. To compute L and U in Line 7.7.5, we also need $\widehat{\sigma}_{SM}$. But s is an estimate of σ, and since $\sigma_{SM} = \sigma/\sqrt{n}$, it follows that $\widehat{\sigma}_{SM} = \widehat{\sigma}/\sqrt{n} = s/\sqrt{n}$. So we can obtain $\widehat{\mu}_{SM}$ and $\widehat{\sigma}_{SM}$ from a simple random sample of size n from the sampled population. Substituting $\widehat{\mu}$ for $\widehat{\mu}_{SM}$ and $\widehat{\sigma}/\sqrt{n}$ for $\widehat{\sigma}_{SM}$ in L and U in Line 7.7.5, we get

$$L = \widehat{\mu} - (T2)\frac{\widehat{\sigma}}{\sqrt{n}} \quad \text{and} \quad U = \widehat{\mu} + (T2)\frac{\widehat{\sigma}}{\sqrt{n}}. \qquad \textbf{Line 7.7.6}$$

As noted in Box 6.3.1, the table value $T2$ obtained from Table T-2 depends on the confidence coefficient specified by the investigator and on the degrees of freedom $n - 1$ used to compute $\widehat{\sigma}$. Thus, L and U in Line 7.7.6 are the same as L and U in Box 6.3.1. We have therefore shown that if a sample size is "sufficiently large," the formulas in Box 6.3.1 can also be used to compute a confidence interval for the mean of a *nonnormal* population.

Computer Simulation

The proof of the central limit theorem requires mathematics beyond what is required for this book, but we can demonstrate its results by using computer simulation. This is discussed in Exploration 7.7.1.

EXPLORATION 7.7.1

In this exploration we will use Minitab to demonstrate the central limit theorem described in Box 7.7.3. We have written the macro **clt.mac** for use in performing this demonstration. If you have access to Minitab, we encourage you to repeat this exercise. We begin our demonstration by considering two populations, each of size $N = 3{,}000$. Relative frequency histograms of the two populations are shown in Figures 7.7.1 and 7.7.2.

As you can see, neither population is *normal*. According to the central limit theorem, the population of sample means for samples of size n selected from either population should be approximately *normal* if n is "sufficiently large," say $n \geq 30$.

To use the macro to demonstrate the central limit theorem, enter the following command in the session window of Minitab:

```
MTB > %a:\macro\clt  n  q popno start
```

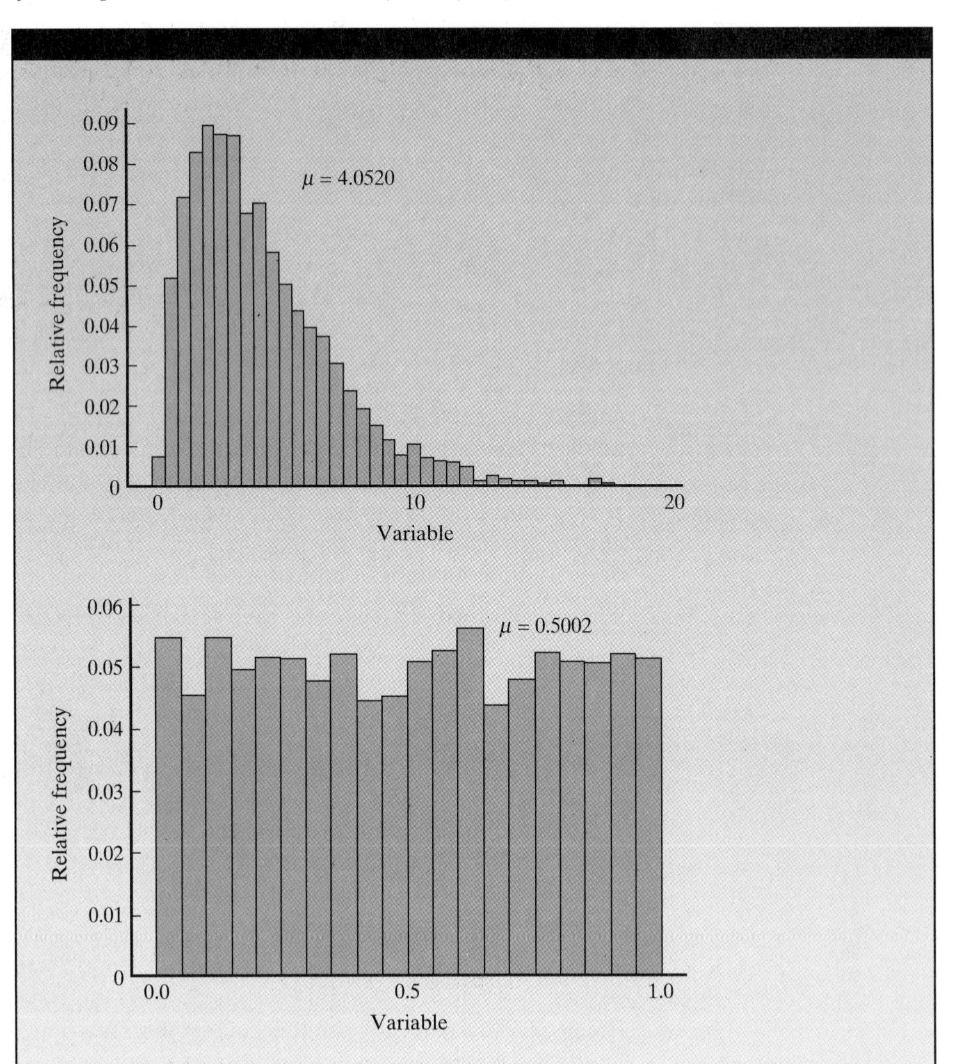

FIGURE 7.7.1
Relative Frequency Histogram for Population 1

FIGURE 7.7.2
Relative Frequency Histogram for Population 2

where `n` is the sample size, `q` is the number of sample means you want to select, `popno` = 1 or 2 designates the population you want to sample, and `start` is any integer (whole number) from 1 to 10,000. The word `start` tells the computer where to get the random numbers to use in the simulation. We suggest that you choose a value for `n` between 30 and 60 and set `q` = 1,000. If you select a value of `q` greater than 1,000, it might take quite a long time to perform the simulation.

We consider samples of size $n = 40$. Here $N > 10n$ as required in Box 7.7.3. Ideally, to verify the central limit theorem, we would like to compute the sample mean for *every* possible sample of size 40 from the sampled population. This collection of sample means is the derived population of sample means. However, the total number of possible samples of size 40 from a population of 3,000 is roughly 1×10^{91}, or the number 1 followed by 91 zeros! This is obviously too large a number for any computer. Thus, we will use 1,000 simple random samples of size $n = 40$, and from these samples we will compute 1,000 sample means. Since we do not use every possible sample of size 40 from the population, the resulting relative frequency histogram will not appear as a smooth curve as shown in Figure 6.2.1. The Minitab command that is used to select $q = 1{,}000$ samples of size $n = 40$ from population 1 is

```
MTB > %a:\macro\clt 40 1000 1 396
```

A relative frequency histogram of the 1,000 sample means obtained by using the previous command is displayed in Figure 7.7.3. This does in fact resemble a bell curve. If we had used n greater than 40, the figure would have more closely resembled a bell curve. The relative frequency histogram you obtain will not appear exactly like the one in Figure 7.7.3 unless you use the same value for `start (396)` that we used.

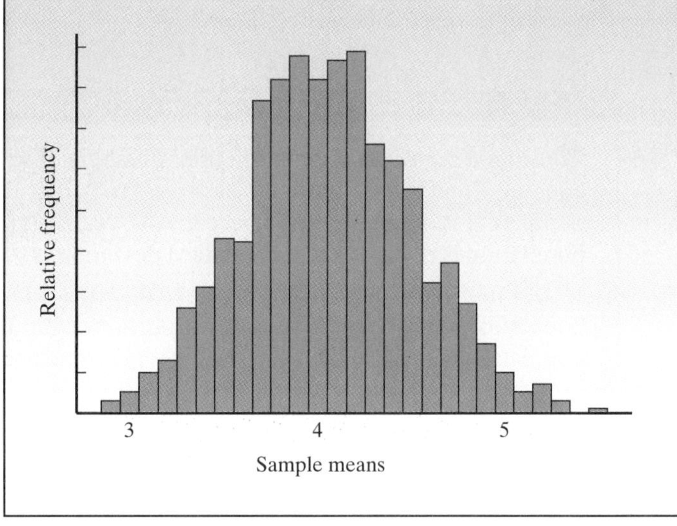

FIGURE 7.7.3
Relative Frequency Histogram of a Population of 1,000 Sample Means of Size 40 from Population 1

Using the Central Limit Theorem to Derive Confidence Intervals for π, a Population Proportion

We now show how the central limit theorem can be used to derive the confidence interval for a population proportion π given in Box 3.2.1. Recall that π is the proportion of population items that have a specified characteristic. For instance, π might be the proportion of individuals in the population who are Republicans, or the proportion of children in a city who have been immunized against diphtheria, or the proportion of bottles of aspirin that contain one or more broken pills. The variable of interest in each case is a qualitative variable with two values—*possesses the characteristic* and *does not possess the characteristic*.

It is sometimes convenient to code the values of this type of qualitative variable by assigning the number 1 to items that possess the characteristic of interest and the number 0 to items that do not possess this characteristic. We then have a population of numbers that are either 0 or 1. Such populations are called **zero–one populations**.

Consider a zero–one population of N items of which K have the value 1 and $N - K$ have the value 0. Thus, the proportion of population items with the characteristic of interest is $\pi = K/N$. If the K items with the characteristic are coded 1 and the remaining $N - K$ items are coded 0, then the population mean is

$$\mu = \frac{1+1+\cdots+1+0+0+\cdots+0}{N} = \frac{(K)(1)+(N-K)(0)}{N} = \frac{K}{N} = \pi.$$

Thus, the mean μ of a zero–one population is equal to π, the proportion of items in the population with the characteristic of interest. The standard deviation of a zero–one population, denoted by σ, is

$$\sigma = \sqrt{\frac{(1-\mu)^2 + (1-\mu)^2 + \cdots + (1-\mu)^2 + (0-\mu)^2 + (0-\mu)^2 + \cdots + (0-\mu)^2}{N}}$$

$$= \sqrt{\frac{(K)(1-\mu)^2 + (N-K)(0-\mu)^2}{N}} = \sqrt{\left(\frac{K}{N}\right)(1-\mu)^2 + \left(1-\frac{K}{N}\right)(0-\mu)^2}$$

$$= \sqrt{\pi(1-\pi)^2 + (1-\pi)\pi^2} = \sqrt{\pi(1-\pi)}.$$

These results are summarized in Box 7.7.4.

BOX 7.7.4: CENTRAL LIMIT THEOREM FOR ZERO–ONE POPULATIONS

Consider a zero–one population of N numbers of which K numbers are equal to 1 and $N - K$ numbers are equal to 0. The proportion of 1's in this population is denoted by π, where $\pi = K/N$. The mean of this zero–one population is $\mu = \pi$, and the standard deviation is $\sigma = \sqrt{\pi(1-\pi)}$.

If the sample size n is sufficiently large and $N > 10n$, then the derived population of all possible sample proportions (sample means) is a *normal* population with mean and standard deviation respectively given by

$$\mu_{\text{SM}} = \mu = \pi \qquad \text{and} \qquad \sigma_{\text{SM}} = \frac{\sigma}{\sqrt{n}} = \sqrt{\frac{\pi(1-\pi)}{n}}.$$

The confidence interval $L \leq \mu_{\text{SM}} \leq U$ in Line 7.7.4, where L and U are defined in Line 7.7.5, can be used to obtain a confidence interval for π since $\mu_{\text{SM}} = \pi$. Substituting

$$\widehat{\pi} \text{ for } \widehat{\mu}_{\text{SM}} \qquad \text{and} \qquad \sqrt{\frac{\widehat{\pi}(1-\widehat{\pi})}{n}} \text{ for } \widehat{\sigma}_{\text{SM}}$$

in L and U in Line 7.7.5, we get

$$L = \widehat{\pi} - T2\sqrt{\frac{\widehat{\pi}(1-\widehat{\pi})}{n}} \qquad \text{and} \qquad U = \widehat{\pi} + T2\sqrt{\frac{\widehat{\pi}(1-\widehat{\pi})}{n}}. \qquad \textbf{Line 7.7.7}$$

It is customary to use $T1$ in place of $T2$ in the formulas for computing confidence intervals for π; we do so in Box 3.2.1. If n is at least 100, as required to use Box 3.2.1, the difference between $T1$ and $T2$ is 0.05 or less for 90%, 95% and 99% confidence intervals. So using $T1$ in place of $T2$ will make little difference in the confidence interval in Line 7.7.7. If we make these replacements, the confidence interval for π is

$$L \leq \pi \leq U,$$

where

$$L = \hat{\pi} - T1\sqrt{\frac{\hat{\pi}(1-\hat{\pi})}{n}} \quad \text{and} \quad U = \hat{\pi} + T1\sqrt{\frac{\hat{\pi}(1-\hat{\pi})}{n}}.$$

These are the formulas for the endpoints of a confidence interval for π given in Box 3.2.1.

EXAMPLE 7.7.3

CALCULUS SCORES. The table in Appendix B reports a set of scores for 2,600 students on a standardized calculus test administered at a large university. The data in Appendix B are also stored in the file **grades.mtp** on the data disk, where column 1 contains the student number, column 2 contains the student's score, and column 3 is a column of zeros and ones. A 1 indicates that the student was a mathematics major, and a 0 indicates that the student was not a math major. The relative frequency histogram of this zero–one population is shown in Figure 7.7.4.

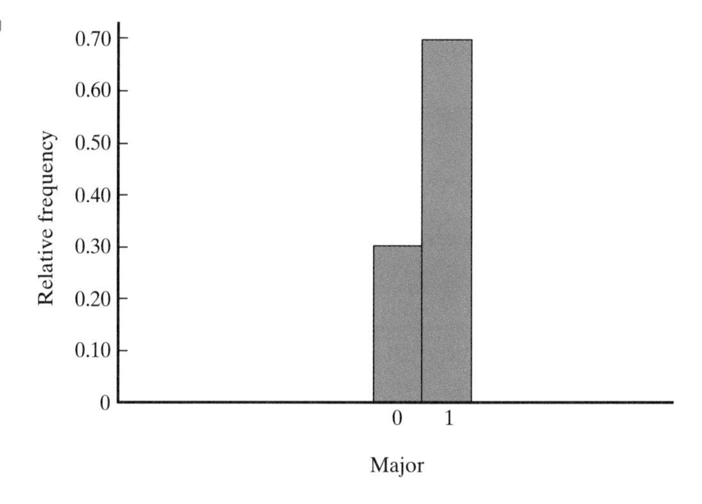

FIGURE 7.7.4
Relative Frequency Histogram of the Zero-One Population of Student Majors

The relative frequency histogram of this zero–one population is vastly different from a bell curve. This is the reason why a simple random sample of size 100 or more is required when using the formulas in Box 3.2.1 to compute a confidence interval for π. However, in cases where π is very close to either 0 or 1, a sample size larger than 100 may be needed for the sample size to be considered "sufficiently large" and for the actual confidence coefficient to be close enough to the specified confidence coefficient.

The following questions, with solutions, can be answered using the material in this section.

7.7.1 Suppose you use a computer to select all possible simple random samples of size 34 from a population $Y_1, Y_2, \ldots, Y_N$ that has a mean μ and a standard deviation σ. The sample mean is computed for each sample, and this population of sample means is denoted by $\bar{y}_1, \bar{y}_2, \ldots, \bar{y}_k$. What are the mean and the standard deviation of this population?

▶**Answer:** *Using Box 7.7.2 the mean of the population of sample means is μ, the same as the mean of the sampled population $Y_1, Y_2, \ldots, Y_N$. The standard deviation of the population of sample means is $f\sigma/\sqrt{n} = f\sigma/\sqrt{34}$, where*

$f = \sqrt{(N-34)/(N-1)}$ *and* σ *is the standard deviation of the sampled population* $Y_1, Y_2, \ldots, Y_N$.

7.7.2 Box 7.7.3 states that if the sample size is sufficiently large, the population of sample means is approximately *normal* with mean μ and standard deviation $\sigma/\sqrt{n}$, where μ and σ are the mean and the standard deviation, respectively, of the sampled population. Use this information to work the following problem. A manufacturer of potato chips ships the chips in boxes that contain 16 bags of chips. A population of the weights of bags of potato chips manufactured in one day (about 10,000 bags) is *normal* with mean $\mu = 567$ grams and standard deviation $\sigma = 10$ grams. A box (consisting of 16 bags of chips) is to be selected at random from all boxes packaged in one day and the 16 bags weighed. What is the probability that the mean of the 16 bags is less than 572 grams? Assume the 16 bags in any box is a simple random sample from the population of bags.

▶**Answer:** *By Box 7.7.2 the derived population of sample means for all boxes packaged in one day has a mean of 567 grams and a standard deviation of* $10/\sqrt{16} = 2.5$ *grams. Since the sampled population is normal, the derived population of sample means is also normal. Let* $\bar{y}$ *represent a simple random sample of one mean to be selected from the derived population of sample means. We need to determine* $Pr(\bar{y} < 572)$. *To do this, follow the instructions in Box 6.2.2, but replace* σ *with the standard deviation of the derived population of sample means. That is, set* $572 = 567 + z(2.5)$ *and solve for* z. *This yields* $z = 2.0$. *Thus,* $Pr(\bar{y} < 572)$ *is equal to the proportion of boxes that are less than 2.0 standard deviations above the mean. From line (4) of Table 6.2.1, the value is 0.977. Thus,* $Pr(\bar{y} < 572) = 0.977$.

7.7.3 An article in the *Denver Post* reported that in 1993 there were 53,000 bee colonies used for honey production in the state of Colorado. A colony is defined as a hive with one queen bee. This population of colonies had an average yield of 73 pounds of honey per year. Assume the population of yield values for the 53,000 colonies is a *normal* population with a standard deviation of 7.126 pounds.

(a) What proportion of the colonies have a yield larger than 79 pounds?

▶**Answer:** *If we let* Y *denote the yield of a simple random sample of size one from all colonies, we must find* $Pr(Y > 79)$. *By Line 6.2.4 we obtain* $Pr(Y > 79) = 1 - Pr(Y \leq 79)$, *so we will find* $Pr(Y \leq 79)$. *Using the procedure in Box 6.2.2, we set* $79 = 73 + z(7.126)$ *and solve for* z. *We obtain* $z = 0.842$. *So* $Pr(Y \leq 79) = Pr(Y \leq \mu + 0.842\sigma)$, *and from line (9) of Table 6.2.1 we note that* $Pr(Y \leq \mu + 0.842\sigma) = 0.80$. *So* $Pr(Y \leq 79) = 0.80$ *and* $Pr(Y > 79) = 0.20$. *So 20% of the colonies have an annual yield of honey greater than 79 pounds.*

(b) A simple random sample of 16 colonies was selected and the sample mean $\bar{y}$ computed. What is the probability that $\bar{y}$ is less than 76.5 pounds?

▶**Answer:** *Since the sampled population is normal, the derived population of sample means of size* $n = 16$ *is* normal *with a mean* $\mu = 73$ *and a standard deviation* $\sigma/\sqrt{n} = 7.126/4 = 1.782$. *We must determine* $Pr(\bar{y} < 76.5)$. *Let* $76.5 = 73 + z(1.782)$, *and solve for* z. *This gives* $z = 1.96$. *From Table 6.2.1 we see that the proportion of values less than* $\mu + 1.96\sigma$ *is 0.975. So the probability that a sample mean of 16 observations will be less than 76.5 pounds of honey is 97.5%.*

CHAPTER 8

STATISTICAL TESTS

8.1 INTRODUCTION

Let us again review the process of statistical inference. In any investigation where one wants to examine a population, the variable of interest must be identified and the target and sampled populations defined. Whenever possible, these two populations should be the same. If they are not, the sampled population should resemble the target population as closely as possible. The next step is to decide what parameter is needed to help answer questions about the population. If every item in the sampled population is measured, the exact value of the parameter can be obtained. However, in most applied problems it is impractical to measure every item in the population, and so the value of the parameter will not be known exactly. In this case one must obtain a sample from the sampled population, and use the sample values to obtain a point estimate and a confidence interval for the parameter of interest.

In some problems the investigator is not primarily interested in estimating the value of a population parameter. Rather, the interest is in determining whether a population parameter is within specified limits. We illustrate with four examples.

EXAMPLE 8.1.1

DISEASED APPLE TREES. A large apple orchard must be sprayed each spring for a certain leaf disease. In years past, the orchard supervisor sprayed all the trees in the orchard with a standard spray used in the fruit-growing industry. The supervisor will use the spray again this year unless he has evidence that π is less than 10%, where π is the proportion of infected trees in the orchard. If he is convinced that

$\pi < 0.10$, he will use a less expensive spray that is known to be less effective. To help decide which spray to use, the supervisor selects a simple random sample of trees from the orchard. If the sample provides sufficient evidence for the supervisor to conclude that $\pi < 0.10$, the less expensive spray will be used to spray the orchard. If the sample evidence is not sufficient to conclude that $\pi < 0.10$, then the supervisor will use the standard spray. The supervisor is primarily interested in determining if $\pi < 0.10$, not in estimating π.

EXAMPLE 8.1.2

EPA REGULATIONS. The Environmental Protection Agency (EPA) notified the manager of a manufacturing plant that carbon monoxide emissions exceeded EPA limits several times last month. In order to avoid a large fine, the manager must provide evidence within 60 days that the plant is in compliance with EPA regulations. Compliance requires that emissions not exceed EPA limits more than 1% of the time over any 30-day period. In order to meet the 60-day deadline for compliance, the company's emissions equipment was repaired within 30 days after receiving the notice. After the repairs, the manager claims that the emissions are now in compliance. The next 30 days are then divided into 8,640 five-minute time periods. To be in compliance with the EPA, π must be less than 0.01, where π is the proportion of the 8,640 time periods in which the EPA limits for carbon monoxide emissions are exceeded. The manager will obtain a simple random sample from the 8,640 time periods. If the sample contains sufficient evidence to convince the EPA that $\pi < 0.01$, the fine will not be imposed. The manager is primarily interested in providing evidence that $\pi < 0.01$, not in estimating the value of π.

EXAMPLE 8.1.3

BOXES OF CEREAL. An operator of a cereal-packaging machine monitors the net weights of the packaged boxes by periodically weighing random samples of boxes. The standard protocol is for the operator to allow the machine to run unless a sample of boxes suggests that the machine is not operating properly. When such a situation occurs, the machine is taken off-line and adjusted. One condition required for the proper operation of the machine is that $\mu = 453$ grams, where μ is the average net weight of the 10,000 most recently packaged boxes. The operator is not primarily interested in estimating μ but is interested in determining if there is sufficient evidence to conclude that $\mu \neq 453$.

EXAMPLE 8.1.4

TO LOAN OR NOT TO LOAN. A company that owns a chain of department stores has requested a loan from a bank to build a store in a Midwestern city. After conducting the required marketing research, the bank loan officer decides to deny the loan unless she is convinced that $\pi > 0.50$, where π is the proportion of households in the city with annual incomes exceeding \$30,000. From previous experience, the loan officer knows that this type of department store tends to be successful in cities where more than 50% of the households have annual incomes exceeding \$30,000. The loan officer informs the company that the loan will be denied unless the company can provide convincing evidence that $\pi > 0.50$. To provide this evidence, the company uses a simple random sample of households from the city. The company is primarily interested in determining if the sample provides sufficient evidence to conclude that $\pi > 0.50$.

As you can see, in each of these examples the primary interest is not in estimating the value of a population parameter. Rather, the primary interest is in determining if the sample provides sufficient evidence to conclude that a population parameter is contained in an interval specified by the investigator.

A **statistical test** is a procedure used to determine the amount of evidence a sample provides for concluding that a population parameter is in an interval specified by an investigator. Statistical tests are sometimes referred to as either **hypothesis tests** or **significance tests**.

Statistical Tests

To conduct a statistical test, an investigator considers two complementary statements about a population parameter. For instance, in Example 8.1.1 the two complementary statements are

$$\text{Statement 1: } \pi \geq 0.10 \qquad \text{versus} \qquad \text{Statement 2: } \pi < 0.10.$$

The supervisor will use the standard spray unless the sample evidence is sufficient for concluding that Statement 1 is false. If the supervisor concludes that Statement 1 is false (and hence that Statement 2 is true), then the less expensive spray will be used to spray the orchard.

In Example 8.1.2 the two complementary statements about the population parameter are

$$\text{Statement 1: } \pi \geq 0.01 \qquad \text{versus} \qquad \text{Statement 2: } \pi < 0.01.$$

The manager must provide sufficient sample evidence for the EPA to conclude that Statement 1 is false and Statement 2 is true. If she can do this, then a fine will *not* be imposed. If the sample evidence is not sufficient to support the claim in Statement 2, the fine will be imposed.

To provide a general description of statistical tests, we use the symbol θ, which is the Greek letter **theta**, to represent any population parameter such as π (a population proportion), M (a population median), σ (a population standard deviation), or μ (a population mean). An investigator considers two complementary statements about θ. Statement 1 is called the **null hypothesis** and is denoted by H_0. Statement 2 is called the **alternative hypothesis** and is denoted by H_a.

BOX 8.1.1: STATEMENT OF HYPOTHESES IN STATISTICAL TESTS

In a statistical test, two complementary statements are made about a population parameter θ:

- Statement 1 is the null hypothesis, H_0.
- Statement 2 is the alternative hypothesis, H_a.

Every possible value of θ must be included in either H_0 or H_a, and no value can be included in both hypotheses.

In Example 8.1.1 $\theta = \pi$ and the supervisor conducts a statistical test using the hypotheses

$$H_0 : \pi \geq 0.10 \qquad \text{versus} \qquad H_a : \pi < 0.10.$$

Every possible value of π is in either H_0 or H_a, and no value appears in both hypotheses. In Example 8.1.3 $\theta = \mu$ and the operator conducts a statistical test using the hypotheses

$$H_0 : \mu = 453 \quad \text{versus} \quad H_a : \mu \neq 453.$$

Every possible value of μ is in either H_0 or H_a, and no value appears in both hypotheses.

How to Choose H_0 and H_a

In conducting a statistical test it is important that the null and alternative hypotheses be chosen correctly. This choice is the responsibility of the investigator. In order to help you properly identify H_0 and H_a, we present guidelines in Box 8.1.2 for two situations in which a statistical test is commonly performed.

BOX 8.1.2: GUIDELINES FOR CHOOSING H_0 AND H_a

1. To solve a problem, suppose that an investigator plans to use a standard course of action unless sample data strongly suggest that an alternative course of action under consideration would be better. The statement that represents the standard course of action is chosen as H_0, the null hypothesis. The statement that represents the alternative course of action is chosen as H_a, the alternative hypothesis.
2. An investigator wants to determine the amount of evidence a sample provides to support a claim or conjecture. This claim is chosen as H_a, the alternative hypothesis.

Example 8.1.1 describes a situation similar to (1) in Box 8.1.2. The orchard supervisor will use the standard spray to treat the orchard (a standard course of action) unless sample data convince him that the proportion of infected trees in the orchard is less than 0.10. In this case, a less expensive spray will be used to spray the orchard (an alternative course of action). The less expensive spray will be used only if the sample provides convincing evidence that $\pi < 0.10$. Thus, the null hypothesis is $H_0 : \pi \geq 0.10$, which is identified with the standard spray (the standard course of action). The alternative hypothesis is $H_a : \pi < 0.10$, which is identified with the less expensive spray (the alternative course of action).

Example 8.1.2 is similar to situation (2) in Box 8.1.2. In this problem, the emissions equipment is repaired and the plant manager claims that the plant is now in compliance with the EPA regulations. The manager must provide sample evidence to support this claim. In this problem, the alternative hypothesis represents the manager's claim, namely $H_a : \pi < 0.01$. The null hypothesis is the complementary statement $H_0 : \pi \geq 0.01$.

Reporting the Result of a Statistical Test

In a statistical test, if the sample evidence is sufficient to convince an investigator that the null hypothesis H_0 is false (and hence the alternative hypothesis H_a is true), the result of the test is stated as "reject H_0." If the sample evidence is not sufficient to convince the investigator that the null hypothesis H_0 is false, the result of the test is stated as "do not reject H_0."

The statement "do not reject H_0" does *not* mean the sample provides convincing evidence that H_0 is true. It simply means that the sample evidence is not sufficient for the investigator to reject H_0.

∗ ***Summary*** In a statistical test, an investigator will conclude that the alternative hypothesis (H_a) is true only if the sample evidence is sufficient to reject the null hypothesis (H_0). Thus, a statistical test is conducted to determine if the amount of evidence contained in a sample is sufficient to reject H_0 (and thereby conclude that H_a is true). So a statistical test can be viewed as "testing a null hypothesis."

It may seem odd to you that a statistical test is used to determine if H_a is true by determining if H_0 can be rejected. However, this is the way statistical tests are typically conducted. Ideally, an investigator would undoubtedly like to "know for sure" whether H_a is true. However, to do this, the entire population (or almost the entire population) must be examined. As usual, it is not practical to observe all of the items in a population, and so decisions must be based on sample values. Since investigators may not be able to "know for sure" whether H_a is true, they will "conclude" that H_a is true whenever the sample evidence is so overwhelmingly against H_0 being true that H_0 must be rejected. We illustrate this thought process in Example 8.1.5.

EXAMPLE 8.1.5

CONCLUDING THAT H_a IS TRUE BY REJECTING H_0. Suppose that there are 1,000 bins identical in outside appearance and that each bin contains 1,000 balls that are indistinguishable except that some are white and some are black. One bin (Bin 1) is on a shelf and 999 bins (Bin 2, Bin 3, Bin 4, . . . , Bin 1,000) are on the floor. The contents of the bins are known, and are shown in Figure 8.1.1, where W stands for a white ball and B stands for a black ball.

FIGURE 8.1.1

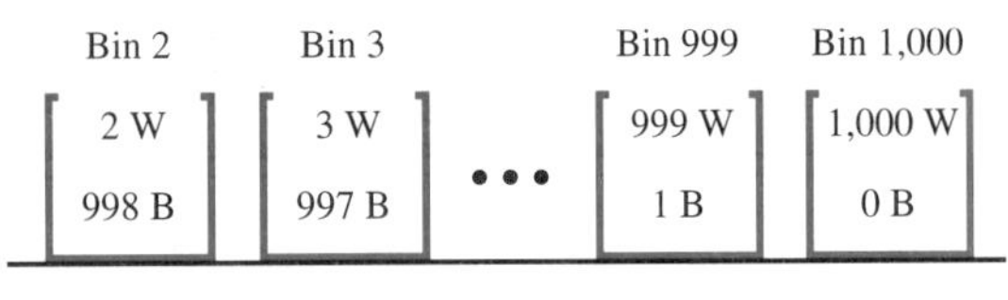

One of the bins has been damaged, and an investigator brings it to me, claiming that the damaged bin came from the floor. I will conduct a statistical test to determine if the sample evidence is sufficient for me to conclude that her claim is true. Using guideline (2) in Box 8.1.2, the claim "The bin came from the floor" is chosen as the alternative hypothesis. Thus, I will conduct a statistical test using the hypotheses

$$H_0 : \text{The bin came from the shelf}$$

versus

$$H_a : \text{The bin came from the floor.}$$

The sample evidence for this test is the color of one ball drawn at random from the bin handed to me. Using the color of only one ball, I cannot "know for sure" whether the claim is true that the bin came from the floor. But the sample data may persuade me to reject H_0 and hence accept that H_a is true. The following discussion illustrates this idea.

There are two possible values of a simple random sample when one ball is selected from the bin given to me: (1) the selected ball is white; (2) the selected ball is black. The possible conclusions, based on the sample values, are described below.

(1) THE BALL DRAWN IS WHITE (W).

The probability of randomly drawing a *white* ball from the bin on the shelf is 0.001. This probability is very small, and it would be extremely unusual to draw a white ball from the bin on the shelf. It would not, however, be unusual to draw a white ball from some of the bins on the floor. Thus, if I draw a white ball, I conclude that I was *not* handed the bin from the shelf. The sample evidence represented by the probability of drawing a white ball from the bin on the shelf is so small (so overwhelmingly against H_0 being true) that I conclude that the ball was not drawn from the bin on the shelf. Therefore, I reject H_0 and accept H_a, that the ball came from a bin on the floor.

(2) THE BALL DRAWN IS BLACK (B).

The probability of drawing a *black* ball from the bin on the shelf is 0.999. This probability is not overwhelmingly small, and so selecting a black ball from the bin on the shelf would not be an unusual occurrence. So this possibility cannot be ruled out without further evidence. Thus, if I draw a black ball, I have no *compelling* reason to reject H_0. This does not mean that I believe H_0 is true, because there are also bins on the floor from which the selection of a black ball would not be an unusual occurrence. The only statement that can be made at this point is that the sample evidence is not sufficient to reject H_0.

THINKING STATISTICALLY: SELF-TEST PROBLEMS

8.1.1 Give some examples where you need to know if a parameter is in a specified interval, but you don't need to estimate the value.

8.1.2 State the hypotheses H_0 and H_a that are appropriate for conducting a statistical test in each of the following situations.

a. An internal auditor for a department store claims that the proportion of overdue customer accounts is greater than 0.10. Evidence based on a simple random sample of accounts will be used to support the claim that $\pi > 0.10$, where π is the proportion of overdue customer accounts.

b. A marketing research firm designs an advertising campaign directed at college students for a client who sells personal computers. Historically, the primary marketing medium used to target college students has been college newspaper advertisements. The research firm plans to continue using newspaper advertising unless evidence can be obtained that suggests this is not an effective medium. If such evidence is found, the marketing firm will use direct mailing in the campaign. In an attempt to determine if newspaper advertising is effective, the company will select a simple random sample of students from the sampled population of all students who attend colleges in Colorado. Let π represent the proportion of students in the sampled population who read the college newspaper at least once a week. If π is less than 0.20, direct mailing will be used instead of college newspaper advertisements.

8.1.3 Suppose the supervisor in Example 8.1.1 used the hypotheses

$$H_0 : \pi \geq 0.10 \qquad \text{versus} \qquad H_a : \pi < 0.15.$$

What is wrong with this formulation?

8.1.4 Suppose the supervisor in Example 8.1.1 used the hypotheses

$$H_0 : \pi \geq 0.10 \qquad \text{versus} \qquad H_a : \pi < 0.08.$$

What is wrong with this formulation?

8.1.5 Suppose the supervisor in Example 8.1.1 used the hypotheses

$$H_0 : \pi < 0.10 \qquad \text{versus} \qquad H_a : \pi \geq 0.10.$$

Is this formulation correct?

8.2 USING CONFIDENCE INTERVALS TO CONDUCT STATISTICAL TESTS

In this section we show how to use a confidence interval to conduct a statistical test. The objective of a statistical test is to make one of the two decisions presented in Box 8.2.1.

BOX 8.2.1: POSSIBLE DECISIONS IN A STATISTICAL TEST

One of the following *two* decisions is made in a statistical test:

1. reject H_0;

2. do not reject H_0.

Since sample data are used to make the decision, we might ask the question, "What is the probability of getting sample data that lead to an erroneous decision?" For example, what is the probability that the sample data will indicate that H_0 should be rejected even though H_0 is true? Or, what is the probability that the sample data will indicate that H_0 should not be rejected even though H_0 is false?

BOX 8.2.2: PROBABILITY OF MAKING MISTAKES WHEN CONDUCTING A STATISTICAL TEST

The following two mistakes (errors) are possible when conducting a statistical test:

1. the mistake of rejecting H_0 when H_0 is true. The probability of making this mistake is denoted by the symbol α, which is the Greek letter **alpha**.
2. the mistake of not rejecting H_0 when H_0 is false. The probability of making this mistake is denoted by the symbol β, which is the Greek letter **beta**.

The probabilities of making either of the two mistakes described in Box 8.2.2 are schematically exhibited in Table 8.2.1.

TABLE 8.2.1 Probability of Making Mistakes in a Statistical Test

		If H_0 is true	**If H_0 is false**
Decision	**reject H_0**	probability of a mistake is α	zero (no mistake)
	do not reject H_0	zero (no mistake)	probability of a mistake is β

When conducting a statistical test, it is desirable to choose small values for both α and β so that the probability of making a mistake is small. However, if values for α and β are both specified (say, both are specified to be small), a certain sample size is required to satisfy these specifications. Moreover, the required sample size depends on the values of population parameters, which typically are not known. Since the population parameters are not known, the investigator will not know the exact sample size that is required in order to attain the specified values for α and β. The procedure that is generally used when conducting a statistical test is to specify α to be a small value (0.05, 0.01, 0.001, etc.) and leave the value of β unspecified. This specified value of α can be attained for any sample size, and the larger the sample size, the smaller the value of β will tend to be.

The selection of a value for α must be made before examining the data. This choice should reflect the practical consequence of making the mistake of rejecting H_0 when H_0 is true. Generally, an investigator wants the probability of making this mistake to be small. In applications, α is typically chosen to be 0.10, 0.05, 0.01, or 0.001, but it can be chosen to be any value between 0 and 1. For example, suppose an investigator chooses $\alpha = 0.05$. This is the risk that this investigator is willing to take in erroneously rejecting H_0 when H_0 is true. If the sample evidence is sufficient to reject H_0 using $\alpha = 0.05$, then the investigator will reject H_0 and conclude that H_a is true, knowing that the probability of making the mistake of erroneously rejecting H_0 is 0.05. If the sample evidence is not sufficient to reject H_0 using $\alpha = 0.05$, then this investigator will conclude that H_0 cannot be rejected. Although the probability of making the mistake of not rejecting H_0 when H_0 is false (β) is not known, it can be estimated. We do not discuss this topic.

EXAMPLE 8.2.1

DISEASED APPLE TREES. In Example 8.1.1 a standard spray will be used unless the sample of trees convinces the supervisor that $\pi < 0.10$, where π is the proportion of infected trees in the orchard. The supervisor will conduct a statistical test using the hypotheses

$$H_0 : \pi \geq 0.10 \qquad \text{versus} \qquad H_a : \pi < 0.10.$$

If $\pi \geq 0.10$ (H_0 is true) and the supervisor erroneously rejects H_0, the supervisor will erroneously use the cheaper, less effective spray. A consequence of this error is that the apple crop may be severely reduced because the trees will not be properly treated. The supervisor wants the probability of making this mistake to be small, so α is chosen to be 0.01. If the sample evidence is sufficient to reject H_0 using $\alpha = 0.01$, then the supervisor will conclude that H_a is true (and use the less effective spray), knowing that the probability of making the mistake of erroneously rejecting H_0 is 0.01.

By now it might have occurred to you that an investigator could use a confidence interval to conduct a statistical test. Actually, this is one way a statistical test can be performed. A confidence interval is used to obtain plausible values for a parameter, and these values are then used to make one of the two decisions shown in Box 8.2.1. An alternative testing procedure employs what is called a **test statistic**. This procedure is described in Appendix 2 of this chapter (Section 8.7) and is equivalent to the confidence interval procedure, which we now discuss.

In this book we use confidence intervals to conduct statistical tests for two reasons: (1) Confidence intervals are used throughout this book, so they are readily available for use in statistical tests (test statistics do not need to be computed and described); (2) in addition to being useful for conducting statistical tests, confidence intervals can be helpful in determining the practical significance of the test results.

Procedures for Conducting a Statistical Test with Confidence Intervals

Recall that we use the symbol θ to denote any population parameter. Three sets of hypotheses that are of interest in applications are

$$H_0 : \theta \leq \theta_0 \qquad \text{versus} \qquad H_a : \theta > \theta_0, \qquad \textbf{Line 8.2.1}$$

$$H_0 : \theta \geq \theta_0 \qquad \text{versus} \qquad H_a : \theta < \theta_0, \qquad \textbf{Line 8.2.2}$$

and

$$H_0 : \theta = \theta_0 \qquad \text{versus} \qquad H_a : \theta \neq \theta_0, \qquad \textbf{Line 8.2.3}$$

where θ_0 is a number specified by the investigator. Box 8.2.3 describes procedures for conducting statistical tests for the hypotheses in Line 8.2.1 and Line 8.2.2. Later in this section we describe a procedure for conducting a statistical test for the hypotheses in Line 8.2.3.

BOX 8.2.3: USING CONFIDENCE INTERVALS TO CONDUCT A STATISTICAL TEST FOR THE HYPOTHESES IN LINES 8.2.1 AND 8.2.2

Let θ denote a population parameter, and consider testing the hypotheses in Line 8.2.1, namely

$$H_0 : \theta \leq \theta_0 \qquad \text{versus} \qquad H_a : \theta > \theta_0,$$

where θ_0 is specified by the investigator. Choose a value for α, and select a simple random sample from the population. Use the sample to compute L, the lower endpoint of a confidence interval for θ with a confidence coefficient equal to $1 - 2\alpha$.

Reject H_0 if the confidence interval is completely to the right of θ_0. That is, reject H_0 and conclude that $\theta > \theta_0$ if

$$\theta_0 < L. \qquad \textbf{Line 8.2.4}$$

If the confidence interval is not completely to the right of θ_0 (that is, if $L \leq \theta_0$), do not reject H_0.

Now consider testing the hypotheses shown in Line 8.2.2, namely

$$H_0 : \theta \geq \theta_0 \qquad \text{versus} \qquad H_a : \theta < \theta_0,$$

where θ_0 is specified by the investigator. Choose a value for α, and select a simple random sample from the population. Use the sample to compute U, the upper endpoint of a confidence interval for θ with a confidence coefficient equal to $1 - 2\alpha$.

Reject H_0 if the confidence interval is completely to the left of θ_0. That is, reject H_0 and conclude that $\theta < \theta_0$ if

$$U < \theta_0. \qquad \textbf{Line 8.2.5}$$

If the confidence interval is not completely to the left of θ_0 (that is, if $\theta_0 \leq U$), do not reject H_0.

If the test procedures described in Box 8.2.3 are used, the probability of obtaining a sample that erroneously rejects H_0 when H_0 is true is no greater than α. We discuss this result in more detail in Appendix 1 (Section 8.6). The test procedure for the hypotheses described in Box 8.2.3 is visually presented in Figure 8.2.1.

FIGURE 8.2.1
Illustration of Testing Procedure for the Hypotheses in Lines 8.2.1 and 8.2.2

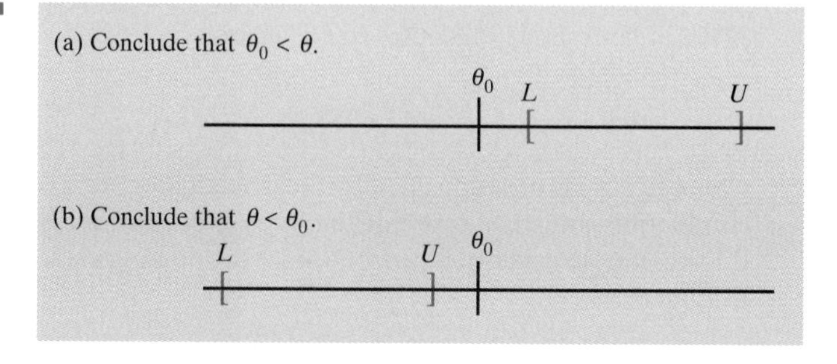

In addition to testing the hypotheses in Line 8.2.1, the procedure in Line 8.2.4 of Box 8.2.3 can be used to test the hypotheses

$$H_0 : \theta < \theta_0 \qquad \text{versus} \qquad H_a : \theta \geq \theta_0$$

or

$$H_0 : \theta = \theta_0 \qquad \text{versus} \qquad H_a : \theta > \theta_0.$$

For both of these sets of hypotheses, H_0 is rejected if $\theta_0 < L$. In addition to testing the hypotheses in Line 8.2.2, the procedure in Line 8.2.5 of Box 8.2.3 can be used to test the hypotheses

$$H_0 : \theta > \theta_0 \qquad \text{versus} \qquad H_a : \theta \leq \theta_0$$

or

$$H_0 : \theta = \theta_0 \qquad \text{versus} \qquad H_a : \theta < \theta_0.$$

For both of these sets of hypotheses, H_0 is rejected if $U < \theta_0$.

Notice that a confidence coefficient of $1 - 2\alpha$ is used in Box 8.2.3. This requires α to be no more than 0.50 (otherwise the confidence coefficient will be negative, an impossible situation). If α is specified to be 0.50 or larger, this means that an investigator is willing to reject H_0 with the probability of at least 0.50 of being wrong. The procedures in Box 8.2.3 must be slightly modified if you choose $\alpha \geq 0.50$. We do not discuss this modification since we think it has little practical value.

The procedure for conducting a statistical test for the hypotheses shown in Lines 8.2.1 and 8.2.2 can be used for any parameter discussed in this book.

EXAMPLE 8.2.2

DISEASED APPLE TREES. This is a continuation of Example 8.1.1, where the supervisor of a large apple orchard conducts a statistical test to decide which of two sprays should be used to treat the orchard. The supervisor selects $\alpha = 0.05$ and tests the hypotheses

$$H_0 : \pi \geq 0.10 \qquad \text{versus} \qquad H_a : \pi < 0.10.$$

Since the hypotheses are in the form of Line 8.2.2, we use the rule in Line 8.2.5 of Box 8.2.3. Here $\alpha = 0.05$, $\theta = \pi$, and $\theta_0 = \pi_0 = 0.10$. Since $\alpha = 0.05$, we must compute U, the upper endpoint of a confidence interval for π with confidence coefficient equal to $1 - 2(0.05) = 0.90 = 90\%$. From a simple random sample of 150 trees, 6 were infected with the leaf disease. Thus, $\widehat{\pi} = 6/150 = 0.04$. The formula for computing U is given in Box 3.2.1. Since the confidence coefficient is 90%, the table value $T1 = 1.65$. The computed upper endpoint is

$$U = 0.04 + 1.65\sqrt{\frac{0.04(0.96)}{150}} = 0.0664.$$

From this result we obtain

$$U = 0.0664 < \theta_0 = \pi_0 = 0.10.$$

Thus, H_0 is rejected and H_a ($\pi < 0.10$) is accepted. The supervisor concludes that $\pi < 0.10$, knowing that the probability of erroneously rejecting H_0 when H_0 is true is no greater than 0.05. The supervisor will take the alternative course of action and use the less expensive spray.

EXAMPLE 8.2.3

LONGEVITY OF RATS. Consider Exploration 7.3.1, where an investigator is interested in the longevity of a certain species of rat when fed a low-fat diet. From past experience the investigator knows that the median number of days that this species of rat lives when fed an ordinary diet is about 580 days. From preliminary studies the investigator conjectures that a low-fat diet she has developed results in a median lifetime that is greater than 580 days. To determine if a sample provides sufficient evidence to support her conjecture, she will select $\alpha = 0.025$ and conduct a statistical test of the hypotheses

$$H_0 : M \leq 580 \qquad \text{versus} \qquad H_a : M > 580.$$

If H_0 is rejected (and hence H_a is accepted), the investigator will conclude that the low-fat diet she has developed contributes to a longer median lifetime for rats. The probability of making the mistake of rejecting H_0 when H_0 is true is 0.025.
A simple random sample of 120 rats was obtained from a colony of 200,000 rats and fed a low-fat diet. The number of days each rat lived was recorded. The data are stored in column 2 in the file **lowfat.mtp** on the data disk. The hypotheses for this test are in the form of Line 8.2.1, where $\theta = M$ and $\theta_0 = M_0 = 580$. Thus, we will use the rule in Line 8.2.4 of Box 8.2.3 to test this set of hypotheses. We must compute L, the lower endpoint of a confidence interval for M with confidence coefficient equal to $1 - 2\alpha = 1 - 2(0.025) = 0.95 = 95\%$. The 95% confidence interval for M was previously computed in Exploration 7.3.1 to be $L = 555$ days to $U = 638$ days. Since the confidence interval is not completely to the right of 580 (that is, L is not greater than $\theta_0 = M_0 = 580$), H_0 is not rejected. This does not mean that the investigator accepts the null hypothesis $H_0 : M \leq 580$. In fact, the point estimate of M is $\widehat{M} = 590$ days. The fact that H_0 is not rejected simply means that the sample evidence is not sufficient to reject H_0 using $\alpha = 0.025$. Even though H_0 is not rejected, the 95% confidence interval may give the investigator useful information.

In Box 8.2.4 we describe a procedure for conducting a statistical test for the hypotheses

$$H_0 : \theta = \theta_0 \qquad \text{versus} \qquad H_a : \theta \neq \theta_0$$

shown in Line 8.2.3.

BOX 8.2.4: USING A CONFIDENCE INTERVAL TO CONDUCT A STATISTICAL TEST FOR THE HYPOTHESES $H_0 : \theta = \theta_0$ VERSUS $H_a : \theta \neq \theta_0$

Let θ denote a population parameter, and consider testing the hypotheses shown in Line 8.2.3, namely

$$H_0 : \theta = \theta_0 \qquad \text{versus} \qquad H_a : \theta \neq \theta_0,$$

where θ_0 is specified by the investigator. Choose a value for α, and select a simple random sample from the population. Use the sample to compute L and U, the endpoints of a confidence interval for θ with a confidence coefficient equal to $1 - \alpha$. (Note that here we use $1 - \alpha$ and not $1 - 2\alpha$ as in Box 8.2.3.)

Reject H_0 if the confidence interval does not contain θ_0. That is, reject H_0 and conclude that $\theta \neq \theta_0$ if

$$\theta_0 < L \quad \text{or} \quad U < \theta_0. \qquad \textbf{Line 8.2.6}$$

Do not reject H_0 if the interval contains θ_0. That is, do not reject H_0 if

$$L \leq \theta_0 \leq U.$$

The probability of erroneously rejecting H_0 when H_0 is true is equal to α for the test procedure described in Box 8.2.4. The procedure in Box 8.2.4 is presented visually in Figure 8.2.2.

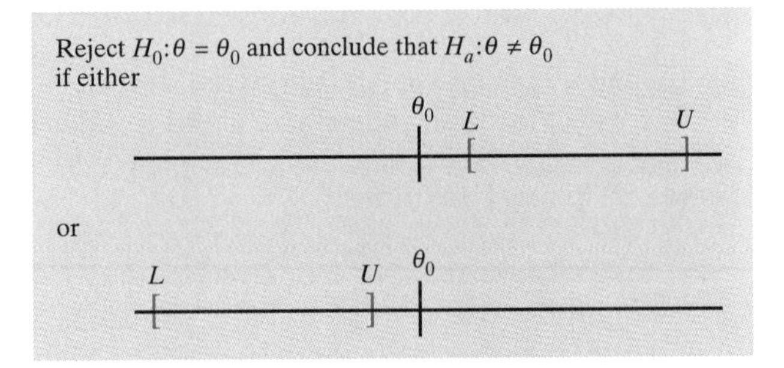

FIGURE 8.2.2
Illustration of Testing Procedure for the Hypotheses in Line 8.2.3

EXAMPLE 8.2.4

BOXES OF CEREAL. Consider Example 8.1.3, where a packaging machine is operating properly if $\mu = 453$ grams, where μ is the average net weight of the 10,000 most recently filled boxes. The operator of the machine will conduct a statistical test to determine if the evidence contained in a simple random sample of boxes is sufficient to conclude that $\mu \neq 453$. If such evidence exists, the machine will be taken off-line and adjusted. The operator selects $\alpha = 0.01$ and conducts a statistical test of the hypotheses

$$H_0 : \mu = 453 \quad \text{versus} \quad H_a : \mu \neq 453.$$

A simple random sample of 50 boxes was selected from the 10,000 most recently filled boxes, and the weight of each box was measured. The sample data were used to compute a confidence interval for μ with a confidence coefficient equal to $1 - \alpha = 1 - 0.01 = 0.99 = 99\%$. The resulting confidence interval is $L = 448.6$ grams to $U = 453.8$ grams. Using the procedure in Box 8.2.4 with $\theta = \mu$, we see that this interval contains $\theta_0 = \mu_0 = 453$, and so H_0 is not rejected. This means that production is not halted to adjust the machine. The fact that $H_0 : \mu = 453$ is not rejected does not mean that the null hypothesis $H_0 : \mu = 453$ is accepted, but only that the sample evidence is not sufficient to reject $H_0 : \mu = 453$ using $\alpha = 0.01$. Even though H_0 is not rejected, the 99% confidence interval gives the investigator valuable information.

A statistical test of the hypotheses

$$H_0 : \mu = \mu_0 \qquad \text{versus} \qquad H_a : \mu \neq \mu_0$$

is used extensively in statistical studies. However, this test is not too realistic. In Example 8.2.4, H_0 states $\mu = 453$. This means that μ is *exactly* equal to 453. It seems quite unrealistic to believe that the operator expects μ, the average net weight of 10,000 boxes of cereal, to be *exactly equal* to 453 grams (say, to several decimal places). But this is what H_0 states. It can be shown that if μ is not *exactly equal* to 453, the hypothesis H_0 can *always be rejected* if the sample size is large enough. What the investigator may really want to test in such a situation is whether μ is *approximately* equal to 453 and whether the difference between μ and 453 can be considered negligible for practical purposes. If so, instead of testing the hypotheses

$$H_0 : \mu = 453 \qquad \text{versus} \qquad H_a : \mu \neq 453,$$

it may be more appropriate to test

$$H_0 : \mu_1 \leq \mu \leq \mu_2 \qquad \text{versus} \qquad H_a : \mu \text{ is not between } \mu_1 \text{ and } \mu_2, \qquad \textbf{Line 8.2.7}$$

where μ_1 and μ_2 are specified by the investigator.

To conduct this test, choose a value for α and compute a confidence interval L to U for μ with a confidence coefficient equal to $1 - \alpha$. Reject H_0 in Line 8.2.7 if *either* (a) or (b) below is satisfied:

$$\text{(a)} \quad \mu_2 < L \qquad \text{or} \qquad \text{(b)} \quad U < \mu_1. \qquad \textbf{Line 8.2.8}$$

Do not reject H_0 if neither (a) nor (b) is satisfied. In this procedure the probability of erroneously rejecting H_0 when H_0 is true is no greater than α. The rule in Line 8.2.8 can be used for any parameter θ discussed in this book.

BOXES OF CEREAL. Suppose the operator in Example 8.2.4 is required to adjust the machine if μ is outside of the interval 452 to 454 grams, where μ is the average net weight of the 10,000 most recently packaged boxes. The operator selects $\alpha = 0.01$ and conducts a statistical test of the hypotheses

$$H_0 : 452 \leq \mu \leq 454 \qquad \text{versus} \qquad H_a : \mu \text{ is not in the interval 452 to 454.}$$

In this example, $\mu_1 = 452$ and $\mu_2 = 454$. Since $1 - \alpha = 1 - 0.01 = 0.99 = 99\%$, a confidence interval for μ with confidence coefficient equal to 99% is reported in Example 8.2.4 as $L = 448.6$ grams to $U = 453.8$ grams. From these values we observe that neither (a) nor (b) in Line 8.2.8 is satisfied, so H_0 is not rejected. The sample does not provide sufficient evidence (using $\alpha = 0.01$) to warrant adjustment of the machine.

P-values

In any study where a statistical test is conducted, the investigator decides on an appropriate value for α based on the practical consequence of rejecting H_0 when H_0 is true. Sometimes many individuals are interested in a study, and each individual may choose a different value for α. For this reason the results of any study should report the smallest value of α for which H_0 can be rejected. Each individual investigator can then use the results of the study to determine whether he or she would reject H_0.

DEFINITION 8.2.1: *P*-VALUE

The smallest value of α for which H_0 can be rejected in a statistical test is called the ***P*-value** (probability value). The P-value measures the amount of sample evidence in favor of rejecting H_0.

We illustrate with an example.

EXAMPLE 8.2.6

GASOLINE ADDITIVE. Based on a series of rigorous tests, it has been determined that the average miles per gallon (mpg) for highway driving for a population of cars of a certain model is 30 mpg. A small research group has developed a new gasoline additive, and it claims that the 30 mpg can be increased by more than 4 mpg if one pint of the additive is used for each 40 gallons of gasoline. The research group plans to contact several oil companies and ask them to market the additive. The group knows it will have to present statistical evidence to support its claim. Thus, the research group conducts a statistical test using the hypotheses

$$H_0 : \mu \leq 34 \qquad \text{versus} \qquad H_a : \mu > 34, \qquad \text{Line 8.2.9}$$

where μ is the average mpg for the population of cars if each car uses the additive. In order to support the research group's claim that $\mu > 34$, the null hypothesis must be rejected. The research group realizes that each oil company might select different values for α in order to evaluate the results of the study. Thus, the research group will report the smallest value of α (the P-value) for which H_0 will be rejected. An oil company can compare its chosen value of α with the computed P-value to determine if it would reject H_0. To conduct the study, 36 new cars of the model were obtained and driven over a specified test route using the additive. These 36 cars are considered to be a simple random sample from a conceptual population of cars of this model that use the additive. The data are in the file **additive.mtp** on the data disk.

The testing procedure in Box 8.2.3 requires a $1 - 2\alpha$ confidence interval for μ. The research group computed a $1 - 2\alpha$ confidence interval for μ for five values of α ($\alpha = 0.20, 0.10, 0.05, 0.01,$ and 0.005). The values of α, $\theta_0 = \mu_0$, and the $1 - 2\alpha$ confidence intervals are shown in Table 8.2.2. Using Line 8.2.4 in Box 8.2.3, H_0 is rejected if the confidence interval is completely to the right of $\theta_0 = \mu_0 = 34$ (that is, H_0 is rejected if $34 < L$). In order to determine the P-value, the research group must find the smallest value of α for which a confidence interval for μ with confidence coefficient equal to $1 - 2\alpha$ is completely to the right of $\mu_0 = \theta_0 = 34$.

TABLE 8.2.2 Results of Five Confidence Intervals for μ

α	μ_0	Confidence interval L to U	Decision
0.200	34	35.06 to 36.17	Reject H_0.
0.100	34	34.76 to 36.46	Reject H_0.
0.050	34	34.51 to 36.71	Reject H_0.
0.010	34	34.03 to 37.19	Reject H_0.
0.005	34	33.85 to 37.38	Do not reject H_0.

From Table 8.2.2, the P-value for this problem is a value between 0.010 and 0.005. If more computing is performed, it can be determined that the exact P-value is 0.0088.

Determining the P-value of a test by obtaining a table similar to Table 8.2.2 requires a large amount of computing. However, most computer software programs print the P-value when conducting a statistical test. To use Minitab to determine the P-value of the statistical test in Example 8.2.6, enter the following command and subcommand in the Session window. (Since a "t interval" is used to conduct a test involving μ, the test is called a "t test." See Section 6.3.)

```
MTB > ttest mu0 ci;
SUBC> alternative=+1.
```

Here `mu0` is the value of $\theta_0 = \mu_0$, and `i` is the column that contains the data. The subcommand `alternative` refers to the alternative hypothesis. If the alternative hypothesis H_a is $\mu > \mu_0$, use the subcommand `alternative=+1`. If H_a is $\mu < \mu_0$, use `alternative=-1`. If H_a is $\mu \neq \mu_0$, the subcommand `alternative` is omitted. In Example 8.2.6, $\mu_0 = 34$, and the alternative hypothesis is $H_a : \mu > 34$ (`alternative=+1`). To obtain the P-value using Minitab for Example 8.2.6, retrieve the data stored in column 2 in the file **additive.mtp** on the data disk and enter the following command and subcommand in the Session window (the output is shown in Exhibit 8.2.1):

```
MTB > ttest 34 c2;
SUBC> alternative=+1.
```

```
-----------------------------------------------------------------------------
T-Test of the Mean

Test of mu = 34.000 vs mu > 34.000

Variable     N      Mean     StDev    SE Mean        T     P-Value
MPG         36    35.611     3.879      0.646     2.49      0.0088
-----------------------------------------------------------------------------
```

EXHIBIT 8.2.1

The P-value to test the hypotheses in Line 8.2.9 is 0.0088. This number is shown in the column labeled `P-Value` in Exhibit 8.2.1. In the second line of Exhibit 8.2.1, Minitab reports

```
Test of mu = 34.000 vs mu > 34.000.
```

This means that the P-value in Exhibit 8.2.1 is the result of the test

$$H_0 : \mu = 34.0 \qquad \text{versus} \qquad \mu > 34.0.$$

However, as noted in the paragraph following Figure 8.2.1, the statistical test (and hence the P-value) is the same if the hypotheses are

$$H_0 : \mu = 34 \qquad \text{versus} \qquad H_a : \mu > 34,$$

$$H_0 : \mu \leq 34 \qquad \text{versus} \qquad H_a : \mu > 34,$$

or

$$H_0 : \mu < 34 \qquad \text{versus} \qquad H_a : \mu \geq 34.$$

How a *P*-Value Can Be Used When Conducting a Statistical Test

One use of a *P*-value when conducting a statistical test is to determine if H_0 should be rejected when a value for α has been specified. In Example 8.2.6 the *P*-value is 0.0088. This means that (before looking at the data) if an oil company had chosen α to be greater than 0.0088, the company would have rejected H_0. If another oil company had chosen α to be 0.0088 or smaller, this company would not have rejected H_0.

Suppose that three observers examine an investigation that involves a statistical test. *Before looking at the data or the P-value*, suppose they decide upon the risks they are willing to take in erroneously rejecting H_0 when H_0 is true. That is, each investigator selects a value for α.

Observer 1 states, "In this investigation, I will be satisfied with a 10% probability of erroneously rejecting H_0. Thus, I will conduct a statistical test using $\alpha = 0.10$."

Observer 2 states, "In this investigation, I will be satisfied with a 5% probability of erroneously rejecting H_0. Thus, I will conduct a statistical test using $\alpha = 0.05$."

Observer 3 states, "In this investigation, I will be satisfied with a 1% probability of erroneously rejecting H_0. Thus, I will conduct a statistical test using $\alpha = 0.01$."

Suppose that the three observers now examine the data and discover that the *P*-value = 0.03. Based on this *P*-value, observers 1 and 2 would reject H_0, but observer 3 would not reject H_0.

Based on this discussion, the *P*-value can be used as described in Box 8.2.5 to perform a statistical test.

BOX 8.2.5: USING A *P*-VALUE TO CONDUCT A STATISTICAL TEST

Before looking at the sample data or the *P*-value associated with a set of hypotheses, select a value of α.

1. If your selected value of α is greater than the computed *P*-value, you will reject H_0.

2. If your selected value of α is less than or equal to the computed *P*-value, you will *not* reject H_0.

The rule in Box 8.2.5 will give you exactly the same test results as the procedures described in Boxes 8.2.3 and 8.2.4. Note that the *P*-value is *not* the probability of erroneously rejecting H_0 when H_0 is true. The probability of erroneously rejecting H_0 when H_0 is true is equal to α, the value of which the investigator chooses before looking at the *P*-value.

Another Interpretation of a *P*-Value

In some studies an investigator may not know enough about the practical consequences of erroneously rejecting H_0 to objectively choose a value for α. However, the investigator may still want to determine what the data indicate about rejecting H_0. The *P*-value has an interpretation that can be used in this situation which we now present.

Remember that the letter P in P-value stands for *probability*. So, a P-value is a number between zero and one, and *the smaller the P-value, the greater is the disagreement between the sample data and the null hypothesis* (the "more unusual" the sample). Thus, if H_0 is true, the smaller the P-value, the "more unusual" the sample. For example, if H_0 is true, a sample that yields a P-value $= 0.01$ is "more unusual" than a sample that yields a P-value $= 0.05$. Suppose a sample is collected and a P-value computed. Further suppose that the investigator concludes that the computed P-value is overwhelmingly small. If H_0 is true, the probability of obtaining such a sample is so overwhelmingly small that one would conclude that H_0 is not true (H_0 would be rejected). This is similar to the situation in Example 8.1.5, where the probability of drawing a white ball from the bin on the shelf is so overwhelmingly small that an investigator would conclude that "such an unusual" sample (a white ball) did not come from the bin on the shelf. Hence, H_0 would be rejected. Many investigators consider a P-value of 0.05, 0.01, or 0.001 to be overwhelmingly small.

THINKING STATISTICALLY: SELF-TEST PROBLEMS

8.2.1 The research group of a company that manufactures light bulbs has developed a new long-lasting bulb. After extensive testing the research group claims that the average life of the new bulb is greater than 700 hours.

a. State a set of hypotheses, H_0 and H_a, that can be used in a statistical test to determine if the sample evidence supports this claim.

b. The research group obtained a simple random sample of 100 light bulbs produced during the past three months and measured the lifetime of each bulb. The data are stored in column 2 in the file **bulbs2.mtp** on your data disk. The average burning time for the bulbs in the sample is 716.53 hours, and the sample standard deviation is 18.81 hours. Use this information to test the hypotheses

$$H_0 : \mu \leq 700 \qquad \text{versus} \qquad H_a : \mu > 700$$

with $\alpha = 0.025$.

c. Use Minitab or another software program to find the P-value for a statistical test of

$$H_0 : \mu \leq 700 \qquad \text{versus} \qquad H_a : \mu > 700.$$

d. Suppose company policy requires the probability of erroneously rejecting H_0 when H_0 is true to be no greater than 0.01. What value of α must be specified? What is your conclusion using the P-value in part (c) with this value of α?

8.2.2 In a public health newsletter it was reported that 30% of hospital patients in the state do not have adequate health insurance. The director of the largest hospital in the state decides to conduct a statistical test to determine the amount of sample evidence to support $\pi > 0.30$, where π represents the proportion of patients who used the facilities in this hospital during the past 12 months that did not have adequate health insurance. The director chose $\alpha = 0.05$ and tested the hypotheses

$$H_0: \pi \leq 0.30 \qquad \text{versus} \qquad H_a: \pi > 0.30.$$

A simple random sample of 150 of last year's patients was contacted by mail and asked if their health insurance was adequate. Of the 150 patients contacted, 90 said their insurance was not adequate.

a. Estimate π, the proportion of patients who did not have adequate health insurance.

b. Use Box 8.2.3 to test the hypotheses with $\alpha = 0.05$. What is your conclusion?

c. Use $\alpha = 0.005$ and perform the test. What is your conclusion?

8.2.3 An investigator conducted a statistical test using $\alpha = 0.05$. The P-value associated with the hypotheses of interest was 0.15.

a. What is the investigator's conclusion?

b. What is the investigator's conclusion if the P-value was 0.02?

8.3 CHAPTER SUMMARY

KEY TERMS

1. statistical test
2. *P*-value
3. null hypothesis
4. alternative hypothesis

SYMBOLS

1. H_0—the null hypothesis
2. H_a—the alternative hypothesis
3. θ_0—a parameter value of interest specified by the investigator
4. α—probability of rejecting the null hypothesis when the null hypothesis is true
5. β—probability of not rejecting the null hypothesis when the null hypothesis is false

KEY CONCEPTS

1. Investigators are interested in statistical tests when they need to know if a population parameter is within specified limits.
2. The null hypothesis and the alternative hypothesis are complementary statements. Every possible value of a population parameter must be included in either the null hypothesis or the alternative hypothesis, and no value can be included in both hypotheses.
3. The statement "reject H_0" means the sample evidence is sufficient to convince the investigator that the null hypothesis is false.
4. The statement "do not reject H_0" means the sample evidence is not sufficient to convince the investigator that the null hypothesis is false.
5. The statement "do not reject H_0" does not mean the sample evidence provides convincing evidence that the null hypothesis is true.

6. Confidence intervals can be used to perform a statistical test.
7. P-values can be used to draw conclusions in a statistical test.
8. The P-value measures the amount of sample evidence in favor of rejecting the null hypothesis.

SKILLS

1. Be able to choose H_0 and H_a in a statistical test.
2. Perform a statistical test using a confidence interval.
3. Perform a statistical test using a P-value.

8.4 CHAPTER EXERCISES

8.1 This is a continuation of Example 8.1.4, where a company has applied to a bank for a loan to build a department store. The bank officials have decided to deny the loan unless the company can provide evidence that $\pi > 0.50$, where π is the proportion of households in the city with annual incomes exceeding $30,000.

a. State the set of hypotheses that must be tested in this problem.

b. A simple random sample of 200 households was obtained to test the hypotheses in part (a). In this sample of 200 households, 136 reported that last year's income exceeded $30,000. Using $\alpha = 0.05$, does this provide sufficient evidence to reject H_0?

8.2 A corporation owns a large apple orchard that is leased to an operator. The lease requires the operator to spray the orchard (at the operator's expense) for a certain leaf disease each spring. However, the requirement will be waived if the operator can provide evidence that π is less than 0.04, where π is the proportion of diseased trees in the orchard. To convince the corporation that π is less than 0.04, the operator conducted a statistical test.

a. Specify H_0 and H_a for this test.

b. The operator selected $\alpha = 0.05$ and conducted the test for the hypotheses stated in part (a). The test was based on a simple random sample of 200 trees selected from the orchard. The resulting P-value was 0.002. Is this sufficient evidence to reject H_0?

8.3 A new fertilizer for watermelons has been developed which is less expensive than the fertilizer commonly used by most farmers. A farmer has read an advertisement about the fertilizer that states it will increase the average weight of watermelons by at least 3 pounds. This sounds great to the farmer, but he still plans to use the standard fertilizer unless he can find sufficient evidence that the claim is correct. Last year the average weight of watermelons grown by the farmer using the standard fertilizer was 15.6 pounds. Growing conditions are expected to be the same this year as last year, and so the farmer decides to see if the claim is true by using the new fertilizer on a small plot of land.

a. State the set of hypotheses that the farmer should use in a statistical test. Let μ represent the average weight of the population of watermelons grown with the new fertilizer on the small plot of land.

b. The farmer selected 35 watermelons at random from the plot of land that was treated with the new fertilizer and weighed each watermelon. This sample was used by the farmer to test the hypotheses in part (a). The computed P-value $= 0.40$. If the farmer had selected $\alpha = 0.05$, what is the conclusion of this test?

8.4 A large state university must implement an employee incentive program for carpooling in order to satisfy a state-mandated environmental law. However, there is a waiver available if the university can present evidence that more than 30% of the employees presently carpool to work. Let π represent the proportion of university employees who presently carpool to work. The university must provide evidence that π is greater than 0.30.

a. State the null and alternative hypotheses that the university will use in an attempt to provide evidence that $\pi > 0.30$.

b. Suppose that a simple random sample of 150 employees was selected and it was determined that 54 of these employees carpooled to work. Use this sample and the procedure in Box 8.2.3 to test the hypotheses in part (a) using $\alpha = 0.05$.

8.5 A state environmental control agency believes that more than 35% of the cars licensed in the state are emitting at least 1.2% carbon monoxide. This level of carbon monoxide is too high for a healthy city environment. The agency has asked a state legislator to introduce a bill requiring all automobiles to be inspected for carbon monoxide once every two years. If an automobile emits more than 1.2% carbon monoxide, the car must be serviced before it is licensed. Before taking action, the legislator wants to be quite certain that the claim of the control agency is correct. A statistician working for the state was asked to collect a simple random sample of automobiles and conduct a statistical test to determine the amount of evidence to support the claim that more than 35% of the cars are emitting at least 1.2% carbon monoxide. If the evidence is sufficient to convince the legislator that the claim is correct, he will introduce the bill.

a. State the null and alternative hypotheses that the statistician should test. Let π represent the proportion of cars licensed in the state that are emitting at least 1.2% carbon monoxide.

b. The statistician obtained a simple random sample of 200 cars that were presently registered in the state and tested each car for the amount of carbon monoxide emitted. Of the 200 cars that were tested, 86 were found to emit more than 1.2% carbon monoxide. Test the hypotheses in part (a) using this sample with $\alpha = 0.025$.

8.6 A state legislator claimed that more than 60% of the registered voters in the state favored a proposition that would place constitutional limits on lawsuits and damage awards. A newspaper decided to hire a consulting agency to conduct a statistical test to determine if the claim is correct that $\pi > 0.60$, where π is the proportion of registered voters who favored the proposition.

a. State the null and alternative hypotheses that the consulting firm should test.

b. The consulting company selected a simple random sample of 200 registered voters and found that 162 of these voters favored the proposition. Conduct the statistical test described by the hypotheses in part (a) using $\alpha = 0.05$.

c. Is the P-value associated with the sample in part (b) greater than 0.05 or less than 0.05?

8.7 A medical company has developed a new cholesterol-lowering drug that it claims reduces heart attacks. Using the records of a large hospital, the company identified a population of men with high cholesterol. Based on past medical studies it is believed that about 21% of these men will have a nonfatal heart attack during the next two years if their condition goes untreated. The medical company states that less than 21% of these men will have a nonfatal heart attack during the next two years if they take the new drug. The director of the hospital decides to test the claim that this drug will reduce heart attacks. In particular, the director wants to determine the amount of evidence that supports the claim that the new drug will result in less than 21% of this population having nonfatal heart attacks.

a. What are the null and alternative hypotheses that should be used in a statistical test to examine this claim? Let π represent the proportion of men in the population who will have a nonfatal heart attack during the next two years if all the men take the new drug.

b. A simple random sample of 100 men from this population was obtained and given the new drug for two years. At the end of the two-year period, seven of these men had suffered a nonfatal heart attack. Use these data to conduct the statistical test in part (a) with $\alpha = 0.005$.

c. Is the P-value associated with the sample in part (b) greater than 0.005 or less than 0.005?

8.8 An office supply store offers a payment period of 30 days to its credit customers but has historically averaged a collection period of 50 days. This collection period is about 10 days greater than the standard recommended by the store's accountant. The accountant suggests the average collection period could be shortened by enclosing a stamped return envelope in the monthly statements mailed to each customer. The manager of the store thinks this idea has merit, but before implementing this new procedure he wants to know if enclosing a stamped return envelope in the monthly statements will shorten the average collection period by at least 10 days. In order to answer this question, the manager will include stamped return envelopes in a simple random sample of 36 customer billing statements next month. Based on this sample, he will determine the amount of evidence to support $\mu < 40$, where μ is the average collection period if all credit customers are sent stamped return envelopes with their billing statements.

a. State the null and alternative hypotheses that the manager needs to test.

b. The simple random sample of 36 customers had a sample average payment period of 37.10 days. The sample standard deviation was 6.14 days. Use this information to test the hypotheses in part (a) with $\alpha = 0.05$. Does it appear that enclosing stamped return envelopes will reduce the average collection period to less than 40 days?

8.9 A state representative wants to know the median salary for the faculty at a state university in her district. She recently read a newspaper article that stated the median salary for faculty in state universities is $45,000. At present, the representative does not plan to support any bill that would increase faculty salaries at the university. However, if she can be convinced that the median salary of the faculty is less than $45,000, she will support such a bill. The representative asked an intern to select a simple random sample of 25 faculty from the list of faculty in the university catalog and use these salaries to conduct a statistical test using $\alpha = 0.025$.

a. State the hypotheses that should be tested in order for the representative to determine if she will support a faculty salary increase. Let M represent the median salary for the population of faculty at the university.

b. The median salary for the sample of 25 faculty is $44,000, and the 95% confidence interval is $L =$ $40,000 to $U =$ $46,000. Use this information to test the hypotheses in part (a). Will the representative support a faculty salary increase?

8.10 The purchasing agent of a manufacturing firm has been purchasing light bulbs from a particular supplier for the past 10 years. A representative of a competing supplier has recently presented the agent an offer to sell her a similar bulb from his company for a slightly lower price. The purchasing agent prefers to stay with the present supplier but will change to the new supplier if she is convinced (by using a statistical test) that at least one-half of the bulbs to be purchased will last over 2,050 hours. The bulb presently in use has a median life of 2,000 hours, and unless the new bulb has better performance, it will not be worth the trouble to change suppliers. The purchasing officer obtained a simple random sample of 20 bulbs from the new supplier and used them until they each burned out. The lifetimes of the 20 bulbs, ordered from least to greatest, are as follows:

2038	2058	2061	2078	2079
2081	2104	2111	2113	2117
2118	2119	2127	2129	2133
2151	2156	2194	2209	2222

a. State the null and alternative hypotheses that the purchasing agent should use in the statistical test. Let M represent the median lifetime for the population of bulbs sold by the potential new supplier.

b. Compute a confidence interval for M to test the hypotheses in part (a) with $\alpha = 0.025$.

8.11 A company sells frozen orange juice in a can labeled "10 fluid ounces." The machine that places the juice in the cans must be monitored to ensure it is not putting either too much or too little juice in the cans. The monitoring process is performed by selecting a simple random sample of 40 cans every hour and measuring the amount of juice in each can.

a. Construct a null hypothesis and an alternative hypothesis that might be used in a statistical test to monitor the machine in a given hour. Let μ represent the average amount of juice for the population of cans filled in one hour.

b. A simple random sample of 40 cans has an average of 10.02 fluid ounces, with a sample standard deviation of 0.06 fluid ounces. Use this sample to test the hypotheses stated in part (a) using $\alpha = 0.01$. Does it appear the machine is operating correctly?

c. Is the P-value associated with the test in part (b) greater than 0.01 or less than 0.01?

8.12 The manufacturing of circuit boards involves a process in which gold plating is placed on each board. During this process, measurements of the gold thickness are obtained for a sample of circuit boards in order to determine if the boards have the required level of gold plating. A simple random sample of 40 boards is selected from the population of boards manufactured on a given day. The thickness of gold plating is measured in microinches for each sampled board using a method called X-ray fluorescence. A microinch is one millionth of an inch. The machine that places the plating on the boards must be adjusted if there is evidence that the average thickness for the population of boards is not equal to 2.0 microinches.

a. State the hypotheses used to monitor this process using a statistical test. Let μ represent the average thickness of the gold plating for the population of circuit boards manufactured on one day.

b. A simple random sample of 40 boards has a sample average gold thickness of 2.20 microinches, with a sample standard deviation of 0.43 microinches. Use this sample with $\alpha = 0.01$ to test the hypotheses in part (a). Does the machine appear to be operating properly?

c. In addition to monitoring the average gold thickness, it is necessary to monitor the standard deviation of the gold thickness. In particular, the machine is operating properly if the standard deviation is less than 0.4 microinches. State the hypotheses that should be tested in order to monitor this condition. Use σ to represent the population standard deviation for the population of circuit boards manufactured in one day.

d. Use the sample described in part (b) to test the hypotheses in part (c). Assume the population of gold thickness values is *normal* and use $\alpha = 0.05$. Does the machine appear to be operating properly with regard to the standard deviation of the gold thickness?

8.5 SOLUTIONS FOR SELF-TEST PROBLEMS IN CHAPTER 8

8.1.1

Example (1). You want to go to the movies and a ticket costs $5.50. You will be able to attend the movie only if you can borrow at least $5.50 from a friend. You don't need to know how much money you can borrow but only if you can borrow at least $5.50.

Example (2). Suppose you want to drive your car to a football game that is 20 miles from your home. You don't need to know exactly how much gas is in your car, but you do need to know if there is enough gas to drive at least 40 miles to the game and back.

Example (3). A citizen's group wants to know if more than 50% of the registered voters in a state favor a proposition that would place constitutional limits on lawsuits and damage awards. If this is the case, then the group will attempt to place this proposition on next year's ballot. The group is not primarily interested in the actual proportion of registered voters who favor the proposition, but only if more than 50% favor the proposition.

Example (4). A company that manufactures light bulbs advertises that 98% of the light bulbs will last more than 700 hours. A consumer protection agency is not interested in the actual number of hours each bulb lasts but only if at least 98% of the bulbs last over 700 hours.

Example (5). Some of the bottles of aspirin a company manufactures contain broken pills. The company vice president decides that if more than 2% of the bottles manufactured last month contain broken pills, this batch will be sold at a reduced price under a different brand name. The vice president is not primarily interested in knowing the proportion of bottles that contain broken pills but only if the proportion is greater than 0.02.

8.1.2

a. The sample will be used to determine if the auditor's claim that $\pi > 0.10$ is true. Following guideline (2) in Box 8.1.2, this claim is chosen as the alternative hypothesis H_a. A statistical test is conducted using the hypotheses

$$H_0 : \pi \leq 0.10 \qquad \text{versus} \qquad H_a : \pi > 0.10.$$

b. The standard course of action is to advertise in the college newspapers. According to guideline (1) in Box 8.1.2, the null hypothesis consistent with this standard course of action is $H_0 : \pi \geq 0.20$. The alternative hypothesis consistent with the alternative course of action (direct mailing) is $H_a : \pi < 0.20$. Thus, a statistical test is conducted using the hypotheses

$$H_0 : \pi \geq 0.20 \qquad \text{versus} \qquad H_a : \pi < 0.20.$$

8.1.3 As stated in Box 8.1.1, every possible value of π must be included in either H_0 or H_a, and no value can be included in both hypotheses. The values of π in the interval $0.10 \leq \pi < 0.15$ are included in both H_0 and H_a.

8.1.4 The values of π in the interval $0.08 \leq \pi < 0.10$ do not appear in either H_0 or H_a.

8.1.5 No. The supervisor wants to determine the sample evidence in favor of rejecting $\pi \geq 0.10$. If the sample provides sufficient evidence, the less expensive spray will be used to spray the orchard. Hence, $\pi \geq 0.10$ is chosen as the null hypothesis.

8.2.1

a. The research group claims that μ is greater than 700 hours. Following guideline (2) in Box 8.1.2 the claim is chosen as H_a. Thus, the hypotheses are

$$H_0 : \mu \leq 700 \qquad \text{versus} \qquad H_a : \mu > 700.$$

b. The hypotheses are stated in Line 8.2.1, where $\theta = \mu$ and $\theta_0 = \mu_0 = 700$. Using Line 8.2.4, H_0 is rejected if $700 < L$. Using the confidence interval for μ in Box 6.3.1, we compute L with a confidence coefficient equal to $1 - 2\alpha = 1 - 2(0.025) = 0.95$. We get

$$L = 716.53 - 1.98\left(\frac{18.81}{\sqrt{100}}\right) = 712.80$$

and

$$700 < L = 712.80.$$

So H_0 is rejected using $\alpha = 0.025$, and $H_a : \mu > 700$ is accepted.

c. To compute the P-value associated with this test, use the Minitab command and subcommand

```
MTB > ttest 700  c2;
SUBC>  alternative=+1.
```

The output from this command is

```
T-Test of the Mean

Test of mu = 700.00 vs mu > 700.00

Variable    N    Mean StDev SE Mean      T P-Value
Lifetime 100 716.53 18.81     1.88 8.78   0.0000
```

The P-value is 0.0000 (to four-decimal-place accuracy).

d. The specified value for α is 0.01. The P-value $= 0.0000$ to four decimal places; since this is less than $\alpha = 0.01$, H_0 is rejected. It is concluded that $\mu > 700$.

8.2.2

a. $\widehat{\pi} = 90/150 = 0.60$.

b. With $\alpha = 0.05$, the confidence coefficient is $1 - 2(0.05) = 0.90 = 90\%$, and the table value is $T1 = 1.65$. Using the formula in Box 3.2.1, we obtain $L = 0.53$. The value of $\theta_0 = \pi_0$ is 0.30, and

$$0.30 < L = 0.53.$$

From this we observe that the interval is completely to the right of 0.30, and hence H_0 is rejected. We conclude that $\pi > 0.30$.

c. With $\alpha = 0.005$, the confidence coefficient is $1 - 2(0.005) = 0.99 = 99\%$, and the table value is $T1 = 2.58$. Using the formula in Box 3.2.1, we obtain $L = 0.49$. The value of $\theta_0 = \pi_0$ is 0.30, and

$$0.30 < L = 0.49.$$

From this we observe that the interval is completely to the right of 0.30, and hence H_0 is rejected. We conclude that $\pi > 0.30$. This is the same conclusion reached when $\alpha = 0.05$.

8.2.3

a. Since α is less than the P-value, H_0 is not rejected, in accordance with Box 8.2.5.

b. Since α is greater than the P-value, H_0 is rejected, in accordance with Box 8.2.5.

8.6 APPENDIX 1: AN OUTLINE OF A PROOF FOR THE PROCEDURE IN BOX 8.2.3

In this appendix we show that if Box 8.2.3 is used to test the hypotheses in Line 8.2.1 or 8.2.2, the probability of erroneously rejecting H_0 when H_0 is true is less than or equal to α. First we discuss one-sided, two-sided, and equal-tailed confidence intervals.

One-sided, Two-sided, and Equal-tailed Confidence Intervals

If you look closely at the testing procedures in Box 8.2.3, you will see that each decision is made using either L or U, but not both. For instance, the upper endpoint U is not used in testing the hypotheses in Line 8.2.1. Similarly, the lower endpoint L is not used in testing the hypotheses in Line 8.2.2. The confidence intervals $L \leq \theta$ and $\theta \leq U$ are called **one-sided confidence intervals**. The upper endpoint for the one-sided confidence interval $L \leq \theta$ is understood to be the maximum possible value for θ. Similarly, the lower endpoint for the one-sided confidence interval $\theta \leq U$ is understood to be the minimum possible value for θ. For example, suppose θ is the population proportion π. Then

$$L \leq \pi \quad \text{is equivalent to} \quad L \leq \pi \leq 1$$

and

$$\pi \leq U \quad \text{is equivalent to} \quad 0 \leq \pi \leq U.$$

The confidence intervals in this book, which we write L to U (or sometimes $L \leq \theta \leq U$), are called **two-sided confidence intervals** because each endpoint is computed using sample data.

Let L to U represent a two-sided confidence interval for θ with a confidence coefficient equal to $1 - 2\alpha$. In your mind's eye, suppose you obtain every possible sample of size n from a population. For each of these samples you compute L and U, the endpoints of a two-sided confidence interval for θ with confidence coefficient equal to $1 - 2\alpha$. In this set of all possible confidence intervals with confidence coefficient $1 - 2\alpha$, a proportion $1 - 2\alpha$ contains θ and a proportion $1 - (1 - 2\alpha) = 2\alpha$ does not contain θ. Suppose that one-half of all the confidence intervals that do not contain θ (a proportion α of all possible confidence intervals) is completely to the left of θ and one-half of all the confidence intervals that do not contain θ (a proportion α of all possible confidence intervals) is completely to the right of θ. Two-sided confidence intervals with this property are called **equal-tailed confidence intervals**. Figure 8.6.1 displays 100 confidence intervals for θ computed from 100 simple random samples drawn from a population. Each confidence interval has a confidence coefficient of $1 - 2\alpha = 0.90$. Ten of the intervals do not contain θ. Of these 10 intervals, 5 have values that are all greater than θ and the other 5 have values that are all less than θ. The 10 intervals that do not contain θ are denoted with asterisks in Figure 8.6.1.

All formulas for two-sided confidence intervals in this book provide equal-tailed confidence intervals. Box 8.6.1 describes how a one-sided confidence interval can be obtained from an equal-tailed confidence interval.

BOX 8.6.1: COMPUTING A ONE-SIDED CONFIDENCE INTERVAL

1. Suppose the confidence interval L to U is an equal-tailed confidence interval for θ with confidence coefficient equal to $1 - 2\alpha$.

2. The lower endpoint of this interval, L, forms a one-sided confidence interval $L \leq \theta$ with confidence coefficient $1 - \alpha$.
3. The upper endpoint of this interval, U, forms a one-sided confidence interval $\theta \leq U$ with confidence coefficient $1 - \alpha$.

The phrase "confidence interval" will always mean "two-sided confidence interval (that is equal tailed)" unless we specifically say "one-sided confidence interval" or specifically say it is not equal tailed.

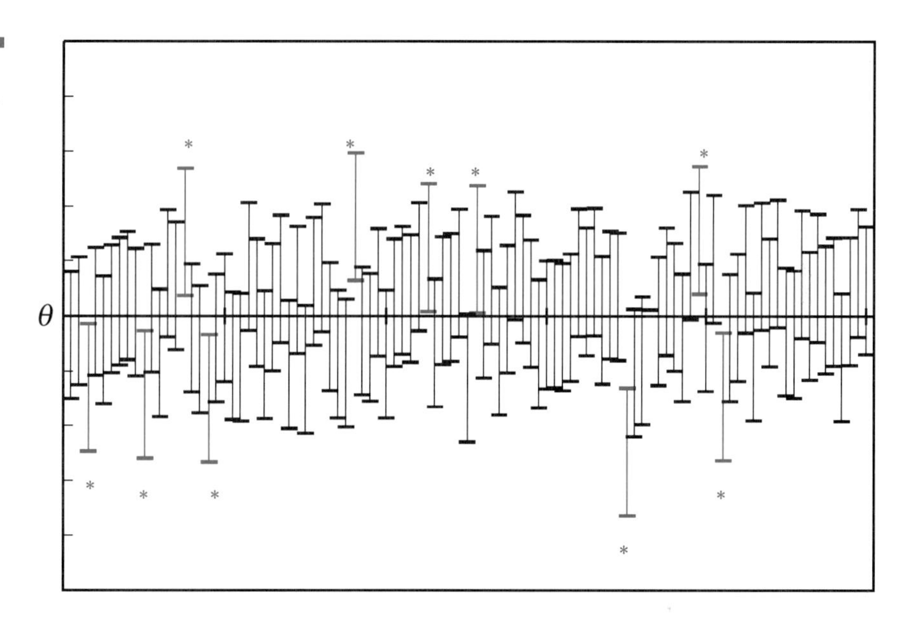

FIGURE 8.6.1
100 90% Equal-tailed Confidence Intervals for θ Where 5% of Lower Bounds Exceed θ and 5% of Upper Bounds Are Less Than θ

Suppose the procedure in Box 8.2.3 is used to test the hypotheses in Line 8.2.1, namely,

$$H_0 : \theta \leq \theta_0 \qquad \text{versus} \qquad H_a : \theta > \theta_0$$

for a given value of α. In your mind's eye, imagine obtaining every possible simple random sample of size n from a population of size N. Suppose there are k such samples, and you compute a $1 - 2\alpha$ (equal-tailed) confidence interval for θ for each sample. These k confidence intervals are denoted by

$$[L_1, U_1],\ [L_2, U_2],\ [L_3, U_3], \ldots, [L_k, U_k].$$

Since these are equal-tailed confidence intervals with confidence coefficient equal to $1 - 2\alpha$, a proportion α of the k intervals is completely to the right of θ. This means a proportion α of the k intervals has a lower endpoint greater than θ, and a proportion $1 - \alpha$ has a lower endpoint less than or equal to θ. We write this statement as

$$1 - \alpha = Pr(L \leq \theta).$$

If H_0 is true, this means that $\theta \leq \theta_0$ and hence

$$1 - \alpha = Pr(L \leq \theta) \leq Pr(L \leq \theta_0).$$

Thus, if H_0 is true,

$$1 - \alpha \le Pr(L \le \theta_0). \qquad \textbf{Line 8.6.1}$$

This states that if H_0 is true, *at least* $1 - \alpha$ of the k confidence intervals have a lower endpoint L that is less than or equal to θ_0. Using Line 8.6.1, we perform some algebra and obtain

$$1 - Pr(L \le \theta_0) \le \alpha \quad \text{if } H_0 \text{ is true.} \qquad \textbf{Line 8.6.2}$$

Using Line 6.2.4, we can write $1 - Pr(L \le \theta_0) = Pr(\theta_0 < L)$ and rewrite Line 8.6.2 as

$$Pr(\theta_0 < L) \le \alpha \quad \text{if } H_0 \text{ is true.} \qquad \textbf{Line 8.6.3}$$

By Line 8.2.4 we reject $H_0 : \theta \le \theta_0$ if $\theta_0 < L$. Thus, Line 8.6.3 states that the probability of rejecting H_0 when H_0 is true is less than or equal to α. A similar argument can be made for testing the hypotheses in Line 8.2.2. Thus, if we use a $1 - 2\alpha$ equal-tailed confidence interval for testing the hypotheses in either Line 8.2.1 or Line 8.2.2, the probability of erroneously rejecting H_0 when H_0 is true is less than or equal to α.

8.7 APPENDIX 2: USING TEST STATISTICS FOR TESTING HYPOTHESES

In Section 8.2 we used confidence intervals to conduct statistical tests. An alternative approach that gives identical results is based on what is called a **test statistic**. In Box 8.2.4 we explained how to use confidence intervals to conduct a statistical test of the hypotheses

$$H_0 : \theta = \theta_0 \qquad \text{versus} \qquad H_a : \theta \ne \theta_0,$$

where θ is any population parameter. For convenience, we describe this procedure again in Box 8.7.1 using $\theta = \mu$, where μ is the mean of a *normal* population.

BOX 8.7.1: USING A CONFIDENCE INTERVAL TO CONDUCT A STATISTICAL TEST FOR THE HYPOTHESES $H_0 : \mu = \mu_0$ VERSUS $H_a : \mu \ne \mu_0$

Let $Y_1, Y_2, Y_3, \ldots, Y_N$ denote a *normal* population with mean μ and standard deviation σ. To test the hypotheses shown in Line 8.2.3, namely

$$H_0 : \mu = \mu_0 \qquad \text{versus} \qquad H_a : \mu \ne \mu_0, \qquad \textbf{Line 8.7.1}$$

where μ_0 is specified by the investigator, choose a value for α and select a simple random sample from the *normal* population. Use the sample data to compute L and U, the endpoints of a confidence interval for μ with a confidence coefficient equal to $1 - \alpha$. Formulas for L and U as shown in Box 6.3.1 are

$$L = \hat{\mu} - (T2)\left(\frac{\hat{\sigma}}{\sqrt{n}}\right) \quad \text{and} \quad U = \hat{\mu} + (T2)\left(\frac{\hat{\sigma}}{\sqrt{n}}\right).$$

Reject H_0 if the confidence interval does not contain μ_0. That is, reject H_0 and conclude that $\mu \neq \mu_0$ if

$$\mu_0 < L \qquad \text{or} \qquad U < \mu_0.$$

Do not reject H_0 if the interval contains μ_0. That is, do not reject H_0 if

$$L \leq \mu_0 \leq U.$$

We now define a test statistic and show how it can be used to test the hypotheses in Line 8.7.1.

BOX 8.7.2: USING A TEST STATISTIC TO CONDUCT A STATISTICAL TEST FOR THE HYPOTHESES IN LINE 8.7.1

Let $Y_1, Y_2, Y_3, \ldots, Y_N$ denote a *normal* population with mean μ and standard deviation σ. To test the hypotheses shown in Line 8.7.1, choose a value for α and select a simple random sample of size n from the *normal* population. Use the sample to compute $\widehat{\mu}$, $\widehat{\sigma}$, and t, where

$$t = \frac{\sqrt{n}(\widehat{\mu} - \mu_0)}{\widehat{\sigma}}. \qquad \textbf{Line 8.7.2}$$

The quantity in Line 8.7.2 is called a **test statistic** for testing the hypotheses in Line 8.7.1.

The null hypothesis H_0 is rejected if the value of t is either less than $-T2$ or larger than $T2$. That is, reject H_0 if

$$t < -T2 \qquad \text{or} \qquad t > T2,$$

where $T2$ is obtained from Table T-2 using a confidence coefficient equal to $1 - \alpha$ and $DF = n - 1$.

The procedure in Box 8.7.2 gives the same result as the confidence interval procedure described in Box 8.7.1. In Box 8.7.3 we discuss the test statistic in Box 8.7.2 in more detail.

BOX 8.7.3: TEST STATISTIC

In your mind's eye, envision *every* possible simple random sample of size n ($n \geq 2$) that can be obtained from a *normal* population with mean μ and standard deviation σ, both unknown. For each of these samples envision that $\widehat{\mu}$, $\widehat{\sigma}$, and t are computed, where

$$t = \frac{\sqrt{n}(\widehat{\mu} - \mu_0)}{\widehat{\sigma}} \qquad \textbf{Line 8.7.3}$$

and μ_0 is specified by the investigator. If the null hypothesis in Line 8.7.1 is true (that is, if $\mu = \mu_0$), the collection of all possible values of t in Line 8.7.3 forms a population called a **t population with $n - 1$ degrees of freedom**.

The relative frequency histogram of a t population is symmetric about zero, and it has a slightly different shape for each different value of the degrees of freedom $n-1$. Figure 6.10.1 shows the curves representing the relative frequency histograms of two t populations, one with 30 degrees of freedom and the other with 3 degrees of freedom. When the degrees of freedom DF is large (say, greater than 30), the relative frequency histogram of the t population is quite similar in appearance to the relative frequency histogram of a *normal* population (a bell curve) with mean 0 and standard deviation 1.

For any specified value for DF, the confidence coefficient in Table T-2 represents the proportion of values in the t population that are in the interval $-T2$ to $T2$. This is represented as the nonshaded area of Figure 8.7.1. Because the t population is symmetric, the proportion of population values larger than $T2$ is equal to the proportion of values smaller than $-T2$.

FIGURE 8.7.1
Proportion of t Population Values Less Than $-T2$ Are Equal to Proportion of t Population Values Greater Than $T2$

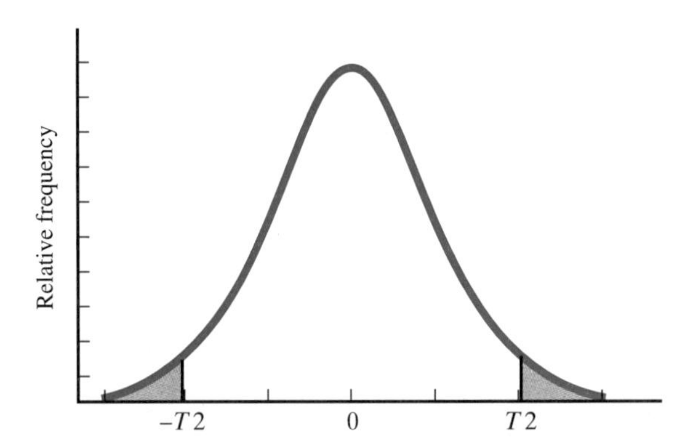

Suppose an investigator selects a simple random sample of size n from a *normal* population and computes $\hat{\mu}$, $\hat{\sigma}$, and t, where

$$t = \frac{\sqrt{n}(\hat{\mu} - \mu_0)}{\hat{\sigma}}.$$

If $H_0 : \mu = \mu_0$ is true, this is equivalent to selecting a value of t at random from the t population with $n-1$ degrees of freedom. The *probability* that the value to be selected will be either smaller than $-T2$ or larger than $T2$ is equal to the *proportion* of values of the t population that are smaller than $-T2$ or larger than $T2$ (the shaded portion of the population shown in Figure 8.7.1). Suppose the proportion of the t population that is between $-T2$ and $T2$ is equal to $1-\alpha$ (this is the nonshaded portion of the relative frequency histogram of the t population shown in Figure 8.7.1). Since the t population is symmetric, the proportion of population values that are smaller than $-T2$ is equal to the proportion that are larger than $T2$. Thus each shaded portion of the relative frequency histogram in Figure 8.7.1 contains a proportion $\alpha/2$ of the population values. The two shaded portions of the t population form the **critical region** of the test. This means that if the computed value of t is in the critical region (is smaller than $-T2$ or larger than $T2$), the null hypothesis is rejected. If $H_0 : \mu = \mu_0$ is true, then the probability that the t value to be selected will be in the critical region is equal to α (since the critical region contains a proportion α of the population values). Thus, if H_0 is true, the probability that H_0 will be rejected is equal to α. So, the probability of making the mistake of rejecting H_0 (when H_0 is true) is equal to α. H_0 will be rejected if either

$$(1)\ t < -T2 \qquad \text{or} \qquad (2)\ t > T2. \qquad \textbf{Line 8.7.4}$$

If we substitute the value of t from Line 8.7.3 into (1) and (2) of Line 8.7.4, we get

$$(1)\ \frac{\sqrt{n}(\widehat{\mu} - \mu_0)}{\widehat{\sigma}} < -T2 \qquad \text{or} \qquad (2)\ \frac{\sqrt{n}(\widehat{\mu} - \mu_0)}{\widehat{\sigma}} > T2. \qquad \textbf{Line 8.7.5}$$

If we solve (1) and (2) for μ_0, we get

$$(1)\ \mu_0 > \widehat{\mu} + T2\left(\frac{\widehat{\sigma}}{\sqrt{n}}\right) \qquad \text{or} \qquad (2)\ \mu_0 < \widehat{\mu} - T2\left(\frac{\widehat{\sigma}}{\sqrt{n}}\right). \qquad \textbf{Line 8.7.6}$$

By Box 8.7.1, the endpoints of a confidence interval for μ are

$$L = \widehat{\mu} - (T2)\left(\frac{\widehat{\sigma}}{\sqrt{n}}\right) \qquad \text{and} \qquad U = \widehat{\mu} + (T2)\left(\frac{\widehat{\sigma}}{\sqrt{n}}\right).$$

So by Line 8.7.6, H_0 is rejected if $\mu_0 > U$ or $\mu_0 < L$. This is exactly the rule given in Box 8.7.1 for rejecting H_0. Thus, using the test statistic in Line 8.7.3 for testing the hypotheses in Line 8.7.1 gives the same result as using the confidence interval procedure in Box 8.2.4.

By examining the test statistic shown in Line 8.7.3, you can see that if H_0 is true (if $\mu = \mu_0$), we would expect the estimate $\widehat{\mu}$ to be somewhat "close" to μ_0 and the value of t would be somewhat close to zero, the center of the t population. However, if μ is not equal to μ_0, we would expect $\widehat{\mu}$ to be somewhat different than μ_0 (either somewhat smaller or somewhat larger). This means that the value of the test statistic would be small (less than $-T2$) or the value of the test statistic would be large (larger than $T2$). If the value of the test statistic is in the shaded portion of Figure 8.7.1, this tells an investigator that $\widehat{\mu}$ is unusually smaller or unusually larger than μ_0. The investigator concludes that this unusual value occurs because $\mu \neq \mu_0$, and so H_0 is rejected.

Test statistics can be used in a similar manner to conduct any of the statistical tests discussed previously using confidence intervals.

CHAPTER 9

SIMPLE LINEAR REGRESSION

9.1 INTRODUCTION

In earlier chapters we have been interested in questions relating to a single variable. Specifically, we have shown how to summarize and display data sets (Chapter 4) and how to make inferences about proportions, medians, means, and standard deviations (Chapters 3, 6, and 7). While these are useful and fundamental tools for making decisions, they form only a small part of what statistics has to offer. Real problems often involve several variables, and investigators may be interested in the interrelationships among them. A question such as, "Is there an association between blood pressure and the amount of salt intake?" is important to answer if one wants to make appropriate lifestyle choices. Public-policy decisions concerning the environment require answers to questions such as, "Is there any association between the level of pollution in the Grand Canyon National Park and the number of visitors who drive their cars in the park each year?" Such questions have led statisticians to develop many useful tools for making inferences about associations among variables based on sample data. In this chapter we focus on statistical methods for examining the association between *two quantitative* variables.

One purpose for investigating the association between two variables is to determine if one variable is useful for *predicting* the other. For example, a physician might be interested in predicting how much a patient's blood pressure can be reduced, based on the amount of an administered drug. A company that manufactures cement blocks may be interested in predicting the strength of the blocks, based on the amount of an administered drug. A company that manufactures cement blocks may be interested in predicting the strength of the blocks, based on the amount of sand used in the mixture. A company that owns a coal-burning power plant may be interested in predicting the amount of sulfur dioxide present in a national park 25 miles away, based on the amount of sulfur dioxide emitted at the plant. You may be interested in predicting the number of pizzas you will need for a party, based on the number of guests who will attend. You can think of many other situations in business, science, and everyday activities where you might want to predict the value of one variable using the value of another variable. We now present several examples to introduce the type of problems discussed in this chapter.

EXAMPLE 9.1.1

SCHOLARSHIP. An educational foundation awards scholarships to high-school graduates to help them with expenses during their first year at a major university. The foundation would like to consider a student for a scholarship only if that student will earn a grade point average (GPA) of at least 2.80 (on a 4.0-point scale) the first year at the university. Since the scholarship is awarded *before* the student enters the university, the first-year GPA must be *predicted*. Each applicant for a scholarship must take an achievement test, the result of which is used to predict his or her first-year GPA.

The director of the foundation has access to the records of the 2,000 students who applied for scholarships during the past five years and completed their first year of college. A simple random sample of 60 students was selected from these records. Their scores on the achievement test and their first-year GPAs are shown in Table 9.1.1.

TABLE 9.1.1 First-year GPA and Achievement Test Score for a Simple Random Sample of 60 Students

Student number	GPA	Score	Student number	GPA	Score	Student number	GPA	Score
1	2.92	0.71	21	3.45	0.86	41	3.19	0.83
2	3.20	0.74	22	3.59	0.88	42	3.09	0.76
3	2.93	0.74	23	2.86	0.71	43	3.16	0.77
4	3.00	0.75	24	3.11	0.76	44	3.41	0.85
5	3.66	0.91	25	3.26	0.81	45	2.94	0.72
6	3.28	0.78	26	2.87	0.72	46	2.73	0.68
7	3.19	0.80	27	3.19	0.79	47	3.07	0.76
8	3.39	0.82	28	3.21	0.78	48	3.25	0.79
9	3.17	0.78	29	3.49	0.83	49	3.39	0.83
10	3.47	0.86	30	3.05	0.76	50	3.07	0.75
11	2.98	0.74	31	3.49	0.86	51	2.78	0.71
12	3.26	0.78	32	3.23	0.80	52	2.82	0.70
13	3.14	0.78	33	3.18	0.79	53	2.47	0.63
14	2.64	0.64	34	3.30	0.83	54	3.63	0.88
15	3.31	0.78	35	3.33	0.85	55	3.15	0.77
16	3.35	0.81	36	3.26	0.75	56	3.24	0.80
17	2.92	0.72	37	3.07	0.76	57	3.19	0.79
18	3.17	0.79	38	2.90	0.72	58	2.54	0.61
19	3.05	0.74	39	3.13	0.78	59	3.06	0.75
20	3.18	0.77	40	3.23	0.78	60	3.08	0.72

Figure 9.1.1 contains a scatter plot of the data in Table 9.1.1.

FIGURE 9.1.1
Scatter Plot of GPA Versus Score on an Achievement Test for a Sample of 60 Students

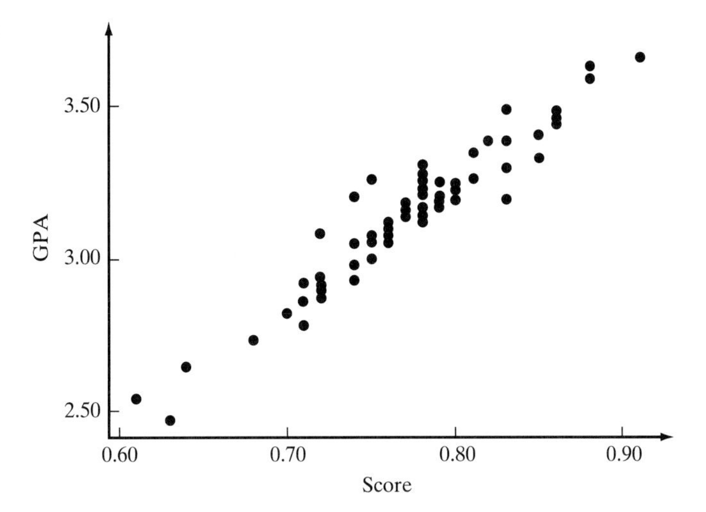

The scatter plot appears to show an association between achievement-test score and first-year GPA. In particular, students with high achievement-test scores tend to have high first-year GPAs, and students with low achievement-test scores tend to have low first-year GPAs.

The variable "Score" is called the **predictor variable** and is typically denoted by the letter X. It is sometimes referred to as the **predictor factor**, the **independent variable**, or the **explanatory variable**. The variable "GPA" is called the **predicted variable** and is typically denoted by the letter Y. The predicted variable is also referred to as the **response variable** or the **dependent variable**. It is desired to predict Y using the value of X. Since the first-year GPA must be predicted in Example 9.1.1, the foundation's director needs a rule or formula (mathematical equation) that can be used to make this prediction. A **prediction function** that can be used for this purpose is defined in Box 9.1.1.

BOX 9.1.1: PREDICTION FUNCTION

A formula that is used to predict the unknown value of the predicted variable Y using the known value of a predictor variable X is called a **prediction function** and is denoted by

$$\widehat{Y}(X).$$

Throughout this book, the symbol $\widehat{\ }$ (hat) is used to indicate an estimate. The reason we use the hat symbol in $\widehat{Y}(X)$ is explained later. Sample data are used to obtain the prediction function $\widehat{Y}(X)$; in Section 9.3 we show the formula used for computing $\widehat{Y}(X)$. Suppose the prediction function in Example 9.1.1 is

$$\widehat{Y}(X) = 0.10 + 3.94X \quad \text{"for } X \text{ from 0.60 to 1.00."} \qquad \textbf{Line 9.1.1}$$

Using this function, the foundation's director predicts that an applicant who received a score of 85% on the achievement test (that is, $X = 0.85$) will earn a first-year GPA of

$$\widehat{Y}(0.85) = 0.10 + 3.94(0.85) = 3.45.$$

An applicant who received a score of 75% on the achievement test is predicted to earn a first-year GPA of

$$\widehat{Y}(0.75) = 0.10 + 3.94(0.75) = 3.06.$$

Based on these predictions and other information, the director of the foundation will decide which applicants receive scholarships. The prediction function $\widehat{Y}(X)$ in Line 9.1.1 includes the statement

"for X from 0.60 to 1.00."

This statement tells us that the formula in Line 9.1.1 can only be used to predict Y for values of X between 0.60 and 1.00, inclusive. If these values of X are apparent from the discussion, this statement is sometimes omitted.

Prediction functions are often used for *forecasting* a future value of Y. However, a prediction function can be used for other purposes, as illustrated in the next two examples.

EXAMPLE 9.1.2

HELICAL SPRING. Helical and spiral springs have long been used for storing energy that can be released to power various devices. One example is the old-fashioned mechanical clock, where a key is used to wind a helical spring. The energy stored in the spring is released in a controlled manner to move the hands of the clock. The amount of energy that can be stored in a "compressed" helical spring depends on a physical property of the spring, referred to as the **spring constant**. We will denote the spring constant by the Greek letter β (beta). One way of determining this spring constant is based on Hooke's law. This law states that, within the limits of elasticity, the amount of compression in a spring (Y) is proportional to the amount of force used to compress it (X). The proportionality constant is the spring constant β. Notationally, Y and X are expected to be related by the equation

$$Y = \beta X. \qquad \text{Line 9.1.2}$$

A laboratory routinely measures spring constants for manufacturers of metallic helical springs by subjecting springs to various forces (X) specified by the investigator. Typical X values for this problem might be 4.00, 4.50, 5.00, . . . , 10.0 Newtons of force. The compression (Y) is measured for each applied amount of force. These data are used to *estimate* the spring constant β by performing appropriate calculations. The data from one such set of measurements on springs of a certain metal are given in Table 9.1.2 and stored in the file **spring.mtp** on the data disk. A scatter plot of these data is shown in Figure 9.1.2.

TABLE 9.1.2 Helical Spring Data

Trial number	Compression Y (mm)	Force X (Newtons)
1	7.8	4.00
2	9.5	4.50
3	10.0	5.00
4	12.2	5.50
5	12.5	6.00
6	12.4	6.50
7	14.6	7.00
8	14.9	7.50
9	16.2	8.00
10	17.0	8.50
11	18.0	9.00
12	18.8	9.50
13	20.2	10.00

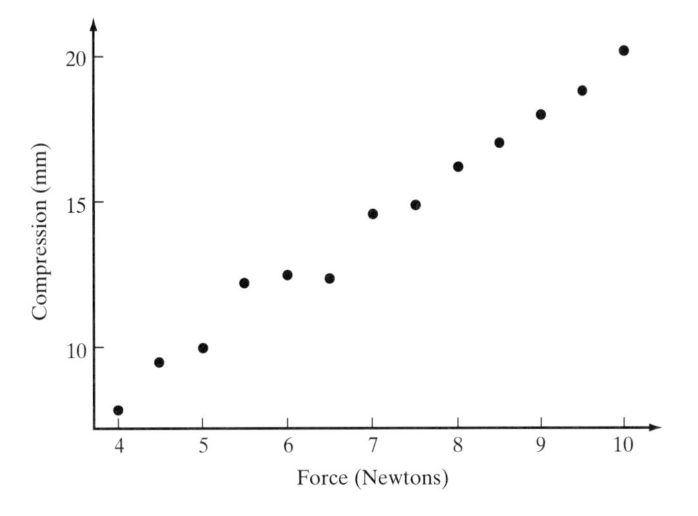

FIGURE 9.1.2
Scatter Plot of Compression Versus Force

It appears from the plot that the straight-line theoretical relationship shown in Line 9.1.2 does indeed hold (to within errors of experimentation). The slope of a straight line drawn through the points will provide an estimate of the spring constant β. The sample data shown in Table 9.1.2 were used to compute the prediction function

$$\widehat{Y}(X) = 2.02X \qquad \text{"for } X \text{ from 4.0 to 10.0."}$$

Using this prediction function, we notice that the estimated value of β is $\widehat{\beta} = 2.02$. The researcher is not interested in using $\widehat{Y}(X)$ to forecast the spring compression based on the force applied. Rather, she is interested in using $\widehat{Y}(X)$ to estimate the spring constant β.

EXAMPLE 9.1.3

VOLUME OF TREES. A corporation that owns several large tree farms must determine the volume of each tree to decide if it should be harvested. In order for the volume of a tree to be measured, the tree must be cut down. This is a very expensive method of measuring volume, since money is lost if a tree is cut down before its volume is large enough to make the maximum profit. The corporation would like to find a method of measuring tree volume that does not require cutting down trees. If there is a relationship between the volume of a tree and its circumference, then circumference could be used to predict volume. Measuring circumference is relatively inexpensive and does not require cutting down the tree. To study the relationship between volume and circumference, a simple random sample of 36 trees was selected from a large farm and cut down. The volume (Y) and the circumference 4.5 feet from the ground (X) were measured for each tree. The data are shown in Table 9.1.3 and stored in the file **tree.mtp** on the data disk. The scatter plot of the data shown in Figure 9.1.3 suggests a *curvilinear* relationship between Y and X.

TABLE 9.1.3 Circumference and Volume of Trees

Tree number	Volume (ft³)	Circumference (in)	Tree number	Volume (ft³)	Circumference (in)
1	191.1	94.9	19	197.5	98.6
2	266.7	111.2	20	85.4	61.9
3	244.0	106.8	21	73.6	54.0
4	103.9	70.1	22	258.2	109.6
5	43.4	35.2	23	183.2	92.0
6	170.4	89.2	24	259.1	110.6
7	54.3	39.6	25	203.2	99.0
8	126.6	77.6	26	234.8	105.6
9	204.5	98.0	27	86.9	56.5
10	181.4	92.4	28	62.7	42.1
11	145.7	82.3	29	272.5	112.5
12	98.1	62.2	30	77.3	52.8
13	88.7	58.1	31	60.2	36.8
14	157.2	85.8	32	59.1	31.4
15	78.2	58.1	33	105.1	68.8
16	125.3	75.4	34	136.7	82.3
17	57.5	46.2	35	231.9	105.9
18	293.9	117.8	36	57.1	38.3

FIGURE 9.1.3
Scatter Plot of Volume Versus Circumference

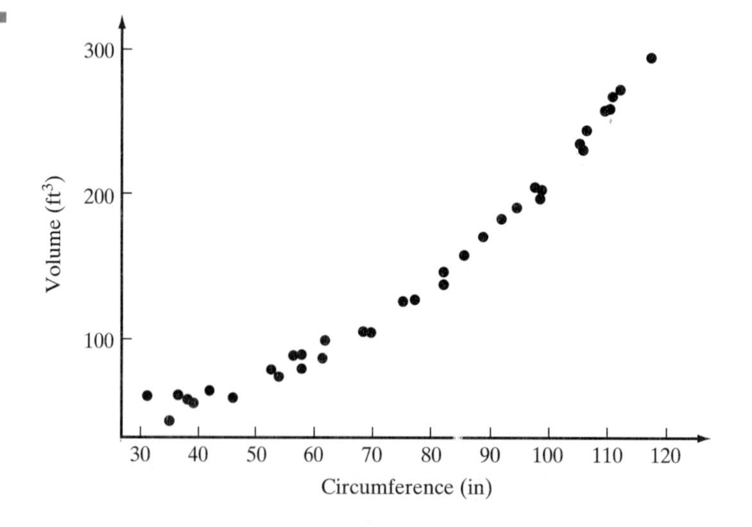

A prediction function based on the 36 sample values shown in Table 9.1.3 is

$$\widehat{Y}(X) = 21.4 + 0.0191X^2 \quad \text{"for } X \text{ from 30.0 to 120.0 inches."}$$

In this example an investigator is not interested in *forecasting* what the volume of a tree will be at some future date. Rather, the interest is in using X (the circumference of a tree) to *predict* Y (the volume of that tree) to decide if the tree should be harvested.

Uses of a Prediction Function

As stated previously, we often think of prediction as forecasting future values of a variable. However as the previous two examples illustrate, prediction functions have additional practical applications. Three uses of a prediction function are summarized below:

1. A prediction function can be used to *forecast* future values of a variable Y using values of X that are known at present. This is the application described in Example 9.1.1.
2. A prediction function can be used to *estimate* the value of a parameter that describes the relationship between Y and X. This is the application described in Example 9.1.2.
3. A prediction function can be used to *predict* the value of a variable Y that is expensive or impractical to measure using a variable X that is inexpensive or easy to measure. This is the application described in Example 9.1.3.

Target and Sampled Populations

To pursue the concept of prediction functions, we need some terminology and notation. First, a target and a sampled population must be defined and the variables of interest identified. We review these concepts in the following paragraphs.

Target population. This is the population for which we desire to make predictions. The target population can be a population that exists at present, existed in the past, will exist in the future, or exists conceptually. For each item in the target population there are *two* variables of interest: the predicted variable (Y) and the predictor variable (X). Many variables may be associated with each population item, but only two are of interest in the investigation. For instance, in Example 9.1.1 the target population is the set of this year's applicants for scholarships. The two variables of interest are "GPA" (Y) and "Score" (X). Many variables are associated with each applicant, including height, age, weight, parents' income, among others. However, only the two variables "GPA" (Y) and "Score" (X) are of interest in that example. It may not always be possible to obtain samples from the target population. If this is the case, a sample from a sampled population must be used to compute the prediction function.

Sampled population. The sampled population is the population from which a sample is obtained to compute $\widehat{Y}(X)$. Clearly, this population must exist when the sample is collected. If possible, the target and sampled populations should be the same. If this is not possible, the sampled population should be chosen to resemble the target population as closely as possible. Associated with each item in the sampled population are two variables: the predicted variable (Y) and the predictor variable (X). These are the same two variables defined for the target population. The function

$\widehat{Y}(X)$ can be used to predict values of Y in the sampled population. If the sampled population is similar to the target population, the prediction function may be useful in the target population. The word "population" will always refer to a sampled population unless we specifically use the term "target population."

You may envision a population of Y and X values as schematically represented in Table 9.1.4. The symbols Y_i and X_i in Table 9.1.4 represent the values of Y and X for population item number i for $i = 1, 2, 3, \ldots, N$. We generally don't know the values of Y_i and X_i for every item in the population, but they do exist, and you can imagine that they are in a table such as Table 9.1.4. It is customary to refer to this entire collection of values (Y_i, X_i) as a two-variable population of numbers, or simply as a **bivariate** population. We use capital letters Y and X to represent two population variables, and the capital letter N represents the number of items in the population. We use the notation

$$(Y_1, X_1), (Y_2, X_2), \ldots, (Y_N, X_N)$$

to represent a bivariate (two-variable) population of N pairs of values of the variables Y and X.

TABLE 9.1.4 Schematic Representation of a Population of Two Variables Y and X

Population item number	Variable Y	Variable X
1	Y_1	X_1
2	Y_2	X_2
3	Y_3	X_3
...	...	...
...	...	...
...	...	...
N	Y_N	X_N

EXAMPLE 9.1.4

SCHOLARSHIP. This example is a continuation of Example 9.1.1, where the foundation's director must decide which applicants will be awarded scholarships. The target population is the set of all applicants who will apply for a scholarship for the next academic year. Here the target population is a future population since the values of Y will not be known until the end of the next academic year. However, the values of X are known at the time a student applies for the scholarship, since the achievement-test score is required on the application. In order to compute a prediction function, one must obtain a sample of both Y and X values. Since the Y values in the target population are not available at the time the prediction must be made, samples of Y values cannot be obtained. Hence, the sample must be obtained from a sampled population where values of *both* Y and X are available. The director of the foundation believes that the set of 2,000 students who applied for scholarships during the past five years and completed their first year is similar to the target population. Samples can be obtained from this population of 2,000 applicants, and both the Y and X values can be measured. This set of 2,000 applicants is chosen as the sampled population. You can imagine that the set of Y and X values in this sampled population appears in a table such as Table 9.1.5. We don't know the values of all population items, but they do exist. The sample in Table 9.1.1 was obtained from this sampled population of 2,000 applicants.

TABLE 9.1.5 Representation of a Sampled Population of First-Year GPA and Achievement-Test Scores

Population item number	Variable Y	Variable X
1	2.76	0.72
2	3.12	0.79
...	...	...
...	...	...
...	...	...
2,000	3.22	0.91

THINKING STATISTICALLY: SELF-TEST PROBLEMS

9.1.1 Consider the prediction function

$$\widehat{Y}(X) = 6 + 2X \qquad \text{“for } X \text{ from } -10.0 \text{ to } 20.0.\text{”}$$

a. What is the value of $\widehat{Y}(-9.7)$?
b. What is the predicted value of Y if $X = 6.1$?
c. If X changes from 11.5 to 12.5, how much does the predicted value of Y change?

9.1.2 Consider the prediction function

$$\widehat{Y}(X) = 3.0 + 1.1X \qquad \text{“for } X \text{ from 80 to 90.”}$$

a. What is the value of $\widehat{Y}(81.9)$?
b. What is the predicted value of Y when $X = 92$?

9.1.3 A company that manufactures tires has developed a new all-weather tread and wants to study the miles per gallon (mpg) of automobiles equipped with these tires. Part of the study involves predicting the mpg of a car equipped with the new tires (Y), based on the car's speed (X). A simple random sample of 100 tires was obtained from the first 5,000 manufactured, and these tires were mounted on 25 cars. The cars were driven over a specified test route at selected speeds. The miles per gallon (Y) and miles per hour (X) were recorded for each test run. These data were used to compute the prediction function

$$\widehat{Y}(X) = 60 - 0.5X \qquad \text{“for } X \text{ from 40 to 75 miles per hour.”}$$

a. What is the predicted mpg of a car if it travels 55 miles per hour?
b. What is the predicted mpg of a car if it travels 60 miles per hour?
c. What is the predicted mpg of a car if it travels 80 miles per hour?

9.2 REGRESSION ANALYSIS

In the previous section we discussed the relationship between two quantitative variables and described various applications of prediction functions. A general approach for obtaining a prediction function using sample data is called **regression analysis**, or simply **regression**. In this section we introduce regression and show how it is used to obtain prediction functions. We begin with an example.

EXAMPLE 9.2.1

WARRANTY FOR VCRS. A company that manufactures videocassette recorders (VCRs) is studying the possibility of instituting a long-term (up to five years) warranty policy for its product. To help determine how much to charge for a policy, the company needs information about the maintenance cost of VCRs. This cost includes the costs of repairing, cleaning, and servicing a VCR. Among other things, the total maintenance cost of a VCR depends on its age. Thus, the company wants to predict the total maintenance cost of a VCR based on age. From the purchase registration records, the company determined that 2,500 VCRs were sold during the past five years. A simple random sample of 30 VCR owners was selected from this population, and each was asked the following two questions:

1. What is the age of your VCR (rounded *down* to the nearest year)? For example, if you have owned your VCR for 3 years, 10 months, and 3 days, the age is recorded as 3 years. If you have owned your VCR for 3 years, 2 months, and 16 days, the age is also recorded as 3 years.
2. During the years recorded in (1), how much money did you spend on maintenance for the VCR?

We use Y to denote the variable "Maintenance cost" (in dollars) and X to denote the variable "Age" (in years). The sample data are given in Table 9.2.1 and stored in the file **vcrsampl.mtp** on the data disk. A scatter plot of these data is shown in Figure 9.2.1.

TABLE 9.2.1 Sample of 30 VCRs

Sample number	Maintenance cost Y (dollars)	Age X (years)	Sample number	Maintenance cost Y (dollars)	Age X (years)
1	88	4	16	39	1
2	77	3	17	74	3
3	91	4	18	50	2
4	33	1	19	110	5
5	51	2	20	97	4
6	88	4	21	31	1
7	33	1	22	108	5
8	76	3	23	91	4
9	87	4	24	87	4
10	23	1	25	53	2
11	65	3	26	67	3
12	112	5	27	109	5
13	69	3	28	72	3
14	71	3	29	49	2
15	114	5	30	67	3

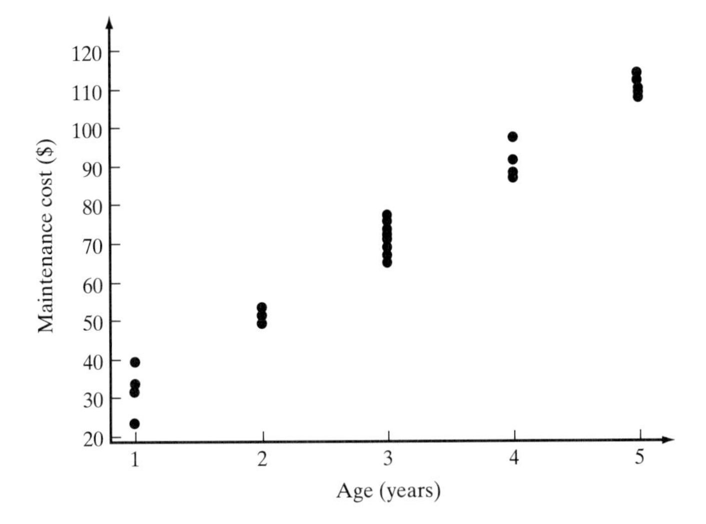

FIGURE 9.2.1
Scatter Plot of Maintenance Cost (Y) Versus Age (X) for a Sample of 30 VCRs

When the company sells a VCR, it would like to predict the maintenance cost for that VCR for the first X years, where $X = 1, 2, 3, 4$, or 5. The prediction function that will be used for this purpose is denoted by $\widehat{Y}(X)$. Using the methods to be discussed in Section 9.3, we obtain the prediction function

$$\widehat{Y}(X) = 11.8 + 19.6X \qquad \text{"for } X = 1, 2, 3, 4, 5.\text{"}$$

Using this function, the two-year maintenance cost of a VCR ($X = 2$) is predicted to be

$$\widehat{Y}(2) = 11.8 + 19.6(2) = \$51.00.$$

The five-year maintenance cost of a VCR ($X = 5$) is predicted to be

$$\widehat{Y}(5) = 11.8 + 19.6(5) = \$109.80.$$

Using predictions along with other information, the company will determine how much to charge for a warranty contract.

We use Example 9.2.1 throughout this section to explain the concepts of regression analysis. For illustration, we have constructed an artificial population of the Y and X values for the 2,500 VCRs sold during the past five years, where X = "Age" and Y = "Maintenance cost." This population is stored in the file **maintvcr.mtp** on the data disk. Column 1 contains the population item number, column 2 contains the Y value, and column 3 contains the X value. The first five and the last five items in this population are shown in Table 9.2.2. A scatter plot of the population data is shown in Figure 9.2.2.

Subpopulations

Suppose the company in Example 9.2.1 wants to predict the three-year maintenance cost for a VCR. This means that the company wants to predict the value of Y when $X = 3$. If you examine Figure 9.2.2, you will notice that there are many different values of Y for $X = 3$. This is because the maintenance cost of a VCR is not solely

TABLE 9.2.2 Population of Maintenance Cost and Age of 2,500 VCRs

Population item number	Maintenance cost Y (dollars)	Age X (years)
1	108	5
2	110	5
3	91	4
4	71	3
5	113	5
...	...	...
...	...	...
...	...	...
2,496	55	2
2,497	113	5
2,498	36	1
2,499	49	2
2,500	74	3

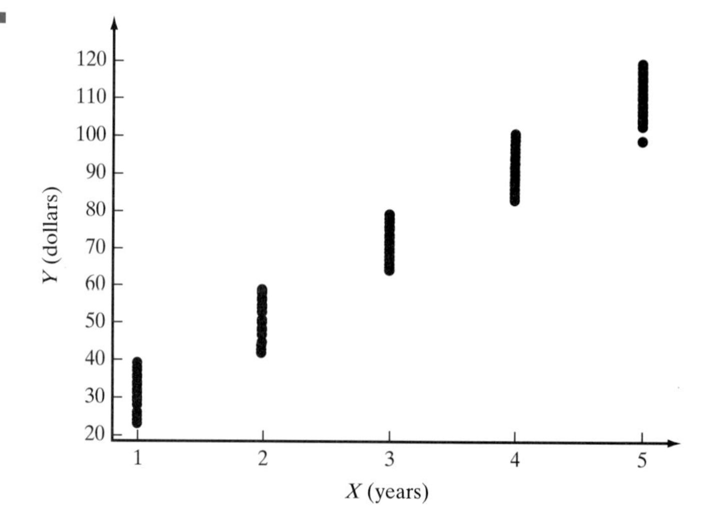

FIGURE 9.2.2
Scatter Plot of Maintenance Cost (Y) Versus Age (X) for the Population of VCRs

determined by age. Other factors such as the amount of usage, quality of videotapes used in the machine, and owner's care will affect maintenance cost. There are 500 values of Y for $X = 3$. The collection of 500 VCRs that are all 3 years old is defined to be a **subpopulation** of the population of all 2,500 VCRs. We denote the mean of this subpopulation by the symbol

$$\mu_Y(3).$$

In this notation, μ represents a mean, the *sub*script Y indicates a *sub*population of Y values, and the number 3 indicates that we are interested in the mean of the subpopulation of Y values where $X = 3$. These ideas are summarized in Box 9.2.1.

BOX 9.2.1: SUBPOPULATIONS AND MEANS OF SUBPOPULATIONS

Consider the bivariate population $(Y_1, X_1), (Y_2, X_2), \ldots, (Y_N, X_N)$. For every distinct value of X in the population, there are one or more values of Y. The set of Y values for a given value of X is called a **subpopulation** of Y values, where the predictor value is equal to X. The mean of this subpopulation of Y values is denoted by

$$\mu_Y(X).$$

For the population shown in Table 9.2.2, there are five distinct values of X: 1, 2, 3, 4, and 5. We partitioned this population of 2,500 items into five subpopulations, one subpopulation for each distinct value of X. Subpopulation 1 contains VCRs that are 1 year old ($X = 1$); subpopulation 2 contains VCRs that are 2 years old ($X = 2$), and so on. There are 500 VCRs in each subpopulation; these data are contained in the file **vcrsub.mtp** on the data disk. Table 9.2.3 shows the first five and last five items in each subpopulation.

TABLE 9.2.3 Subpopulations of VCR Maintenance

Item	**Subpopulation 1**		**Subpopulation 2**		**Subpopulation 3**		**Subpopulation 4**		**Subpopulation 5**	
	Y	X	Y	X	Y	X	Y	X	Y	X
1	31	1	53	2	71	3	91	4	108	5
2	36	1	50	2	72	3	87	4	110	5
3	35	1	49	2	69	3	89	4	113	5
4	33	1	48	2	75	3	92	4	119	5
5	28	1	54	2	65	3	92	4	112	5
...	...	...	...	...	...	...	...	...	...	...
...	...	...	...	...	...	...	...	...	...	...
...	...	...	...	...	...	...	...	...	...	...
496	29	1	55	2	70	3	96	4	106	5
497	30	1	47	2	70	3	91	4	114	5
498	32	1	53	2	71	3	91	4	113	5
499	27	1	55	2	66	3	91	4	110	5
500	36	1	49	2	74	3	91	4	113	5

The mean and the standard deviation of the Y values in each of the five subpopulations in Table 9.2.3 are shown in Table 9.2.4 (rounded to whole numbers).

TABLE 9.2.4 The Means and Standard Deviations of the Y Values for the Five Subpopulations in Table 9.2.3

Subpopulation	$X =$	1	2	3	4	5
Mean of Y	$\mu_Y(X) =$	31	51	71	91	111
Standard Deviation of Y	$\sigma =$	3	3	3	3	3

You will notice that the standard deviation of the Y values is equal to 3 for each subpopulation. We denote this common standard deviation by σ. Another interesting feature of these subpopulations is that $\mu_Y(X)$ can be represented as

$$\mu_Y(X) = 11 + 20X \qquad \text{“for } X = 1, 2, 3, 4, 5.\text{”} \qquad \textbf{Line 9.2.1}$$

Using $\mu_Y(X)$, we see that $\mu_Y(1) = 31$, $\mu_Y(2) = 51$, and so on, as shown in Table 9.2.4. This is demonstrated graphically in Figure 9.2.3.

FIGURE 9.2.3
Relative Frequency Histograms of Maintenance Cost for Each of Five Subpopulations of VCRs

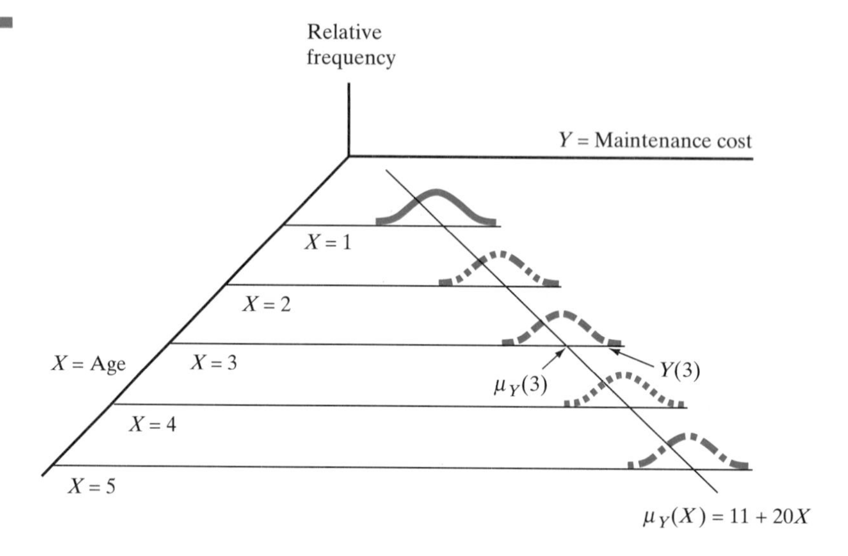

We now define the **regression** of Y on X.

BOX 9.2.2: REGRESSION OF Y ON X

Let $(Y_1, X_1), (Y_2, X_2), \ldots, (Y_N, X_N)$ be a bivariate population where an investigator is interested in studying the relationship between Y and X.

a. For every distinct value of X, there is a subpopulation of Y values.

b. The mean of the subpopulation of Y values for any specified value of X is a function of X. This mean is denoted by $\mu_Y(X)$ and is called the **population regression function of Y on X.**

Simple Linear Regression

In many cases $\mu_Y(X)$ is a linear function of X. This is the case for the subpopulations shown in Table 9.2.3. In such cases the *form* of the population regression function is written as

$$\mu_Y(X) = \beta_0 + \beta_1 X \qquad \text{“for specified values of } X,\text{”} \qquad \textbf{Line 9.2.2}$$

where β_0 and β_1 are population parameters. The function in Line 9.2.2 is referred to as a straight-line population regression function. Since there is only one predictor variable X in this model, it is called a **simple linear population regression function.** The parameter β_0 in Line 9.2.2 is called the **intercept** of the line, and the parameter β_1 is called the **slope** of the line. The value of the population regression function changes by an amount equal to β_1 when X increases by one unit. This is indicated by the relationship

$$\mu_Y(X + 1) = \mu_Y(X) + \beta_1.$$

Models

The data in Table 9.2.3 were artificially constructed to create a population regression function where the subpopulation means lie *exactly* on a straight line and the standard deviations of all subpopulations are *exactly* equal. In practice, it would be rare, if ever, that we encountered a problem where the subpopulation means lie exactly on a straight line or where the standard deviations of all subpopulations are exactly equal. For this reason, the function in Line 9.2.2 represents a *model* of an *idealized* population regression function. Models are used frequently in everyday activities to help make decisions, as the next example illustrates.

EXAMPLE 9.2.2

MODELS. You may recall Example 6.2.1, where an individual wants to put new carpet on a living room floor and needs to know how many square yards of carpet to purchase. Although the floor resembles a theoretical rectangle, it is not an *exact* theoretical rectangle since the angles formed by intersecting sides of the room are not *exactly* 90 degrees (if measured to many decimals). However, the room is approximately a theoretical rectangle, so the individual uses a theoretical rectangle as a *model* and uses the formula *length* × *width* to compute the square yards of carpet needed. In a similar manner, the function shown in Line 9.2.2 is used as a model to describe the regression function of Y on X. We don't expect to encounter any practical problems where the regression function can be *exactly* represented by the function shown in Line 9.2.2, but this model can be used to approximate a population regression function in many practical problems.

Convention. When we write the statement, "The population regression function is $\mu_Y(X) = \beta_0 + \beta_1 X$," or when we write the statement, "The population regression function is linear," we mean that the actual population regression function for the problem under study is adequately approximated by the function shown in Line 9.2.2.

The function in Line 9.2.2 is used by theoretical statisticians to develop formulas for computing point estimates and confidence intervals for the unknown quantities. If the actual regression function under study is adequately approximated by the function shown in Line 9.2.2, then the results obtained using these formulas will be useful. In any application, one of the investigator's tasks is to assess whether the approximation is adequate. Statistical methods are available to help determine this; we discuss some of these methods in Section 9.5. In the remainder of this chapter we will use the population regression function shown in Line 9.2.2.

Sample Regression Function

In most applications the values of β_0 and β_1 in the population regression function

$$\mu_Y(X) = \beta_0 + \beta_1 X$$

are not known. In this case a sample is collected from the bivariate population

$$(Y_1, X_1), (Y_2, X_2), \ldots, (Y_N, X_N)$$

and used to estimate β_0 and β_1. We denote these estimates by $\widehat{\beta}_0$ and $\widehat{\beta}_1$, respectively. The estimates $\widehat{\beta}_0$ and $\widehat{\beta}_1$ are used to estimate $\mu_Y(X)$. This estimate is

$$\widehat{\mu}_Y(X) = \widehat{\beta}_0 + \widehat{\beta}_1 X.$$

For illustration, the sample in Table 9.2.1 is used to obtain the estimates $\widehat{\beta}_0 = 11.8$ and $\widehat{\beta}_1 = 19.6$. Using these estimates, we find that the estimate of $\mu_Y(X)$ in Example 9.2.1 is

$$\widehat{\mu}_Y(X) = 11.8 + 19.6X.$$

The estimate $\widehat{\mu}_Y(X)$ is referred to as the **sample regression function**. The sample regression function estimates the population regression function, $\mu_Y(X) = \beta_0 + \beta_1 X$. In Section 9.3 we show how to compute $\widehat{\beta}_0$, $\widehat{\beta}_1$, and $\widehat{\mu}_Y(X)$ using sample data.

Predicting the Value of a Future Observation

When studying the regression of Y on X, we use the notation $Y(X)$ to denote the Y value of an item that is to be chosen at random from the subpopulation, where all values of the predictor variable are equal to a specified value of X. For instance, in Example 9.2.1 suppose I plan to buy a VCR. If I want to predict the three-year maintenance cost for this particular VCR, I can envision that I selected it at random from the subpopulation of three-year-old VCRs and measured its maintenance cost. The symbol $Y(3)$ represents the three-year maintenance cost of the randomly selected VCR that I will purchase (see $Y(3)$ in Figure 9.2.3). The value of $Y(3)$ is not known at the time the VCR is purchased, but it can be estimated using sample values. The estimate of $Y(3)$ is denoted by $\widehat{Y}(3)$, where the symbol $\widehat{\ }$ (hat) denotes an estimate. Instead of calling $\widehat{Y}(3)$ the "estimated value" of $Y(3)$, it is customary to call it the **predicted value** of $Y(3)$. We will use this latter terminology. For a general value X, the symbol $Y(X)$ represents a value of Y that will be chosen at random from the subpopulation of Y values that all have the same value for the predictor value X. How should one obtain a predicted value for $Y(X)$? Recall from Chapter 5 that the mean of a population is a single number that is often used to represent all the values in the population. In regression it is customary to use $\mu_Y(X)$, the mean of a subpopulation, to represent all the values in that subpopulation. Thus, if $\mu_Y(X) = \beta_0 + \beta_1 X$ is known, it would be used to estimate (predict) $Y(X)$, the value of an item to be chosen at random from that subpopulation. In general, we will not know the values of β_0 and β_1, so we estimate them and denote the estimates by $\widehat{\beta}_0$ and $\widehat{\beta}_1$, respectively. These are substituted into $\mu_Y(X) = \beta_0 + \beta_1 X$ to obtain the sample regression function

$$\widehat{\mu}_Y(X) = \widehat{\beta}_0 + \widehat{\beta}_1 X.$$

Thus, $\widehat{\mu}_Y(X) = \widehat{\beta}_0 + \widehat{\beta}_1 X$ is the estimate of $\mu_Y(X) = \beta_0 + \beta_1 X$ and is also the estimate of $Y(X)$. That is, we use $\widehat{\mu}_Y(X) = \widehat{\beta}_0 + \widehat{\beta}_1 X$ for the prediction function $\widehat{Y}(X)$ and write

$$\widehat{Y}(X) = \widehat{\mu}_Y(X) = \widehat{\beta}_0 + \widehat{\beta}_1 X.$$

For example, the value of the sample regression function when $X = 3$ is $\widehat{\mu}_Y(3) = \widehat{\beta}_0 + \widehat{\beta}_1(3)$, and this will be used as $\widehat{Y}(3)$, the predicted value of Y when $X = 3$. These results are summarized in Box 9.2.3.

BOX 9.2.3: SUMMARY OF THE RELATIONSHIP BETWEEN SAMPLE REGRESSION FUNCTIONS AND PREDICTION FUNCTIONS

Consider a bivariate sampled population $(Y_1, X_1), (Y_2, X_2), \ldots, (Y_N, X_N)$, where the population regression function is denoted by

$$\mu_Y(X) = \beta_0 + \beta_1 X \qquad \text{"for specified values of } X\text{."}$$

A simple random sample is obtained from the population and used to compute the estimates $\widehat{\beta}_0$ and $\widehat{\beta}_1$. Using these estimates, we obtained functions (1) and (2) below.

1. $\widehat{\mu}_Y(X) = \widehat{\beta}_0 + \widehat{\beta}_1 X$.
This is the **sample regression function**. It is an estimate of the population regression function $\mu_Y(X) = \beta_0 + \beta_1 X$.

2. $\widehat{Y}(X) = \widehat{\beta}_0 + \widehat{\beta}_1 X$.
This is the **prediction function**. It is the predicted value of Y for an item that will be chosen at random from a subpopulation of Y values, where the predictor value is equal to X for all items in the subpopulation.

The fact that we use $\widehat{\mu}_Y(X)$ as the prediction function should not be too surprising, since we would use $\mu_Y(X)$ to predict $Y(X)$ if the values of β_0 and β_1 were known.

Suppose I plan to purchase a VCR and want to predict its four-year maintenance cost using estimates based on a simple random sample from the population. Thus, $\widehat{Y}(4) = \widehat{\beta}_0 + \widehat{\beta}_1(4)$ is the predicted value of $Y(4)$. We do not expect that $\widehat{Y}(4)$ will be exactly equal to $Y(4)$, and a confidence interval for $Y(4)$ can be used to determine "how close" $\widehat{Y}(4)$ is to $Y(4)$. In general, for any value of X the predicted value of $Y(X)$ is denoted by $\widehat{Y}(X)$ and a confidence interval for $Y(X)$ can be used to determine "how close" this predicted value is to $Y(X)$. All the concepts of confidence intervals discussed in Chapter 3 apply to a confidence interval for $Y(X)$. It is customary to call a confidence interval for $Y(X)$ a **prediction interval**. In the next section we discuss procedures for using sample data to compute the prediction function, $\widehat{Y}(X)$, and prediction intervals for $Y(X)$.

EXPLORATION 9.2.1

In this exploration we study Example 9.1.1 in more detail. Recall that the foundation's director must decide which applicants will be awarded scholarships. To help make this decision, the director would like to know the applicant's first-year GPA. However, the decision to award a scholarship must be made before the applicant's first-year GPA is known, so the director will have to predict this GPA. The score on an achievement test (X) is used to predict the first-year GPA (Y). The target population is the set of all scholarship applicants for the next academic year. The values of Y in the target population will not be known until the end of the academic year, and so data must be obtained from a sampled population.

The sampled population is the set of 2,000 students who, during the past five years, applied for scholarships and completed their first year. The data for the 2,000 applicants in the sampled population are available in the foundation office records. These data are stored in the file **gpa.mtp** on the data disk, where column 1 contains the student identification number, column 2 contains the first-year GPA (Y), and column 3 contains the score on the achievement test (X).

1. Does it appear that score on the achievement test is useful in predicting GPA in the sampled population of 2,000 applicants?

Answer: *One way to answer this question is to examine a scatter plot of the data. This plot is shown in Figure 9.2.4.*

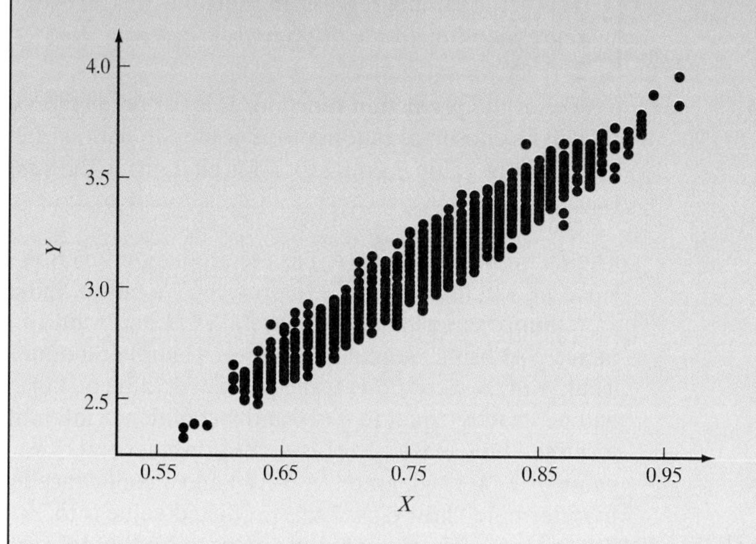

FIGURE 9.2.4
Scatter Plot of GPA (Y) Versus Grade on an Achievement Test (X) for the Sampled Population

From this plot you can see that Y tends to increase as X increases (that is, the first-year GPA tends to increase as the score on the achievement test increases). So it appears that X is useful for predicting Y in the sampled population.

2. Naomi, one of this year's applicants, received a score of 70% on the achievement test. In order to decide if Naomi will be awarded a scholarship, the director would like to predict her first-year GPA. What value might be used to predict her GPA?

Answer: *We might consider predicting Naomi's first-year GPA by using the first-year GPA of a student in the sampled population who received 70% on the achievement test. However, there are many applicants in this population who received a score of 70%, and they do not all have the same first-year GPA. Since the mean of a subpopulation is a good predictor of the value of an item that is to be chosen at random from that subpopulation, we could use the mean of the subpopulation where* $X = 0.70$ *to predict the first-year GPA of a student in that subpopulation. This mean is denoted by* $\mu_Y(0.70)$. *If the entire sampled population of 2,000 students is available,* $\mu_Y(0.70)$ *will be used to predict the first-year GPA of a student in that population who received a score of 70% on the achievement test.*

3. Suppose it is impractical to collect the necessary information for all 2,000 applicants in the sampled population. If this is the case, a sample of size n is obtained from the sampled population and used to compute $\widehat{\mu}_Y(X)$. The prediction function is now $\widehat{Y}(X) = \widehat{\mu}_Y(X)$, and $\widehat{Y}(0.70)$ could be used to predict Naomi's first-year GPA.

   To illustrate, a simple random sample of 60 students was selected from the 2,000 students in the sampled population. Using this sample, we can make a statistical inference to the sampled population of 2,000 students. However, since Naomi is not a member of the sampled population, a judgment inference is also required. See Figure 9.2.5.

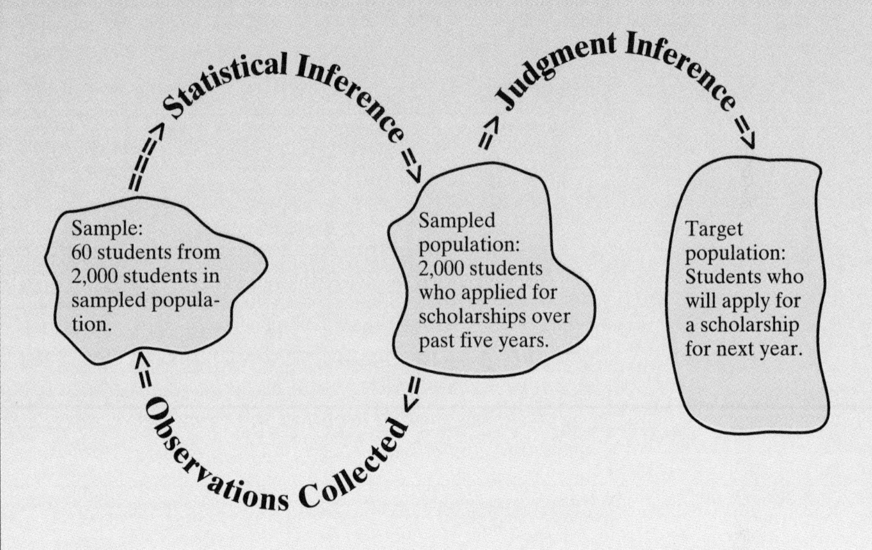

FIGURE 9.2.5
Relationships Among Sample, Sampled Population, and Target Population in Exploration 9.2.1

The sample data are shown in Table 9.1.1. This sample was used to compute the prediction function

$$\widehat{Y}(X) = 0.10 + 3.94X \qquad \text{"for } X \text{ from 0.60 to 1.00."} \qquad \textbf{Line 9.2.3}$$

Construct a scatter plot of the sample data in Table 9.1.1. Draw the prediction function on the same plot.

Answer: *The result is shown in Figure 9.2.6. To use Minitab to produce this figure, retrieve the data in the file* **gpasampl.mtp** *and enter the following command in the Session window. The variable Y is in column 2, and X is in column 3.*

```
MTB > %fitline c2 c3
```

4. Compare the scatter plot (of the sample values) in Figure 9.2.6 with the scatter plot (of population values) in Figure 9.2.4. Is the sample a good representation of the sampled population?

Answer: *Yes. Of course, in most real problems the entire population will not be known and we will have to rely on the sample only.*

FIGURE 9.2.6
Plot of Sample Data Given in Table 9.1.1

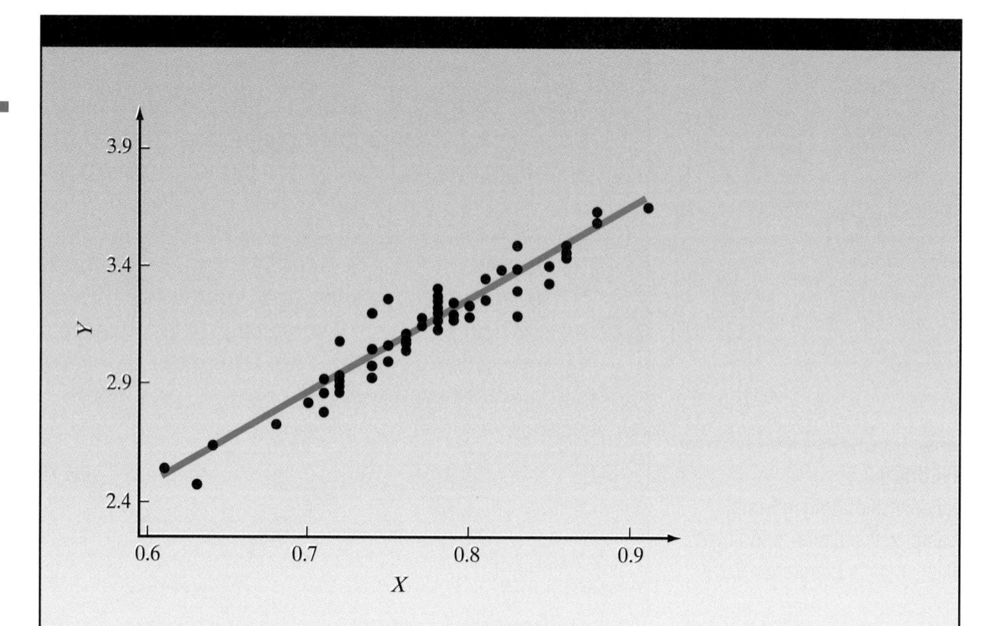

5. Use the prediction function shown in Line 9.2.3 to predict Naomi's first-year GPA. Recall she received a score of 70% on the achievement test.

Answer: *From Line 9.2.3 we obtain* $\widehat{Y}(0.70) = 0.10 + 3.94(0.70) = 2.86$. *You must remember that the predicted GPA of 2.86 is the predicted GPA of an applicant in the sampled population who received a score of 70% on the achievement test. If the sampled population resembles the target population, then 2.86 will be a valid prediction of Naomi's first-year GPA.*

THINKING STATISTICALLY: SELF-TEST PROBLEMS

9.2.1 A car rental company purchases several hundred new cars each year. Most of these cars are driven between 20,000 and 60,000 miles during the first year of service. The company manager wants to study the relationship between Y and X, where Y is the maintenance cost of a car during the first year and X is the miles driven during the first year. The population regression function is

$$\mu_Y(X) = \beta_0 + \beta_1 X \qquad \text{"for } X \text{ from 20,000 to 60,000."}$$

a. What symbol represents the change in the average maintenance cost if mileage increases by 1,000?
b. A simple random sample of 60 cars was selected from the population of rental cars that were one year old. The computed sample regression function is

$$\widehat{\mu}_Y(X) = -60.1 + 0.0318X \qquad \text{"for } X \text{ from 20,000 to 60,000."}$$

What is the estimate of β_1?
c. What is the estimate of the change in the average maintenance cost if mileage increases by 1,000?

d. What is the predicted maintenance cost of a car that was driven 29,000 miles in the first year?
e. What is the estimate of the average maintenance cost of all cars that were driven 25,000 miles in the first year?
f. What is the estimate of the average maintenance cost of all cars that were driven 75,000 miles in the first year?

9.2.2 Which of the following statements are true, and which are false?
a. If $\mu_Y(3)$ is known, then $Y(3)$ is known, where $Y(3)$ is a Y value to be chosen at random from the subpopulation where the predictor variable is equal to 3.
b. If $\widehat{Y}(X) = 3.2 - 6.1X$ "for $X = 6$ to 18," then $\widehat{Y}(8) = -45.6$.
c. $\widehat{Y}(X) = \widehat{\mu}_Y(X)$.
d. $\widehat{Y}(X) = Y(X)$.

9.3 STATISTICAL INFERENCE IN SIMPLE LINEAR REGRESSION

In this section we discuss prediction functions and prediction intervals for

$$Y(X) \qquad \text{Line 9.3.1}$$

and point estimates and confidence intervals for

$$\beta_0, \quad \beta_1, \quad \text{and} \quad \mu_Y(X). \qquad \text{Line 9.3.2}$$

For the statistical inference procedures described in this chapter to be valid, certain assumptions must be satisfied. These assumptions are given in Box 9.3.1.

BOX 9.3.1: ASSUMPTIONS FOR STATISTICAL INFERENCE IN REGRESSION

Suppose an investigator is interested in studying the regression of Y on X in a bivariate population

$$(Y_1, X_1), (Y_2, X_2), \ldots, (Y_N, X_N).$$

For statistical inference procedures described in this chapter to be valid, the bivariate population must satisfy assumptions described in (1), (2), and (3) below. In addition, the sample must be obtained by either method (4a) or (4b) described below.

1. The subpopulation of Y values for each distinct value of X has a mean denoted by $\mu_Y(X)$, where

$$\mu_Y(X) = \beta_0 + \beta_1 X \qquad \text{"for specified values of } X\text{."}$$

The term "for specified values of X" defines the values of X for which this linear regression function is valid.

2. For each distinct value of X, the subpopulation of Y values must have the same standard deviation. This common standard deviation is denoted by σ (that is, the standard deviation of every subpopulation is equal to σ).

3. The subpopulation of Y values for each distinct value of X is a *normal* population. This assumption is not needed for point estimation but is required for confidence intervals and prediction intervals.

4a. A simple random sample of n items is selected from the sampled population, and Y and X are observed for each sampled item. The sample values are denoted as $(y_1, x_1), (y_2, x_2), \ldots, (y_n, x_n)$.

4b. An investigator chooses any desired set of X values. A simple random sample of one or more Y values is then selected from each subpopulation determined by these X values to obtain a sample size of n. As in (4a), the sample values are denoted as $(y_1, x_1), (y_2, x_2), \ldots, (y_n, x_n)$.

The sample data in Examples 9.1.1 and 9.1.3 were obtained by the sampling method described in assumption (4a) in Box 9.3.1. The sample data in Example 9.1.2 were obtained by the sampling method described in assumption (4b). In that example, the set of X values was specified to be $x_1 = 4.00, x_2 = 4.50, \ldots, x_{13} = 10.00$. Assumptions (1), (2), and (3) in Box 9.3.1 concern the parameters and relative frequency histogram of a population—quantities that are seldom known in practical applications. In this case sample data must be used to help decide if these assumptions are satisfied (or approximately satisfied). This topic is discussed in Section 9.5.

Computer Calculations Using Minitab

Computers are almost a necessity for performing calculations in regression analysis. In this section we show the Minitab commands that can be used for this purpose. If you do not have Minitab, ignore the commands and simply refer to the output shown in the exhibits. This output is similar to output that any statistical software package would produce. In the appendix to this chapter (Section 9.9) we present formulas that can be used to perform the calculations using a hand-held calculator.

To obtain the prediction function $\widehat{Y}(X)$ and point estimates of $\beta_0, \beta_1, \mu_Y(X)$, and σ using Minitab, enter the following command in the Session window:

```
MTB > regress ci 1  cj
```

where `ci` stands for column `i`, which contains the Y variable, and `cj` stands for column `j`, which contains the X variable. The number `1` that appears between `ci` and `cj` means that there is only *one* predictor variable. You might think of this command as stating, "Regress the predicted variable Y, which is in column `i`, on the single (`1`) predictor variable X, which is in column `j`." We illustrate the command with an example.

EXAMPLE 9.3.1

WARRANTY FOR VCRS. Recall Example 9.2.1, where a company is considering a warranty policy for its VCRs and wants to predict the maintenance cost of VCRs for 1, 2, 3, 4, or 5 years. Since the entire sampled population is contained in the file **maintvcr.mtp**, we could compute the *exact* values of the population parameters shown in Line 9.3.2 using all the data. However, for illustrative purposes, we selected a simple random sample of 30 items from this population, which we will use to obtain a prediction function $\widehat{Y}(X)$ and point estimates for the quantities in Line 9.3.2. The assumptions in Box 9.3.1 are presumed to be satisfied. The sample data are in Table 9.3.1 and stored in the file **vcrsampl.mtp** on the data disk, where column 1 contains the sample number, column 2 contains the maintenance cost in dollars (Y), and column 3 contains the age in years (X). You should always construct a scatter plot to see if it is appropriate to assume a linear relationship between Y and X. The scatter plot of the data in Table 9.3.1 is shown in Figure 9.3.1. From this plot, a linear population regression function seems appropriate for this example.

TABLE 9.3.1 Sample of 30 VCRs

Sample number	Cost Y (dollars)	Age X (years)	Sample number	Cost Y (dollars)	Age X (years)
1	88	4	16	39	1
2	77	3	17	74	3
3	91	4	18	50	2
4	33	1	19	110	5
5	51	2	20	97	4
6	88	4	21	31	1
7	33	1	22	108	5
8	76	3	23	91	4
9	87	4	24	87	4
10	23	1	25	53	2
11	65	3	26	67	3
12	112	5	27	109	5
13	69	3	28	72	3
14	71	3	29	49	2
15	114	5	30	67	3

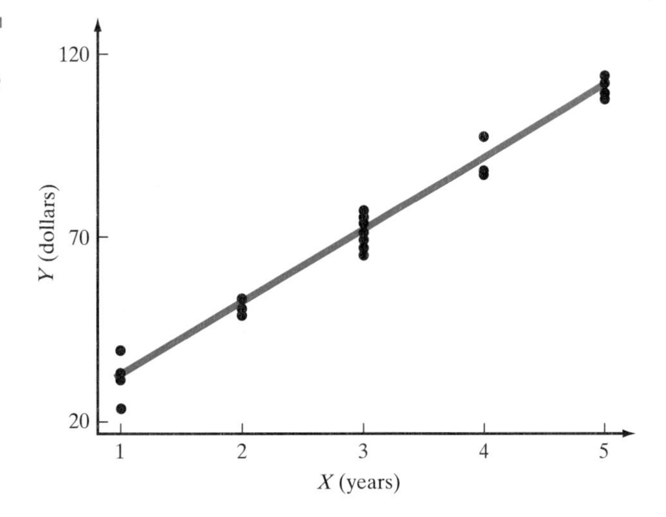

FIGURE 9.3.1
Scatter Plot of Cost (Y) Versus Age (X) for a Sample of 30 VCRs

To perform the computations using Minitab, retrieve the data and enter the following command in the Session window:

```
MTB > regress c2 1 c3
```

A portion of the resulting output is shown in Exhibit 9.3.1.

```
-----------------------------------------------------------------
Regression Analysis

The regression equation is
Y = 11.8 + 19.6 X

Predictor        Coef        Stdev     t-ratio        p
Constant       11.846        1.760        6.73    0.000
X             19.6410       0.5237       37.50    0.000

S = 3.729        R-sq = 98.0%     R-sq(adj) = 98.0%
-----------------------------------------------------------------
```

EXHIBIT 9.3.1

The third line in Exhibit 9.3.1 contains the prediction function, $\widehat{Y}(X) = 11.8 + 19.6X$. Minitab uses the letter `Y` to represent $\widehat{Y}(X)$. Remember that $\widehat{Y}(X)$ is equal to the sample regression function, $\widehat{\mu}_Y(X)$. Under the column labeled `Coef` are two numbers, 11.846 (in the row labeled `Constant`) and 19.641 (in the row labeled `X`). The number 11.846 is $\widehat{\beta}_0$, and the number 19.641 is $\widehat{\beta}_1$. These are the same estimates of β_0 and β_1 shown in $\widehat{Y}(X)$, but with greater decimal accuracy. The last line in Exhibit 9.3.1 contains the estimate of σ, which is labeled `S`. Hence,

$$\widehat{\beta}_0 = 11.846, \qquad \widehat{\beta}_1 = 19.641, \qquad \widehat{\sigma} = 3.729,$$
$$\widehat{Y}(X) = \widehat{\mu}_Y(X) = 11.846 + 19.641X.$$

Suppose we want to estimate $\mu_Y(4)$, the *average* maintenance cost of the subpopulation of four-year-old VCRs. This estimate is $\widehat{\mu}_Y(4) = 11.846 + 19.641(4) = \$90.41$.

Suppose I plan to purchase a VCR and I want to predict its two-year maintenance cost. Using the prediction function in Exhibit 9.3.1, we obtain

$$\widehat{Y}(2) = 11.846 + 19.641(2) = \$51.13.$$

So the predicted two-year maintenance cost of the VCR I will purchase is \$51.13. Other quantities shown in Exhibit 9.3.1 will be discussed as needed.

Using Minitab to Compute Prediction Intervals and Confidence Intervals

We now discuss how to use Minitab to compute prediction intervals for

$$Y(X) \qquad \textbf{Line 9.3.3}$$

and confidence intervals for

$$\beta_0, \quad \beta_1, \quad \text{and} \quad \mu_Y(X). \qquad \textbf{Line 9.3.4}$$

Enter the following Minitab command and subcommands in the Session window, where the Y variable is in column `i` and the X variable is in column `j`:

```
MTB > regress ci 1 cj;
SUBC> predict X;
SUBC> confidence = p.
```

The quantity X is the value of X you want to use in $\widehat{\mu}_Y(X)$ and $\widehat{Y}(X)$. The quantity p is the desired confidence coefficient in percentage (for example, 90, 95, or 99). We illustrate these commands with an example.

EXAMPLE 9.3.2

WARRANTY FOR VCRS. This is a continuation of Example 9.3.1, where a company wants to predict the maintenance cost of VCRs for 1, 2, 3, 4, and 5 years. Suppose an investigator wants a 95% prediction interval for $Y(4)$ using the simple random sample shown in Table 9.3.1. The data are stored in the file **vcrsampl.mtp** on the data disk, where column 1 contains the sample number, column 2 contains the Y variable, and column 3 contains the X variable. Retrieve the data and enter the following command and subcommands in the Session window:

```
MTB > regress c2 1 c3;
SUBC> predict 4;
SUBC> confidence = 95.
```

The output is shown in Exhibit 9.3.2.

```
Regression Analysis

The regression equation is
Y = 11.8 + 19.6 X

Predictor       Coef       Stdev    t-ratio        p
Constant      11.846       1.760       6.73    0.000
X            19.6410      0.5237      37.50    0.000

S = 3.729       R-sq = 98.0%      R-sq(adj) = 98.0%

     Fit  Stdev.Fit        95.0% C.I.            95.0% P.I.
  90.410      0.828   ( 88.714,  92.107)   ( 82.584,  98.236)
```

EXHIBIT 9.3.2

The quantities in the first seven lines of Exhibit 9.3.2 are identical to those in the first seven lines of Exhibit 9.3.1. The number 90.410 under the column labeled Fit in the last row of Exhibit 9.3.2 is $\widehat{\mu}_Y(4)$. Of course, this value is also $\widehat{Y}(4)$, the predicted value of $Y(4)$. Under the column labeled 95.0% C.I. is the 95% confidence interval for $\mu_Y(4)$. Thus, we can state, "We are 95% confident that $\mu_Y(4)$ is contained in the interval \$88.71 to \$92.11." Under the column labeled 95.0% P.I. is the 95% prediction interval for $Y(4)$. We say, "We are 95% confident that $Y(4)$ will be contained in the interval \$82.58 to \$98.24." Remember that even though $\widehat{Y}(X) = \widehat{\mu}_Y(X)$, it is generally not true that $Y(X) = \mu_Y(X)$. That is, a Y value selected at random from a subpopulation is generally not equal to the subpopulation mean. Since $Y(X)$ might not be equal to $\mu_Y(X)$, the 95% prediction interval for $Y(X)$ is not equal to the 95% confidence interval for $\mu_Y(X)$. The 95% prediction interval for $Y(X)$ is always wider than the 95% confidence interval for $\mu_Y(X)$.

The number listed in the column labeled `p` and the row labeled `X` in Exhibit 9.3.2 is a P-value used to test the hypotheses

$$H_0 : \beta_1 = 0 \quad \text{versus} \quad H_a : \beta_1 \neq 0.$$

This value is 0.000 (to three decimals) in Exhibit 9.3.2. Investigators sometimes use this test to decide if X is useful for predicting Y with the model $\mu_Y(X) = \beta_0 + \beta_1 X$. However, we recommend that one consider prediction intervals rather than statistical tests in deciding if the regression model provides useful predictions.

Confidence Intervals for β_0 and β_1

Since $\mu_Y(0) = \beta_0$, a confidence interval for β_0 can be obtained from the previous Minitab command and subcommands by using the subcommand

```
predict 0;
```

Many software packages, including Minitab, do not compute confidence intervals for β_1. However, a confidence interval for β_1 can be computed using the standard output provided by most software packages. Box 9.3.2 shows how to perform this computation using the Minitab output in Exhibit 9.3.2.

BOX 9.3.2: HOW TO COMPUTE A CONFIDENCE INTERVAL FOR β_1

A confidence interval for β_1 using the output in Exhibit 9.3.2 is L to U, where

$$L = \widehat{\beta}_1 - (T2)\texttt{Stdev}(\widehat{\beta}_1) \qquad \text{and} \qquad U = \widehat{\beta}_1 + (T2)\texttt{Stdev}(\widehat{\beta}_1). \quad \textbf{Line 9.3.5}$$

The quantity $T2$ is a table value found in Table T-2 with $DF = n - 2$. The quantity `Stdev`$(\widehat{\beta}_1)$ is located in the column labeled `Stdev` and the row labeled `X` in Exhibit 9.3.2. The term `Stdev` is commonly called the **standard error**. Thus, `Stdev`$(\widehat{\beta}_1)$ is the standard error of $\widehat{\beta}_1$. Standard errors are discussed in more detail in Section 9.9.

To demonstrate the use of the formulas in Box 9.3.2, we will construct a 95% confidence interval for β_1 in Example 9.3.2. For a sample size of $n = 30$, $DF = 28$ and from Table T-2 we get $T2 = 2.05$. From Exhibit 9.3.2 we get `Stdev`$(\widehat{\beta}_1) = 0.5237$. The endpoints of a 95% confidence interval for β_1 using the results in Exhibit 9.3.2 are

$$L = 19.641 - (2.05)(0.5237) = \$18.56 \text{ per year}$$

and

$$U = 19.641 + (2.05)(0.5237) = \$20.72 \text{ per year.}$$

EXPLORATION 9.3.1

Over the past three years, 7,031 students have completed the first course in statistics at a major university. The department head is interested in knowing the relationship between the midterm and final examination scores in this course. Specifically, the interest is in predicting a student's final exam score (Y) using the student's midterm exam score (X). Using these predictions, help sessions can be provided for students who are predicted to receive a "low" final-exam score. All of the exam scores for the past three years are written in grade books stored in the department. However, it would be too time consuming to enter all these scores into a computer database. The department head decides to obtain the exam scores for a simple random sample of 50 students. The data are presented in Table 9.3.2 and stored in the file **statgrde.mtp** on the data disk. Column 1 in the file contains the student number, column 2 contains the variable "Final" (Y), and column 3 contains the variable "Midterm" (X). We suppose the assumptions in Box 9.3.1 are satisfied, where the population regression function is given by

$$\mu_Y(X) = \beta_0 + \beta_1 X \qquad \text{"for } X \text{ from 50 to 100."}$$

A scatter plot of these data is shown in Figure 9.3.2.

TABLE 9.3.2 Sample of Midterm and Final Grades

Student number	Final-test score = Y	Midterm test score = X	Student number	Final-test score = Y	Midterm test score = X
1	67	73	26	58	63
2	78	80	27	74	79
3	72	75	28	61	69
4	77	79	29	71	77
5	61	65	30	65	72
6	77	81	31	82	87
7	78	84	32	79	82
8	75	81	33	65	70
9	70	74	34	83	85
10	62	67	35	76	79
11	78	82	36	70	72
12	94	95	37	78	79
13	50	56	38	79	80
14	62	66	39	74	76
15	79	83	40	80	84
16	67	72	41	76	81
17	72	74	42	58	65
18	85	86	43	81	84
19	79	85	44	62	69
20	80	85	45	63	67
21	72	75	46	78	80
22	85	85	47	68	71
23	81	84	48	62	65
24	77	79	49	75	76
25	79	83	50	84	85

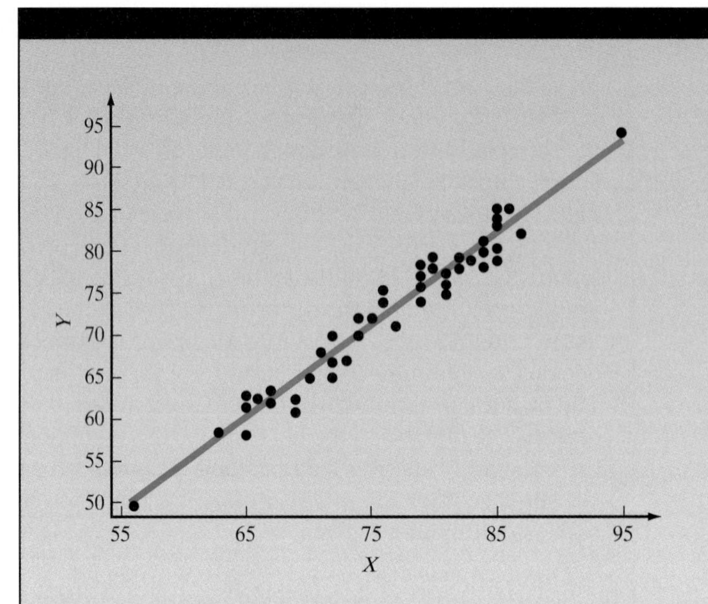

FIGURE 9.3.2
Scatter Plot of a Sample of Final Scores (Y) Versus Midterm Scores (X)

Some of the calculations that will be needed are presented in Exhibit 9.3.3, which was produced using Minitab.

```
------------------------------------------------------------
The regression equation is
Y = - 11.5 + 1.10 X

Predictor        Coef        Stdev      t-ratio        p
Constant      -11.486        2.525        -4.55    0.000
X             1.10070      0.03267        33.69    0.000

S = 1.777         R-sq = 95.9%      R-sq(adj) = 95.9%
------------------------------------------------------------
```

EXHIBIT 9.3.3

1. Does the following population regression function seem appropriate for this study?

$$\mu_Y(X) = \beta_0 + \beta_1 X \qquad \text{“for } X \text{ from 50 to 100.”}$$

Answer: *Yes. Based on Figure 9.3.2, a linear population regression function seems appropriate.*

2. What is the estimate of the population regression function?
Answer: *The estimate of the population regression function obtained from Exhibit 9.3.3 is*

$$\widehat{\mu}_Y(X) = \widehat{\beta}_0 + \widehat{\beta}_1 X = -11.5 + 1.1X \qquad \text{“for } X \text{ from 50 to 100.”}$$

This is the sample regression function.

3. What symbol represents the predicted value of Y for a specified value of X?
Answer: *The symbol is* $\widehat{Y}(X)$. *Since* $\widehat{Y}(X)$ *is equal to the sample regression function* $\widehat{\mu}_Y(X)$,

$$\widehat{Y}(X) = \widehat{\mu}_Y(X) = \widehat{\beta}_0 + \widehat{\beta}_1 X = -11.5 + 1.1X \qquad \text{“for } X \text{ from 50 to 100.”}$$

4. What symbol represents the average final-exam score of all students who received a midterm exam score of 75?
Answer: *The symbol is* $\mu_Y(75)$.

5. What is the value of $\mu_Y(75)$?
Answer: $\mu_Y(75)$ *is a population parameter, and its value cannot be determined without collecting the exam scores for all 7,031 students.*

6. Estimate the average final-exam score of all students who received a score of 75 on the midterm exam.
Answer: *This estimate is*

$$\widehat{\mu}_Y(75) = \widehat{\beta}_0 + \widehat{\beta}_1(75) = -11.5 + 1.1(75) = 71.$$

7. A student received a score of 75 on the midterm exam. Predict this student's final-exam score.
Answer: *This student's predicted final-exam score is* $\widehat{Y}(75)$, *where*

$$\widehat{Y}(75) = \widehat{\mu}_Y(75) = 71.$$

8. Obtain a 95% prediction interval for $Y(75)$.
Answer: *We use the following Minitab command and subcommands to compute this prediction interval:*

```
MTB > regress c2 1 c3;
SUBC> predict 75;
SUBC> confidence = 95.
```

A portion of the output is shown in Exhibit 9.3.4. From this exhibit we observe that the 95% prediction interval for $Y(75)$ *is from 67.45 to 74.68.*

9. Find a 95% confidence interval for $\mu_Y(75)$, the average final-exam score of all students who received a score of 75 on the midterm exam.
Answer: *The 95% confidence interval for* $\mu_Y(75)$ *obtained from Exhibit 9.3.4 is 70.54 to 71.59. Note that the 95% prediction interval for* $Y(75)$ *obtained in part (8) is wider than the 95% confidence interval for* $\mu_Y(75)$. *As stated earlier, this will always be true since there is greater uncertainty in predicting a single* Y *value to be selected at random from a subpopulation than there is in estimating the mean of the subpopulation.*

10. This semester's midterm exam was given yesterday, and a student received a grade of 85. What score is she predicted to receive on the final exam?

```
-----------------------------------------------------------------
Regression Analysis

The regression equation is
Y = - 11.5 + 1.10 X

Predictor        Coef        Stdev     t-ratio        p
Constant      -11.486        2.525       -4.55    0.000
X             1.10070      0.03267       33.69    0.000

S = 1.777         R-sq = 95.9%      R-sq(adj) = 95.9%

     Fit  Stdev.Fit       95.0% C.I.          95.0% P.I.
  71.067      0.259   (70.546,71.588)   (67.455,74.679)
-----------------------------------------------------------------
```

EXHIBIT 9.3.4

Answer: *If a student from the sampled population of 7,031 students scored an 85 on the midterm exam, her predicted final-exam score would be* $\hat{Y}(85) = -11.5 + 1.1(85) = 82$. *However, the student in question is not a member of the sampled population. Rather, she belongs to a target population of students who are presently enrolled in the course and have not yet taken the final exam. If the sampled population is similar to the target population, the score of 82 can be used to predict the final-exam score for this student.*

THINKING STATISTICALLY: SELF-TEST PROBLEMS

9.3.1 The vice president of a national chain of restaurants wants to study the impact of local advertising on sales. To study this relationship, a simple random sample of 30 restaurants was selected from the population of restaurants in the chain. For each restaurant in the sample, the total sales (Y) and the amount of money spent on local advertising (X) for the past month were recorded. These data are stored in the file **advertis.mtp** on the data disk. Column 1 contains the sample number, column 2 contains Y, and column 3 contains X. A scatter plot of the data is shown in Figure 9.3.3.

a. From the scatter plot does it appear that the following population regression function is appropriate?

$$\mu_Y(X) = \beta_0 + \beta_1 X \qquad \text{“for } X \text{ from 900 to 6,000.”}$$

b. Suppose that the assumptions in Box 9.3.1 are satisfied. The data stored in the file **advertis.mtp** were retrieved, and the following Minitab command and subcommands were executed to produce the output in Exhibit 9.3.5.

```
MTB > regress c2 1 c3;
SUBC> predict 2500;
SUBC> confidence = 95.
```

FIGURE 9.3.3
Scatter Plot of Sales (Y) Versus Advertising (X)

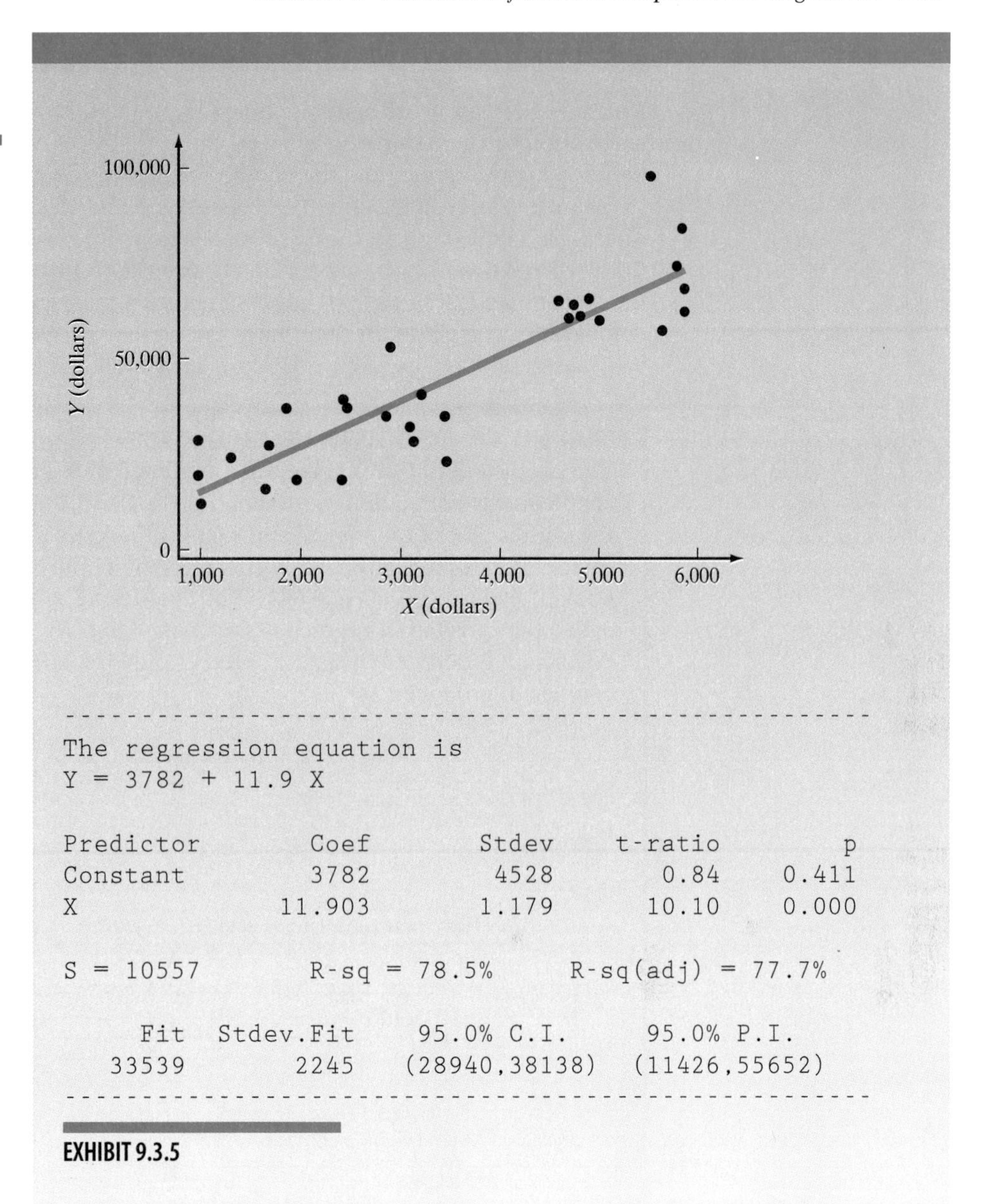


```
The regression equation is
Y = 3782 + 11.9 X

Predictor        Coef        Stdev    t-ratio        p
Constant         3782         4528       0.84    0.411
X              11.903        1.179      10.10    0.000

S = 10557        R-sq = 78.5%      R-sq(adj) = 77.7%

     Fit  Stdev.Fit     95.0% C.I.         95.0% P.I.
   33539       2245   (28940,38138)    (11426,55652)
```

EXHIBIT 9.3.5

What symbol represents the difference between the average monthly sales of restaurants that spent \$3,000 per month in advertising and those that spent \$2,000 per month?

c. Estimate $1{,}000\beta_1$.

d. Predict sales for a restaurant that spent \$2,500 on local advertising and construct a 95% prediction interval.

e. What is the estimate of the average sales of all restaurants that spent \$2,500 for advertising last month?

f. Find a 95% confidence interval for the average sales of all restaurants that spent \$2,500 for advertising last month.

9.4 CORRELATION

You have perhaps heard statements similar to the following: "I wonder if there is an association between smoking and lung cancer?" Or, "I believe there is an association between highway speed and the number of highway deaths." As stated in Section 9.1, there is often interest in studying the association between two quantitative variables. There are many ways to measure association; we discuss one of these measures, **linear correlation**. Linear correlation is also called **Pearson correlation**, in honor of the statistician Karl Pearson, who studied this measure extensively. A dictionary definition of correlation is "mutually associated" or "mutually related." Definition 9.4.1 describes how linear correlation is measured in a bivariate population.

DEFINITION 9.4.1: COEFFICIENT OF LINEAR CORRELATION

The coefficient of linear correlation of Y and X is a measure of how well the simple linear population regression function $\mu_Y(X) = \beta_0 + \beta_1 X$ represents the scatter plot of the population values $(Y_1, X_1), (Y_2, X_2), \ldots, (Y_N, X_N)$. That is, it measures how tightly clustered the population values are about the straight line $\mu_Y(X) = \beta_0 + \beta_1 X$. The coefficient of linear correlation is often called the **correlation coefficient** between Y and X. This correlation coefficient is denoted by $\rho_{Y,X}$, where ρ is the Greek letter **rho**. The correlation coefficient is defined as

$$\rho_{Y,X} = \frac{\sigma_{Y,X}}{\sigma_Y \sigma_X},$$

where

$$\sigma_{Y,X} = \frac{(Y_1 - \mu_Y)(X_1 - \mu_X) + (Y_2 - \mu_Y)(X_2 - \mu_X) + \cdots + (Y_N - \mu_Y)(X_N - \mu_X)}{N},$$

$$\sigma_Y = \sqrt{\frac{(Y_1 - \mu_Y)^2 + (Y_2 - \mu_Y)^2 + \cdots + (Y_N - \mu_Y)^2}{N}},$$

$$\sigma_X = \sqrt{\frac{(X_1 - \mu_X)^2 + (X_2 - \mu_X)^2 + \cdots + (X_N - \mu_X)^2}{N}},$$

$$\mu_Y = \frac{Y_1 + Y_2 + \cdots + Y_N}{N},$$

$$\mu_X = \frac{X_1 + X_2 + \cdots + X_N}{N}.$$

You will recognize μ_Y and σ_Y as the mean and the standard deviation of the population of Y values and μ_X and σ_X as the mean and the standard deviation of the population of X values. The quantity $\sigma_{Y,X}$ is called the **covariance** between Y and X. The correlation coefficient $\rho_{Y,X}$ has the following properties:

1. $\rho_{Y,X}$ is a number between -1 and $+1$.
2. Positive values of $\rho_{Y,X}$ indicate that as the X values increase, there is a *tendency* for the Y values to increase. We say that there is a "positive association" between Y and X.

3. Negative values of $\rho_{Y,X}$ indicate that as the X values increase, there is a *tendency* for the Y values to decrease. We say that there is a "negative association" between Y and X.

4. If $\rho_{Y,X} = +1$ or $\rho_{Y,X} = -1$, then all the population values lie exactly on a straight line. This straight line has a positive slope if $\rho_{Y,X} = +1$ and a negative slope if $\rho_{Y,X} = -1$.

5. The larger the magnitude (absolute value) of $\rho_{Y,X}$, the tighter the scatter of the population values about a straight line.

6. If $\rho_{Y,X} = 0$, there is no *linear* association between Y and X. In this case either Y and X are *unrelated* or Y and X are associated in a *nonlinear* fashion.

Figures 9.4.1 through 9.4.5 show scatter plots from several different (artificial) populations with different correlation coefficients.

The correlation coefficients in Figures 9.4.1 and 9.4.2 have the same magnitude but different signs. This means the lines shown in each figure have equal merit for describing the scatter plot. Notice how the tightness of the points around the straight line in each scatter plot differs in accordance with the value of the correlation coefficient. The swarm of points in Figure 9.4.3 is much closer to a straight line than in any of the other figures. Notice that Figures 9.4.4 and 9.4.5 both correspond to

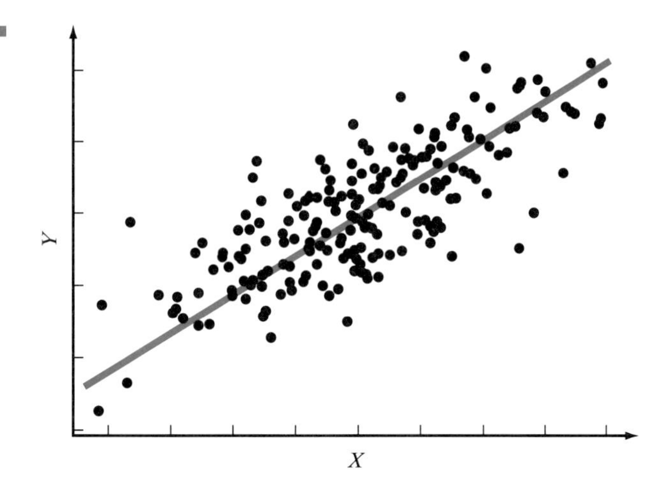

FIGURE 9.4.1
Scatter Plot of Y Versus X ($\rho_{Y,X} = 0.78$)

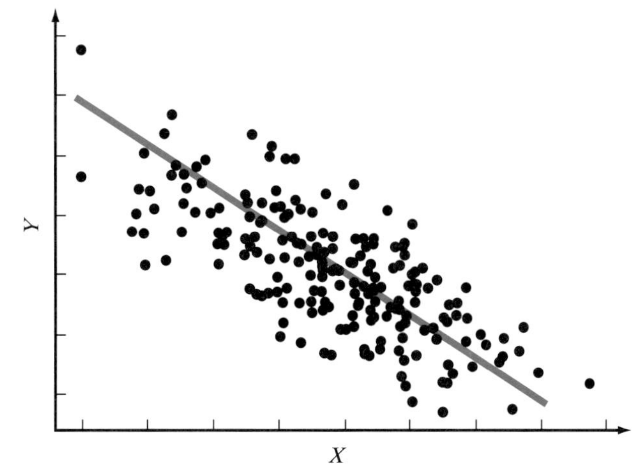

FIGURE 9.4.2
Scatter Plot of Y Versus X ($\rho_{Y,X} = -0.78$)

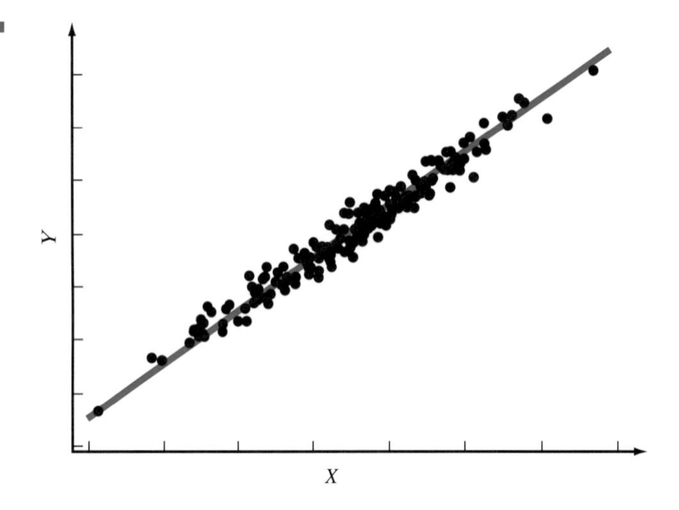

FIGURE 9.4.3
Scatter Plot of Y Versus X
($\rho_{Y,X} = 0.93$)

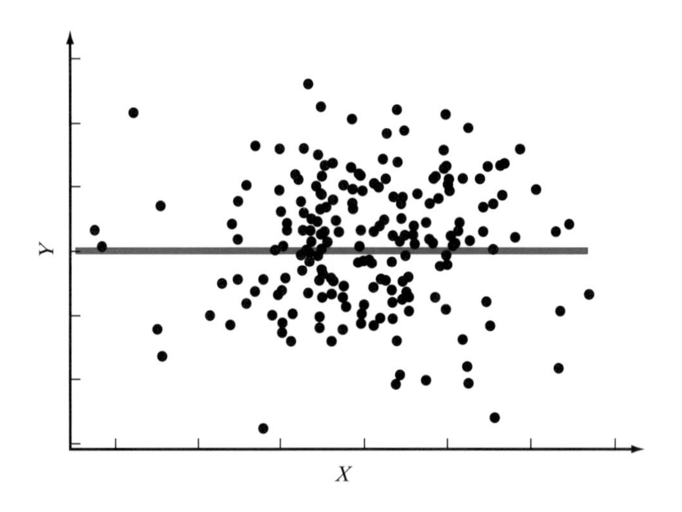

FIGURE 9.4.4
Scatter Plot of Y Versus X
($\rho_{Y,X} = 0$)

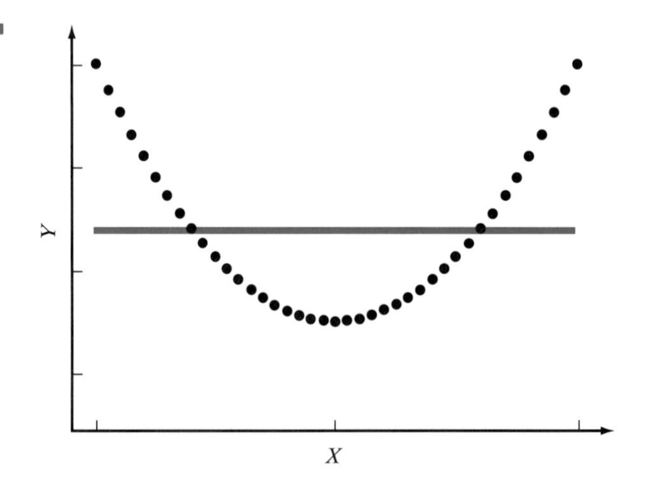

FIGURE 9.4.5
Scatter Plot of Y Versus X
($\rho_{Y,X} = 0$)

populations with $\rho_{Y.X} = 0$. The variables appear to have no association in Figure 9.4.4. However, in Figure 9.4.5 there is a perfect relationship between Y and X although it is not linear. Figure 9.4.5 serves as a reminder that $\rho_{Y.X}$ only measures how well a *straight* line represents the points in a scatter plot.

The larger the magnitude of the *correlation* coefficient $\rho_{Y.X}$, the tighter the swarm of points is to the linear population *regression* function $\mu_Y(X) = \beta_0 + \beta_1 X$. This suggests that there is a relationship between correlation and regression. This relationship is displayed in Box 9.4.1.

BOX 9.4.1: RELATIONSHIP BETWEEN CORRELATION AND REGRESSION

When assumptions (1) and (2) in Box 9.3.1 are satisfied (that is, if the population regression function is $\mu_Y(X) = \beta_0 + \beta_1 X$, and if all subpopulation standard deviations are equal), then

$$\beta_1 = \rho_{Y.X}\left(\frac{\sigma_Y}{\sigma_X}\right) \quad \text{and} \quad \beta_0 = \mu_Y - \beta_1\mu_X.$$

From Box 9.4.1 we see that if $\rho_{Y.X} = 0$, then $\beta_1 = 0$ and $\beta_0 = \mu_Y$. It follows that if $\rho_{Y.X} = 0$, then $\mu_Y(X) = \beta_0 + \beta_1 X = \mu_Y + 0(X) = \mu_Y$, and X is of no value in predicting Y using $\mu_Y(X) = \beta_0 + \beta_1 X$. The square of the correlation coefficient, $\rho^2_{Y.X}$, provides an interesting interpretation of how well X predicts Y when the population regression function is $\mu_Y(X) = \beta_0 + \beta_1 X$. We illustrate this interpretation in Example 9.4.1.

EXAMPLE 9.4.1

WEIGHT-LOSS CLINIC. A man who is interested in losing weight goes to a weight-loss clinic and asks, "If I complete your 12-month program, how much weight can I expect to lose?" The director replies, "The average weight loss was 54 pounds and the standard deviation was 7.5 pounds for the 2,000 men who participated in the program over the past two years." Then the director added, "However, the weight loss of an individual depends on his initial weight when he enters the program. The heavier the man, the more weight he tends to lose. How much do you weigh?" The man replied, "I weigh 200 pounds." The director examined his computer screen and said, "The average weight loss was 40 pounds and the standard deviation was 3 pounds for all men who weighed 200 pounds when they entered the program."
To put this conversation into a regression context, the population of items consists of the 2,000 men who completed the program during the past two years. In the weight-loss program, the amount of weight loss is related to the initial weight of each participant, so there are two variables of interest for each man: "Weight loss" (Y) and "Initial weight" (X). Suppose the population regression function is

$$\mu_Y(X) = -0.575 + 0.203X \qquad \textbf{Line 9.4.1}$$

and the standard deviation (σ) of each subpopulation of Y values for every value of X is 3 pounds. There are two possible answers to the question, "If I complete the 12-month program, how much weight can I expect to lose?"

SOLUTION Answer (1): If initial weight X is not considered, a person's weight loss can be viewed as a simple random sample of size 1 from the population of weight

losses $Y_1, Y_2, \ldots, Y_{2000}$ with mean $\mu_Y = 54$ pounds and standard deviation $\sigma_Y = 7.5$ pounds.
Answer (2): If the initial weight of $X = 200$ pounds is considered, a person's weight loss can be viewed as a simple random sample of size 1 from a subpopulation of the weight losses of all men with an initial weight of 200 pounds. Using Line 9.4.1, this subpopulation has mean $\mu_Y(200) = -0.575 + 0.203(200) = 40.0$ pounds and standard deviation $\sigma = 3$ pounds.

As stated in Chapter 5, the standard deviation measures how well the *mean* of a population (or subpopulation) represents a value selected at random from that population (or subpopulation). The smaller the standard deviation, the better the mean represents the population (or subpopulation). Thus, since the standard deviation σ in answer (2) of Example 9.4.1 is less than the standard deviation σ_Y in answer (1), we can conclude that $\mu_Y(X)$ will predict Y better than μ_Y. When assumptions (1) and (2) in Box 9.3.1 are satisfied, it can be shown that $\sigma \le \sigma_Y$. This means that μ_Y will never be better than $\mu_Y(X)$ for predicting Y. Given this information, you might wonder why $\mu_Y(X)$ is not always used instead of μ_Y for predicting Y. The reason is that knowledge of X is required when using $\mu_Y(X)$. If X is not available, then μ_Y must be used to predict Y because it does not require knowing X. The cost of obtaining X and the relative sizes of σ_Y and σ must be considered in deciding whether to use $\mu_Y(X)$ or μ_Y to predict Y. We now define a quantity that compares the relative sizes of σ_Y and σ.

DEFINITION 9.4.2: COEFFICIENT OF DETERMINATION

Consider a bivariate population $(Y_1, X_1), (Y_2, X_2), \ldots, (Y_N, X_N)$. If assumptions (1) and (2) in Box 9.3.1 are satisfied (remember, these assumptions say that the population regression function is $\mu_Y(X) = \beta_0 + \beta_1 X$ and the standard deviation of each subpopulation is equal to σ), then the ratio

$$\frac{\sigma_Y^2 - \sigma^2}{\sigma_Y^2}$$

is a measure of how much better $\mu_Y(X)$ is than μ_Y for predicting Y. This ratio is called the **coefficient of determination** of Y with X. It can be shown that the coefficient of determination is equal to the *square of the linear correlation coefficient.* That is,

$$\rho_{Y,X}^2 = \frac{\sigma_Y^2 - \sigma^2}{\sigma_Y^2}. \qquad \textbf{Line 9.4.2}$$

In Example 9.4.1 we have $\sigma = 3.0$ and $\sigma_Y = 7.5$. Using Definition 9.4.2, we get

$$\rho_{Y,X}^2 = \frac{(7.5)^2 - (3)^2}{(7.5)^2} = 0.84.$$

The coefficient of determination $\rho_{Y,X}^2$ can take on values from 0 to +1. The larger the value of $\rho_{Y,X}^2$, the smaller σ^2 is relative to σ_Y^2, and the better the population regression function $\mu_Y(X) = \beta_0 + \beta_1 X$ is than μ_Y for predicting Y. In Figures 9.4.1

through 9.4.5 we showed the scatter plot of Y versus X and the corresponding value of $\rho_{Y,X}$ for five different bivariate populations. As you can see, the larger the magnitude of $\rho_{Y,X}$, the larger the value of $\rho^2_{Y,X}$, and the closer the swarm of points is to the line $\mu_Y(X) = \beta_0 + \beta_1 X$. The correlation coefficient $\rho_{Y,X}$ will always have the same sign as β_1, the slope of the linear population regression function. Finally, we previously noted that if $\rho_{Y,X} = 0$, then $\beta_1 = 0$ and $\mu_Y(X) = \mu_Y$. You can also see from Line 9.4.2 that if $\rho_{Y,X} = 0$, then $\rho^2_{Y,X} = 0$ and $\sigma^2_Y = \sigma^2$.

Estimating $\rho_{Y,X}$ Using Minitab

Since $\rho_{Y,X}$ is a population parameter, in most practical problems it is unknown and may be estimated using a simple random sample of n values, $(y_1, x_1), \ldots, (y_n, x_n)$. The estimate of $\rho_{Y,X}$ is denoted by $\widehat{\rho}_{Y,X}$.

In Section 9.9 we present formulas that can be used to compute $\widehat{\rho}_{Y,X}$. The Minitab command to perform this calculation is

```
MTB > correlate ci cj
```

where the Y values are in column `i` and the X values are in column `j`.

EXAMPLE 9.4.2

WARRANTY FOR VCRS. Consider Example 9.3.1, where a simple random sample of 30 VCRs was used to compute the sample regression function of maintenance cost (Y) on the age of a VCR (X). The data are shown in Table 9.3.1 and stored in the file **vcrsampl.mtp** on the data disk, where Y is in column 2 and X is in column 3. To obtain $\widehat{\rho}_{Y,X}$, retrieve the data and enter the following Minitab command in the Session window:

```
MTB > correlate c2 c3
```

The output is

```
    Correlations (Pearson)

Correlation of y and x = 0.990
```

The estimate of $\rho_{Y,X}$ is $\widehat{\rho}_{Y,X} = 0.99$, which is very close to $+1$. This indicates that the population regression function $\mu_Y(X) = \beta_0 + \beta_1 X$ is very good relative to μ_Y for predicting Y. The last line of the Minitab regression output in Exhibit 9.3.1 contains the quantity

```
R-sq=98.0%
```

The term `R-sq` is $\widehat{\rho}^2_{Y,X}$, the estimate of the population coefficient of determination. Notice that $(0.990)^2 = 0.98 = 98\%$.

The estimate $\widehat{\rho}_{Y,X}$ is a number between -1 and $+1$. Positive values of $\widehat{\rho}_{Y,X}$ indicate that as the sample values of X increase, the sample values of Y tend to increase. Negative values of $\widehat{\rho}_{Y,X}$ indicate that as the sample values of X increase, the sample values of Y tend to decrease. The larger the magnitude of $\widehat{\rho}_{Y,X}$, the tighter the scatter of the sample values is about a straight line.

THINKING STATISTICALLY: SELF-TEST PROBLEMS

9.4.1 A scientist wants to identify factors that determine the energy level of third-graders at the end of a typical school day. The population consists of 1,500 third-grade students in a large school district. Let Y, X, and

Z represent the energy level, blood pressure, and weight, respectively, of these students. An investigator uses the population to compute the population regression functions for Y on X and Y on Z, namely,

$$\mu_Y(X) = -23.1 + 11.7X \qquad \text{and} \qquad \mu_Y(Z) = 308.3 - 1.9Z.$$

The population correlation coefficients are $\rho_{Y,X} = 0.37$ and $\rho_{Y,Z} = -0.59$.

a. Is $\mu_Y(X)$ or $\mu_Y(Z)$ better for predicting energy level?

b. Suppose that the two population correlation coefficients are $\rho_{Y,X} = 0.57$ and $\rho_{Y,Z} = -0.57$. Is $\mu_Y(X)$ or $\mu_Y(Z)$ better for predicting energy level?

9.4.2 Suppose that the population regression function in an investigation is

$$\mu_Y(X) = \beta_0 + \beta_1 X.$$

If $\rho_{Y,X} = 0$, what does this tell us about β_1?

9.4.3 Figure 9.4.6 contains a scatter plot of a simple random sample from the bivariate population $(Y_1, X_1), (Y_2, X_2), \ldots, (Y_N, X_N)$. Which of the following statements is true?

a. $\widehat{\rho}_{Y,X} > 0$

b. $\widehat{\rho}_{Y,X} = 0$

c. $\widehat{\rho}_{Y,X} < 0$

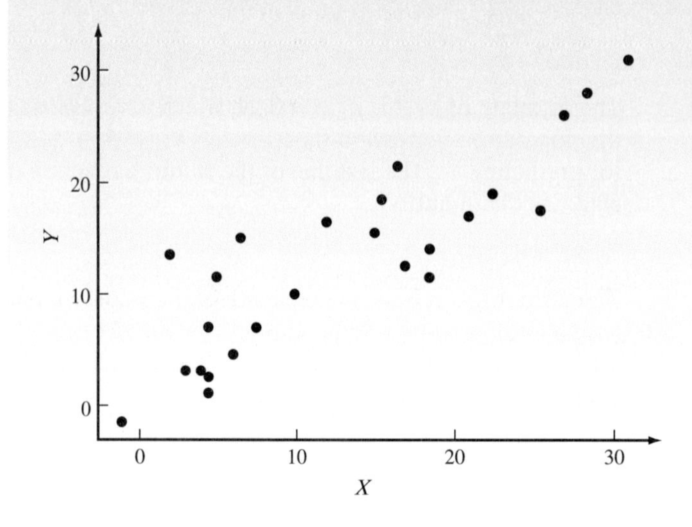

FIGURE 9.4.6
Scatter Plot of a Simple Random Sample from a Bivariate Population

9.5 CHECKING REGRESSION ASSUMPTIONS

For the statistical inference procedures discussed in Section 9.3 to be valid in regression problems, certain assumptions must be satisfied. These assumptions are given in Box 9.3.1 and for convenience are repeated in Box 9.5.1.

BOX 9.5.1: ASSUMPTIONS FOR STATISTICAL INFERENCE IN REGRESSION

Suppose an investigator is interested in studying the regression of Y on X in a bivariate population

$$(Y_1, X_1), (Y_2, X_2), \ldots, (Y_N, X_N).$$

For statistical inference procedures described in this chapter to be valid, the bivariate population must satisfy assumptions described in (1), (2), and (3) below. In addition, the sample must be obtained by either method (4a) or (4b) described below.

1. The subpopulation of Y values for each distinct value of X has a mean denoted by $\mu_Y(X)$, where

$$\mu_Y(X) = \beta_0 + \beta_1 X \qquad \text{"for specified values of } X\text{."}$$

The term "for specified values of X" defines the values of X for which this linear regression function is valid.

2. For each distinct value of X, the subpopulation of Y values must have the same standard deviation. This common standard deviation is denoted by σ (that is, the standard deviation of every subpopulation is equal to σ).

3. The subpopulation of Y values for each distinct value of X is a *normal* population. This assumption is not needed for point estimation but is required for confidence intervals and prediction intervals.

4a. A simple random sample of n items is selected from the sampled population and Y and X are observed for each sampled item. The sample values are denoted as $(y_1, x_1), (y_2, x_2), \ldots, (y_n, x_n)$.

4b. An investigator chooses any desired set of X values. A simple random sample of one or more Y values is then selected from each subpopulation determined by these X values to obtain a sample size of n. As in (4a), the sample values are denoted as $(y_1, x_1), (y_2, x_2), \ldots, (y_n, x_n)$.

Statisticians have developed an array of techniques for using sample values to help determine if the assumptions in Box 9.5.1 are satisfied. We recommend the following four steps when examining the assumptions in Box 9.5.1. If it appears that some or all of these assumptions are violated, you should consult advanced textbooks on regression analysis or seek the aid of a professional statistician.

1. **Determine the procedure used to select the sample.** You should determine if the sample was collected using either method (4a) or (4b) in Box 9.5.1.

2. **Construct a scatter plot.** A scatter plot of y_i versus x_i is one of the most helpful tools for checking regression assumptions. By examining the scatter plot, you should be able to decide if the population regression function is linear as stated in assumption (1) of Box 9.5.1. When the population regression function appears to be nonlinear, you may consider more complex population regression functions such as the curvilinear function described in Example 9.1.3. Methods for using more complex population regression functions can be found in advanced textbooks on regression.

3. **Examine the residuals.** If the scatter plot of the data suggests that assumption (1) in Box 9.5.1 is satisfied, obtain the prediction function $\widehat{Y}(X) = \widehat{\beta}_0 + \widehat{\beta}_1 X$. Using this function, compute the predicted values $\widehat{Y}(x_i) = \widehat{\beta}_0 + \widehat{\beta}_1 x_i$ for each item in the sample. The difference $y_i - \widehat{Y}(x_i)$ (observed Y value minus predicted Y value of the ith sample item) is denoted by $\widehat{e}_i$ and called the **residual** for the ith sample item. The behavior of the residuals is an important indicator for deciding if the assumptions in Box 9.5.1 are satisfied. Most investigators transform the residuals using the formula

$$r_i = \frac{\widehat{e}_i}{\hat{\sigma}\sqrt{1 - \frac{1}{n} - \frac{(x_i - \bar{X})^2}{\text{SSD}x}}}$$

where

$$\text{SSD}x = (x_1 - \bar{X})^2 + (x_2 - \bar{X})^2 + \cdots + (x_n - \bar{X})^2.$$

The r_i are called **standardized residuals**. If the assumptions in Box 9.5.1 are satisfied, the standardized residuals will behave as a simple random sample from a *normal* population with mean 0 and standard deviation 1. So, to help decide if the assumptions in Box 9.5.1 are satisfied, the following graphs should be constructed to see whether if the standardized residuals appear to be as a simple random sample from a *normal* population with $\mu = 0$ and $\sigma = 1$:

a. *a frequency histogram of the standardized residuals (r_i).* If the assumptions in Box 9.5.1 are satisfied, the frequency histogram of the standardized residuals should look like the frequency histogram of a simple random sample from a *normal* population with mean 0 and standard deviation 1.

b. *a normal rankit-plot of the standardized residuals.* If the assumptions in Box 9.5.1 are satisfied, this plot should appear to follow the pattern of a straight line through the origin with a slope equal to 1.

c. *a scatter plot with the standardized residuals on the vertical axis and the predicted values $\widehat{Y}(x_i)$ on the horizontal axis.* If the assumptions in Box 9.5.1 are satisfied, the points in this scatter plot should be randomly scattered about the horizontal line through zero. There should be no pattern in the scatter plot.

4. **Look for outliers.** As mentioned earlier in the book, outliers are sample values that do not appear to be from the same population as the rest of the sample. Outliers sometimes occur because mistakes were made during data collection, transcription, or computer entry. Sometimes they occur because a sample item is not a member of the sampled population. For example, consider a sampled population of men between the ages of 40 and 50 for a study of skin cancer. A sample was selected by calling random telephone numbers and asking to speak to any household male member between the ages of 40 and 50. If a person responding to the telephone call answers that he is 50 when in fact he is 65, he will incorrectly be included in the sample even though he is not a member of the sampled population.

   There are a couple of ways to identify outliers. One way is to examine the values of the standardized residuals. Since these residuals are supposed to behave as a simple random sample from a *normal* population with mean 0 and standard deviation 1, an outlier can be defined as any item with a standardized residual that exceeds 3 in magnitude. Outliers can also be detected in a normal rankit-plot of the standardized residuals as points that do not conform to the straight-line pattern.

Each sample value that appears to be an outlier should be thoroughly investigated to determine the reason for its occurence. If you cannot determine why a sample value is an outlier, it may be wise to carry out two analyses—one with the outlier included and one with it excluded. If the two analyses lead to identical decisions, then the outlier does not affect the result. If the two analyses lead to different decisions, you must look further into the matter.

We now show you how to use Minitab to check regression assumptions using the data in Example 9.1.1.

EXAMPLE 9.5.1

CHECKING REGRESSION ASSUMPTIONS FOR THE SCHOLARSHIP DATA. In Example 9.1.1, the director of the foundation obtained a simple random sample of 60 students from a population of 2,000 college students who applied for scholarships and completed the first year during the past five years. The variables of interest are "Score" (X) and "GPA" (Y). The sample values are shown in Table 9.1.1 and stored in the file **gpasampl.mtp**. We now examine steps 1 through 4 for checking the regression assumptions in Box 9.5.1.

1. **Determine the procedure used to select the sample.** The sample was obtained using method (4a).
2. **Construct a scatter plot.** A scatter plot is shown in Figure 9.5.1. By examining the scatter plot, we have no reason to doubt that the population regression function is linear. Thus, the regression model $\mu_Y(X) = \beta_0 + \beta_1 X$ appears to be appropriate.

FIGURE 9.5.1
Scatter Plot of GPA (Y) Versus Achievement-Test Scores (X)

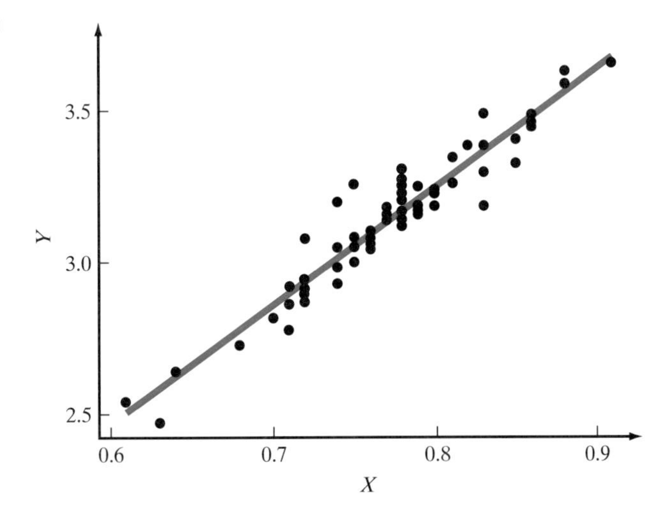

3. **Examine the residuals.** The Minitab command `regress` can be used to obtain a prediction function $\widehat{Y}(X)$, the standardized residuals r_i, and the predicted values $\widehat{Y}(x_i)$. The command is

```
MTB > regress ci 1 cj (std res in) ck (pred values in) cm
```

where Y is in column `i` and X is in column `j`. The phrases in parentheses are not needed but are useful in remembering the command. Minitab will compute and store the standardized residuals in column k and the predicted values in column m. For the scholarship data, enter the following command in the Session window:

```
MTB > regress c2 1 c3 c4 c5
```

The standardized residuals will be stored in column 4 and the predicted values in column 5. The resulting prediction function is

$$\widehat{Y}(X) = 0.10 + 3.94X \qquad \text{“for } X \text{ from 0.60 to 1.00.”}$$

Figure 9.5.2 shows a frequency histogram of the standardized residuals.

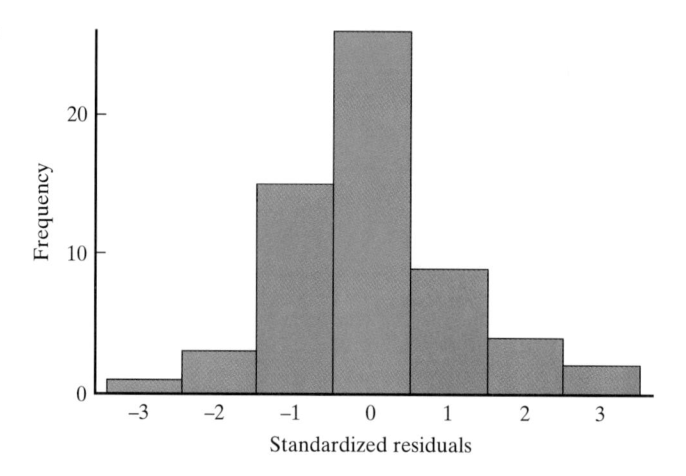

FIGURE 9.5.2
Frequency Histogram of Standardized Residuals of GPA

To obtain the nscores for the standardized residuals and construct a normal rankit-plot, enter the following Minitab commands in the Session window:

```
MTB > nscores c4 c6
MTB > %fitline c4 c6
```

The nscores are placed in column 6. The normal rankit-plot is produced by the macro **fitline** and shown in Figure 9.5.3. Based on the normal rankit-plot, there

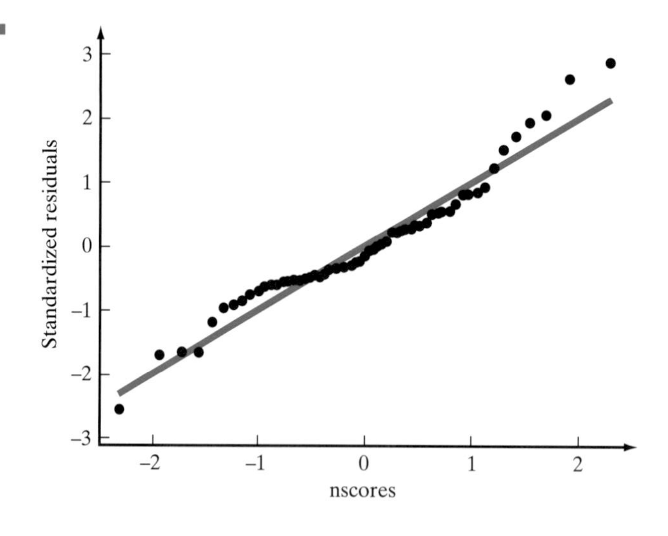

FIGURE 9.5.3
Normal Rankit-plot of the Standardized Residuals

is some reason to doubt that the standardized residuals came from a *normal* population since the points in the northeast corner veer away from the straight line. If this is true, statistical inferences may be affected, and the advice of a statistician is warranted.
Figure 9.5.4 contains a scatter plot of the standardized residuals against the predicted values.

FIGURE 9.5.4
Scatter Plot of the Standardized Residuals Versus Predicted Values

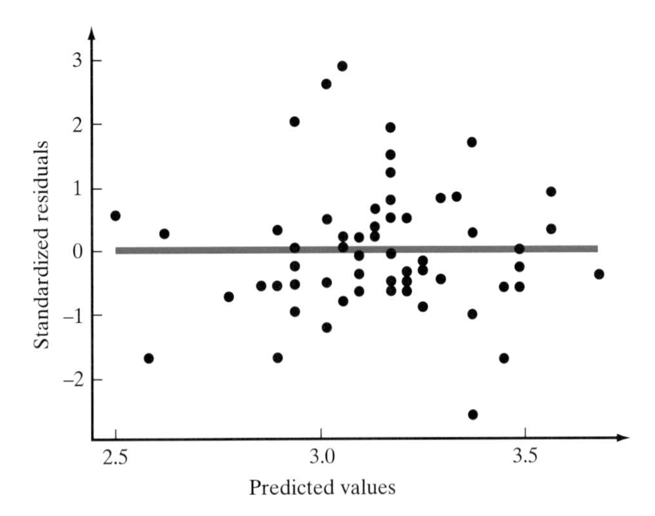

The scatter plot in Figure 9.5.4 shows no obvious pattern and appears to exhibit a random scatter about the horizontal line at zero.

4. **Outliers.** By examining the scatter plot, we find that no outliers appear to be in the sample. However, we can examine the sample more closely for outliers using the standardized residuals. We observe in Figure 9.5.4 that the standardized residual with the largest magnitude is less than 3, so there does not appear to be any outliers in the sample.

THINKING STATISTICALLY: SELF-TEST PROBLEMS

9.5.1 Consider the sample of 30 VCRs presented in Table 9.3.1 and stored in the file **vcrsampl.mtp**. A scatter plot of y_i and x_i is shown in Figure 9.5.5. A scatter plot of the standardized residuals against the predicted values $\widehat{Y}(x_i)$ is shown in Figure 9.5.6. A normal rankit-plot of the standardized residuals is shown in Figure 9.5.7.

a. Was the sample selected using either (4a) or (4b) in Box 9.5.1? If so, which one?

b. Does a linear regression function appear to be appropriate for this sample?

c. Do there appear to be any outliers in the sample?

d. Does it appear that the assumptions in Box 9.5.1 are satisfied?

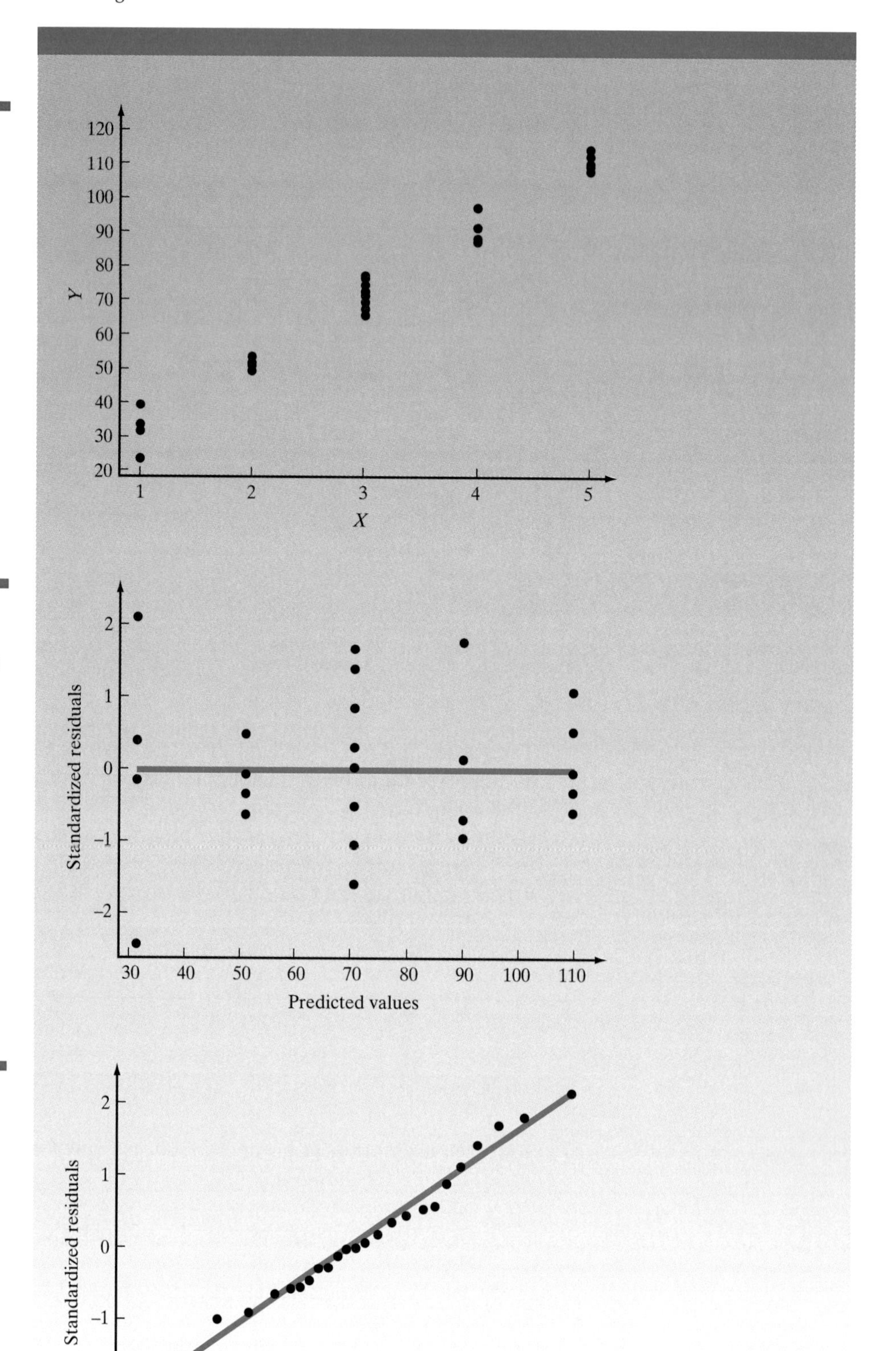

FIGURE 9.5.5
Scatter Plot of y_i Versus x_i for VCR Sample

FIGURE 9.5.6
Scatter Plot of Standardized Residuals Versus $\widehat{Y}(x_i)$ for VCR Sample

FIGURE 9.5.7
Normal Rankit-Plot of Standardized Residuals for VCR Sample

9.6 CHAPTER SUMMARY

KEY TERMS

1. predictor variable (X)
2. predicted variable (Y)
3. subpopulation
4. population regression function
5. sample regression function
6. prediction function
7. prediction interval
8. coefficient of linear correlation
9. coefficient of determination
10. residuals and standardized residuals

SYMBOLS

1. $\mu_Y(X) = \beta_0 + \beta_1 X$—population regression function
2. β_0—the intercept of a population regression function
3. β_1—the slope of a population regression function
4. $\widehat{\mu}_Y(X) = \widehat{\beta}_0 + \widehat{\beta}_1 X$—sample regression function
5. $\widehat{Y}(X) = \widehat{\beta}_0 + \widehat{\beta}_1 X$—prediction function
6. $Y(X)$—a value of Y to be selected at random from a subpopulation where all items have the same value of X
7. $\rho_{Y,X}$—population correlation coefficient between Y and X
8. $\widehat{\rho}_{Y,X}$—sample estimate of a population correlation coefficient between Y and X
9. $\widehat{e}_i$—residual of ith sample item
10. r_i—standardized residual of ith sample item

KEY CONCEPTS

1. A prediction function is used to predict the unknown value of a variable Y using the known value of a variable X.
2. The mean of a subpopulation can be used to predict the value of an item to be selected at random from the subpopulation.
3. The sample regression function is used as a prediction function when the entire population is not available.
4. The assumptions in Box 9.3.1 (Box 9.5.1) must be (approximately) satisfied in order to make valid statistical inferences in regression analysis.
5. A confidence interval for $\mu_Y(X)$ will always be shorter than a prediction interval for $Y(X)$.
6. $\mu_Y(X)$ is much better than μ_Y for predicting Y when σ is much less than σ_Y.
7. If $\rho_{Y,X} = 0$, then $\beta_1 = 0$.

SKILLS

1. Use a prediction function to predict values of Y.
2. Examine a scatter plot of sample values y_i and x_i, and determine if a linear population regression function is appropriate.
3. Make proper inferences concerning a regression of Y on X using information provided by a computer software package.
4. Use sample data to decide if the assumptions in Box 9.3.1 (Box 9.5.1) are (approximately) satisfied.

9.7 CHAPTER EXERCISES

9.1 An investigation was performed in which a simple random sample was used to compute the prediction function

$$\widehat{Y}(X) = 213.9 - 47.2X \qquad \text{“for } X \text{ from 35 to 92.”}$$

a. Compute $\widehat{Y}(42)$.

b. Compute $\widehat{Y}(90)$.

c. Compute $\widehat{Y}(30)$.

9.2 Indicate whether each of the following statements is true or false.

a. $\mu_Y(X) = \beta_0 + \beta_1 X$ is a population regression function.

b. $Y(X)$ is a prediction function.

c. $\widehat{\mu}_Y(X) = \widehat{\beta}_0 + \widehat{\beta}_1 X$ is a sample regression function.

d. $\mu_Y(X) = Y(X)$.

e. If $\mu_Y(X) = \beta_0 + \beta_1 X$ and $\beta_1 < 0$, then $\sigma_Y = \sigma$.

f. If $\mu_Y(X) = \beta_0 + \beta_1 X$, then it is possible that $\rho_{Y,X} = -1$.

g. If $\rho_{Y,X} = 0$, this always means that there is no association between Y and X.

9.3 A company that removes trash from beaches wants to predict the amount of trash that accumulates each day so that an adequate number of trucks can be dispatched to remove it by 6:00 a.m. the next morning. Past experience indicates that the daily high temperature is a good predictor of the amount of trash that will accumulate on a given day. The population regression function is

$$\mu_Y(X) = \beta_0 + \beta_1 X$$
$$\text{“for } X \text{ from 50 degrees to 90 degrees.”}$$

For the past three years the company has recorded the amount of trash collected in thousands of pounds (Y) and the daily high temperature in degrees Fahrenheit (X). From these data a simple random sample of 60 days was selected and used to compute the sample regression function

$$\widehat{\mu}_Y(X) = 23.10 + 0.21X$$
$$\text{“for } X \text{ from 50 degrees to 90 degrees.”}$$

We suppose throughout that the assumptions in Box 9.3.1 are satisfied.

a. What is the predicted amount of trash that must be removed by 6:00 a.m. if the high temperature of the previous day was 79 degrees?

b. What is the predicted amount of trash that must be removed by 6:00 a.m. if the high temperature the previous day was 100 degrees?

c. Give a reason why the population regression function might not be appropriate if the high temperature is 100 degrees.

9.4 An economics professor wants to study a person's incentive to find work during unemployment. Specifically, he wants to examine the relationship between the number of job applications filed during the first two months of unemployment (Y) and the most recent hourly wage rate before unemployment (X). A simple random sample of 25 unemployment insurance claimants in the state was obtained from the several hundred who applied for insurance during the past year. The data are stored in the file **unemploy.mtp** on the data disk. Column 1 contains the sample number, column 2 contains the number of job applications (Y), and column 3 contains the most recent hourly wage (X). A scatter plot is shown in Figure 9.7.1.

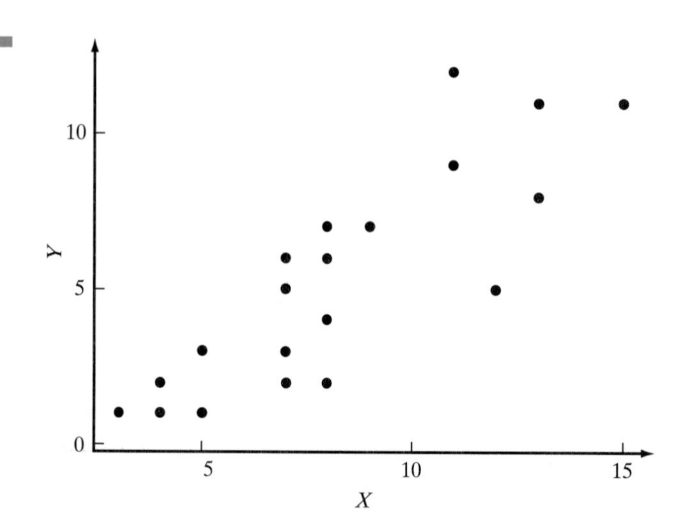

FIGURE 9.7.1
Scatter Plot of Number of Job Applications (Y) Versus Hourly Wage (X)

```
The regression equation is
Y = - 1.84 + 0.886 X

Predictor        Coef        Stdev     t-ratio        p
Constant       -1.840        1.045       -1.76    0.091
X              0.8862       0.1212        7.31    0.000

S = 1.814        R-sq = 69.9%       R-sq(adj) = 68.6%

    Fit   Stdev.Fit         95.0% C.I.              95.0% P.I.
  7.021       0.431    (  6.130,    7.913)   (   3.164,   10.879)
```

EXHIBIT 9.7.1

Exhibit 9.7.1 shows the Minitab output resulting from the following command and subcommands:

```
MTB > regress c2 1 c3;
SUBC> predict 10;
SUBC> confidence=95.
```

a. Does it appear that the following population regression function is appropriate?

$$\mu_Y(X) = \beta_0 + \beta_1 X \qquad \text{"for } X \text{ from \$3 to \$15"}$$

b. Suppose the assumptions in Box 9.3.1 are satisfied. Estimate $\mu_Y(10)$, the average number of job applications for the subpopulation of claimants who were earning $10 per hour when they became unemployed. What is the 95% confidence interval for $\mu_Y(10)$?

c. Predict $Y(10)$, the number of job applications for a claimant selected at random from the subpopulation of claimants who were earning $10 per hour when they became unemployed. What is the 95% prediction interval for $Y(10)$? Based on this interval, would you consider it unusual if the claimant made only four applications during the period of unemployment?

d. Interpret β_1 in the context of this problem. Construct a 95% confidence interval for β_1.

e. Estimate the correlation coefficient between number of job applications and hourly wage.

9.5 Officers of a large oil company want to examine the relationship between the cost to drill an oil well (Y) and the depth of the well (X). This relationship will be used to predict the cost to drill a new well. A simple random sample of 100 oil wells was obtained from a population of oil wells presently in operation. All of these wells were drilled to depths between 4,000 and 12,000 feet. The data are stored in the file **oilfield.mtp** on the data disk. Column 1 contains the sample number, column 2 contains Y, and column 3 contains X. Cost is reported in thousands of dollars and depth is reported in thousands of feet. A scatter plot is shown in Figure 9.7.2.

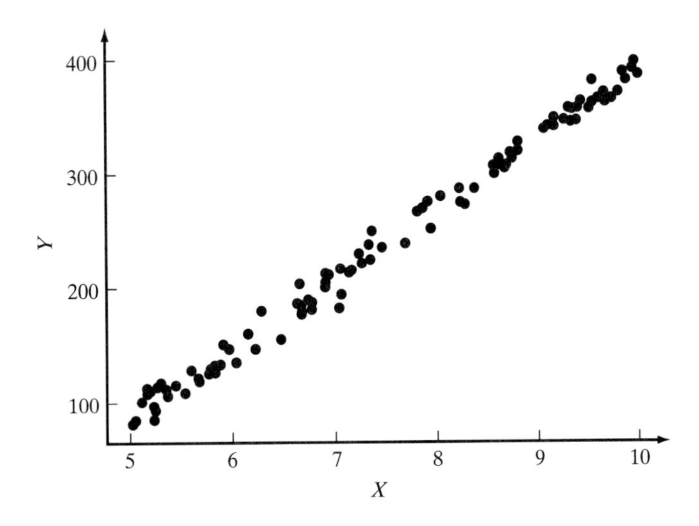

FIGURE 9.7.2
Scatter Plot of Cost (Y) Versus Depth (X)

```
The regression equation is
Y = - 224 + 61.4 X

Predictor        Coef        Stdev     t-ratio        p
Constant     -223.730        4.251      -52.63    0.000
X             61.4428       0.5505      111.61    0.000

S = 8.777         R-sq = 99.2%      R-sq(adj) = 99.2%
     Fit   Stdev.Fit          95.0% C.I.            95.0% P.I.
 267.812       0.911   ( 266.004, 269.621)   ( 250.296, 285.328)
```

EXHIBIT 9.7.2

A Minitab output that resulted from the following command and subcommands is shown in Exhibit 9.7.2:

```
MTB > regress c2 1 c3;
SUBC> predict 8;
SUBC> confidence = 95.
```

a. Does it appear that the population regression function $\mu_Y(X) = \beta_0 + \beta_1 X$ is appropriate for the range of X values from 4,000 to 12,000 feet?

b. What are the estimates of β_0 and β_1?

c. What is the sample regression function?

d. What population parameter represents the change in the average drilling cost when depth increases by 1 (thousand feet)? What is the point estimate of this parameter?

e. What is the predicted cost of a well that was drilled to a depth of 8 (thousand feet)? What is the 95% prediction interval?

f. What is the estimate of the average cost of all wells that were drilled to a depth of 8 (thousand feet)?

g. Estimate the population correlation coefficient between cost and depth.

h. Estimate the proportion that expresses how much better $\mu_Y(X)$ is than μ_Y for predicting cost.

9.6 A large utility company must predict electrical consumption for the population of residential customers. One variable that affects consumption is the size of the customer's residence. A simple random sample of 40 customers was selected from the population of residential customers. The average electrical consumption for the past month measured in kilowatt hours per day (Y) and the size of the residence measured in square feet (X) were recorded for each sample item. The sample data are stored in the file **elecusag.mtp** on the data disk, where column 1 contains the sample number, column 2 contains Y, and column 3 contains X. A scatter plot of the data is shown in Figure 9.7.3.

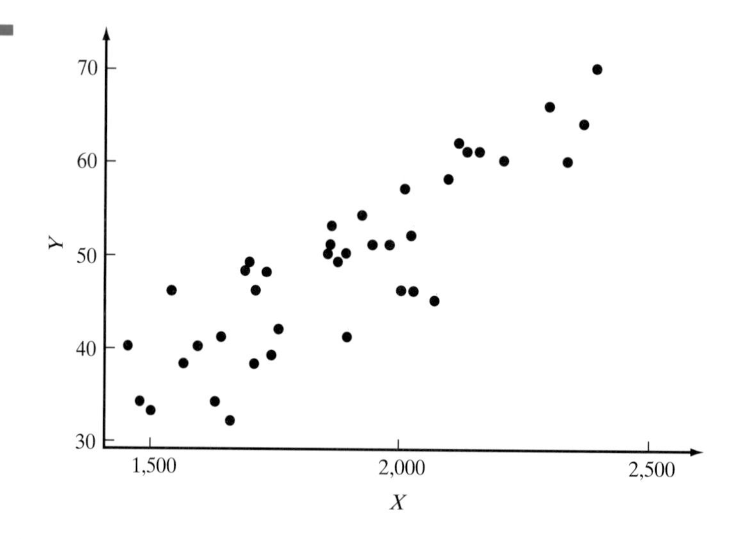

FIGURE 9.7.3
Scatter Plot of Electric Consumption (Y) Versus Size of Residence (X)

```
--------------------------------------------------------------------
The regression equation is
Y = - 13.6 + 0.0331 X

Predictor        Coef        Stdev     t-ratio        p
Constant      -13.577        5.678       -2.39    0.022
X            0.033085     0.002982       11.10    0.000

S = 4.722        R-sq = 76.4%       R-sq(adj) = 75.8%

     Fit   Stdev.Fit          95.0% C.I.            95.0% P.I.
  52.592       0.818    ( 50.935,  54.249)   ( 42.888,   62.296)
--------------------------------------------------------------------
```

EXHIBIT 9.7.3

Suppose the assumptions in Box 9.3.1 are satisfied. The following Minitab command and subcommands produced Exhibit 9.7.3:

```
MTB > regress c2 1 c3;
SUBC> predict 2000;
SUBC> confidence = 95.
```

a. Does it appear that an appropriate population regression function is

$\mu_Y(X) = \beta_0 + \beta_1 X$ "for X from 1,400 to 2,500"?

b. What is the parameter that represents the change in average electrical consumption for a 1-square-foot increase in size of residence? Estimate this parameter, and construct a 95% confidence interval.

c. What is the predicted electrical consumption for a residence that contains 2,000 square feet? What is the 95% prediction interval?

9.7 Many college students find it necessary to work part-time in order to finance their education. There are two arguments concerning how working part-time affects a student's academic performance. Some people argue that time spent working takes away from study time and thereby negatively affects a student's performance. Other people argue that the increased demand on a student's time forces the student to better manage time and thereby positively affects a student's performance. A group of students in an elementary statistics course decided to address this question in their class project. To do this, they received permission from their instructor to select a simple random sample of 50 students from the 800 students enrolled in the course. The sample was obtained at the end of the semester, and the 50 sampled students were given confidential questionnaires in which they reported their grade point average (GPA) for the current semester (Y) and the average number of part-time work hours each week (X). The results of the survey are stored in the file **work.mtp**, where column 1 contains the sample number, column 2 contains GPA (Y), and column 3 contains part-time work hours (X). The scatter plot of the data is shown in Figure 9.7.4.

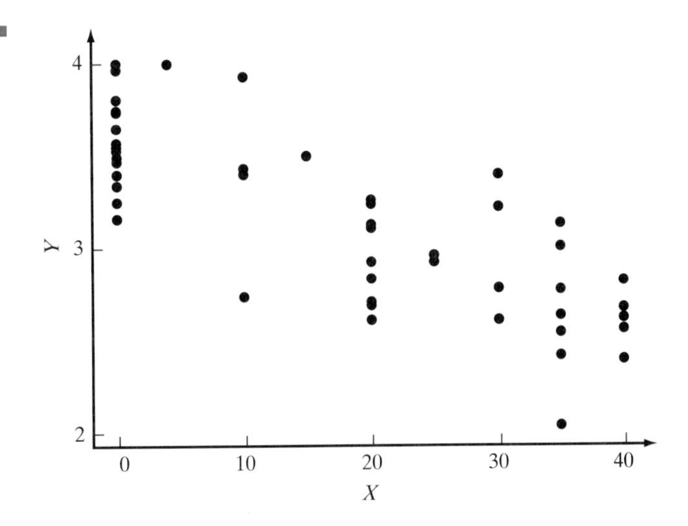

FIGURE 9.7.4
Scatter Plot of GPA (Y) Versus Part-Time Work Hours (X)

```
The regression equation is
Y = 3.61 - 0.0261 X

Predictor        Coef        Stdev     t-ratio        p
Constant      3.61029      0.06361       56.76    0.000
X           -0.026094     0.002808       -9.29    0.000

S = 0.2955        R-sq = 64.3%     R-sq(adj) = 63.5%

    Fit  Stdev.Fit         95.0% C.I.            95.0% P.I.
 3.0884     0.0426   ( 3.0028,  3.1741)   ( 2.4879,  3.6889)
```

EXHIBIT 9.7.4

Suppose that the assumptions in Box 9.3.1 are satisfied. The following Minitab command and subcommands were used to produce Exhibit 9.7.4:

```
MTB > regress c2 1 c3;
SUBC> predict 20;
SUBC> confidence = 95.
```

a. Does it appear that a linear population regression function is appropriate for values of X between 0 and 40 hours? If so, write the population regression function.

b. What is the sample regression function?

c. Construct the 95% confidence interval for β_1. Which of the two arguments does this confidence interval support?

d. Predict the GPA for a student to be selected at random from the subpopulation of students who work 20 hours per week. What is the 95% prediction interval?

9.8 In manufacturing it is often assumed that costs of repairing machines that are used in a production process are linearly related to the hours of production. An accountant for a company that manufactures circuit boards wants to examine this relationship. He selects a simple random sample of 35 machines from the population of 400 machines that are used in the manufacturing of circuit boards. Measurements are obtained for the cost of repairs in dollars (Y) and the hours of operation (X) for each machine for the past 12 months. The sample data are stored in the file **machine.mtp** on the data disk. Column 1 contains the sample number, column 2 contains the repair cost (Y), and column 3 the number of hours of operation (X). Figure 9.7.5 contains a scatter plot of the data.

FIGURE 9.7.5
Scatter Plot of Repair Cost (Y) Versus Hours of Operation (X)

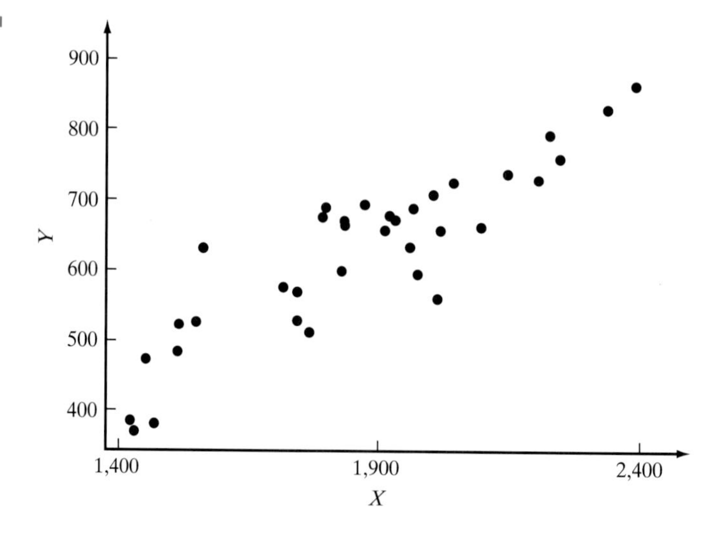

```
The regression equation is
Y = - 135 + 0.407 X

Predictor        Coef       Stdev    t-ratio        p
Constant      -135.31       67.99      -1.99    0.055
X             0.40703     0.03608      11.28    0.000

S = 55.55        R-sq = 79.4%      R-sq(adj) = 78.8%

     Fit  Stdev.Fit         95.0% C.I.            95.0% P.I.
  678.74      10.55    ( 657.27,  700.22)   ( 563.68,  793.81)
```

EXHIBIT 9.7.5

Exhibit 9.7.5 was produced using the following Minitab command and subcommands:

```
MTB > regress c2 1 c3;
SUBC> predict 2000;
SUBC> confidence = 95.
```

a. Does it appear that an appropriate population regression function is

$\mu_Y(X) = \beta_0 + \beta_1 X$ "for X from 1,400 to 2,400"?

b. Estimate and construct a 95% confidence interval for the average change in repair costs for a 100-hour increase in operation.

c. The company has recently purchased a new machine to replace an existing machine that is no longer operational. Predict the repair cost for the new machine after 2,000 hours of operation. What is the 95% prediction interval? What assumption did you make about the new machine in order to answer this question?

9.9 Consider Exploration 9.3.1, where the department head wants to use midterm exam scores to predict final-exam scores in a statistics course. The scores for a simple random sample of 50 students are shown in Table 9.3.2 and stored in the file **statgrde.mtp** on the data disk. Figures 9.7.6 through 9.7.8 present three graphs used to determine if the assumptions in Box 9.3.1 are satisfied.

a. Does a linear population regression function seem appropriate for this problem?

b. Was the sample selected using either method (4a) or (4b) in Box 9.3.1? Which one?

c. Does there appear to be any outliers?

d. Do the assumptions in Box 9.3.1 appear to be satisfied?

FIGURE 9.7.6
Scatter Plot of Final-Exam Score (Y) Versus Midterm Exam Score (X)

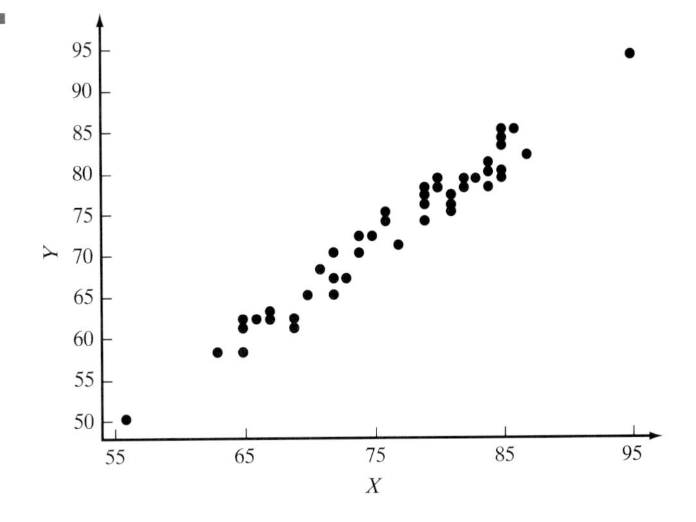

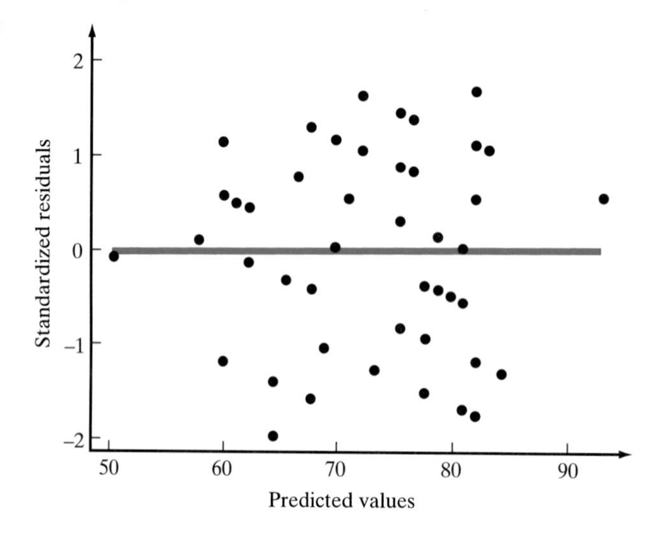

FIGURE 9.7.7
Scatter Plot of Standardized Residuals Versus Predicted Values

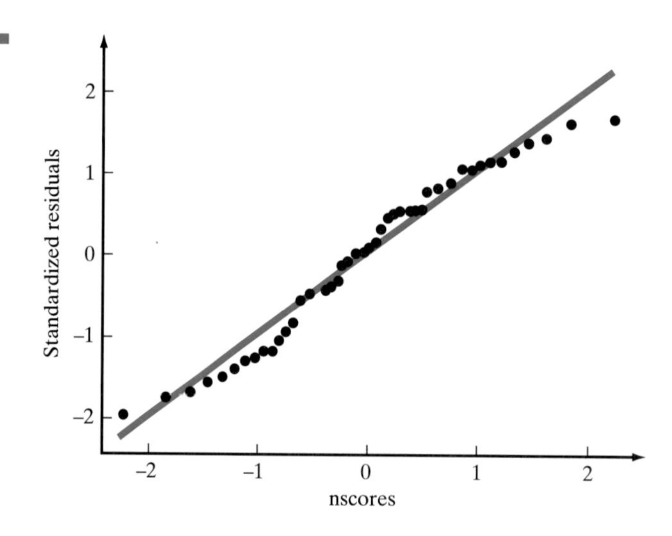

FIGURE 9.7.8
Normal Rankit-Plot of Standardized Residuals

9.8 SOLUTIONS FOR SELF-TEST PROBLEMS IN CHAPTER 9

9.1.1

a. $\widehat{Y}(-9.7) = 6 + 2(-9.7) = -13.4$.

b. $\widehat{Y}(6.1) = 6 + 2(6.1) = 18.2$.

c. The predicted value of Y changes by 2 units. This is the coefficient of X in the prediction function.

9.1.2

a. $\widehat{Y}(81.9) = 3.0 + 1.1(81.9) = 93.09$.

b. The prediction function $\widehat{Y}(X)$ is not defined for $X = 92$.

9.1.3

a. $\widehat{Y}(55) = 60 - 0.5(55) = 32.5$ mpg.

b. $\widehat{Y}(60) = 30$ mpg.

c. The prediction function is not defined for $X = 80$.

9.2.1

a. If the mileage increases from X to $X + 1{,}000$, the change in the average maintenance cost is

$$\mu_Y(X + 1{,}000) - \mu_Y(X) = 1{,}000\beta_1.$$

b. The estimate of β_1 is $\widehat{\beta}_1 = 0.0318$, the coefficient of X in the sample regression function.

c. If the mileage increases from X to $X + 1{,}000$, the estimate of the change in the average maintenance cost is

$$\widehat{\mu}_Y(X + 1{,}000) - \widehat{\mu}_Y(X) = 1{,}000\widehat{\beta}_1 = \$31.80.$$

d. Since $\widehat{Y}(X) = \widehat{\mu}_Y(X)$, the predicted maintenance cost is $\widehat{Y}(29{,}000) = -60.1 + 0.0318(29{,}000) = \$862.10$.

e. $\widehat{\mu}_Y(25{,}000) = -60.1 + 0.0318(25{,}000) = \$734.90$.

f. The sample regression function is not defined for $X = 75{,}000$.

9.2.2

a. false

b. true

c. true

d. false

9.3.1

a. Yes. While the sample exhibits a considerable amount of scatter, it appears that the linear population regression function is reasonable.

b. The symbol is $1{,}000\beta_1$.

c. From Exhibit 9.3.5, $\widehat{\beta}_1 = 11.903$. Hence, $1{,}000\widehat{\beta}_1 = \$11{,}903$.

d. We must predict $Y(2{,}500)$ and obtain a 95% prediction interval for $Y(2{,}500)$. From Exhibit 9.3.5 we see that the predicted sales *last* month for a restaurant selected at random from all restaurants that spent $2,500 on local advertising is $\widehat{Y}(2{,}500) = \$33{,}539$. If the target population of next month's advertising and sales is similar to the sampled population, this can be used as the predicted sales next month for a restaurant that spends $2,500 on advertising. The 95% prediction interval for $Y(2{,}500)$ shown in Exhibit 9.3.5 is $11,426 to $55,652.

e. From the last line in Exhibit 9.3.5, $\widehat{\mu}_Y(2{,}500) = \$33{,}539$.

f. From the last line in Exhibit 9.3.5, we see that the 95% confidence interval for $\mu_Y(2{,}500)$ is from $28,940 to $38,138.

9.4.1

a. Since the magnitude of $\rho_{Y,Z}$ is larger than the magnitude of $\rho_{Y,X}$ (0.59 is larger than 0.37), $\mu_Y(Z)$ is better than $\mu_Y(X)$ for predicting energy level (Y).

b. Both regression functions predict Y equally well since the magnitudes of the correlation coefficients are equal.

9.4.2 From Box 9.4.1, if $\rho_{Y,X} = 0$, then $\beta_1 = 0$.

9.4.3 Statement **(a)** is true.

9.5.1

a. Yes. The sample was selected according to assumption (4a) of Box 9.5.1. The values of X were not preselected.

b. Yes. Based on Figure 9.5.5, a straight-line population regression function seems quite appropriate.

c. No. There is no evidence of outliers since no standardized residual is larger than 3 in magnitude.

d. Yes.

9.9 APPENDIX: COMPUTATIONAL FORMULAS IN SIMPLE LINEAR REGRESSION

In this section we provide formulas that can be used to compute quantities in simple linear regression using a hand-held calculator. These formulas provide the same answers as Minitab or any other statistical computing package (except perhaps for rounding error).

Least Squares

The method of **least squares** is used to compute the estimates $\widehat{\beta}_0$, $\widehat{\beta}_1$, and $\widehat{\mu}_Y(X)$ from a simple random sample of size n obtained from a bivariate population using either sampling method (4a) or (4b) described in Box 9.3.1. The least-squares method determines the equation of the straight line that best fits a scatter plot of the points $(y_1, x_1), (y_2, x_2), \ldots, (y_n, x_n)$. For a simple example, consider a simple random sample of size 6 selected from a bivariate population. The sample data are shown in Table 9.9.1. The scatter plot of the six sample values is shown in Figure 9.9.1.

Figure 9.9.2 shows two straight lines superimposed on the points in Figure 9.9.1. Each line seems to provide a good fit to the data. You can probably draw other straight lines that also fit the data quite well, so how can one determine which line fits best? That is, how can we determine the prediction function that will yield the "best" predictions? To answer this question we consider the sample residuals

TABLE 9.9.1 A Small Sample

Sample number	Y	X
1	11.9	2
2	15.3	3
3	19.1	5
4	25.4	7
5	30.7	9
6	31.1	10

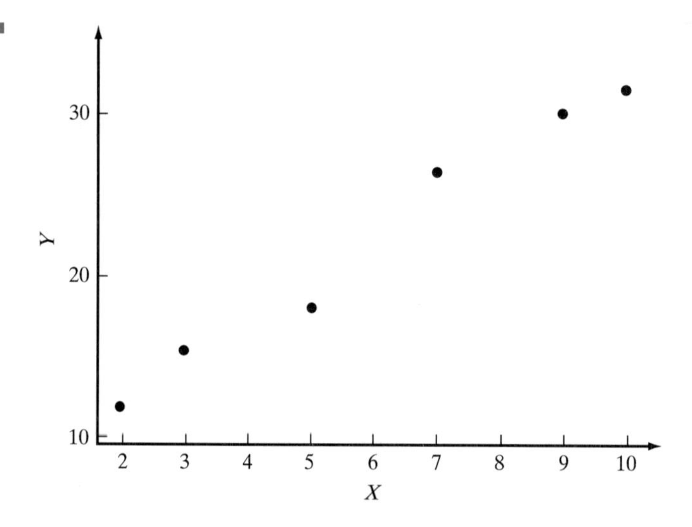

FIGURE 9.9.1
Scatter Plot of Six Data Values

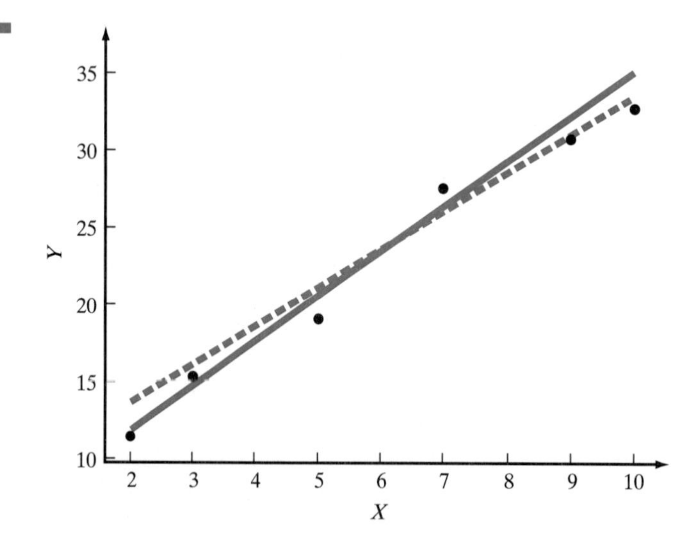

FIGURE 9.9.2
Two Lines Fit to the Six Points

discussed in Section 9.5. Recall that a residual is defined as $\hat{e}_i = y_i - \widehat{Y}(x_i)$. A good prediction function will produce residuals that are "small" in magnitude. Thus, we desire a prediction function that minimizes the "combined magnitude" of the residuals. Although there are several ways to measure the combined magnitude of the residuals, a common measure is the sum of the squared residuals, $\hat{e}_1^2 + \hat{e}_2^2 + \cdots + \hat{e}_n^2$. The method of least squares provides the prediction function that minimizes this quantity.

BOX 9.9.1: LEAST SQUARES

The method of least squares selects values for $\widehat{\beta}_0$ and $\widehat{\beta}_1$ in the prediction function $\widehat{Y}(X) = \widehat{\beta}_0 + \widehat{\beta}_1 X$ such that the sum of the squared residuals is as small as possible. These values of $\widehat{\beta}_0$ and $\widehat{\beta}_1$ are called the **least-squares estimates** of β_0 and β_1, respectively.

The formulas for $\widehat{\beta}_0$ and $\widehat{\beta}_1$ are given in Box 9.9.2 and Box 9.9.3, respectively.

BOX 9.9.2: LEAST-SQUARES ESTIMATE OF β_1 IN SIMPLE LINEAR REGRESSION

The least-squares estimate of β_1 is

$$\widehat{\beta}_1 = \frac{\text{SPD}yx}{\text{SSD}x},$$

where

$$\text{SPD}yx = (y_1 - \bar{y})(x_1 - \bar{X}) + (y_2 - \bar{y})(x_2 - \bar{X}) + \cdots + (y_n - \bar{y})(x_n - \bar{X}),$$
$$\text{SSD}x = (x_1 - \bar{X})^2 + (x_2 - \bar{X})^2 + \cdots + (x_n - \bar{X})^2,$$
$$\bar{x} = \frac{x_1 + x_2 + \cdots + x_n}{n} \quad \text{and} \quad \bar{y} = \frac{y_1 + y_2 + \cdots + y_n}{n}.$$

The notation "SPDyx" means "**S**um of the **P**roducts of **D**eviations of y and x from their respective sample means." The notation "SSDx" means "**S**um of the **S**quared **D**eviations of the x values from $\bar{x}$."

BOX 9.9.3: LEAST-SQUARES ESTIMATE OF β_0 IN SIMPLE LINEAR REGRESSION

The least-squares estimate of β_0 is

$$\widehat{\beta}_0 = \bar{y} - \widehat{\beta}_1 \bar{x}.$$

The quantities $\bar{y}$, $\bar{x}$, and $\widehat{\beta}_1$ are defined in Box 9.9.2.

We use the least-squares estimates, $\widehat{\beta}_0$ and $\widehat{\beta}_1$, to obtain the estimate $\widehat{\mu}_Y(X)$ and the prediction function $\widehat{Y}(X)$. These estimates are given in Box 9.9.4 and Box 9.9.5, respectively. A point estimate of σ is given in Box 9.9.6.

BOX 9.9.4: POINT ESTIMATE OF $\mu_Y(X)$ IN SIMPLE LINEAR REGRESSION

The point estimate of $\mu_Y(X)$ is

$$\widehat{\mu}_Y(X) = \widehat{\beta}_0 + \widehat{\beta}_1 X.$$

BOX 9.9.5: PREDICTED VALUE OF $Y(X)$ IN SIMPLE LINEAR REGRESSION

The predicted value of $Y(X)$ is

$$\widehat{Y}(X) = \widehat{\beta}_0 + \widehat{\beta}_1 X.$$

BOX 9.9.6: POINT ESTIMATE OF σ IN SIMPLE LINEAR REGRESSION

The point estimate of σ is

$$\widehat{\sigma} = \sqrt{\frac{\hat{e}_1^2 + \hat{e}_2^2 + \cdots + \hat{e}_n^2}{n-2}},$$

where

$$\hat{e}_i = y_i - \widehat{Y}(x_i) = y_i - (\widehat{\beta}_0 + \widehat{\beta}_1 x_i).$$

The formula for computing an estimate of $\rho_{Y,X}$ is presented in Box 9.9.7.

BOX 9.9.7: POINT ESTIMATE OF $\rho_{Y,X}$

The point estimate of $\rho_{Y,X}$ is

$$\widehat{\rho}_{Y,X} = \frac{\text{SPD}yx}{\sqrt{(\text{SSD}X)(\text{SSD}y)}},$$

where

$$\text{SPD}yx = (y_1 - \overline{y})(x_1 - \overline{x}) + (y_2 - \overline{y})(x_2 - \overline{x}) + \cdots + (y_n - \overline{y})(x_n - \overline{x}),$$
$$\text{SSD}y = (y_1 - \overline{y})^2 + (y_2 - \overline{y})^2 + \cdots + (y_n - \overline{y})^2,$$
$$\text{SSD}x = (x_1 - \overline{x})^2 + (x_2 - \overline{x})^2 + \cdots + (x_n - \overline{x})^2.$$

We illustrate the computations in Boxes 9.9.2 through 9.9.7 in Example 9.9.1.

EXAMPLE 9.9.1

A SMALL SAMPLE. Table 9.9.1 contains a simple random sample of size 6 from a bivariate population. The data are stored in columns 2 and 3 in the file **smalsmpl.mtp** on the data disk. For convenience we repeat Table 9.9.1 here.

TABLE 9.9.1 (repeated) A Small Sample

Sample number	Y	X
1	11.9	2
2	15.3	3
3	19.1	5
4	25.4	7
5	30.7	9
6	31.1	10

Using these data we first compute $\bar{x} = 6.00$ and $\bar{y} = 22.25$. To obtain $\widehat{\beta}_1$, $\widehat{\beta}_0$, and $\widehat{\sigma}$, it is convenient to organize the computations as shown in Table 9.9.2.
The first three columns of Table 9.9.2 are the columns shown in Table 9.9.1. Column 4 is obtained by subtracting $\bar{y} = 22.25$ from each of the y values in column 2. Column 5 is obtained by subtracting $\bar{x} = 6.00$ from each of the x values in column 3. Column 6 is obtained by multiplying the corresponding entries in columns 4 and 5. Column 7 is obtained by squaring each entry in column 5. The sum of the entries in column 6 is 129.3, so we have

$$\text{SPD}yx = (y_1 - \bar{y})(x_1 - \bar{X}) + (y_2 - \bar{y})(x_2 - \bar{X}) + \cdots + (y_6 - \bar{y})(x_6 - \bar{X}) = 129.30.$$

Also, by summing column 7, we obtain

$$\text{SSD}x = (x_1 - \bar{X})^2 + (x_2 - \bar{X})^2 + \cdots + (x_6 - \bar{X})^2 = 52.$$

We now use Boxes 9.9.2 through 9.9.5 to obtain

$$\widehat{\beta}_1 = \frac{\text{SPD}yx}{\text{SSD}x} = \frac{129.30}{52} = 2.4865,$$
$$\widehat{\beta}_0 = \bar{y} - \widehat{\beta}_1\bar{x} = 22.25 - 2.4865(6) = 7.33,$$
$$\widehat{\mu}_Y(X) = 7.33 + 2.4865X,$$
$$\widehat{Y}(X) = 7.33 + 2.4865X.$$

TABLE 9.9.2 Organizing the Computations for Least-Squares Estimates

(1)	(2)	(3)	(4)	(5)	(6)	(7)	(8)	(9)	(10)
Sample number	y_i	x_i	$(y_i - \bar{y})$	$(x_i - \bar{X})$	$(y_i - \bar{y})(x_i - \bar{X})$	$(x_i - \bar{X})^2$	$\widehat{y}_i$	$\hat{e}_i$	$\hat{e}_i^2$
1	11.9	2	−10.35	−4	41.40	16	12.3030	−0.4030	0.16241
2	15.3	3	−6.95	−3	20.85	9	14.7895	0.5105	0.26061
3	19.1	5	−3.15	−1	3.15	1	19.7625	−0.6625	0.43891
4	25.4	7	3.15	1	3.15	1	24.7355	0.6645	0.44156
5	30.7	9	8.45	3	25.35	9	29.7085	0.9915	0.98307
6	31.1	10	8.85	4	35.40	16	32.1950	−1.0950	1.19903
Total	133.5	36			129.30	52			3.48559

Square the numbers in columns 4 and 5 and obtain the total of each column. Then substitute these totals and the total of column 6 into the formula for $\widehat{\rho}_{Y,X}$ in Box 9.9.7 to obtain

$$\widehat{\rho}_{Y,X} = \frac{129.30}{\sqrt{(324.995)(52)}} = 0.9946.$$

Column 8 in Table 9.9.2 contains the predicted y values for the sample items. For instance, for sample item 1 we have $\widehat{Y}(x_1) = \widehat{Y}(2) = 7.33 + 2.4865(2) = 12.3030$. Column 9 contains the residuals. Finally, column 10 is obtained by squaring the corresponding entries in column 9. The sum of the entries in column 10 is 3.48559. Using this we obtain

$$\widehat{\sigma} = \sqrt{\frac{3.48559}{n-2}} = \sqrt{\frac{3.48559}{4}} = 0.9335.$$

The formulas presented in this section are useful when computations are performed by hand. Computer software packages generally use alternative (but equivalent) formulas that minimize round-off error.

Confidence Intervals for Regression Quantities

We now present formulas for obtaining prediction intervals for

$$Y(X) \qquad \textbf{Line 9.9.1}$$

and confidence intervals for

$$\beta_0, \quad \beta_1, \quad \mu_Y(X). \qquad \textbf{Line 9.9.2}$$

Let θ (the Greek letter **theta**) represent any one of the quantities in Lines 9.9.1 and 9.9.2, and let $\widehat{\theta}$ represent the corresponding estimate (prediction). The general form of a confidence (prediction) interval for θ is

$$\widehat{\theta} - (T2)SE(\widehat{\theta}) \le \theta \le \widehat{\theta} + (T2)SE(\widehat{\theta}), \qquad \textbf{Line 9.9.3}$$

where the table value $T2$ is found in Table T-2 with $DF = n - 2$. The quantity $SE(\widehat{\theta})$ is called the **standard error of** $\widehat{\theta}$. The term $(T2)SE(\widehat{\theta})$ is the margin of error for the estimate $\widehat{\theta}$.

The point estimates of θ (which are denoted by $\widehat{\theta}$) are given in Boxes 9.9.2 through 9.9.5, and the quantity $SE(\widehat{\theta})$ is calculated from sample data using the following formulas:

$$SE(\widehat{\beta}_0) = \widehat{\sigma}\sqrt{\frac{1}{n} + \frac{\overline{x}^2}{\text{SSD}x}}$$

$$SE(\widehat{\beta}_1) = \widehat{\sigma}\sqrt{\frac{1}{\text{SSD}x}}$$

$$SE(\widehat{\mu}_Y(X)) = \widehat{\sigma}\sqrt{\frac{1}{n} + \frac{(X - \overline{x})^2}{\text{SSD}x}}$$

$$SE(\widehat{Y}(X)) = \widehat{\sigma}\sqrt{1 + \frac{1}{n} + \frac{(X - \overline{x})^2}{\text{SSD}x}}.$$

We illustrate how to compute the four standard errors using the data in Example 9.9.1. We obtain

$$SE(\widehat{\beta}_0) = 0.9335\sqrt{\frac{1}{6} + \frac{(6)^2}{52}} = 0.8652,$$

$$SE(\widehat{\beta}_1) = 0.9335\sqrt{\frac{1}{52}} = 0.1295,$$

$$SE(\widehat{\mu}_Y(X)) = 0.9335\sqrt{\frac{1}{6} + \frac{(X - 6)^2}{52}},$$

$$\text{For } X = 4, \quad SE(\widehat{\mu}_Y(4)) = 0.9335\sqrt{\frac{1}{6} + \frac{(4 - 6)^2}{52}} = 0.4607,$$

$$SE(\widehat{Y}(X)) = 0.9335\sqrt{1 + \frac{1}{6} + \frac{(X - 6)^2}{52}},$$

$$\text{For } X = 4, \quad SE(\widehat{Y}(4)) = 0.9335\sqrt{1 + \frac{1}{6} + \frac{(4 - 6)^2}{52}} = 1.041.$$

Exhibit 9.9.1 presents Minitab output for predicting Y when $X = 4$. You can compare this output to the values computed in this section.

```
---------------------------------------------------------------------
Regression Analysis

The regression equation is
Y = 7.33 + 2.49 X

Predictor       Coef       Stdev    t-ratio       p
Constant      7.3308      0.8652       8.47   0.001
X             2.4865      0.1295      19.21   0.000

S = 0.9335      R-sq = 98.9%     R-sq(adj) = 98.7%

     Fit  Stdev.Fit       95.0% C.I.             95.0% P.I.
  17.277      0.461   ( 15.997,  18.556)  (  14.386,  20.168)
---------------------------------------------------------------------
```

EXHIBIT 9.9.1

In the column labeled `Stdev` are the two numbers 0.8652 and 0.1295. These are the values for $SE(\widehat{\beta}_0)$ and $SE(\widehat{\beta}_1)$, respectively. In the column labeled `Fit` are $\widehat{\mu}_Y(4)$ and $\widehat{Y}(4)$. These are equal, so $\widehat{\mu}_Y(4) = \widehat{Y}(4) = 17.277$. In the column labeled `Stdev.Fit` is the number 0.461. This is the value of $SE(\widehat{\mu}_Y(4))$. The point estimates in Boxes 9.9.2 through 9.9.4 and the standard errors can be substituted into Line 9.9.3 to obtain confidence intervals for β_0, β_1, and $\mu_Y(X)$, respectively. A prediction interval for $Y(X)$ is obtained in the same manner using $\widehat{Y}(X)$ in Box 9.9.5 and $SE(\widehat{Y}(X))$.

Here are some questions, with solutions, that can be answered using the material in this section.

9.9.1. Exercise scientists recommend that you monitor your pulse rate during exercise to avoid excessive stress to the cardiovascular system. This recommendation is based on the observation that a pulse rate increases with more strenuous exercise. A competitive runner wants to investigate this relationship. She kept a record of her pulse rate (beats per minute) after a one-mile run at different paces. The target and sampled populations in this problem are the same conceptual population. You can imagine that this individual ran the mile at 10 different times each day for two years and recorded her pulse rate in beats per minute (Y) and the time to run a mile in seconds (X). The population consists of the 7,300 pairs of (Y, X) measurements. You can view the sample as being six of the pairs selected at random from this population. The sample data are given in Table 9.9.3 and stored in the file **pulse.mtp** on the data disk.

TABLE 9.9.3 Pulse Rate Data

Trial	Pulse y (beats per minute)	Time x (seconds)
1	202	313
2	187	356
3	175	402
4	150	490
5	141	525
6	135	602

Suppose the assumptions in Box 9.3.1 are satisfied. Suppose the population regression function is

$$\mu_Y(X) = \beta_0 + \beta_1 X \qquad \text{“for } X = 300 \text{ to } 610.\text{”}$$

a. Use the formulas in this section to compute estimated values for β_0, β_1, and σ.

▶**Answer:** *The required intermediate calculations are organized in the following table. The sample means are* $\overline{y} = 165$ *and* $\overline{x} = 448$.

Trial	y_i	x_i	$y_i - \bar{y}$	$x_i - \bar{x}$	$(y_i - \bar{y})(x_i - \bar{X})$	$(x_i - \bar{X})^2$	$\hat{y}_i$	$\hat{e}_i$	$\hat{e}_i^2$
1	202	313	37	−135	−4,995	18,225	197.682	4.318	18.647
2	187	356	22	−92	−2,024	8,464	187.272	−0.272	0.074
3	175	402	10	−46	−460	2,116	176.136	−1.136	1.291
4	150	490	−15	42	−630	1,764	154.832	−4.832	23.352
5	141	525	−24	77	−1,848	5,929	146.359	−5.359	28.722
6	135	602	−30	154	−4,620	23,716	127.719	7.281	53.018
Total					−14,577	60,214			125.104

From this table we obtain SPDyx $= -14{,}577$, *SSDx* $= 60{,}214$, $\hat{\beta}_1 = -0.242087$, $\hat{\beta}_0 = 273.455$, $\hat{e}_1^2 + \hat{e}_2^2 + \cdots + \hat{e}_6^2 = 125.104$, $\hat{\sigma} = 5.593$, *and* $\widehat{Y}(X) = \widehat{\mu}_Y(X) = 273.455 - 0.242087X$.

b. Predict the pulse rate at the end of a one-mile run if the athlete ran the mile in 440 seconds.

▶**Answer:** *The predicted value of the pulse rate at the end of a one-mile run in 440 seconds is* $\widehat{Y}(440) = 273.455 - 0.242087(440) = 166.937$.

c. Compute a 95% prediction interval for the pulse rate at the end of a one-mile run if the athlete ran the mile in 440 seconds.

▶**Answer:** *We compute* $SE(\widehat{Y}(440)) = 6.044$. *The table value obtained from Table T-2 for 95% confidence and 4 degrees of freedom is* $T2 = 2.78$. *So the 95% prediction interval for* $Y(440)$ *is L to U, where* $L = 166.937 - 16.802 = 150.135$ *and* $U = 166.937 + 16.802 = 183.739$.

d. Compute $\widehat{\rho}_{Y,X}$, the estimate of the population correlation coefficient of Y and X.

▶**Answer:** $\widehat{\rho}_{Y,X} = -0.983$.

INTERLUDE

CRITICAL ASSESSMENT OF STATISTICAL STUDIES

I.1 INTRODUCTION

In the preceding nine chapters, we have presented the central concepts of statistical inference. In the chapters that follow, we provide extensions of this basic knowledge. Before beginning Chapter 10 we want to break from the standard format of the text and discuss some concepts on a less technical level. In this interlude we provide practical guidelines to help you better understand and interpret the results of a statistical study.

In Section I.2 we provide two lists of questions that can be used to critically analyze the results of a statistical study. Section I.3 discusses the topic of cause-and-effect relationships. This is an important issue to understand when interpreting the results of a statistical study in which the primary interest concerns the nature of the relationship between two variables.

I.2 EVALUATING STATISTICAL AND JUDGMENT INFERENCES

Although you may not plan to be a professional statistician, there will be many occasions when you are presented with the report of a statistical analysis. Such a report may be presented to you at work, described in a newspaper article, or referenced in a television commercial. The conclusions of the statistical analysis may be based on a well-planned experiment that took many years to complete or on a collection of phone calls to a radio talk show during a two-hour period. The importance you place on the results of a statistical report should be based on the appropriateness of the statistical and judgment inferences. In this section we present two sets of questions that will help you make this evaluation.

Statistical inference concerns the ability to draw conclusions about a sampled population based on a sample. Judgment inference concerns the ability to draw conclusions about a target population based in part on a statistical inference about a related sampled population. Accordingly, we have developed a set of questions to help determine the appropriateness of each type of inference. We begin with the set of questions in Box I.2.1. These questions can be used to help determine the appropriateness of a *statistical* inference.

BOX I.2.1: QUESTIONS TO HELP DETERMINE THE APPROPRIATENESS OF A STATISTICAL INFERENCE

1. What is the objective of the study?
2. What is the sampled population?
3. Why is it necessary to make a statistical inference?
4. How was the sample collected?
5. How are the variables defined and measured?
6. Do the variables provide the information necessary to accomplish the study objective?
7. Are there any items in the sample that were not measured? If so, what effect does the omission of these items have on the conclusions?
8. Are appropriate statistical methods used to perform the analysis?
9. How large is the margin of error?

We now discuss each question in more detail.

1. *What is the objective of the study?*

It is important that the study objective be stated clearly. Typically, the objective is to make a decision. Before making the decision, it is desired to have some specific information. Often the desired information is the value of a population parameter. Thus, in order to state the objective, it is necessary to define the target population, the variables, and the parameters needed to make the decision.

2. *What is the sampled population?*

Remember that any statistical inference pertains only to the sampled population. Require an exact definition of the sampled population, including the time period in which the data were collected. Request a list of all items in the sampled population or determine how the list was compiled for selecting the sample.

3. *Why is it necessary to make a statistical inference?*

In some cases it may be possible to measure every item in the sampled population. If so, then the exact value of a parameter can be determined and no statistical inference is necessary. If a statistical inference is presented, ask why the population was not completely enumerated.

4. *How was the sample collected?*

It is necessary that the sample be collected using a valid statistical sampling procedure. There are many different types of such procedures. The simple random sampling procedure was described in Chapter 2. Some other commonly used procedures are presented in Chapter 14. One characteristic common to all these procedures is that observations are selected using a well defined random process. If this is not the case, then one cannot make a statistical inference even if the sample appears to be representative of the population.

5. *How are the variables defined and measured?*

This question addresses many issues. It is important that variables be carefully defined. As an illustration, consider a study to determine the relationship between eating habits and performance in school for a population of high-school students. One variable of interest might be a response to the question, "Do you eat a healthy breakfast every school day?" Since there are many ways to define the term "healthy breakfast," it is necessary that a precise definition be given to the respondents before they answer the question. Definitions of all variables should be included in the report of the statistical analysis.

After all the variables of interest have been carefully defined, the variables must be measured. If a questionnaire is used for this purpose, request a copy and examine it to determine if the questions are relevant and if they measure the variables accurately. If interviewers were used to collect data, determine how they were trained and what safeguards were used to minimize interviewer errors. Determine the process used to calibrate and operate measuring instruments for variables such as "Time," "Length," "Voltage," and "Optical density." Measurements must have little or no **bias**, and good **precision**. We list some examples to illustrate these concepts.

Bias. When one measures the value of a continuous variable, the measured value is never *exactly* equal to the true value. The difference between the measured value and the true value is called **measurement error**. We say that a measurement procedure is **unbiased** if there is no systematic tendency for the procedure to underestimate or overestimate the true value. For example, if the weight of 100 people was measured using an unbiased scale, we would expect the reported weight to be greater than the true value for some people and less than the true value for others. On average, the error in measurement would be essentially zero. However, if the scale is not properly calibrated, it might consistently yield readings that are lower (or higher) than the true weight. In such a case, the scale is said to be **biased**.

Precision. This property concerns the reliability of a measurement procedure. If one measures the same item several times, it is hoped that the measurements are more or less the same each time and do not fluctuate wildly. Measurement procedures that exhibit little variability when measuring the same item are said to be **precise**. Measurements of variables such as "Height," "Weight," and "Temperature" that use well-calibrated instruments are likely to be very precise. Responses of subjects to questions about attitudes or opinions are unlikely to be precise unless the questions are worded clearly and are not open to multiple interpretations.

6. *Do the variables provide the information necessary to accomplish the study objective?*

This question refers to the concept of **validity**. A measurement is valid if it measures the characteristic of interest. For instance, suppose we want to measure an individual's intelligence before deciding whether to admit the person to a graduate program. It is quite common to use the scores on one of several standardized achievement tests to make this evaluation. However, the characteristic called "intelligence" is a vague concept that cannot be measured with a single number. It has been contended that even intelligence quotient (IQ) scores do not actually measure this characteristic. Examples abound where an individual tagged with a low IQ score shows exceptional talent in some area of the arts or sciences. Before any measure of "intelligence" is used, one must critically evaluate if it is measuring characteristics related to the study objective.

In some situations it may not be possible to measure the variable that appears to have the greatest validity. For example, the faculty of a business college performed a study to determine criteria for predicting success in a graduate business program. It was argued that the most valid measures of success reflect a student's performance in the business world after graduation. Many variables can be used to measure this performance. These might include annual salary, job title, level of personal satisfaction, and civic contributions. Unfortunately, the time frame for completing the study did not afford an opportunity to locate graduates and collect this information. Rather, the variable used to measure success was the student's grade point average at graduation. Measurements of this variable were easily obtained from college records. The importance that someone places on the conclusions of this study would depend on whether the person believes grade point average is a valid measure of future success in the business world.

7. *Are there any items in the sample that were not measured? If so, what effect does the omission of these items have on the conclusions?*

Quite often, particularly for data collected from human populations, measurements are not obtained for every item in the sample. Sample items that are not measured are called **nonrespondents**. In many cases, values of variables of interest may differ greatly between nonrespondents and respondents. Thus, a study with a large proportion of nonrespondents can provide misleading results. In order to avoid this situation, it is important to make every possible effort to collect measurements from every item in the sample. One must consider how nonrespondents and respondents differ and what effect these differences might have on the sample estimates.

8. *Are appropriate statistical methods used to perform the analysis?*

Determine the statistical methods and the computer software used to perform computations. Ask if all required assumptions are reasonably satisfied. If results are described as "adjusted" or "weighted," determine how and why the adjustments were performed.

9. *How large is the margin of error?*

An investigator generally needs to use a point estimate of a population parameter in place of that parameter when making decisions. The investigator will be comfortable in doing this if the margin of error is small enough. If the margin of error is too large to accomplish the study objective, one must collect additional data.

It is unfortunate that instead of asking, "How large is the margin of error?" most people ask, "How large is the sample?" It is even more unfortunate that the latter question is often the *only* question asked about a study. Sample size affects the margin of error because the margin of error decreases when sample size increases (for a given level of confidence). However, knowledge of sample size alone is not sufficient to determine if the study objective can be accomplished. The margin of error is needed for this purpose. If the margin of error is small enough to accomplish the study objective, then the sample size is as large as it needs to be. Although

P-values may be useful when conducting statistical tests to determine the value of a parameter, point estimates with the accompanying margins of error are generally more useful.

By asking the questions in Box I.2.1, you will have a much better chance of determining the appropriateness of any reported statistical inference. Unfortunately, in many published reports there is insufficient information to answer many of these questions. However, by asking these questions, you will better understand the limitations that should be considered in evaluating the study's conclusions.

As noted previously, statistical inference is only one consideration in evaluating a statistical report. In many practical applications, the target population and the sampled population are not identical, and so it is necessary to make a judgment inference. If you have answered the questions in Box I.2.1 and believe the statistical inference is appropriate, then you should ask the two questions shown in Box I.2.2.

BOX I.2.2: QUESTIONS TO HELP DETERMINE THE APPROPRIATENESS OF A JUDGMENT INFERENCE

1. How does the target population differ from the sampled population, and how will the values of the population parameters differ in the two populations?
2. What information is used to make the judgment inference?

We now discuss each question in Box I.2.2 in more detail.

1. *How does the target population differ from the sampled population, and how will the values of the population parameters differ in the two populations?*

It is important to determine how the sampled population and the target population differ and how these differences impact the parameter values of each population. Often the target population is the sampled population as it will exist in the future. If so, one needs to determine future events that could dramatically change the values of the population parameters. It might be desirable to make several judgment inferences, each one based on a different scenario of future events.

Often the sampled population is either a subset of the target population or contains items that do not belong in the target population. In either case, the possible effect on parameter values of the items not contained in both populations must be considered in making any judgment inference.

2. *What information is used to make the judgment inference?*

Judgment inferences are based on statistical inferences, expert opinion, personal feelings, and data external to the sample. It is difficult, if not impossible, to quantify the uncertainty associated with this combined information, but it is necessary to evaluate the quality of each component. The appropriateness of a statistical inference can be determined using the questions presented in Box I.2.1. If an expert is used to help make the judgment inference, one should examine the expert's qualifications and motivation for making the judgment. In some situations an expert may have a cause to promote, and this may bias the expert's view of the situation. The value of personal feelings can be validated by past performance in similar situations. Data external to the sample can be evaluated in much the same manner as a statistical inference.

We now present two examples that demonstrate how the questions in Boxes I.2.1 and I.2.2 can be used to determine the validity of a statistical study.

EXAMPLE I.2.1

MARKETING SURVEY. A marketing research firm designed an advertising campaign directed at college students for a client who sells personal computers. The firm proposed to use advertisements in college newspapers as the primary marketing medium. Unsure that this method would effectively reach the target market, the client asked the research firm to estimate the proportion of college students who read their college newspaper. The client agreed to advertise in college newspapers if evidence suggests that at least 30% of college students read their college newspaper at least three days a week. The research firm selected a simple random sample of 400 students from the student directory of a large state university in Arizona. Each sampled student was to be interviewed by telephone and asked if they read their college newspaper at least three days a week. Responses were obtained from 220 of the 400 students in the sample. Of the remaining 180 students, 120 were unavailable at the time the phone call was made and the other 60 refused to respond to the interviewer. Of the 220 respondents, 88 stated that they read the college newspaper at least three days a week. The research firm estimated the proportion π of students who read the paper at least three days a week as $\widehat{\pi} = 88/220 = 0.40$ and reported the 95% margin of error to be 0.065. The research firm claimed that these results could be applied to all college students in the United States and recommended that the client advertise in college newspapers throughout the country. We will now evaluate the validity of the conclusions for this study using the list of questions in Box I.2.1.

1. *What is the objective of the study?*

The research firm wants to estimate the proportion of college students who read their college newspaper at least three days a week. Their client will use this information to determine if college newspaper advertising will effectively reach the target market.

2. *What is the sampled population?*

A simple random sample of 400 names was obtained from the population of all students listed in the university's student directory. The sampled population is the set of students listed in the student directory.

3. *Why is it necessary to make a statistical inference?*

It is impractical to contact every student listed in the student directory.

4. *How was the sample collected?*

Each name in the student directory was numbered sequentially. There was a total of $N = 30{,}000$ names in the directory. A simple random sample of $n = 400$ integers from 1 to 30,000 was selected using a random number generator. These numbers were matched with the line numbers in the directory to select the sample of 400 students.

5. *How are the variables defined and measured?*

The variable of interest is the answer to the question, "Do you read the college newspaper at least three days a week?" Responses to this question should be relatively precise, but there is some chance for misinterpretation and measurement error. For example, suppose a student reads the paper five days on some weeks and one day on other weeks. How should she answer this question? It might be expected that the student would answer in accordance with her general behavior, but that is not stated in the question. Perhaps an alternative wording might be more precise. Think of an alternative wording that might provide a better question.

6. *Do the variables provide the information necessary to accomplish the study objective?*

Responses to the question, "Do you read the college newspaper at least three days a week?" will enable the firm to estimate the proportion of students in the sampled population who read the college newspaper at least three days a week. However, responses to this question will provide no information concerning the likelihood that a student will read an advertisement placed in the newspaper. Questions that might provide more useful information are "Do you read advertisements in the college newspaper at least three days a week?" or "Have you ever purchased an item based on an advertisement you read in the college newspaper?"

7. *Are there any items in the sample that were not measured? If so, what effect does the omission of these items have on the conclusions?*

Yes. Although 400 student names were selected in the sample, responses were obtained from only 220 students. Of the 180 nonrespondents, 120 were unavailable at the time the phone call was made (additional effort should be made to contact these students) and the other 60 refused to respond to the interviewer. Thus, the nonresponse rate is $180/400 = 45\%$. Based on discussions with the interviewers, it was conjectured that most of the 60 students who did not respond to the interviewers did not read the newspaper and were unwilling to admit this fact. Students who could not be contacted were believed to be neither more nor less likely to read the newspaper than the 220 respondents. If this is true, the proportion of students in the sampled population who read the newspaper at least three days a week is likely to be less than the proportion of 220 respondents who read the newspaper at least three days a week.

8. *Are appropriate statistical methods used to perform the analysis?*

The margin of error reported by the research firm was computed using the formula in Box 3.2.1 which requires a simple random sample from the sampled population. Because of nonresponses, the sample of 220 students is not a simple random sample from the sampled population. Therefore the computed margin of error is not valid for making inferences about the sampled population. However, the reported margin of error is valid for making inferences about a different population, namely, the subset of students in the sampled population who could have been contacted and who would have responded.

9. *How large is the margin of error?*

Using Box 3.2.1 with $\hat{\pi} = 88/220 = 0.40$ and $T1 = 1.96$, we find that the 95% margin of error is 0.065. As noted in the response to the previous question, this margin of error is not valid for the sampled population, but is valid for the subpopulation consisting of all students listed in the student directory who could have been contacted and who would have responded to the interviewer.

We now focus on the judgment inference that the results of the study could be applied to all college students in the United States.

1. *How does the target population differ from the sampled population, and how will the values of the population parameters differ in the two populations?*

Because of the nonresponses, the statistical inference applies only to the subset of the subpopulation described in the answer to question 8 on the previous page. Is this subset of the sampled population representative of the target population of all college students in the United States? Probably not. In fact, as suggested earlier, there is some possibility that this subset of the sampled population is not even representative of the sampled population. Furthermore, student attitudes and behaviors vary between public and private universities and across different regions of the country. Whether these differences affect the likelihood of reading a college newspaper is a question the client will have to answer in order to make a good decision.

2. *What information is used to make the judgment inference?*

The research firm used the statistical inference and knowledge of the target market of college students to make the judgment inference.

After evaluating the results of the survey and examining the procedures used to conduct it, the client will decide if college newspapers are an effective advertising medium.

EXAMPLE I.2.2

LONGEVITY OF RATS. In Example 8.2.3 an investigator wanted to study the longevity of a certain species of rat fed a low-fat diet. Many previous studies have confirmed that the median lifetime for this species of rat when fed an ordinary diet is about 580 days. The investigator believes that the low-fat diet she has developed will result in a median lifetime that is greater than 580 days. To examine this, she conducts a preliminary study using a simple random sample of 120 rats selected from a colony of 200,000 rats. Based on this sample, a 95% confidence interval for the population median was computed to be from 555 days to 638 days. Based on this interval, she concluded that the data did not demonstrate that the median lifetime is greater than 580 days. She then made a judgment inference that the results of this study could be applied to all rats of this species.

1. *What is the objective of the study?*

The objective is to determine the median lifetime of a certain species of rat when fed a low-fat diet. Additionally, the investigator wants to know if this value is greater than 580 days, and if so, by how much.

2. *What is the sampled population?*

The sampled population is conceptual. In your mind's eye, imagine that each of the 200,000 rats in the colony is fed the low-fat diet and its lifetime is determined. This collection of 200,000 numbers is the sampled population. In actuality, the diet was fed only to the sample of 120 rats and their lifetimes were recorded. These values can be regarded as a simple random sample from the sampled population.

3. *Why is it necessary to make a statistical inference?*

It is impractical to feed the low-fat diet to all 200,000 rats in the colony. Thus, a statistical inference based on a sample of rats is needed to accomplish the study objective.

4. *How was the sample collected?*

A simple random sampling procedure is used to select a sample of 120 rats from the 200,000 rats in the colony. Each rat in the colony has an identification number from 1 to 200,000. A random sample of 120 integers between 1 and 200,000 is selected, and the identified rats are placed in the sample.

5. *How are the variables defined and measured?*

The variable "Days lived" is measured by a research assistant. At the beginning of each day, the research assistant performs an inventory of all rats used in the experiment. The assistant records the lifetime in days of any rat that died since the previous inventory.

6. *Do the variables provide the information necessary to accomplish the study objective?*

Yes. The variable measured in this study is "Days lived." It provides the information needed to accomplish the study objective.

7. *Are there any items in the sample that were not measured? If so, what effect does the omission of these items have on the conclusions?*

No. Lifetimes were measured for all 120 rats in the study.

8. *Are appropriate statistical methods used to perform the analysis?*

Yes. The confidence interval for the median was computed using Box 7.3.2. To use this formula, it is required that the sample size be at least 10.

9. *How large is the margin of error?*

The 95% confidence interval for the population median was computed in Exploration 7.3.1 to be from $L = 555$ days to $U = 638$ days. Note that the estimated median $\widehat{M} = 590$ days is not in the middle of this interval. When this is the case, we define the margin of error to be the maximum of the two numbers $U - \widehat{M}$ and $\widehat{M} - L$. In this problem, the margin of error is $638 - 590 = 48$ days. We now examine the judgment inference made by the investigator using the two questions given in Box I.2.2.

1. *How does the target population differ from the sampled population, and how will the values of the population parameters differ in the two populations?*

The target population is conceptual and consists of all rats of this species, where every rat is fed the low-fat diet. So to answer this question, we need to know more about the colony of 200,000 rats. The owner of the colony might be able to tell us about any breeding techniques unique to the colony that might affect the median lifetime of the rats. The owner of the colony might have other information about this species of rat that could help us discover differences between the sampled population and the target population.

2. *What information is used to make the judgment inference?*

In addition to the statistical inference, information from the owner of the colony is used to help make the judgment inference.

I.3 UNDERSTANDING CAUSE-AND-EFFECT RELATIONSHIPS

As discussed in Chapter 9, in many statistical studies the main interest is in determining the relationship between two quantitative variables. Often one is interested in predictive relationships between the variables. This is the primary use of regression analysis, described in Chapter 9. For simplicity let us consider predicting a variable Y using the value of a predictor variable X. Suppose the prediction function is given by $\widehat{Y}(X) = \widehat{\beta}_0 + \widehat{\beta}_1 X$. This means that if we know the value of X, we can predict the value of Y.

Sometimes the fact that X is a useful predictor of Y is misinterpreted to mean that X and Y are causally related, that is, that X causes Y. For instance, you may have read or heard a statement such as "researchers have found an association between coffee drinking and liver disease." This statement might suggest to you that coffee drinking *causes* liver disease. Such an interpretation is often incorrect and should not be made without learning more about the nature of the relationship. The following example illustrates why it is not appropriate to make a causal interpretation based solely on an observed association between two variables.

EXAMPLE I.3.1

CHURCHES AND VIOLENT CRIME. Table I.3.1 presents the number of violent crimes reported during the past year (Y) and the number of churches (X) for a simple random sample of 20 cities selected from a collection of U.S. cities with populations between 150,000 and 300,000 people. The data are stored in the file **violence.mtp** on the data disk, where the sample number is in column 1, Y is in column 2, and X is in column 3. A scatter plot of Y versus X is presented in Figure I.3.1.

TABLE I.3.1 Number of Violent Crimes and Number of Churches for a Simple Random Sample of Cities

City number	Number of violent crimes reported Y	Number of churches X
1	531	127
2	395	120
3	182	80
4	126	83
5	155	46
6	610	126
7	353	92
8	653	124
9	578	80
10	263	68
11	171	31
12	425	144
13	345	70
14	458	41
15	582	142
16	309	81
17	635	98
18	417	122
19	613	154
20	562	150

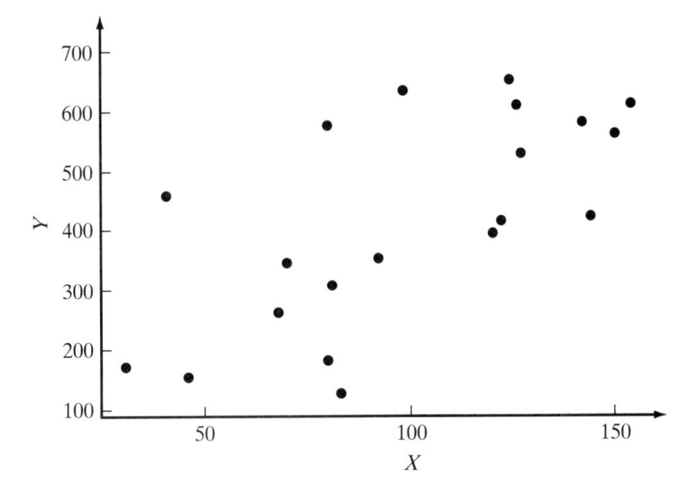

FIGURE I.3.1
Scatter Plot of Number of Violent Crimes (Y) Versus Number of Churches (X)

It appears that the population regression function of Y on X may be adequately represented by

$$\mu_Y(X) = \beta_0 + \beta_1 X \quad \text{"for } X \text{ between 30 and 160."}$$

We suppose that the assumptions in Box 9.3.1 are satisfied. The Minitab output for the data in Table I.3.1 is shown in Exhibit I.3.1.

```
The regression equation is
Y = 110 + 3.12 X

Predictor        Coef        Stdev     t-ratio       p
Constant       109.64        86.78        1.26   0.223
X              3.1178       0.8230        3.79   0.001

S = 134.1        R-sq = 44.4%      R-sq(adj) = 41.3%
```

EXHIBIT I.3.1

Suppose we want a 95% confidence interval for β_1. For 95% confidence and DF $= 20 - 2 = 18$, enter Table T-2 and obtain $T2 = 2.10$. The endpoints of a 95% interval for β_1 are L and U, where

$$L = 3.1178 - (2.10)(0.8230) = 1.38, \qquad U = 3.1178 + (2.10)(0.8230) = 4.85.$$

Consider all cities that have a specified number of churches (say, 120 churches). We are 95% confident that cities that have one additional church (say, 121 churches) had an average of between 1.38 and 4.85 more violent crimes last year.

Suppose a report of the study in Example I.3.1 states, "There is a linear relationship between the number of churches and the number of violent crimes." How should one interpret this statement? In particular, is it appropriate to conclude that a change in X will *cause* Y to change? Is it appropriate to conclude that closing churches in a city will decrease the number of violent crimes? Or that building more churches will increase the number of violent crimes?

If it can be demonstrated that changing the value of X will *cause* the value of Y to change, the relationship between X and Y is said to be **causal**. However, as Example I.3.1 might suggest, the existence of a relationship between Y and X is not sufficient to conclude that Y and X are causally related. Before explaining how to determine if a causal relationship exists, we describe two types of studies often used to examine the relationships between variables.

Experimental Versus Observational Studies

It is very difficult to demonstrate that two variables are causally related. Certain types of studies provide stronger evidence for concluding causality than other types of studies. For this purpose, studies can be broadly categorized into two types: (1) observational studies, and (2) experimental studies.

Observational studies are studies in which the investigator merely observes and records information on the items in the sample. **Experimental studies** are studies in which the investigator compares two or more **treatments** or conditions after *randomly* assigning the sample items (subjects) to the different treatment groups.

To illustrate the difference between an observational and an experimental study, consider a study to investigate the association between smoking and the incidence of lung cancer. An experimental study can be designed using laboratory animals such as rats. There have been numerous experimental studies in which rats (sample items) are randomly assigned to different treatments (various levels of tobacco smoke exposure for a predetermined period of time). For each rat in the sample, both the level of tobacco smoke exposure (X) and the severity of damage to the lungs (Y) are recorded. The data are then examined to see if there is an association between the level of exposure to tobacco smoke and the severity of lung damage. All rats are treated identically except for the exposure to tobacco smoke.

Obviously, ethical considerations prevent us from conducting such an experimental study using human subjects. An observational study that addresses the same question might select a sample of adults and record whether each person has lung cancer and whether the person is (or has been) a smoker. The incidence of lung cancer in the two groups can then be compared to see if a difference exists. This is an example of an observational study because the investigator has no control over which of the subjects in the sample smoke and which do not.

The reason many studies are conducted is that one wants to understand the mechanism underlying various phenomena and ultimately put this understanding to use by devising methods for controlling such phenomena. The key issue of contention in most situations is whether an observed relationship is *causal.* For instance, does smoking *cause* cancer? Does exposure to asbestos *cause* cancer? Do vaccinations *cause* a reduction in the incidence of various diseases? Do emission control laws *cause* a reduction in environmental pollution? While it is relatively straightforward to demonstrate relationships or associations between different variables, it is very difficult to demonstrate that such relationships are causal. Observational studies are useful only to the extent that they allow us to discover patterns or relationships that may exist between variables. However, observational studies do not provide evidence toward establishing a causal connection. Properly conducted experimental studies are almost essential for a scientific demonstration of causality.

Demonstration of Causality

To establish causality, one must first demonstrate the existence of a relationship between the variables for the population of interest. At this stage, one should critically evaluate the appropriateness of any statistical and judgment inferences using the list of questions in Boxes I.2.1 and I.2.2. However, there are many more questions that must be answered to demonstrate causality. Box I.3.1 reports several such questions.

BOX I.3.1: QUESTIONS TO HELP DECIDE IF TWO VARIABLES ARE CAUSALLY RELATED

1. Is the observed association between the variables "real," or can it be attributed to an unusual sample due to chance?
2. Is the study an experimental study or an observational study?
3. Can the claimed causal relationship be explained by a plausible mechanism?
4. Is the magnitude of the claimed effect large enough to be of practical consequence?
5. Did the proposed cause precede the effect?
6. Are there other variables that might explain the association?
7. Is the association consistent over many studies?

We now present an example that demonstrates how the questions in Box I.3.1 can be used to assess whether the association between two variables is causal.

EXAMPLE I.3.2

CHURCHES AND VIOLENT CRIME. We consider the relationship between the number of churches and the incidents of violent crime reported in Example I.3.1. The data demonstrated (at the 95% confidence level) that, on average, there are between 1.38 and 4.85 additional violent crimes for each additional church in a city. We now answer each question in Box I.3.1 to assess whether this association could be causal.

1. *Is the observed association between the variables "real," or can it be attributed to an unusual sample due to chance?*

We can be quite confident that the observed association is not due to chance since a 95% confidence interval for β_1 indicates that the slope is positive. This implies that cities with more churches tend to have a greater number of violent crimes.

2. *Is the study an experimental study or an observational study?*

This is an observational study since the number of churches in a city (X) was not controlled by the investigator.

3. *Can the claimed causal relationship be explained by a plausible mechanism?*

There does not seem to be a good explanation for why increasing the number of churches might result in an increased number of violent crimes. Perhaps one could argue that as the number of violent crimes increases in a city, the citizens build more churches in an effort to combat the increasing crime rate. However, this explanation suggests that the number of violent crimes causes changes in the number of churches, and not vice versa.

4. *Is the magnitude of the claimed effect large enough to be of practical consequence?*

The data demonstrate (at the 95% confidence level) that, on average, there are between 1.38 and 4.85 additional violent crimes for each additional church. If the relationship is causal, this would mean that closing down one church would, on average, result in a decrease of 1.38 to 4.85 violent crimes. This is probably large enough to be of practical importance.

5. *Did the proposed cause precede the effect?*

This question asks whether the number of churches increased before the number of violent crimes increased. This seems unlikely. Most probably, both numbers change in the same direction as the population of a city changes. It would be useful to examine how the number of violent crimes and the number of churches have changed over the years.

6. *Are there other variables that might explain the association?*

Yes. One variable that could explain the association is the *population* of each city. Such a variable is called a **confounding** variable. Larger cities are expected to have more churches as well as more violent crimes than smaller cities. So the apparent relationship between the number of violent crimes and the number of churches in a city could be explained by the population size. For this reason, it would perhaps be better to examine the relationship between the number of churches and the number of violent crimes on a **per capita** basis. That is, one might consider the association between the number of violent crimes per 1,000 people living in a city (Y) and the number of churches per 1,000 people living in a city (X). Other statistical techniques can also be used to adjust for the effect of confounding variables when studying the association between two variables. These techniques are discussed in more-advanced textbooks.

7. *Is the association consistent over many studies?*

This is an especially useful question when the answers to the previous questions support the claim of a causal relationship. That is not the case in the present example. However, for the sake of completeness, we will answer this question. If one were to examine data collected over different time periods, there would probably be evidence of an association between the number of churches and the number of violent crimes in every data set.

CONCLUSION The answers to the questions in Box I.3.1 do not appear to support the claim of a causal relationship between the number of churches and the number of violent crimes. A major difficulty to proving causality is that other possible explanations for the observed association exist (question 6). In particular, it appears that population size can be used to explain the values of both variables. Another problem with the study is that no convincing mechanism is evident to explain the causal relationship (question 3).

We close this section with an exploration that concerns a major environmental study called the "Winter Haze Intensive Tracer Experiment" (WHITEX). This study was conducted in the late 1980s to determine whether a particular power plant in Arizona was causing visibility problems at the Grand Canyon National Park. This is an extremely complex study with many interconnected components that cannot be explained thoroughly within the context of this book. However, we hope that this material will provide you with a more practical example of a claimed causal relationship.

EXPLORATION 1.3.1

As the front cover of this book demonstrates, the Grand Canyon National Park is one of the most spectacular natural wonders in the world. In some places, the canyon is 1.6 km deep and 30 km wide. One attraction of the Grand Canyon is that on a clear day, one can overlook the canyon and see the details of over 100 km of breathtaking landscape features. One's ability to see and enjoy the scenic beauty depends on the clearness of the air. A coal-fired power plant called the Navajo Generating Station (NGS) is located about 110 km northeast of the Grand Canyon Village tourist area. In 1987 the U.S. National Park Service suspected that emissions from the NGS were causing significant loss in visibility in the Grand Canyon. If the National Park Service could demonstrate this to be true, the Environmental Protection Agency was going to ask the power plant to install pollution-reducing devices in its smoke stacks.

It is well known that coal-fired power plants release large amounts of sulfur dioxide gas that can be transported by the wind. During transport, some of the sulfur dioxide gas molecules undergo chemical transformations in the atmosphere and become sulfate particles. Various studies have established that sulfate particles in the atmosphere scatter sunlight and cause the haze that results in visibility degradation.

In January and February of 1987, a large-scale study was carried out to determine how much of the reduction in visibility at the Grand Canyon was due to the emissions from the NGS. The study was called the "Winter Haze Intensive Tracer Experiment" (WHITEX). The project was sponsored in part by the National Parks Service, the Environmental Protection Agency, and a consortium of electrical utilities. A detailed report of the WHITEX results may be obtained by contacting the National Park Service.

The WHITEX study attempted to determine if emissions from the NGS contributed to the level of haze in the Grand Canyon. To do this, an inert gas not normally found in nature or emitted by other sources was released continuously through the smoke stacks at the NGS in quantities roughly proportional to the amount of sulfur dioxide emitted by the plant. This inert gas, called deuterated methane (chemical symbol CD_4), was referred to as the **tracer**. When the tracer was detected at the Grand Canyon, it was concluded that a proportional amount of sulfur dioxide (particles and gas) from the plant was also present at the Grand Canyon. One objective of the WHITEX study was to estimate the fraction of the sulfate particles at the Grand Canyon that originated at the NGS. If one could estimate the total amount of sulfur (sulfate particles + sulfur dioxide gas) at the Grand Canyon that came from the NGS, then one might be able to estimate the amount of sulfate particles and the accompanying loss in visibility at the Grand Canyon attributable to the NGS. Thus, the success of the study in estimating the impact of the NGS on the visibility at the Grand Canyon depended on establishing several links in a complex chain of events. Both judgment inference and statistical inference were needed to form conclusions.

We now consider a small data set collected during the study. The concentration of the tracer (X) and the total sulfur concentration (Y) were measured by collecting air samples at Hopi Point near the Grand Canyon Village tourist area. The data are presented in Table I.3.2 and stored in the file **grndcnyn.mtp** on the data disk. The sample number is in column 1, Y is in column 2, and X is in column 3. These data represent only a small subset of the entire WHITEX database. Each measurement in this table is based on one of the many air samples collected over different 12-hour intervals during the study period. The concentrations of sulfur are given in micrograms per cubic meter ($\mu g/m^3$), and the concentrations of the tracer are in parts per billion (ppb).

TABLE I.3.2 Grand Canyon Tracer Data*

Sample number	Sulfur ($\mu g/m^3$) Y	Tracer (ppb) X	Sample number	Sulfur ($\mu g/m^3$) Y	Tracer (ppb) X
1	0.24917	0.004420	19	0.91159	0.005950
2	1.49678	0.010400	20	1.32867	0.007700
3	0.71102	0.003280	21	0.75229	0.005140
4	0.38131	0.004090	22	0.70710	0.003590
5	0.32583	0.003400	23	0.54845	0.004510
6	0.59321	0.003680	24	0.23977	0.003740
7	0.78407	0.005680	25	0.34218	0.004820
8	0.56541	0.004120	26	0.54682	0.003320
9	0.33719	0.003100	27	0.78823	0.002950
10	0.19850	0.003260	28	0.64884	0.002680
11	0.07781	0.002270	29	0.48953	0.002650
12	0.07416	0.002030	30	0.26603	0.003300
13	0.08969	0.002150	31	0.54113	0.005575
14	0.12012	0.002410	32	0.46071	0.004810
15	1.00689	0.002610	33	0.40007	0.005850
16	0.49704	0.002360	34	0.25450	0.004110
17	4.42405	0.004060	35	0.16249	0.001630
18	0.60679	0.005640	36	0.14716	0.002110

*Data courtesy of Dr. William C. Malm of the U.S. National Park Service

We now pose several questions to help you analyze the data in Table I.3.2.

1. *Obtain a scatter plot of the total sulfur concentration (Y) against the tracer concentration (X). Does there appear to be an association between sulfur levels and tracer levels?*

   The scatter plot is shown in Figure I.3.2. The sulfur concentrations appear to increase with increasing values of the tracer. One of the plotted points is an outlier. An examination of Table I.3.2 reveals that the outlier corresponds to observation number 17, which has an unusually high value for the sulfur concentration.

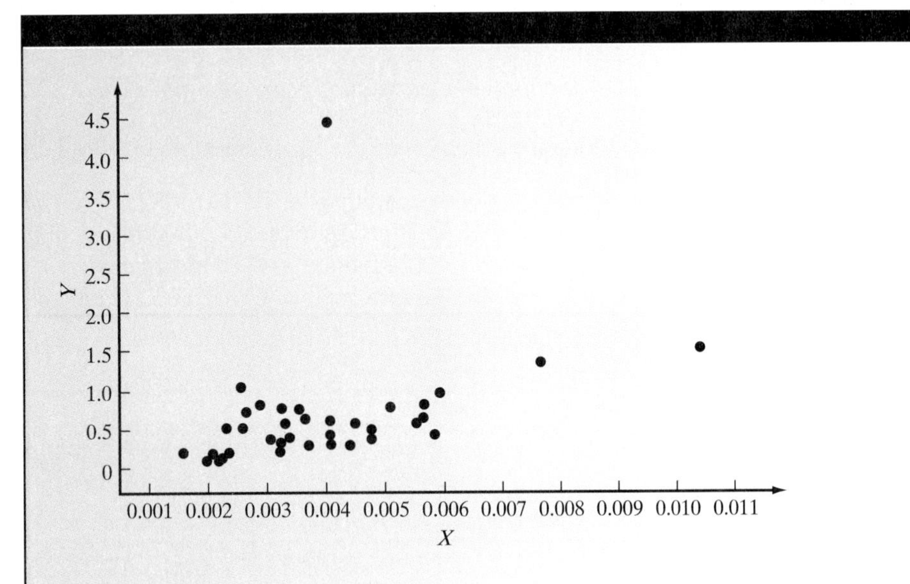

FIGURE I.3.2
Scatter Plot of Total Sulfur Concentration (Y, in $\mu g/m^3$) Versus Tracer Concentration (X, in ppb)

2. *An examination of the measurements for observation number 17 indicated that the meteorological conditions for this time period were atypical. For this reason, we will omit this observation in all further analyses of the data. For the remaining 35 observations, obtain a scatter plot of Y versus X. Does there appear to be an association between sulfur levels and tracer levels?*
The scatter plot of the data after excluding the outlier (observation 17) is given in Figure I.3.3. There does appear to be an association between the sulfur levels and the tracer levels. As the tracer values increase, it appears that the sulfur values tend to increase.

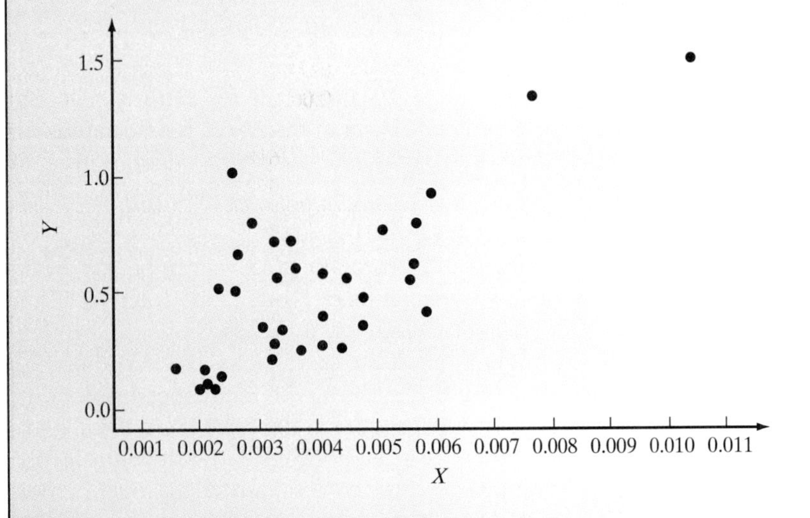

FIGURE I.3.3
Scatter Plot of Total Sulfur Concentration (Y, in $\mu g/m^3$) Versus Tracer Concentration (X, in ppb) Without Observation 17

3. *Does a straight line appear to be a reasonable description of the relationship between the tracer and the total sulfur?*
Yes. The scatter plot in Figure I.3.3 suggests that a straight line is a reasonable summary of the relationship between the tracer and the total sulfur.

4. *Suppose the population regression function of Y on X is given by*

$$\mu_Y(X) = \beta_0 + \beta_1 X.$$

and the other assumptions in Box 9.3.1 are satisfied. Compute the sample regression function using the 35 observations that remain after deleting observation 17. Compute a 90% confidence interval for β_1.
Using Minitab, we obtained Exhibit I.3.2 to help perform these computations.

```
------------------------------------------------------------------
The regression equation is
Y = - 0.028 + 134 X

Predictor        Coef        Stdev     t-ratio        p
Constant      -0.0283       0.1005       -0.28    0.780
X              133.79        23.11        5.79    0.000

S = 0.2399       R-sq = 50.4%      R-sq(adj) = 48.9%
------------------------------------------------------------------
```

EXHIBIT I.3.2

From Exhibit I.3.2 we obtain the sample regression function

$$\hat{\mu}_Y(X) = -0.028 + 134X.$$

To compute a 90% confidence interval for β_1 using $DF = 35 - 2 = 33$, we obtain $T2 = 1.69$. The 90% confidence interval for β_1 is

$$L = 133.79-(1.69)(23.11) = 94.73, \quad U = 133.79+(1.69)(23.11) = 172.85.$$

Thus, with 90% confidence we can say that, on average, each additional one part per billion of the tracer is associated with an additional amount of sulfur between 94.73 and 172.85 micrograms per cubic meter.

5. *Is there an association between total sulfur concentrations and tracer concentrations at Grand Canyon?*
Using the results of part (4), we can be 90% confident that the slope of the regression function of Y on X is between 94.73 and 172.85. In particular, the data indicate that the slope is positive, confirming what we found by a visual examination of the scatter plot. Hence, there appears to be a relationship between total sulfur and tracer concentrations at the Grand Canyon.
By now we seem to have sufficient graphical and statistical evidence to suggest there is a relationship between sulfur levels and tracer levels at Hopi Point, Grand Canyon. In particular, higher amounts of the tracer are associated with higher amounts of the total sulfur. However, some of the sulfur will exist in the form of sulfur dioxide gas and the rest of it will exist as sulfate particles. Therefore, it seems reasonable to expect that higher amounts of the tracer are associated with higher amounts of sulfate particles originating from the NGS, which in turn affects the visibility at the Grand Canyon. This was the spirit of the argument put forth by investigators who claimed that

emissions from the NGS "caused" as much as 60% of the loss in visibility attributable to aerosol particles at the Grand Canyon. The Environmental Protection Agency reviewed all available information and, based on a combination of statistical and judgment inferences, concluded that the power plant was at least partially responsible for the visibility degradation at the Grand Canyon. Based on this conclusion, the power plant owners were asked to install pollution-reducing technology.

Although the data present a basis for these actions, some people have argued that a causal relationship between visibility levels at the Grand Canyon and emission levels at the NGS was not demonstrated satisfactorily. The primary argument against a causal relationship comes in response to question 6 in Box I.3.1. In particular, there are other factors that might explain the observed association between tracer and sulfur concentrations. For the sake of argument, suppose all emissions from the NGS are shut down but the tracer gas continues to be injected into the stacks and released into the atmosphere. Even though none of the sulfur at the Grand Canyon could conceivably be from the NGS under these circumstances, one would still find an association between sulfur and tracer concentrations if sulfur were transported to the Grand Canyon from another source that is "upwind" from the NGS. That is, air mass already containing sulfur from other power plants might pass over the NGS and pick up the tracer. Then air samples at the Grand Canyon would contain both tracer and sulfur, and a relationship would exist between the two variables. Thus, an association between the tracer and sulfur at the Grand Canyon is not definitive evidence to claim that the sulfur came from the NGS.

In conclusion, the evidence regarding causality seems to be mixed. It appears that the plausibility of a causal relationship between emissions at the NGS and Grand Canyon visibility has been demonstrated, particularly given that the NGS is very close to the Grand Canyon. The presence of the tracer gas at the Grand Canyon provides further evidence that emissions from the NGS travel to the Grand Canyon. However, the actual amount of sulfate particles at the Grand Canyon attributable to the NGS has not been adequately estimated. The association between sulfur and tracer concentrations at the Grand Canyon does not provide definitive evidence that the sulfur actually came from the NGS. The possibility that the sulfur came from other power plants cannot be ruled out based solely on the data. One reason for the lack of more definitive conclusions is the fact that WHITEX is an observational study and not an experimental study.

I.4 CHAPTER SUMMARY

KEY TERMS

1. validity
2. bias
3. precision
4. nonrespondents
5. causal relationship
6. observational studies
7. experimental studies

KEY CONCEPT

1. If evidence suggests that two variables are related, it does not necessarily imply that the two variables are causally related.

SKILL

1. Use Boxes I.2.1, I.2.2, and I.3.1 to critically examine the results of a statistical study.

CHAPTER 10

COMPARING POPULATION PROPORTIONS; CONTINGENCY TABLES

10.1 DIFFERENCE BETWEEN TWO POPULATION PROPORTIONS

In Chapter 3 we discussed point estimation and confidence intervals for a population proportion π. In some situations an investigator may be interested in comparing two different population proportions. In this case, $\pi_1 - \pi_2$ is of interest, where π_1 is the proportion of items in population 1 with a certain characteristic and π_2 is the proportion of items in population 2 with the same characteristic. We illustrate with three examples.

EXAMPLE 10.1.1

SPANKING CHILDREN. A child protection agency conducted a poll of adults in the state and asked them the question, "Is a hard spanking sometimes necessary to discipline a child?" The agency wants to determine the difference between π_1, the proportion of men who believe that a hard spanking is sometimes necessary, and π_2, the proportion of women who have this belief. That is, the agency wants to estimate $\pi_1 - \pi_2$, the difference between the proportion of men and women in the state who believe that a hard spanking is sometimes necessary to discipline a child. There are two populations in this problem. One population consists of adult men and the other consists of adult women. The question of interest is whether there is a difference between two population proportions.

EXAMPLE 10.1.2

CIGARETTE ADVERTISING. A poll of adults in a large city was conducted to collect information on attitudes concerning cigarette advertising. One objective of the study was to estimate $\pi_1 - \pi_2$, where π_1 is the proportion of nonsmokers who believe that all cigarette advertising should be banned, and π_2 is the proportion of smokers who believe that all cigarette advertising should be banned. As in Example 10.1.1, interest concerns the difference between two population proportions.

EXAMPLE 10.1.3

STATE OF THE ECONOMY. A candidate for the state legislature wants to know what the voters think about economic conditions in the state. There are two populations of interest: (1) voters who are registered as Republicans; and (2) voters who are registered as Democrats. The candidate wants to know $\pi_1 - \pi_2$, where π_1 is the proportion of Republicans who think the economy is better now than it was two years ago, and π_2 is the proportion of Democrats who have this belief.

These examples demonstrate that the difference between two population proportions is of interest in many applications. Since entire populations are seldom known, $\pi_1 - \pi_2$ is generally unknown. A sample must be collected from each of the two populations, and the sample data used to obtain a point estimate and confidence interval for $\pi_1 - \pi_2$.

Point Estimate of $\pi_1 - \pi_2$

Consider two sampled populations, which we denote as population 1 and population 2. It is desired to estimate $\pi_1 - \pi_2$, where π_1 is the proportion of items in population 1 with a certain characteristic, and π_2 is the proportion of items in population 2 with the same characteristic. To illustrate, in Example 10.1.1 we arbitrarily designate the population of men as population 1 and the population of women as population 2. The parameter π_1 represents the proportion of men who believe that a hard spanking is sometimes necessary to discipline a child, and π_2 represents the proportion of women who believe that a hard spanking is sometimes necessary to discipline a child. In Box 10.1.1 we describe the procedure for estimating $\pi_1 - \pi_2$.

BOX 10.1.1: PROCEDURE FOR ESTIMATING THE DIFFERENCE BETWEEN TWO POPULATION PROPORTIONS

A simple random sample of size n_1 from population 1 contains k_1 items that have a characteristic of interest. The estimate of π_1, the proportion of items in population 1 with the characteristic, is

$$\widehat{\pi}_1 = \frac{k_1}{n_1}.$$

A simple random sample of size n_2 from population 2 contains k_2 items that have the characteristic of interest. The estimate of π_2, the proportion of items in population 2 with the characteristic, is

$$\widehat{\pi}_2 = \frac{k_2}{n_2}.$$

The point estimate of $\pi_1 - \pi_2$ is

$$\widehat{\pi}_1 - \widehat{\pi}_2 = \frac{k_1}{n_1} - \frac{k_2}{n_2}.$$

EXAMPLE 10.1.4

SPANKING CHILDREN. Consider Example 10.1.1, where a child protection agency wants to know $\pi_1 - \pi_2$, where π_1 is the proportion of men who believe that a hard spanking is sometimes necessary to discipline a child, and π_2 is the proportion of women who believe that a hard spanking is sometimes necessary to discipline a child. The results of a simple random sample of 571 men and a simple random sample of 611 women are shown in Table 10.1.1.

TABLE 10.1.1 Attitudes on Spanking Children

		Men	Women
Is spanking necessary?	Yes	421	367
	No	150	244
	Total	571	611

From Table 10.1.1 we obtain $n_1 = 571$, $k_1 = 421$, $n_2 = 611$, and $k_2 = 367$. From these results $\widehat{\pi}_1 = k_1/n_1 = 421/571 = 0.7373$, and $\widehat{\pi}_2 = k_2/n_2 = 367/611 = 0.6007$. The estimate of $\pi_1 - \pi_2$ is $\widehat{\pi}_1 - \widehat{\pi}_2 = 0.7373 - 0.6007 = 0.1366$. Thus, we estimate that 13.7% more men than women believe that a hard spanking is sometimes necessary to discipline a child.

Confidence Intervals for the Difference Between Two Population Proportions

Confidence intervals can be used to determine "how close" $\widehat{\pi}_1 - \widehat{\pi}_2$ is to $\pi_1 - \pi_2$. The computations for obtaining a confidence interval for $\pi_1 - \pi_2$ are shown in Box 10.1.2.

BOX 10.1.2: CONFIDENCE INTERVALS FOR THE DIFFERENCE BETWEEN TWO PROPORTIONS

A simple random sample of size n_1 from population 1 contains k_1 items that have a characteristic of interest. A simple random sample of size n_2 from population 2 contains k_2 items that have the same characteristic of interest. The endpoints L and U of an approximate confidence interval for $\pi_1 - \pi_2$ are

$$L = \widehat{\pi}_1 - \widehat{\pi}_2 - \text{ ME},$$
$$U = \widehat{\pi}_1 - \widehat{\pi}_2 + \text{ ME},$$

where

$$\text{ME } = (T1)\sqrt{\frac{\widehat{\pi}_1(1-\widehat{\pi}_1)}{n_1} + \frac{\widehat{\pi}_2(1-\widehat{\pi}_2)}{n_2}}.$$

The table value $T1$ is found in Table T-1. The approximation is satisfactory if n_1 and n_2 are both greater than 100, if both k_1 and $n_1 - k_1$ are greater than or equal to 5, and if both k_2 and $n_2 - k_2$ are greater than or equal to 5.

The formulas in Box 10.1.2 can be derived using the central limit theorem discussed in Section 7.7.

EXAMPLE 10.1.5

SPANKING CHILDREN. Consider Example 10.1.4, where a child protection agency wants to know the difference between the proportion of men and women who believe that a hard spanking is sometimes necessary to discipline a child. The results of a simple random sample of 571 men and a simple random sample of 611 women are displayed in Table 10.1.1. From this table we obtain $n_1 = 571$, $k_1 = 421$, $n_2 = 611$, $k_2 = 367$, $\widehat{\pi}_1 = 0.7373$, $\widehat{\pi}_2 = 0.6007$, and $\widehat{\pi}_1 - \widehat{\pi}_2 = 0.1366$. To compute a 95% confidence interval for $\pi_1 - \pi_2$, the value $T1 = 1.96$ is obtained from Table T-1. Substituting these values into the formula in Box 10.1.2, we obtain

$$\text{ME} = 1.96\sqrt{\frac{0.7373(0.2627)}{571} + \frac{0.6007(0.3993)}{611}} = 0.05302,$$

$$L = 0.1366 - 0.05302 = 0.0835, \quad \text{and} \quad U = 0.1366 + 0.05302 = 0.1897.$$

We are 95% confident that $\pi_1 - \pi_2$ is between 0.0835 and 0.1897.
As stated in Box 10.1.2, this confidence interval is approximate. However, since n_1 and n_2 are both greater than 100 and since k_1, $n_1 - k_1$, k_2, and $n_2 - k_2$ are all greater than 5, the approximation is satisfactory.

Sampling Considerations

In each of the examples discussed so far in this section, a separate simple random sample was obtained from each of two populations. The sampling is conducted by randomly selecting n_1 items from the first population and n_2 items from the second population. If either n_1 or n_2 is too small, the resulting confidence interval for $\pi_1 - \pi_2$ is likely to be so wide that the results of the study will be inconclusive. To avoid this situation, before collecting data the investigator should determine the sample sizes n_1 and n_2 that will be needed in order to obtain a confidence interval for $\pi_1 - \pi_2$ whose width will be small enough to be useful. This will not be discussed in this book, but a professional statistician should be consulted for guidance on choosing the sample sizes n_1 and n_2, taking into consideration factors such as cost, convenience, and the width of the resulting confidence interval.

Suppose values for n_1 and n_2 have been determined. If two separate lists are available, one for each population, the sampling is conducted by randomly selecting n_1 items from the list containing the first population and randomly selecting n_2 items from the list containing the second population. However, there are many situations where separate lists of items are not available for each population. Rather, the only list available is a combined list that contains the items belonging to *both* populations. In this situation investigators will sometimes select a simple random sample of a predetermined size n from the combined list. After the sample is obtained, the population membership is determined for each item in the sample. The values of

n_1 and n_2 are thus known only after the data collection is completed. Of course, $n_1 + n_2 = n$. We illustrate with an example.

EXAMPLE 10.1.6

HEALTH PROVIDER. An insurance company that provides health insurance to several thousand individuals is interested in ways to decrease health costs. One topic of interest to the company is comparing the proportion of smokers and nonsmokers who get annual physical examinations. Let π_1 represent the proportion of smokers who had a physical exam during the past 12 months, and let π_2 represent the proportion of nonsmokers who had a physical exam during the past 12 months. The interest is in $\pi_1 - \pi_2$, the difference between the proportions of smokers and nonsmokers who had a physical examination during the past 12 months. There are two populations: (1) the population of individuals insured by the company who are smokers; and (2) the population of individuals insured by the company who are nonsmokers. If 500 individuals are to be randomly selected from the combined list of those insured, this list contains individuals from both populations. Before selecting the sample the investigator has no idea whether or not a person is a smoker. Each individual in the sample will be asked two questions: (1) "Do you smoke?" and (2) "Did you have a physical examination during the past 12 months?" Here the investigator can choose the value of n, the total sample size to use, but the values of n_1 and n_2 (the number of individuals in the sample who do and do not smoke, respectively) will be known only after the sample has been obtained.

In Example 10.1.3 the two populations are voters registered as Republicans (population 1) and voters registered as Democrats (population 2). If one can obtain a list of registered Republicans and a separate list of registered Democrats, then a sample of n_1 Republicans can be obtained from the first list and a sample of n_2 Democrats can be obtained from the second list. However, if the sample is selected from a list of registered voters in which party affiliation is not designated, then one would select a simple random sample of size n from the combined list. In this case the number of Republicans (n_1) and the number of Democrats (n_2) in the sample are determined only after the data collection is completed.

In some applications, selecting a sample of size n from a combined list may not be a wise procedure. For example, suppose the number of items in population 1 is much smaller than the number of items in population 2. In this situation, it is possible that the sample may contain very few items, or even no items, from population 1. As a result, the confidence interval for $\pi_1 - \pi_2$ may turn out to be too wide to be useful. The following example provides an illustration.

EXAMPLE 10.1.7

STUDENTS' GRADES. The registrar of a major university wants to determine the difference between π_1, the proportion of international students who have a GPA of 3.5 or greater, and π_2, the proportion of U.S. students who have a GPA of 3.5 or greater. A simple random sample of 300 students was obtained from the combined list of all students in the university. The records of these 300 students were examined; 10 out of the 300 students were international students, and 2 of these 10 had a GPA of 3.5 or higher. There were 290 U.S. students in the sample, and 22 of them had a GPA of 3.5 or higher. In this case we get a good estimate of π_2 since the sample size for estimating π_2 is 290. But we have a poor estimate of π_1 since it is based on a sample of size 10. As a consequence, the confidence interval for $\pi_1 - \pi_2$ would undoubtedly be too wide to be useful. It would have been better to specify the values of n_1 and n_2 before collecting the data, so that a tighter confidence interval for $\pi_1 - \pi_2$ could have been obtained.

With some additional effort, the registrar in Example 10.1.7 might have been able to obtain two separate lists for the two populations of interest. When this is not possible, the investigator should determine a suitable plan for selecting the samples with the help of a professional statistician so that the sample sizes n_1 and n_2 are large enough to provide a confidence interval for $\pi_1 - \pi_2$ that has a small-enough width to be useful.

THINKING STATISTICALLY: SELF-TEST PROBLEMS

10.1.1 Give two examples where the difference between two proportions is of interest.

10.1.2 Find two newspaper articles where the difference between two proportions is discussed.

10.1.3 Last year a total of 1,121 graduate students applied for financial aid at a large university. Of these students, 710 were males and 411 were females. The university awarded financial aid to 492 male students and 358 female students. The graduate student council wants to know if students of each gender were treated equally in awarding financial assistance last year.

a. What are the two target populations of interest?
b. Let π_1 represent the proportion of male applicants who received financial aid last year. Compute π_1.
c. Let π_2 represent the proportion of female applicants who received financial aid last year. Compute π_2.
d. Compute $\pi_1 - \pi_2$.
e. Last year the proportion of female applicants who received financial aid was greater than the proportion of male applicants who received financial aid. The graduate student council decides to determine if this situation has been true for the past five years. A simple random sample of 1,569 students was selected from a list of all financial aid applicants (male and female) during the past five years. The results are shown in Table 10.1.2.

TABLE 10.1.2 Sample of Financial Aid Applicants

		Male	**Female**
Received aid	Yes	489	529
	No	354	197
	Total	843	726

Use Table 10.1.2 to estimate π_1, π_2, and $\pi_1 - \pi_2$, where π_1 represents the proportion of male applicants who received financial aid during the past five years and π_2 represents the proportion of female applicants who received financial aid during the past five years.
f. Compute and interpret a 95% confidence interval for $\pi_1 - \pi_2$.

10.2 DIFFERENCES AMONG SEVERAL POPULATION PROPORTIONS

In the previous section we restricted our attention to comparing two population proportions where each population had one variable of interest, and this variable had two values. For instance, in Example 11.1.1 the two populations are Men and Women. The variable of interest is labeled "Is spanking necessary?" with two values, *yes* and *no*. It is common to encounter studies in which the variable of interest has many possible values (more than two). For instance, suppose a company is marketing two different pesticides to farmers. The farmers are asked to evaluate the effectiveness of the pesticide using one of three possible values—*ineffective, partially effective*, and *very effective*. There are two populations (two pesticides), and the variable of interest, "Effectiveness," has three possible values. The company is interested in comparing the two pesticides with respect to their effectiveness. In this section we provide methods that can be used to determine the difference between two population proportions for several values of a variable. The methods are simple extensions of the methods introduced in the previous section. We begin our discussion with an example.

EXAMPLE 10.2.1

FOOT-ROT IN PIGS. A company has developed a new drug for treatment of foot-rot in pigs. An experiment is conducted to determine the effectiveness of the drug by comparing it with a standard drug for treating this disease. A simple random sample of 100 pigs is obtained from about 50,000 pigs that are suffering from foot-rot. Each pig in this sample is treated with the new drug. A second simple random sample of 100 pigs is obtained from the same population of 50,000 pigs, and each pig in this sample is treated with a standard drug. Thus, there are two sampled populations and both are conceptual. Sampled population 1 is the set of 50,000 pigs where every pig is treated with the new drug. Sampled population 2 is the set of 50,000 pigs where every pig is treated with the standard drug. After one month of treatment, a veterinarian assesses the degree of recovery for each of the 200 pigs in the samples. Without knowing which treatment a pig received, the veterinarian assigns one of the following four values to each pig: (1) *No recovery*; (2) *Slight recovery*; (3) *Substantial recovery*; and (4) *Complete recovery* (*cure*). The variable of interest, "Recovery," has four possible values. The sample data are shown in Table 10.2.1 and stored in columns 1 and 2 in the file **foot-rot.mtp** on the data disk.

TABLE 10.2.1 Veterinarian's Assessment of Foot-Rot for Two Samples of 100 Pigs (frequencies)

		New drug	Standard drug
Recovery	No	6	12
	Slight	14	10
	Substantial	30	18
	Complete	50	60
	Total	100	100

Table 10.2.2 contains the proportion of pigs in each column of Table 10.2.1 that were assigned to each of the four recovery values. To obtain the entries of Table 10.2.2, each number in Table 10.2.1 is divided by the total of the column in which it

appears. For example, the number in row 1 and column 2 in Table 10.2.1 is 12. If we divide this number by the column total—100—we get the proportion $12/100 = 0.12$, which appears in row 1 of column 2 in Table 10.2.2.

TABLE 10.2.2 Veterinarian's Assessment of Foot-Rot for Two Samples of 100 Pigs (proportions)

		New drug	Standard drug
Recovery	No	0.06	0.12
	Slight	0.14	0.10
	Substantial	0.30	0.18
	Complete	0.50	0.60
	Total	1.00	1.00

Table 10.2.2 has four rows (excluding the row of totals) and two columns. It is of interest to determine how much the proportions differ in each row of the table. For example, the new drug has a smaller proportion of cases than the standard drug in the rows labeled *No* and *Complete*. The new drug has a larger proportion of cases than the standard drug in the rows labeled *Slight* and *Substantial*. Since the data in Table 10.2.1 come from two samples of pigs, the proportions in Table 10.2.2 are estimates. Confidence intervals are needed to determine "how close" these estimates are to the corresponding population proportions.

Contingency tables. A table such as Table 10.2.1 that summarizes the frequencies of values of a variable for each sample from two populations is called a **contingency table**. When used in this manner, the word "contingency" means that the table entries were obtained "at random." It is conventional to describe a contingency table by the number of rows and columns it contains. For example, Table 10.2.1 has four rows and two columns and is described as a "4 by 2 contingency table." When describing a contingency table, we give the number of rows first and then the number of columns.

Convention. In this chapter the variable of interest will be placed in the row, and the row labels will identify the values of the variable. The column labels will identify the two populations. Since we consider only two populations, the general form of a contingency table is r by 2, where $r \geq 2$.

We now present a method for obtaining confidence intervals for the difference between the proportions in every row of an r by 2 table of population proportions. For ease of explanation, we will use a 4 by 2 table. Obvious modifications will allow you to analyze any r by 2 contingency table where $r > 2$. If $r = 2$, you should use the method described in Section 10.1.

Confidence Intervals for the Difference Between Population Proportions in an r by 2 Table ($r > 2$)

Suppose an investigator is interested in two sampled populations, where population 1 consists of N_1 items, and population 2 consists of N_2 items. In Example 10.2.1 population 1 is a conceptual population of pigs that received a new drug, and population 2 is a conceptual population of pigs that received a standard drug. Using our convention, the column labels identify the two populations. The row labels (the values of the variable) are the veterinarian's assessment of "Recovery" for each drug using the four values *No*, *Slight*, *Substantial*, and *Complete*. The population frequencies (counts) for each of these four values are denoted by the uppercase letters $K_{11}, K_{21}, K_{31}, K_{41}$ and $K_{12}, K_{22}, K_{32}, K_{42}$, where K_{ij} denotes the number of population items in the ith row and the jth column in the table of population frequencies. Although the K_{ij} values exist, they are not known unless both sampled populations are observed in their entirety. You can envision the populations schematically as shown in Table 10.2.3.

TABLE 10.2.3 Schematic Representation of the Frequencies of Sampled Populations in a 4 by 2 Table

		Population	
		New drug	**Standard drug**
Recovery	No	K_{11}	K_{12}
	Slight	K_{21}	K_{22}
	Substantial	K_{31}	K_{32}
	Complete	K_{41}	K_{42}
	Total	N_1	N_2

The column totals in Table 10.2.3 are N_1 and N_2, respectively. We are interested in comparing the proportions of items in columns 1 and 2 for each of the four rows in the table. To obtain the proportions, each value in column 1 is divided by N_1 and each value in column 2 is divided by N_2. Table 10.2.4 reports the values of π_{ij}, the proportion of the total items in column j that appear in row i. These values are computed using the formula $\pi_{ij} = K_{ij}/N_j$, for $i = 1, 2, 3, 4$ and $j = 1, 2$.

TABLE 10.2.4 Schematic Representation of the Proportions of Sampled Populations in a 4 by 2 Table

		Population	
		New drug	**Standard drug**
Recovery	No	π_{11}	π_{12}
	Slight	π_{21}	π_{22}
	Substantial	π_{31}	π_{32}
	Complete	π_{41}	π_{42}
	Total	1.00	1.00

If the two proportions in columns 1 and 2 are equal for every row of Table 10.2.4, then the new drug is no better (nor worse) than the standard drug for treating foot-rot in pigs. This is true in Table 10.2.4 if

$$\pi_{11} = \pi_{12}, \quad \pi_{21} = \pi_{22}, \quad \pi_{31} = \pi_{32}, \quad \text{and} \quad \pi_{41} = \pi_{42}.$$

We illustrate such a situation with the hypothetical 4 by 2 table shown in Table 10.2.5.

Using the frequencies in Table 10.2.5, we calculate the proportions π_{ij} and display them in Table 10.2.6. For example, $\pi_{11} = 10/75 = 0.133$ and $\pi_{32} = 16/60 = 0.267$.

From Table 10.2.6 we see that the two proportions are equal in every row. That is, $\pi_{11} = \pi_{12} = 0.133$, $\pi_{21} = \pi_{22} = 0.200$, $\pi_{31} = \pi_{32} = 0.267$, and $\pi_{41} = \pi_{42} = 0.400$.

TABLE 10.2.5 Hypothetical Frequencies of Sampled Populations in a 4 by 2 Table

		Population 1	Population 2
Row variable	Category 1	10	8
	Category 2	15	12
	Category 3	20	16
	Category 4	30	24
	Total	75	60

TABLE 10.2.6 Hypothetical Proportions of Sampled Populations in a 4 by 2 Table

		Population 1	Population 2
Row variable	Category 1	0.133 (π_{11})	0.133 (π_{12})
	Category 2	0.200 (π_{21})	0.200 (π_{22})
	Category 3	0.267 (π_{31})	0.267 (π_{32})
	Category 4	0.400 (π_{41})	0.400 (π_{42})
	Total	1.000	1.000

In such a case we say that the proportions of the two populations are equal with respect to the values of the variable labeled "Row variable." In practice, it is highly unlikely that the population proportions will be *exactly* equal in *every* row. However, an investigator might decide that if the magnitude of the difference between the two population proportions is small in every row, then the differences are of no practical importance. For example, the investigator might state that if $|\pi_{i1} - \pi_{i2}| < d$ for every row in the table (d is specified by the investigator), then any difference between the two populations with respect to the row variable is of no practical importance.

If all the data in both populations are completely known, then all of the proportions can be computed exactly. As you know by now, this seldom occurs in practice and we must rely on sample data to make inferences about the differences. This can be done by examining confidence intervals for the difference between proportions in each row of the table. For a 4 by 2 table, confidence intervals are needed for

$$\pi_{11} - \pi_{12}, \quad \pi_{21} - \pi_{22}, \quad \pi_{31} - \pi_{32}, \quad \text{and} \quad \pi_{41} - \pi_{42}. \qquad \textbf{Line 10.2.1}$$

Computing Simultaneous Confidence Intervals

We now outline an approach for computing confidence intervals for the differences in Line 10.2.1. Sample data are obtained and displayed in the 4 by 2 contingency table shown in Table 10.2.7, where n_1 and n_2 are the column totals and $n_1 + n_2 = n$ (note that we use lowercase n_1 and n_2 for sample column totals).

TABLE 10.2.7 Schematic Representation of Sample Frequencies

		1	**2**
Row variable	Category 1	k_{11}	k_{12}
	Category 2	k_{21}	k_{22}
	Category 3	k_{31}	k_{32}
	Category 4	k_{41}	k_{42}
	Total	n_1	n_2

In Box 10.2.1 we outline a procedure for computing confidence intervals for all the differences in Line 10.2.1 such that, with a confidence of at least 95% (or any confidence coefficient chosen by the investigator), *all* intervals cover their respective parameters.

BOX 10.2.1: COMPUTATION OF SIMULTANEOUS CONFIDENCE INTERVALS FOR DIFFERENCES BETWEEN POPULATION PROPORTIONS IN EVERY ROW OF AN *r* BY 2 TABLE

An investigator wants to examine the differences between two population proportions for each of the r values of a variable where $r > 2$. For purposes of illustration, we select $r = 4$. Table 10.2.7 contains a schematic representation of k_{ij}, the sample frequencies (counts) when $r = 4$. The following steps can be used to compute confidence intervals for the four differences

$$\pi_{11} - \pi_{12}, \quad \pi_{21} - \pi_{22}, \quad \pi_{31} - \pi_{32}, \quad \text{and} \quad \pi_{41} - \pi_{42}. \qquad \textbf{Line 10.2.1}$$

(1) Compute the estimates of π_{ij} using the formula $\widehat{\pi}_{ij} = k_{ij}/n_j$ for every value of i and j.

(2) Compute the estimates for all the differences in Line 10.2.1 using the formula $\widehat{\pi}_{i1} - \widehat{\pi}_{i2} = k_{i1}/n_1 - k_{i2}/n_2$.

(3) Compute the margin of error $(\text{ME})_i$ for each of the differences in Line 10.2.1 using the formula

$$(\text{ME})_i = (TB)\sqrt{\frac{\widehat{\pi}_{i1}(1-\widehat{\pi}_{i1})}{n_1} + \frac{\widehat{\pi}_{i2}(1-\widehat{\pi}_{i2})}{n_2}},$$

for $i = 1, 2, 3, 4$, where TB is a **Bonferroni table value** found in Table 10.2.8.

(4) Compute the endpoints L_i and U_i of the confidence interval for each difference in Line 10.2.1 using the formulas

$$L_i = \widehat{\pi}_{i1} - \widehat{\pi}_{i2} - (\text{ME})_i,$$
$$U_i = \widehat{\pi}_{i1} - \widehat{\pi}_{i2} + (\text{ME})_i,$$

for $i = 1, 2, 3, 4$.

The formula for $(\text{ME})_i$ in Box 10.2.1 is the same one shown for the margin of error in Box 10.1.2 except that the table value TB replaces $T1$. For any value of r ($r > 2$), it follows that $TB > T1$ for any confidence coefficient. Thus, an interval computed using Box 10.2.1 will be wider than if computed using Box 10.1.2. This is because it is necessary to widen each interval in a set of r intervals in order to ensure that all r intervals *simultaneously* cover their respective parameters with the specified level of confidence. The values of TB are derived using the **Bonferroni method**, which is discussed in the appendix to this chapter (Section 10.7).

Inspection of Table 10.2.8 shows that the value of TB depends on both the desired confidence coefficient and r, the number of confidence intervals obtained. If TB corresponds to a 95% confidence coefficient, we are at least 95% confident that *all r intervals* are simultaneously correct. That is, the confidence we have that *all r* intervals are correct is the same confidence we have that the ball hidden in our hand is black if it was randomly drawn from a bin that contains 100 balls of which at least 95 are black. A similar result holds for any confidence coefficient (in particular, for the other two confidence coefficients shown in Table 10.2.8).

TABLE 10.2.8 Table Values of *TB* for Several Values of *r* with 90%, 95%, and 99% Confidence Coefficients

	Confidence Coefficient		
r	**0.90**	**0.95**	**0.99**
3	2.13	2.39	2.94
4	2.24	2.50	3.02
5	2.33	2.58	3.09
6	2.39	2.64	3.14
7	2.45	2.69	3.19
8	2.50	2.73	3.23
9	2.54	2.78	3.26
10	2.58	2.81	3.29

The formulas in Box 10.2.1 are approximate and can be used if each cell in the table contains at least five observations and if n_1 and n_2 are both greater than 100. After examining the confidence intervals for the differences in Line 10.2.1, the investigator can determine if the magnitude of any difference is large enough to be of practical importance. We illustrate the procedure in Box 10.2.1 using the foot-rot example.

EXAMPLE 10.2.2

FOOT-ROT IN PIGS. We now compute confidence intervals for all four differences in Line 10.2.1 using the samples of pigs shown in Table 10.2.1. For convenience, this table is repeated in Table 10.2.9.

Suppose we want to be at least 95% confident that all four intervals are simultaneously correct. To do this, we follow the steps in Box 10.2.1.

1. The computed values of $\widehat{\pi}_{ij} = k_{ij}/n_j$ are shown in Table 10.2.10.
2. The estimates of the $r = 4$ differences in Line 10.2.1 are shown in the last column of Table 10.2.10.
3. Next we calculate $(\text{ME})_i$ for each of the differences. To obtain at least a 95% confidence with $r = 4$, select $TB = 2.50$ from Table 10.2.8.

TABLE 10.2.9 Veterinarian's Assessment of Foot-Rot for Two Samples of 100 Pigs (frequencies)

		New drug	Standard drug
Recovery	No	6	12
	Slight	14	10
	Substantial	30	18
	Complete	50	60
	Total	100	100

TABLE 10.2.10 Sample Estimates for Foot-Rot Data

		New drug	Standard drug	Estimates of differences
Recovery	No	0.060 ($\hat{\pi}_{11}$)	0.120 ($\hat{\pi}_{12}$)	−0.060 ($\hat{\pi}_{11} - \hat{\pi}_{12}$)
	Slight	0.140 ($\hat{\pi}_{21}$)	0.100 ($\hat{\pi}_{22}$)	0.040 ($\hat{\pi}_{21} - \hat{\pi}_{22}$)
	Substantial	0.300 ($\hat{\pi}_{31}$)	0.180 ($\hat{\pi}_{32}$)	0.120 ($\hat{\pi}_{31} - \hat{\pi}_{32}$)
	Complete	0.500 ($\hat{\pi}_{41}$)	0.600 ($\hat{\pi}_{42}$)	−0.100 ($\hat{\pi}_{41} - \hat{\pi}_{42}$)
	Total	1.000	1.000	

The computation for $(\text{ME})_1$ is

$$(\text{ME})_1 = (2.50)\sqrt{\frac{0.060(1-0.060)}{100} + \frac{0.120(1-0.120)}{100}} = 0.101.$$

The other margins of error are $(\text{ME})_2 = 0.115$, $(\text{ME})_3 = 0.150$, and $(\text{ME})_4 = 0.175$.

4. The point estimates, margins of error, and confidence intervals for the four differences in Line 10.2.1 are shown in Table 10.2.11.

TABLE 10.2.11 Summary Results of Foot-Rot Investigation

Difference	Point estimate	Margin of error	Lower bound	Upper bound
$\pi_{11} - \pi_{12}$	−0.06	0.101	−0.161	0.041
$\pi_{21} - \pi_{22}$	0.04	0.115	−0.075	0.155
$\pi_{31} - \pi_{32}$	0.12	0.150	−0.030	0.270
$\pi_{41} - \pi_{42}$	−0.10	0.175	−0.275	0.075

We are at least 95% confident that all four confidence intervals in Table 10.2.11 are simultaneously correct. Suppose the investigator decides that any difference in Line 10.2.1 is of practical importance if the magnitude (absolute value) is 0.10 or greater. Since all four intervals contain values that are both larger and smaller than

0.10 in magnitude, the results are inconclusive. That is, the data do not provide sufficient evidence to conclude that any of the differences in Line 10.2.1 exceed 0.10 in absolute value, nor do they rule out such a possibility. More sample data are required for the results to be conclusive.

A Minitab Macro for Performing the Computations in Box 10.2.1

Minitab has no single command that can be used to perform the calculations in Box 10.2.1. However, we have written a macro for the case $r > 2$ that is contained in the file **pairprop.mac** on the data disk. To use this macro, place the frequencies of an r by 2 contingency table in two columns of a Minitab worksheet. Enter the following command in the Session window:

```
MTB > %a:\macro\pairprop cp cq confidence
```

where the frequencies are in columns `p` and `q`, and `confidence` is the desired confidence coefficient expressed as a percentage (for example, 90, 95, or 99). To use the macro for Example 10.2.2, retrieve the data in the file **foot-rot.mtp** and enter the command

```
MTB > %a:\macro\pairprop c1 c2 95
```

The output is shown in Exhibit 10.2.1.

```
The estimate of pi_11 minus pi_12 is -0.06000

The estimate of pi_21 minus pi_22 is 0.040000

The estimate of pi_31 minus pi_32 is  0.12000

The estimate of pi_41 minus pi_42 is -0.10000

The confidence interval for pi_11 minus pi_12 is -0.16053 to 0.040531

The confidence interval for pi_21 minus pi_22 is -0.07457 to  0.15457

The confidence interval for pi_31 minus pi_32 is -0.02936 to  0.26936

The confidence interval for pi_41 minus pi_42 is -0.27484 to 0.074839

* All intervals are simultaneously correct with at least 95% confidence
```

EXHIBIT 10.2.1

Within rounding error, the results are the same as those shown in Table 10.2.11.

10.2.1 The student government of a university with over 18,000 students wants to determine student attitudes concerning the student recreation center. A simple random sample of 208 graduate students was selected from all graduate students enrolled in the university. A second simple random sample of 252 undergraduate students was selected from all undergraduate students enrolled in the university. Each student in both samples was asked to rate their level of satisfaction with the recreation center. The results are shown in Table 10.2.12. The frequencies shown in Table 10.2.12 are stored in columns 1 and 2 in the file **stcenter.mtp** on the data disk.

TABLE 10.2.12 Results of a Survey About the Student Recreation Center

		Graduate	Undergraduate
Rating	Very satisfactory	120	148
	Satisfactory	44	45
	Unsatisfactory	32	52
	Very unsatisfactory	12	7
	Total	208	252

a. What is the value of k_{12}? What is the interpretation of k_{12}?
b. What are the values of n_1 and n_2?
c. Compute $\widehat{\pi}_{ij}$ for all i and j, and display them in a 4 by 2 table.
d. Estimate π_{21} and explain its meaning.
e. What differences of population proportions must be examined to determine if the ratings of graduate and undergraduate students are the same?
f. Estimate each difference in part (e).
g. Suppose we want a confidence interval for all four differences in part (e) such that we are at least 90% confident that all four intervals simultaneously cover their respective difference. What is the required table value TB?
h. Compute confidence intervals for all four differences in (e) so that the confidence is at least 95% that all the intervals are simultaneously correct.
i. For all values of the variable "Rating," the student government is interested in knowing if there is any difference between two population proportions that has a magnitude larger than 0.10. From the results in part (h), does it appear that any difference satisfies this criterion? If so, which one(s)?

10.3 A STATISTICAL TEST OF ASSOCIATION

The interest in the two previous sections was in comparing proportions of the values of a variable in two populations. In some applications it is of interest to determine if two variables measured on items in a single population are related. If two variables are related, then knowledge of the value of one variable can be used to help predict the value of the other variable.

EXAMPLE 10.3.1

VOTERS. The two candidates in a city election for mayor were a Democrat and a Republican. A social scientist is interested in determining if there is a relationship between how people voted in the election and their annual income. The two variables are labeled "Vote" and "Income group." If there is no relationship, it means that knowing a person's annual income is of no help in determining how the person voted in the election. Equivalently, knowing how a person voted is of no help in determining the person's income category. A simple random sample of 2,000 voters was obtained, and each voter was asked two questions: (1) "Did you vote for the Democrat or the Republican?" and (2) "What was last year's income for the household in which you reside?" The answer to the second question was classified into one of three income groups: *lower* (less than $25,000); *middle* ($25,000 to $75,000); or *upper* (more than $75,000). The results of the survey are presented in Table 10.3.1 and stored in columns 1 and 2 in the file **vote.mtp** on the data disk.

TABLE 10.3.1 Sample of 2,000 Voters (frequencies)

		Vote	
		Democrat	**Republican**
	Lower	511	202
Income group	Middle	387	401
	Upper	196	303
	Total	1,094	906

Of the people who voted for the Democrat, let π_{i1} denote the proportion who belong to the ith income group. Of the people who voted for the Republican, let π_{i2} denote the proportion who belong to the ith income group. The estimates of π_{i1} and π_{i2} for $i = 1, 2, 3$ are shown in Table 10.3.2.

TABLE 10.3.2 Sample of 2,000 Voters (estimated proportions)

		Vote	
		Democrat	**Republican**
	Lower	0.467	0.223
Income group	Middle	0.354	0.443
	Upper	0.179	0.334
	Total	1.000	1.000

For example, from Table 10.3.2 we estimate that $\widehat{\pi}_{32} = 0.334 = 33.4\%$ of the people who voted for the Republican belong to the upper income group. We estimate that $\widehat{\pi}_{11} = 0.467 = 46.7\%$ of the people who voted for the Democrat belong to the lower income group.

The social scientist in Example 10.3.1 wants to determine if two variables ("Income group" and "Vote") are related in a population of voters. If "Income group" is the row variable in the table, then the variables are *not* related if

$$\pi_{11} = \pi_{12}, \qquad \pi_{21} = \pi_{22} \qquad \text{and} \qquad \pi_{31} = \pi_{32}.$$

In Section 10.2 we used confidence intervals to estimate the differences between population proportions in the same row of an r by 2 contingency table. Confidence intervals can be used in a similar manner to determine if two variables are related. However, many investigators prefer to use a statistical test for this purpose. To illustrate how to conduct this test, we will use a 3 by 2 contingency table. However, this test can be conducted for any contingency table using obvious modifications. The hypotheses tested are

H_0 : The row and column variables are unrelated

versus

H_a : The row and column variables are related.

An equivalent way to state these hypotheses using the population parameters π_{ij} is

$$H_0 : \pi_{11} = \pi_{12}, \quad \pi_{21} = \pi_{22} \quad \text{and} \quad \pi_{31} = \pi_{32}$$

versus **Line 10.3.1**

$$H_a : \pi_{11} \neq \pi_{12} \quad \text{or} \quad \pi_{21} \neq \pi_{22} \quad \text{or} \quad \pi_{31} \neq \pi_{32}.$$

A schematic representation of a sample of size n appears in Table 10.3.3.

TABLE 10.3.3 Schematic Representation of a 3 by 2 Contingency Table with Sample Size n

		Column variable		**Row**
		Category 1	**Category 2**	**total**
	Category 1	k_{11}	k_{12}	r_1
Row variable	Category 2	k_{21}	k_{22}	r_2
	Category 3	k_{31}	k_{32}	r_3
	Column total	c_1	c_2	n

The numbers $k_{11}, k_{12}, k_{21}, k_{22}, k_{31}$, and k_{32} are called **observed cell frequencies**; r_1, r_2, and r_3 are the row totals; c_1 and c_2 are the column totals; and n is the sample size. To conduct a statistical test of the hypotheses in Line 10.3.1, choose a value for α, the probability of erroneously rejecting H_0. The null hypothesis is rejected if the computed P-value is less than the chosen value of α. The P-value can be computed using the instructions in Box 10.3.1.

BOX 10.3.1: COMPUTATION OF *P*-VALUE TO DETERMINE IF TWO QUALITATIVE VARIABLES ARE RELATED

The following steps can be used to compute the P-value for the statistical test of the hypotheses in Line 10.3.1.

1. *Compute the expected cell frequencies.* The expected cell frequencies are defined by the formula $e_{ij} = r_i c_j / n$, where $i = 1, \ldots, r$ and $j = 1, 2$. They represent the frequencies one would expect to observe in each cell if the null hypothesis were true.
2. *Compute the cell discrepancies.* The cell discrepancies are defined by the formula $d_{ij} = (k_{ij} - e_{ij})^2 / e_{ij}$, where $i = 1, \ldots, r$ and $j = 1, 2$. A cell discrepancy is a standardized measure of the difference between the expected cell frequencies and the observed cell frequencies.

3. *Obtain the chi-squared statistic.* The chi-squared statistic is the sum of all the cell discrepancies. For the 3 by 2 contingency table, the chi-squared statistic is $d_{11} + d_{12} + d_{21} + d_{22} + d_{31} + d_{32}$. The chi-squared statistic measures the total discrepancy in the sample.
4. *Obtain the P-value for the test using the chi-squared statistic and Table 10.3.4.* To obtain the P-value for the chi-squared statistic, determine the degrees of freedom given by the formula $DF =$ (number of rows − 1) × (number of columns − 1). The P-values for several degrees of freedom are provided in Table 10.3.4. To use this table, read across the row corresponding to the value of DF and find two consecutive numbers that sandwich the value of the chi-squared statistic computed in step (3). Denote the values at the top of the columns that contain these numbers by P_1 and P_2, where $P_1 < P_2$. The P-value for the test is a number between P_1 and P_2. If the chi-squared statistic is equal to any number in the row, then the value at the top of the column that contains the number is the P-value. If the chi-squared statistic is greater than every number in the row, then the P-value is less than 0.001. If the chi-squared statistic is less than every value in the row, then the P-value is greater than 0.25.

TABLE 10.3.4 *P-values for the Chi-squared Statistic*

	P-value							
DF	**0.25**	**0.20**	**0.15**	**0.10**	**0.05**	**0.025**	**0.01**	**0.001**
2	2.77	3.22	3.79	4.61	5.99	7.38	9.21	13.82
3	4.11	4.64	5.32	6.25	7.81	9.35	11.34	16.27
4	5.39	5.99	6.74	7.78	9.49	11.14	13.28	18.47
5	6.63	7.29	8.12	9.24	11.07	12.83	15.09	20.52
6	7.84	8.56	9.45	10.64	12.59	14.43	16.81	22.45
7	9.04	9.80	10.75	12.02	14.07	16.01	18.48	24.32
8	10.22	11.03	12.03	13.36	15.51	17.53	20.09	26.12
9	11.38	12.24	13.29	14.68	16.92	19.02	21.67	27.87
10	12.55	13.44	14.53	15.99	18.31	20.48	23.21	29.59

A few comments are in order to help explain the procedure in Box 10.3.1. In step (1) the expected cell frequencies are computed using the formula $e_{ij} = r_i c_j / n$. This formula can be explained in the following manner: If $H_0 : \pi_{i1} = \pi_{i2} = \pi_i$ for all i, is true, then r_i/n is the estimate for π_i. The number of items we would expect in the ijth cell (the expected cell frequency) is r_i/n times the number of items in the column, c_j. Thus, $e_{ij} = (r_i/n)c_j = r_i c_j/n$. If the row and column variables are unrelated (if H_0 is true), then the observed cell frequency, k_{ij}, should be close to e_{ij} for every cell in the table. For illustration, consider the voters data in Table 10.3.1. We observe that $511 + 202 = 713$ of the 2,000 voters live in households with an annual income below \$25,000. Thus, it is estimated that a proportion $713/2{,}000 = 0.357$ of voters in the population live in households with an annual income below \$25,000. If there is no relationship between "Income group" and "Vote," we would expect the proportion of voters that live in households with annual incomes below \$25,000 to be 35.7% both for people who voted for the Democrat and people who voted for the Republican.

So we would expect 35.7% of the 1,094 individuals who voted for the Democrat ($e_{11} = 390$) to live in households with annual incomes below \$25,000. Likewise, we would expect 35.7% of the 906 individuals who voted for the Republican ($e_{12} = 323$) to live in households with annual incomes less than \$25,000.

In step (2) the discrepancy was computed for each cell using the formula $d_{ij} = (k_{ij} - e_{ij})^2/e_{ij}$. The term d_{ij} is a standardized measure of discrepancy. You might justifiably think that the discrepancy between observed and expected frequencies should be computed as $k_{ij} - e_{ij}$ or perhaps $(k_{ij} - e_{ij})^2$. However, in order to combine the individual discrepancy measures into a meaningful overall measure of discrepancy, it is necessary to use the standardized form defined by d_{ij}.

In step (3) the chi-squared statistic is obtained by summing the cell discrepancies. Notice that as the difference between the expected and observed frequencies in any cell increases, the value of the chi-squared statistic increases. Large differences between the observed and expected frequencies are evidence against the null hypothesis. Thus, a large chi-squared statistic provides evidence that the variables are related. If the observed and expected frequencies are equal in every cell, the chi-squared statistic is equal to zero.

In step (4) the P-value is determined. The following intuitive explanation of the formula for computing the degrees of freedom (DF) might be helpful. It can be verified that the table of differences $k_{ij} - e_{ij}$ has all row sums and all column sums equal to zero. So, one only needs values of $k_{ij} - e_{ij}$ for the cells not contained in either the last row or the last column. The number of such cells is equal to DF = (number of rows − 1) × (number of columns − 1). The degrees of freedom, DF tells us how many pieces of information are used to compute the chi-squared statistic.

Once the P-value is computed, it is compared to the chosen value of α. If the P-value $< \alpha$, then H_0 is rejected and one concludes that the row and column variables are related. In this case one should construct confidence intervals for the differences between the proportions in each row, as described in Section 10.2. This enables one to determine the magnitude of the differences to see if they are of practical importance. If the P-value $\geq \alpha$, H_0 is not rejected and there is not sufficient evidence in the sample to support the claim that the row and column variables are related. Even though the null hypothesis is not rejected, confidence intervals for the differences between proportions in Line 10.3.1 should be obtained to see if the magnitudes of the differences is small enough to be declared unimportant for the practical problem being investigated. So, regardless of whether or not the null hypothesis is rejected, confidence intervals should be studied.

VOTERS. In this example we illustrate the computations described in Box 10.3.1 using the voter data in Example 10.3.1. The sample is shown in Table 10.3.5.

TABLE 10.3.5 Observed Frequencies for Sample of 2,000 Voters (k_{ij})

		Vote		Row
		Democrat	**Republican**	**total**
Income group	Lower	511	202	713
	Middle	387	401	788
	Upper	196	303	499
	Total	1,094	906	2,000

We select $\alpha = 0.05$ to test

$$H_0: \pi_{11} = \pi_{12}, \quad \pi_{21} = \pi_{22} \quad \text{and} \quad \pi_{31} = \pi_{32}$$

$$\text{versus}$$

Line 10.3.2

$$H_a: \pi_{11} \neq \pi_{12} \quad \text{or} \quad \pi_{21} \neq \pi_{22} \quad \text{or} \quad \pi_{31} \neq \pi_{32}.$$

Box 10.3.1 is used to calculate the P-value for this test. Table 10.3.6 reports the expected frequencies, and Table 10.3.7 reports the discrepancies.

TABLE 10.3.6 Expected Frequencies for Sample of 2,000 Voters (e_{ij})

		Vote		
		Democrat	**Republican**	**Row total**
Income group	Lower	713(1,094)/2,000 = 390.01	713(906)/2,000 = 322.99	713
	Middle	788(1,094)/2,000 = 431.04	788(906)/2,000 = 356.96	788
	Upper	499(1,094)/2,000 = 272.95	499(906)/2,000 = 226.05	499
	Total	1,094	906	2,000

TABLE 10.3.7 Discrepancies for Sample of 2,000 Voters (d_{ij})

		Vote	
		Democrat	**Republican**
Income group	Lower	$(511 - 390.01)^2/390.01 = 37.53$	$(202 - 322.99)^2/322.99 = 45.32$
	Middle	$(387 - 431.04)^2/431.04 = 4.50$	$(401 - 356.96)^2/356.96 = 5.43$
	Upper	$(196 - 272.95)^2/272.95 = 21.69$	$(303 - 226.05)^2/226.05 = 26.19$

The chi-squared statistic is computed by summing the discrepancies in Table 10.3.7. This value is $37.53 + 45.32 + 4.50 + 5.43 + 21.69 + 26.19 = 140.66$. To determine the P-value for the test, refer to Table 10.3.4, with $DF = (3 - 1)(2 - 1) = 2$. Read across the row with $DF = 2$ and look for two values that sandwich the value 140.66. The value 140.66 is greater than any number in this row, so the P-value is less than 0.001. Since the P-value is less than $\alpha = 0.05$, the null hypothesis H_0 is rejected and we conclude that the row and column variables are related. Hence, there is a relationship between how individuals voted and the annual income of their household. Confidence intervals for the differences between each pair of proportions in Line 10.3.2 should be constructed to see if they are of practical importance.

Using Minitab to Compute a *P*-Value

Computer programs can be used to compute the P-value needed to test the hypotheses in Line 10.3.2. The Minitab command used to compute this P-value is

```
MTB > chisquared ci cj
```

where the numbers in the contingency table are in columns `i` and `j`, respectively. To use this command for Example 10.3.2, retrieve the data in **vote.mtp**, and enter the following Minitab command in the Session window:

```
MTB > chisquared c1 c2
```

The output is shown in Exhibit 10.3.1.

```
------------------------------------------------------------------------
Chi-Square Test

Expected counts are printed below observed counts

       Democrat Republcn    Total
    1       511      202      713
         390.01   322.99

    2       387      401      788
         431.04   356.96

    3       196      303      499
         272.95   226.05

Total      1094      906     2000

ChiSq =  37.533 + 45.321 +
          4.499 +  5.432 +
         21.695 + 26.197 = 140.678
df = 2, p = 0.000
------------------------------------------------------------------------
```

EXHIBIT 10.3.1

Exhibit 10.3.1 shows the 3 by 2 table of observed frequencies (k_{ij}), the expected frequencies (e_{ij}), the chi-squared statistic (140.678), the degrees of freedom ($DF = 2$), and the P-value ($p = .000$). In this case the P-value is 0.000 to three decimals.

THINKING STATISTICALLY: SELF-TEST PROBLEM

10.3.1 This is a continuation of Self-Test Problem 10.1.3, where the graduate council of a university decided to investigate the relationship between the proportion of male and female students who received financial aid over the past five years. A simple random sample of 1,569 students was selected from all students who applied for financial aid during the past five years. The sample data are shown in Table 10.3.8. The variable "Gender" was used in Section 10.1 to define two populations of interest. The graduate council will conduct a test to determine if the variables "Gender" and "Received aid" are related for the population of all students who applied for financial aid during the past five years.

a. Compute the expected frequencies, e_{11}, e_{12}, e_{21}, and e_{22}.
b. Compute d_{11}, the discrepancy for the cell in row 1 and column 1.
c. Compute the chi-squared statistic that can be used to test if the two variables "Gender" and "Received aid" are related.

TABLE 10.3.8 Results of Financial Aid Study

		Gender		
		Male	**Female**	**Total**
Received aid	Yes	489	529	1,018
	No	354	197	551
	Total	843	726	1,569

10.3.2 In a 4 by 2 contingency table, suppose an investigator wants to test

$$H_0: \pi_{11} = \pi_{12}, \quad \pi_{21} = \pi_{22}, \quad \pi_{31} = \pi_{32}, \quad \text{and} \quad \pi_{41} = \pi_{42}$$

versus

$$H_a: \text{ at least one pair of proportions is not equal.}$$

a. What value of DF is used to obtain the P-value for this test?
b. What is the P-value of the test if the chi-squared statistic is 6.25?
c. What is the P-value of the test if the chi-squared statistic is 14.6?

10.4 CHAPTER SUMMARY

KEY TERMS

1. table of population proportions
2. contingency table
3. simultaneous confidence intervals
4. Bonferroni method
5. association
6. expected cell frequency
7. chi-squared statistic

SYMBOLS

1. $\pi_1 - \pi_2$—a difference between two population proportions
2. π_{ij}—proportion of items in row i and column j in a table of population proportions
3. $(ME)_i$—margin of error for the estimate $\hat{\pi}_{i1} - \hat{\pi}_{i2}$
4. e_{ij}—expected cell frequency
5. k_{ij}—observed cell frequency
6. d_{ij}—cell discrepancy

KEY CONCEPTS

1. Confidence intervals constructed using the Bonferroni method are all simultaneously correct with at least a specified level of confidence.
2. If the P-value is less than α for a chi-squared statistic, the null hypothesis is rejected and we conclude that the two variables are related.

SKILLS

1. Compute an estimate and a confidence interval for the difference between two population proportions.
2. Use a Bonferroni table value to compute simultaneous confidence intervals for the differences between proportions in each row of an *r* by 2 table of population proportions.
3. Compute a chi-squared statistic to test if two variables in an *r* by 2 contingency table are related.

10.5 CHAPTER EXERCISES

10.1 A car manufacturer purchases bolts from two suppliers, supplier 1 and supplier 2. A quality control engineer for the company obtained a simple random sample of bolts from the most recent lots purchased from each supplier. The sample from supplier 1 contained 462 bolts, 32 of which were defective, and the sample from supplier 2 contained 348 bolts, 10 of which were defective.

a. What are the two sampled populations?

b. Estimate π_1 and π_2, the proportion of defective bolts manufactured by supplier 1 and 2, respectively.

c. Estimate $\pi_1 - \pi_2$.

d. Compute a 95% confidence interval for $\pi_1 - \pi_2$.

10.2 A pesticide manufacturer is testing two different pesticides to determine which one is more successful in killing a type of insect that attacks apple trees. A scientist that works for the company carries out a laboratory experiment, using insects purchased from an insect farm, to determine which pesticide has the higher proportion of "kills." He purchases a random sample of 240 insects from the farm and randomly divides the sample into two groups. One group is sprayed with pesticide 1 and the other with pesticide 2. The number of insects killed with each pesticide is reported in Table 10.5.1. These data are stored in the file **pesticid.mtp** on the data disk.

TABLE 10.5.1 Insects Killed with Two Pesticides

		Pesticide 1	Pesticide 2
Killed	Yes	98	110
	No	22	10
	Total	120	120

a. What are the two sampled populations?

b. Let π_i represent the proportion of all insects on the farm that would have been killed if they had all been sprayed with pesticide *i* for $i = 1, 2$. Estimate π_1 and π_2.

c. Estimate $\pi_2 - \pi_1$.

d. Compute a 95% confidence interval for $\pi_2 - \pi_1$.

10.3 A telephone poll was conducted to obtain information concerning attitudes about cigarette advertising. A simple random sample of 2,003 adults was obtained; each person was asked two questions: (1) "Would you ban all cigarette advertising?" and (2) "Are you a smoker?" The responses are reported in Table 10.5.2.

TABLE 10.5.2 Attitudes About Cigarette Advertising

		Nonsmokers	Smokers
Ban advertising	Yes	757	240
	No	671	335
	Total	1,428	575

a. Estimate π_1, the proportion of nonsmokers who would ban all cigarette advertising.

b. Estimate π_2, the proportion of smokers who would ban all cigarette advertising.

c. Compute a 99% confidence interval for $\pi_1 - \pi_2$.

10.4 Researchers have long been interested in the connection, if any, between diet and longevity. Some researchers have conjectured that low-fat diets generally increase longevity in animals. To help evaluate this conjecture, an investigator planned an experiment using rats of a certain species. She obtained 240 rats from a rat breeder who has 200,000 rats. The investigator randomly assigned 120 of the rats to a low-fat diet, and the remaining 120 rats were assigned to a high-fat diet. The experiment lasted until all the rats died. The results of the study are summarized in the contingency table shown in Table 10.5.3.

TABLE 10.5.3 Days Lived for Rats on Two Diets

		Low-fat diet	High-fat diet
Days lived	< 365	12	84
	≥ 365	108	36
	Total	120	120

There are two parameters of interest in this problem. One is π_1, the proportion of the 200,000 rats that would survive at least 365 days if all the rats were fed the low-fat diet. The other is π_2, the proportion of the 200,000 rats that would survive at least 365 days if all the rats were fed the high-fat diet.

a. Estimate π_1 and π_2.

b. Estimate and interpret $\pi_1 - \pi_2$.

c. Compute a 95% confidence interval for $\pi_1 - \pi_2$.

d. Can we conclude that a low-fat diet will increase the longevity of the 200,000 rats? Can we conclude that a low-fat diet will increase longevity in humans?

10.5 This exercise refers to Example 10.1.3, where a candidate for the state legislature wants to know if the voters think that the state economy is better than it was two years ago. The candidate wants a point estimate and confidence interval for $\pi_1 - \pi_2$, where π_1 is the proportion of Republicans who believe the economy is better than it was two years ago, and π_2 is the proportion of Democrats with this belief. A simple random sample of 3,515 voters was collected from a single list that contained the names of both Republicans and Democrats. Each person was asked two questions: (1) "What is your party affiliation, Republican or Democrat?" and (2) "Do you think the state economy is better than it was two years ago?" The sample data are summarized in Table 10.5.4.

TABLE 10.5.4 Sample from Population of Voters

		Republican	Democrat
Is economy better?	Yes	640	582
	No	875	1,418
	Total	1,515	2,000

a. Estimate π_1, the proportion of Republicans who believe the economy is better than it was two years ago.

b. Estimate π_2, the proportion of Democrats who believe the economy is better than it was two years ago.

c. Estimate $\pi_1 - \pi_2$.

d. Compute a 95% confidence interval for $\pi_1 - \pi_2$.

10.6 This exercise refers to Example 10.2.1, where a new drug has been developed to treat foot-rot in pigs. A second study was conducted to determine how the treatment using the new drug compares to "no treatment." In this study, the new drug was used to treat a simple random sample of 200 pigs from a population of 50,000 pigs that have foot-rot. A second simple random sample of pigs is selected from the same population but receives no treatment for the disease. The sample data are shown in Table 10.5.5 and stored in columns 1 and 2 in the file **footrot2.mtp** on the data disk.

TABLE 10.5.5 Frequencies of Recovery in Second Foot-rot Study

		New drug	No drug
Recovery	No	10	132
	Slight	22	32
	Substantial	58	30
	Complete	110	6
	Total	200	200

a. What are the sampled populations?

b. Let π_{i1} represent the proportion of pigs receiving the new drug that are in "Recovery" category i, and let π_{i2} represent the proportion of pigs receiving no treatment that are in "Recovery" category i. Estimate π_{i1} and π_{i2} for $i = 1, 2, 3, 4$.

c. Estimate the proportion of pigs in the population of 50,000 pigs that would have recovered completely if all the pigs were given the new drug.

d. Compute confidence intervals for all differences $\pi_{i1} - \pi_{i2}$ for $i = 1, 2, 3, 4$ such that the confidence is at least 95% that all intervals are simultaneously correct.

10.7 An exercise scientist wants to compare two weight-loss programs. She selected a simple random sample of 200 members of a health club and randomly assigned 100 of them to weight-loss program 1 and the remaining 100 to weight-loss program 2. At the end of two months, she found that 72 of the participants in program 1 achieved their weight-loss goal, and 50 of the participants in program 2 achieved their weight-loss goal.

a. What are the sampled populations?

b. Let π_1 represent the proportion of people in the population who would achieve their weight-loss goal if they used program 1. Let π_2 represent the proportion of people in the population who would achieve their weight-loss goal if they used program 2. Estimate $\pi_1 - \pi_2$.

c. Compute a 95% confidence interval for $\pi_1 - \pi_2$ and interpret the result. Which program would you recommend?

10.8 An instructor for a statistics course offers a "help session" before each exam given in the course. He would like to determine if there is any relationship between a student's exam grade and whether the student attends the help session. To do this, he selected a simple random sample of 100 students who completed the last hourly exam and sent each one an e-mail message asking if he or she had attended the help session. The frequencies for the values of the two variables "Exam grade" and "Attended help session" are shown in Table 10.5.6.

TABLE 10.5.6 Grade on Exam and Help Session Attendance

		Attended help session	
		Yes	No
Grade	A or B	22	28
	C	14	21
	D or F	4	11
	Total	40	60

a. Compute the P-value that tests whether exam grade and attendance at the help session are related. What is your conclusion if $\alpha = 0.05$?
b. Compute confidence intervals for the three differences $\pi_{11} - \pi_{12}$, $\pi_{21} - \pi_{22}$, and $\pi_{31} - \pi_{32}$ such that the confidence is at least 90% that all three intervals are simultaneously correct.

10.9 A newspaper conducted a telephone survey to determine voter attitudes on two propositions concerning gun control that will appear on the ballot in the forthcoming November election. One proposition would require a mandatory waiting period of seven days before the purchase of a handgun. The second proposition would require the mandatory registration of handguns. A simple random sample of 400 registered voters was contacted by telephone using voter registration lists. Each sampled voter was asked how he or she planned to vote on each proposition and whether he or she owned a handgun. The frequencies for the values of the two variables "Propositions" and "Own handgun" are shown in Table 10.5.7.

TABLE 10.5.7 Support of Propositions and Handgun Ownership

		Own handgun	
		Yes	No
Propositions	Support both	45	175
	Oppose both	90	26
	Support one but not both	15	49
	Total	150	250

a. Compute the P-value that tests whether support of the propositions and ownership of a handgun are related. What is your conclusion if $\alpha = 0.05$?
b. Compute confidence intervals for the three differences $\pi_{11} - \pi_{12}$, $\pi_{21} - \pi_{22}$, and $\pi_{31} - \pi_{32}$, such that the confidence is at least 95% that all three intervals are simultaneously correct.

10.6 SOLUTIONS FOR SELF-TEST PROBLEMS IN CHAPTER 10

10.1.3

a. The target populations of interest are (1) the set of all male graduate students who applied for financial aid last year and (2) the set of all female graduate students who applied for financial aid last year. Complete information is available for both target populations, and so it is not necessary to select a sample.
b. $\pi_1 = 492/710 = 0.693$.
c. $\pi_2 = 358/411 = 0.871$.
d. $\pi_1 - \pi_2 = 0.693 - 0.871 = -0.178$.
e. The estimates are $\widehat{\pi}_1 = 489/843 = 0.580$, $\widehat{\pi}_2 = 529/726 = 0.729$, and $\widehat{\pi}_1 - \widehat{\pi}_2 = -0.149$. It is estimated that over the past five years, 14.9% more female applicants than male applicants received financial aid.
f. Using Box 10.1.2 with $T1 = 1.96$, we obtain

$$L = -0.149 - 1.96\sqrt{\frac{0.580(0.420)}{843} + \frac{0.729(0.271)}{726}} = -0.196,$$

$$U = -0.149 + 1.96\sqrt{\frac{0.580(0.420)}{843} + \frac{0.729(0.271)}{726}} = -0.102.$$

The confidence is 95% that from 10.2% to 19.6% more female applicants than male applicants received financial aid during the past five years.

10.2.1

a. $k_{12} = 148$. This means that of the 252 undergraduate students in the sample, $k_{12} = 148$ rated the student center as *very satisfactory*.
b. $n_1 = 208$ and $n_2 = 252$.
c. Using the formula $\widehat{\pi}_{ij} = k_{ij}/n_j$, we obtain the results in the following table.

		Graduate	Undergraduate
Rating	Very satisfactory	0.577	0.587
	Satisfactory	0.212	0.179
	Unsatisfactory	0.154	0.206
	Very unsatisfactory	0.058	0.028

d. $\widehat{\pi}_{21} = 0.212$. It is estimated that 21.2% of the graduate students in the university rate the student center as *satisfactory*.

e. One must examine $\pi_{11}-\pi_{12}, \pi_{21}-\pi_{22}, \pi_{31}-\pi_{32}$, and $\pi_{41}-\pi_{42}$. If each of these differences is equal to zero, the ratings of graduate and undergraduate students are the same.

f. $\widehat{\pi}_{11}-\widehat{\pi}_{12} = -0.010$, $\widehat{\pi}_{21}-\widehat{\pi}_{22} = 0.033$, $\widehat{\pi}_{31}-\widehat{\pi}_{32} = -0.052$, and $\widehat{\pi}_{41}-\widehat{\pi}_{42} = 0.030$.

g. Using Table 10.2.8, we find that the value is $TB = 2.24$.

h. We will use the Minitab macro **pairprop.mac**. Retrieve the data in the file **stcenter.mtp** and enter the following in the Session window:

```
a%:\macro\pairprop c1 c2 95
```

The output is as follows:

```
-----------------------------------------------------------------------------------------
The estimate of pi_11 minus pi_12 is -0.01038

The estimate of pi_21 minus pi_22 is 0.032967

The estimate of pi_31 minus pi_32 is -0.05250

The estimate of pi_41 minus pi_42 is 0.029915

The confidence interval for pi_11 minus pi_12 is -0.12580 to  0.10504

The confidence interval for pi_21 minus pi_22 is -0.05995 to  0.12589

The confidence interval for pi_31 minus pi_32 is -0.14171 to 0.036708

The confidence interval for pi_41 minus pi_42 is -0.01803 to 0.077863

* All intervals are simultaneously correct with at least 95%  confidence
-----------------------------------------------------------------------------------------
```

i. Since all confidence intervals include zero and all except the last one include at least one value greater in magnitude than 0.10, only the last interval is conclusive. It indicates that $\pi_{41}-\pi_{42}$ has a magnitude less than 0.10 (in fact, less than 0.078).

10.3.1

a. The expected cell frequencies are $e_{11} = 1018(843)/1569 = 546.96$, $e_{12} = 1018(726)/1569 = 471.04$, $e_{21} = 551(843)/1569 = 296.04$, and $e_{22} = 551(726)/1569 = 254.96$.

b. $d_{11} = (489 - 546.96)^2/546.96 = 6.142$.

c. The chi-squared statistic $= 6.142 + 7.132 + 11.348 + 13.176 = 37.798$.

10.3.2

a. The value of DF $= (4-1)(2-1) = 3$.

b. Find the number 3 in the column labeled *DF* in Table 10.3.4. Read across this row, and find two consecutive numbers that sandwich the value of the chi-squared statistic 6.25. The number 6.25 is located in the column labeled 0.10. So the *P*-value = 0.10.

c. Find the number 3 in the column labeled DF in Table 10.3.4. Read across this row until you find two consecutive numbers that sandwich the value of the chi-squared statistic, 14.6. These two values are 11.34 and 16.27. The column that contains 11.34 is labeled $P_1 = 0.01$, and the column that contains 16.27 is labeled $P_2 = 0.001$. So the *P*-value is between 0.01 and 0.001.

10.7 APPENDIX: THE BONFERRONI METHOD

In Box 10.2.1 we use a Bonferroni table value (TB) to construct a set of r confidence intervals that are simultaneously correct with a specified level of confidence. In this section we explain in more detail how to interpret a set of simultaneous confidence intervals and we show how the Bonferroni table values are derived. We illustrate these concepts for a set of $r = 4$ confidence intervals, but the results hold for all values of r. In fact, the results of this section hold for any set of r confidence intervals for any population parameters, not just for population proportions. Consider the four differences of interest in a 4 by 2 table of population proportions:

$$\theta_1 = \pi_{11} - \pi_{12}, \quad \theta_2 = \pi_{21} - \pi_{22}, \quad \theta_3 = \pi_{31} - \pi_{32}, \quad \text{and} \quad \theta_4 = \pi_{41} - \pi_{42}.$$

In your mind's eye, suppose that every simple random sample of size n is collected from a combined list of all items in both populations and used to compute a 99% confidence interval for $\theta_1, \theta_2, \theta_3$, and θ_4. Each confidence interval is computed using Box 10.1.2 with $T1 = 2.58$, the table value for a 99% confidence interval. The value 99% is called the **individual confidence coefficient** because it describes the performance of only one confidence interval. For the M possible samples of size n, Table 10.7.1 presents a schematic representation of the performance of each of the four confidence intervals in each of the M samples. The letter Y (yes) means that the confidence interval covers the parameter (θ_i). The letter N (no) means the confidence interval does not cover the parameter.

TABLE 10.7.1 Confidence Intervals for $\theta_1, \theta_2, \theta_3$, and θ_4

Column 1	Column 2	Column 3	Column 4	Column 5
Samples of size n	**Does confidence interval for θ_1 cover θ_1?**	**Does confidence interval for θ_2 cover θ_2?**	**Does confidence interval for θ_3 cover θ_3?**	**Does confidence interval for θ_4 cover θ_4?**
1	Y	Y	Y	Y
2	Y	N	N	Y
3	Y	Y	Y	Y
4	N	Y	N	Y
5	Y	N	Y	N
6	Y	Y	Y	Y
...	...	...	...	
M	Y	Y	Y	Y

Since the confidence interval for θ_1 has a confidence coefficient of 99%, it follows that 99% of the letters in column 2 of Table 10.7.1 are Y and 1% of the letters are N. A similar statement can be made for columns 3, 4, and 5.

We now wish to determine the confidence that all four confidence intervals are simultaneously correct. That is, we want to know what proportion of rows (samples) in Table 10.7.1 have all Y (have the letter Y in every column). For example, in sample 1 the letter Y appears in all four columns. This means that all four confidence intervals cover the respective parameter. However, in sample 2 the confidence intervals for θ_2 and θ_3 do not cover the parameters. In this row, all four intervals are *not* simultaneously correct. We can determine the proportion of samples in which all four confidence intervals are simultaneously correct by using the Bonferroni method described in Box 10.7.1.

BOX 10.7.1: BONFERRONI METHOD

Suppose a set of r confidence intervals are computed using an individual confidence coefficient of I, where I is written as a proportion. For example, if each confidence interval has a confidence coefficient of 99%, then $I = 0.99$. The confidence that all r confidence intervals are simultaneously correct is at least S, where

$$S = 1 - r(1 - I). \qquad \textbf{Line 10.7.1}$$

Alternatively, if you want to construct a set of r confidence intervals that are simultaneously correct with a confidence of at least S, then each confidence interval should have an individual confidence coefficient of

$$I = 1 - \frac{(1 - S)}{r}. \qquad \textbf{Line 10.7.2}$$

Since $I = 0.99$, we use Line 10.7.1, and state that the proportion of rows in Table 10.7.1 that contain the letter Y in every column is at least $S = 1-4(1-0.99) = 0.96 = 96\%$. This proportion is the confidence that all four confidence intervals are simultaneously correct. To verify this result, in your mind's eye use an asterisk (*) to mark all rows in Table 10.7.1 that contain the letter N. We will show that *at most* $100\% - 96\% = 4\%$ of the rows contain the letter N, meaning that *at least* 96% of the rows are not marked with an asterisk and contain only the letter Y. To do this, proceed through the following four steps.

1. Use an asterisk to mark all rows in Table 10.7.1 that contain the letter N in column 2. Since the confidence interval for θ_1 has a confidence coefficient of 99%, this means that 99% of the letters in column 2 are Y and 1% of the letters are N. Thus, 1% of the rows in Table 10.7.1 are marked with an asterisk.
2. Use an asterisk to mark all rows in Table 10.7.1 that contain the letter N in column 3. Since the confidence interval for θ_2 has a confidence coefficient of 99%, this means that 99% of the letters in column 3 are Y and 1% of the letters are N. But some of the rows that contain an N in column 3 might also contain an N in column 2, and these rows have already been marked in step (1). Hence, *at most* 1% of the rows are marked in this step. The number of marked rows in the table is now at most $1\% + 1\% = 2\%$.
3. We continue in the same manner by marking all rows in Table 10.7.1 that contain the letter N in column 4. Since the confidence interval for θ_3 has a confidence coefficient of 99%, this means that 99% of the letters in column 4 are Y and 1% of the letters are N. But some of the rows that contain an N may have already been marked in steps (1) and (2). Hence, we now have marked at most $1\% + 1\% + 1\% = 3\%$ of the rows in Table 10.7.1.
4. Finally, mark all rows in Table 10.7.1 that contain the letter N in column 5. Since the confidence interval for θ_4 has a confidence coefficient of 99%, this means that 99% of the letters in column 5 are Y and 1% of the letters are N. But some of the rows that contain an N may have been marked in steps (1),

(2), or (3). Hence, we now have marked at most $1\% + 1\% + 1\% + 1\% = 4\%$ of the rows in Table 10.7.1.

All of the rows that contain the letter N in Table 10.7.1 have now been marked. Since this includes at most 4% of the rows, it follows that at least 96% of the rows are not marked. Each of the nonmarked rows contains the letter Y in every column. This is what we set out to show, that at least 96% of all possible samples will yield confidence intervals that simultaneously cover their respective θ_i.

We can now demonstrate how the TB values were obtained for Table 10.2.8. Consider the case where it is desired to have a simultaneous confidence coefficient of $S = 0.90$ for a set of confidence intervals for the differences of proportions in a 5 by 2 table ($r = 5$). One way to proceed is to use Box 10.1.2 to construct a confidence interval for each of the five differences. Using Line 10.7.2, we see that this means the confidence coefficient for each individual interval is $I = 1 - (1 - 0.90)/5 = 0.98$, and the appropriate value for $T1$ is 2.33. This approach will result in the same intervals you will obtain using Box 10.2.1 with a simultaneous confidence of 90%, because the value of TB in Table 10.2.8 with $r = 5$ is also 2.33. The values of TB in Table 10.2.8 were computed so that these two approaches will always be equivalent.

Here are some questions, with solutions, that can be answered using the material in this appendix.

10.7.1 Suppose an investigator wants confidence intervals for three parameters θ_1, θ_2, and θ_3 such that the confidence is at least 91% that all intervals simultaneously cover their respective parameters. What confidence coefficient must be used for each individual interval?

▶**Answer**: Using Line 10.7.2, we obtain $I = 1 - (1 - 0.91)/3 = 1 - 0.03 = 0.97$. Hence, we use a confidence coefficient of 97% for each of the three intervals.

10.7.2 Suppose an investigator wants confidence intervals for five parameters θ_1, θ_2, θ_3, θ_4, and θ_5, such that the confidence is at least 95% that all intervals simultaneously cover their respective parameters. What confidence coefficient must be used for each individual interval?

▶**Answer**: Using Line 10.7.2, we obtain $I = 1 - (1 - 0.95)/5 = 1 - 0.01 = 0.99$. Hence, we use a confidence coefficient of 99% for each of the five intervals.

10.7.3 A set of four confidence intervals is computed using a 95% confidence coefficient for each interval. What is the confidence that all four intervals will be simultaneously correct?

▶**Answer**: Using Line 10.7.1, we obtain $S = 1 - 4(1 - 0.95) = 0.80$. Thus, the confidence is at least 80% that all four confidence intervals are simultaneously correct.

CHAPTER 11

COMPARING POPULATION MEANS

11.1 COMPARING TWO POPULATION MEANS

The vast majority of statistical investigations involve comparing the means of two populations. In these investigations, $\mu_1 - \mu_2$ is generally of interest, where μ_1 is the mean of one population and μ_2 is the mean of the other. For instance, in medical studies one might want to compare the effectiveness of two treatments for cancer to determine which one produces longer average life expectancy. In an industrial experiment one might want to compare two sets of operating conditions to determine which one produces smaller average downtime for machinery. A consumer agency might compare two different types of cars to determine which one has a lower average annual maintenance cost. An agricultural scientist might compare two different varieties of wheat to determine which one has the higher average yield. We present two other applications in the following examples.

EXAMPLE 11.1.1

LIGHT BULBS. A company that owns several large office buildings considers replacing its present brand of fluorescent light bulb with a less expensive brand. To determine if the company should switch brands, it wants to compare μ_1 with μ_2, where μ_1 is the average lifetime (in hours) of the less expensive brand and μ_2 is the average lifetime of its present brand. The company would like to know the value of $\mu_1 - \mu_2$, the difference between the average lifetimes of the two brands.

EXAMPLE 11.1.2

INCOMES OF TWO CITIES. A company that owns department stores plans to build a new store in one of two cities, which we call city 1 and city 2. Let μ_1 represent the average household income last year in city 1 and μ_2 represent the average household income last year in city 2. To help determine in which city to build the store, the company would like to know $\mu_1 - \mu_2$, the difference between last year's average household incomes in the two cities.

Point Estimate of $\mu_1 - \mu_2$

Consider two sampled populations, which we denote as population 1 and population 2. In Box 11.1.1 we describe the procedure for estimating $\mu_1 - \mu_2$, where μ_1 is the mean of population 1 and μ_2 is the mean of population 2.

BOX 11.1.1: PROCEDURE FOR ESTIMATING THE DIFFERENCE BETWEEN TWO POPULATION MEANS

Consider two populations of numbers, population 1 and population 2. Let μ_1 and σ_1 denote the mean and the standard deviation of population 1. Let μ_2 and σ_2 denote the mean and the standard deviation of population 2.

Compute $\bar{y}_1$ and s_1 for a simple random sample of n_1 items from population 1 using the formulas in Definitions 6.3.1 and 6.3.2, respectively. The estimates of μ_1 and σ_1 are

$$\hat{\mu}_1 = \bar{y}_1 \qquad \text{and} \qquad \hat{\sigma}_1 = s_1.$$

Compute $\bar{y}_2$ and s_2 for a simple random sample of n_2 items from population 2 using the formulas in Definitions 6.3.1 and 6.3.2, respectively. The estimates of μ_2 and σ_2 are

$$\hat{\mu}_2 = \bar{y}_2 \qquad \text{and} \qquad \hat{\sigma}_2 = s_2.$$

The estimate of $\mu_1 - \mu_2$, the difference between the means of populations 1 and 2, is

$$\hat{\mu}_1 - \hat{\mu}_2 = \bar{y}_1 - \bar{y}_2.$$

EXAMPLE 11.1.3

INCOMES OF TWO CITIES. This is a continuation of Example 11.1.2, where a company wants to determine $\mu_1 - \mu_2$, the difference between last year's average household incomes in two cities. A simple random sample of 30 households was obtained from each city, and last year's income was obtained for each sampled household. The data are shown in Table 11.1.1 and stored in columns 1 and 2 in the file **2cityinc.mtp** on the data disk.

Exhibit 11.1.1 reports some descriptive measures for the two samples. This exhibit was obtained using the Minitab command `describe c1 c2`. From Exhibit 11.1.1 we obtain $\hat{\mu}_1 = 39.19$ and $\hat{\mu}_2 = 35.43$. Thus, the estimate of $\mu_1 - \mu_2$ is $\hat{\mu}_1 - \hat{\mu}_2 = 3.76$. So we estimate that the difference between the average household incomes is \$3,760.

TABLE 11.1.1 Samples of Household Incomes from Two Cities

Sample number	Income for city 1 (thousands of dollars)	Sample number	Income for city 2 (thousands of dollars)
1	46.14	1	36.72
2	46.50	2	35.04
3	30.48	3	38.46
4	46.56	4	26.58
5	43.20	5	31.62
6	34.38	6	36.54
7	23.52	7	41.94
8	50.22	8	28.74
9	32.88	9	18.06
10	25.80	10	38.76
11	46.32	11	38.28
12	43.86	12	17.52
13	50.34	13	40.38
14	46.50	14	49.14
15	34.02	15	45.00
16	36.42	16	37.50
17	31.68	17	28.14
18	41.34	18	27.30
19	43.26	19	42.18
20	32.16	20	30.12
21	48.12	21	45.00
22	36.18	22	47.34
23	39.24	23	35.94
24	38.88	24	25.02
25	47.70	25	40.20
26	28.14	26	40.20
27	36.12	27	30.12
28	31.68	28	26.70
29	40.92	29	48.54
30	43.14	30	35.76

```
------------------------------------------------------------------------
Descriptive Statistics

Variable        N      Mean    Median    TrMean     StDev    SEMean
City 1         30     39.19     40.08     39.45      7.51      1.37
City 2         30     35.43     36.63     35.75      8.30      1.52

Variable      Min       Max        Q1        Q3
City 1      23.52     50.34     32.70     46.36
City 2      17.52     49.14     28.59     40.77
------------------------------------------------------------------------
```

EXHIBIT 11.1.1

Confidence Intervals for the Difference Between Two Population Means

As you know by now, we do not expect a point estimate of $\mu_1 - \mu_2$ to be *exactly* equal to $\mu_1 - \mu_2$, so we use confidence intervals to determine "how close" the estimate $\hat{\mu}_1 - \hat{\mu}_2$ is to the true value of $\mu_1 - \mu_2$. Instructions for computing a confidence interval for $\mu_1 - \mu_2$ are given in Box 11.1.2.

BOX 11.1.2: COMPUTING A CONFIDENCE INTERVAL FOR THE DIFFERENCE BETWEEN TWO POPULATION MEANS

Let $\bar{y}_1$ and s_1 be the mean and the standard deviation of a simple random sample of size n_1 selected from population 1, whose mean and standard deviation are μ_1 and σ_1, respectively. Let $\bar{y}_2$ and s_2 be the mean and the standard deviation of a simple random sample of size n_2 selected from population 2, whose mean and standard deviation are μ_2 and σ_2, respectively. The estimate of $\mu_1 - \mu_2$ is $\hat{\mu}_1 - \hat{\mu}_2 = \bar{y}_1 - \bar{y}_2$.

The endpoints L and U of a confidence interval for $\mu_1 - \mu_2$ are

$$L = \hat{\mu}_1 - \hat{\mu}_2 - \text{ME},$$
$$U = \hat{\mu}_1 - \hat{\mu}_2 + \text{ME},$$

where ME, the margin of error, is

$$\text{ME} = T2\sqrt{\frac{s_1^2}{n_1} + \frac{s_2^2}{n_2}}.$$

The value of $T2$ is obtained from Table T-2 using the degrees of freedom DF given by

$$\text{DF} = \frac{(h_1 + h_2)^2}{h_1^2/(n_1 - 1) + h_2^2/(n_2 - 1)}, \qquad \textbf{Line 11.1.1}$$

where $h_1 = s_1^2/n_1$ and $h_2 = s_2^2/n_2$. If DF is not an integer, delete the decimal portion and use the resulting value for DF.

The formulas in Box 11.1.2 can be used if each population is *normal*, or when n_1 and n_2 are "sufficiently large" (see Section 7.2 for a discussion of the term "sufficiently large"). In general, the formulas can be used if n_1 and n_2 are both greater than 30. If either n_1 or n_2 is less than 30, the formulas can still be used if the sample (or samples) of size less than 30 is selected from a *normal* population. Normal rankit-plots can be used to help decide if a sample appears to be selected from a *normal* population (see Section 6.6).

EXAMPLE 11.1.4

INCOMES OF TWO CITIES. We continue with Example 11.1.3 and compute a 95% confidence interval for $\mu_1 - \mu_2$. We first construct normal rankit-plots to decide if the samples appear to be from *normal* populations. These plots are shown in Figures 11.1.1 and 11.1.2.

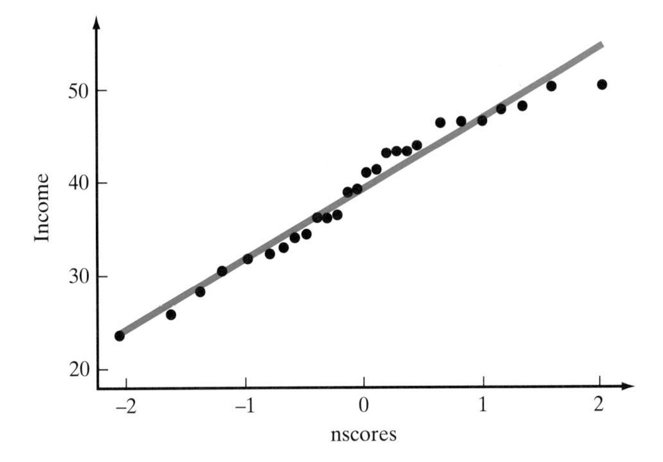

FIGURE 11.1.1
Normal Rankit-plot for City 1 Incomes (in thousands of dollars)

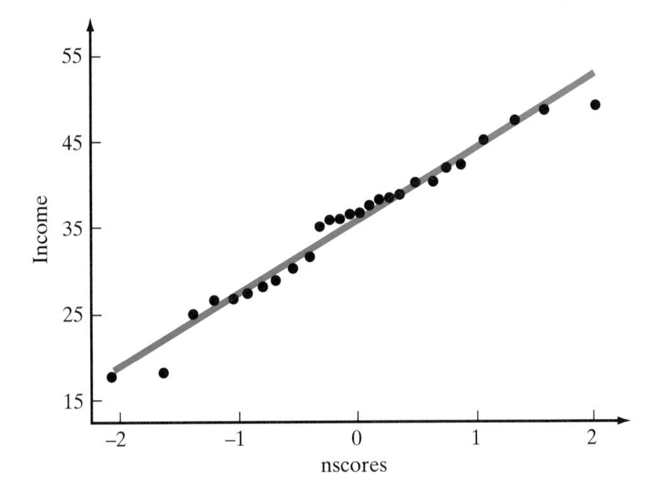

FIGURE 11.1.2
Normal Rankit-plot for City 2 Incomes (in thousands of dollars)

From examining Figures 11.1.1 and 11.1.2 it appears reasonable to assume that the samples came from *normal* populations. In addition, since n_1 and n_2 are both equal to 30, it is appropriate to use the formulas in Box 11.1.2. From Exhibit 11.1.1 we obtain $\bar{y}_1 = 39.19$, $s_1 = 7.51$, $\bar{y}_2 = 35.43$, and $s_2 = 8.30$. To determine DF, we need $h_1 = (7.51)^2/30 = 1.88$ and $h_2 = (8.30)^2/30 = 2.30$. Substituting these values into Line 11.1.1, we obtain

$$\text{DF} = \frac{(1.88 + 2.30)^2}{(1.88)^2/29 + (2.30)^2/29} = 57.42.$$

We delete the decimal portion of 57.42 and use DF = 57 in Table T-2 with 95% confidence to obtain $T2 = 2.00$. Using the formulas in Box 11.1.2, we obtain

$$\text{ME} = 2.00\sqrt{\frac{(7.51)^2}{30} + \frac{(8.30)^2}{30}} = 4.09,$$

$$L = 39.19 - 35.43 - 4.09 = -0.33, \qquad U = 39.19 - 35.43 + 4.09 = 7.85.$$

Thus, we have 95% confidence that the difference between last year's average household incomes in the two cities was between −\$330 and \$7,850. Since both negative and positive values are contained in this confidence interval, the data provide inconclusive information for deciding which city had the larger average household income last year.

Using Minitab to Compute a Confidence Interval for $\mu_1 - \mu_2$

The computations described in Box 11.1.2 can be performed using Minitab by entering the following command in the Session window after the data have been retrieved:

```
MTB > twosample confidence ci cj
```

The quantity `confidence` is the desired confidence coefficient in percent (e.g., 90, 95, 99), and the data are located in columns `i` and `j`. To illustrate, we will compute a 95% confidence interval for $\mu_1 - \mu_2$ in Example 11.1.4. The sample data in Example 11.1.4 are in columns 1 and 2 in the file **2cityinc.mtp**. Retrieve the data and enter the following command in the Session window:

```
MTB > twosample 95 c1 c2
```

The portion of the Minitab output that is of interest to us is shown in Exhibit 11.1.2.

```
---------------------------------------------------------------------
Twosample T for city 1 vs city 2
          N      Mean      StDev    SE Mean
City 1   30     39.19       7.51        1.4
City 2   30     35.43       8.30        1.5

95% C.I. for mu City 1 - mu City 2: ( -0.3,  7.9)
---------------------------------------------------------------------
```

EXHIBIT 11.1.2

From this output we obtain $\hat{\mu}_1 = 39.19$, $\hat{\sigma}_1 = 7.51$, $\hat{\mu}_2 = 35.43$, and $\hat{\sigma}_2 = 8.30$. So $\hat{\mu}_1 - \hat{\mu}_2 = 3.76$, and the 95% confidence interval reported in the last line of Exhibit 11.1.2 is –0.3 to 7.9.

Pooling Estimates of the Standard Deviations

In some problems the investigator may know from past experience that the two standard deviations, σ_1 and σ_2, are equal (or nearly equal). If $\sigma_1 = \sigma_2$, the formulas for computing a confidence interval for $\mu_1 - \mu_2$ given in Box 11.1.2 can be simplified. More importantly, these simpler formulas sometimes result in a shorter confidence interval. If the standard deviations are equal, we denote the common value by σ (that is, $\sigma_1 = \sigma_2 = \sigma$). Instead of using s_1 and s_2 as the estimates of σ_1 and σ_2, respectively, we **pool** (combine) these estimates to obtain a single estimate of the common value σ. This pooled estimate of σ is denoted by s_p and is given by the formula

$$s_p = \sqrt{\frac{(n_1 - 1)s_1^2 + (n_2 - 1)s_2^2}{(n_1 - 1) + (n_2 - 1)}}. \qquad \textbf{Line 11.1.2}$$

The estimate, s_p, replaces s_1 and s_2 in the formulas for L and U in Box 11.1.2; the formula for the margin of error becomes

$$\text{ME} = (T2)s_p\sqrt{\frac{1}{n_1} + \frac{1}{n_2}}. \qquad \textbf{Line 11.1.3}$$

When we pool the estimates s_1 and s_2, the degrees of freedom used to obtain $T2$ is

$$\text{DF} = (n_1 - 1) + (n_2 - 1) = n_1 + n_2 - 2. \qquad \textbf{Line 11.1.4}$$

We recommend that the formulas in Box 11.1.2 be used (that is, do not pool) unless the investigator has reliable information indicating that $\sigma_1 = \sigma_2$.

EXAMPLE 11.1.5

INCOMES OF TWO CITIES. For illustration, we use the pooled estimate s_p to construct the confidence interval required in Example 11.1.4. From Exhibit 11.1.1 we obtain $s_1 = 7.51$, $s_2 = 8.30$, $n_1 = 30$, $n_2 = 30$, and

$$s_p = \sqrt{\frac{(29)(7.51)^2 + (29)(8.30)^2}{(29) + (29)}} = 7.92.$$

For DF $= 29 + 29 = 58$ and a 95% confidence coefficient, we obtain $T2 = 2.00$ from Table T-2. Using the formula in Line 11.1.3, we obtain

$$\text{ME} = 2.00(7.92)\sqrt{(1/30 + 1/30)} = 4.09,$$

$$L = 39.19 - 35.43 - 4.09 = -0.33, \qquad U = 39.19 - 35.43 + 4.09 = 7.85.$$

In this case, the interval using the pooled estimate of σ is virtually identical (to the accuracy reported) to the interval computed in Example 11.1.4 in which there is no pooling.

The Minitab subcommand `pooled` can be used to compute a confidence interval for $\mu_1 - \mu_2$ using the pooled estimate of σ. After retrieving the data, which we assume are in columns `i` and `j`, enter the following command and subcommand in the Session window:

```
MTB > twosample confidence ci cj;
SUBC> pooled.
```

The quantity `confidence` is the desired confidence coefficient in percent (90, 95, 99), and the data are in columns `i` and `j`.

THINKING STATISTICALLY: SELF-TEST PROBLEMS

11.1.1 Give two examples where it is of interest to compare two population means.

11.1.2 Find two newspaper articles where statistical methods are used to compare two or more population means. Summarize these articles in your own words.

11.1.3 Find two articles from technical journals in your field of interest where two or more population means are compared. Summarize these articles in your own words.

11.1.4 A large corporation that develops computer software employs several hundred computer programmers. A financial officer wants to determine if there is a difference between the average salaries earned by male and female programmers. A simple random sample of 12 male programmers and a simple random sample of 8 female programmers were selected from the company payroll records and their salaries were obtained. Using these samples, the financial officer will estimate μ_1, μ_2, and $\mu_1 - \mu_2$, where μ_1 is the average salary for male programmers and μ_2 is the average salary for female programmers. The samples are shown in Table 11.1.2, where the salaries are reported in thousands of dollars. These data are stored in columns 1 and 2 in the file **programr.mtp** on the data disk.

TABLE 11.1.2 Salaries of Programmers

Male (thousands of dollars)	Female (thousands of dollars)
42	38
38	28
45	44
51	41
37	48
29	51
44	39
45	46
53	
58	
49	
47	

a. What are the target and sampled populations?
b. What is the variable of interest?
c. Estimate $\mu_1 - \mu_2$.
d. Construct the normal rankit-plot for each random sample. Based on these plots, does it appear reasonable to assume that the samples were selected from *normal* populations?
e. Compute a 95% confidence interval for $\mu_1 - \mu_2$ using the formulas in Box 11.1.2. Assume that both populations are *normal*.
f. Repeat part (e), assuming that $\sigma_1 = \sigma_2$.

11.2 COMPARING MORE THAN TWO POPULATION MEANS

In Section 11.1 we restricted our attention to studies for comparing two population means. However, it is very common to encounter studies in which one wants to compare more than two population means. For example, one might be interested in comparing average life expectancies for four different treatments for cancer, the effects of five different pesticides on the average yield of tomato plants, or the average salaries of three different ethnic groups, and so on. In this section we present methods for computing point estimates and confidence intervals for comparing more than two population means. We begin with an example.

EXAMPLE 11.2.1

CHOLESTEROL IN CHICKEN EGGS. In poultry science it is well known that the cholesterol content of the yolks of chicken eggs can be controlled by adding certain chemicals to the chicken feed. For health reasons, most people want to eat eggs with low-cholesterol yolks. A poultry scientist wishes to compare the effect of three different chemicals on yolk cholesterol. Three feed rations are prepared using three different chemicals. The rations are denoted ration 1, ration 2, and ration 3. The investigator wants to estimate the average cholesterol content of eggs for each ration and compare the differences among the three averages.

When comparing more than two population means, an investigator should obtain a confidence interval for the difference between **each** *pair.*

If there are k populations of interest, the number of pairs of means is equal to $r = k(k-1)/2$. In Example 11.2.1, suppose an investigator wants to compare the average cholesterol of eggs for $k = 3$ different rations. Thus, there are $r = 3(2)/2 = 3$ pairs of means that must be compared. The differences for these three pairs are $\mu_1 - \mu_2$, $\mu_1 - \mu_3$, and $\mu_2 - \mu_3$. In order to help make reliable decisions, it is necessary to construct confidence intervals for this set of differences in such a manner that these intervals are all simultaneously correct with a specified confidence. One method for doing this is the **Bonferroni method**, which is described in Box 11.2.1.

BOX 11.2.1: BONFERRONI METHOD

An investigator wants to compute a set of confidence intervals for r population quantities (say, r pairwise differences of population means). Suppose that each confidence interval is computed using an individual confidence coefficient of I, where I is written as a proportion. For example, if each confidence interval has a confidence coefficient of 95%, then $I = 0.95$. The confidence that all r confidence intervals are simultaneously correct is at least S, where

$$S = 1 - r(1 - I). \qquad \textbf{Line 11.2.1}$$

Alternatively, suppose you want to construct a confidence interval on each of r population parameters. If you want the confidence to be at least S that all intervals are simultaneously correct, then, using the Bonferroni method, each confidence interval must have an individual confidence coefficient of

$$I = 1 - \frac{1 - S}{r}. \qquad \textbf{Line 11.2.2}$$

We demonstrate the Bonferroni method by considering the case of $k = 3$ population means. The number of pairwise differences of k population means is $r = 3(2)/2 = 3$. These differences are $\mu_1 - \mu_2$, $\mu_1 - \mu_3$, and $\mu_2 - \mu_3$. If each of the three confidence intervals has a confidence coefficient of 95%, then $I = 0.95$, and by Line 11.2.1 the confidence is at least $S = 1 - 3(1 - 0.95) = 0.85$ (85%) that all three intervals are simultaneously correct. If it is desired for the confidence to be at least 95% that all three intervals are simultaneously correct, then by Line 11.2.2 each confidence interval requires a confidence coefficient of $I = 1 - (1 - 0.95)/3 = 0.983$, or 98.3%. An explanation of the Bonferroni method is given in Section 10.7 for any set of r population parameters.

We now describe the method for comparing $k = 3$ population means. The population means are denoted μ_1, μ_2, and μ_3 and the standard deviations are denoted σ_1, σ_2, and σ_3. Suppose a simple random sample of size n_i is available from population i for $i = 1, 2, 3$. To obtain confidence intervals for $\mu_1 - \mu_2$, $\mu_1 - \mu_3$, and $\mu_2 - \mu_3$, first compute $\hat{\mu}_i = \bar{y}_i$ and $\hat{\sigma}_i = s_i$ for $i = 1, 2, 3$. Each confidence interval is then computed using the formula in Box 11.1.2. For us to be at least 95% confident that all three intervals are simultaneously correct, Line 11.2.2 indicates that a confidence coefficient of $I = 1 - (1 - 0.95)/3 = 0.983$ is required for each confidence interval. For us to be (at least) 99% confident that all three intervals are simultaneously correct, a confidence coefficient of $I = 1 - (1 - 0.99)/3 = 0.997$ is required for each interval. The following example illustrates this procedure.

EXAMPLE 11.2.2

CHOLESTEROL IN CHICKEN EGGS. This is a continuation of Example 11.2.1, where a poultry scientist wishes to compare the effects of three different rations on cholesterol in chicken egg yolks. There are three sampled populations in this study, and all three are conceptual. The first sampled population consists of all the eggs laid by chickens on a chicken farm on a specified day where each chicken has been fed ration 1. The second and third populations are defined in the same manner except every chicken is fed ration 2 and ration 3, respectively. We will refer to the three populations as ration 1, ration 2, and ration 3. An investigator plans to use simple random samples of 33 eggs from each population to estimate each population mean and compute confidence intervals for the three pairwise differences

$$\mu_1 - \mu_2, \quad \mu_1 - \mu_3, \quad \mu_2 - \mu_3.$$

Suppose the investigator wants to be at least 97% confident that all three confidence intervals are simultaneously correct. The data are presented in Table 11.2.1 and stored in columns 1, 2, and 3 in the file **eggyolk.mtp** on the data disk. Before computing the required confidence intervals, one should construct normal rankit-plots for each sample and decide if the samples were selected from *normal* populations. Although we do not show these plots, we did construct them; they indicate that it is reasonable to assume the three samples were obtained from *normal* populations.

TABLE 11.2.1 Cholesterol of Egg Yolks in Milligrams (mg) for Three Rations

Ration 1	Ration 2	Ration 3
223	196	205
221	194	209
222	194	201
220	190	220
216	199	204
216	191	204
222	190	211
217	192	202
222	192	201
223	194	196

Ration 1	Ration 2	Ration 3
213	195	208
227	192	209
230	197	203
218	208	204
215	194	204
219	185	212
227	196	210
215	198	202
225	199	208
219	189	196
216	193	211
223	191	204
224	190	205
223	194	211
211	194	206
221	198	205
214	192	205
214	210	210
210	201	210
217	184	210
221	190	202
217	200	213
221	188	211

The means and the standard deviations for the three samples were obtained using Minitab and are shown in Exhibit 11.2.1.

```
-----------------------------------------------------------------------
Descriptive Statistics

Variable        N      Mean    Median   Tr Mean     StDev   SE Mean
Ration1        33    219.45    220.00    219.45      4.72      0.82
Ration2        33    194.24    194.00    193.90      5.52      0.96
Ration3        33    206.42    205.00    206.45      4.99      0.87

Variable      Min       Max        Q1        Q3
Ration1    210.00    230.00    216.00    223.00
Ration2    184.00    210.00    190.50    197.50
Ration3    196.00    220.00    203.50    210.00
-----------------------------------------------------------------------
```

EXHIBIT 11.2.1

From Exhibit 11.2.1 we observe that $n_1 = 33$, $n_2 = 33$, $n_3 = 33$, $s_1 = 4.72$ mg, $s_2 = 5.52$ mg, $s_3 = 4.99$ mg, $\bar{y}_1 = 219.45$ mg, $\bar{y}_2 = 194.24$ mg, and $\bar{y}_3 = 206.42$ mg. The investigator wants confidence intervals for $\mu_1 - \mu_2$, $\mu_1 - \mu_3$, and $\mu_2 - \mu_3$ so that the confidence is at least 97% that all three intervals are simultaneously correct. By Line 11.2.2, each confidence interval for the pairwise differences must be computed using a confidence coefficient of $I = 1 - (1 - 0.97)/3 = 0.99$. We begin by computing the confidence interval for $\mu_1 - \mu_2$. Using the information in

Exhibit 11.2.1 and the formulas in Box 11.1.2, we obtain the following:

$$\hat{\mu}_1 - \hat{\mu}_2 = 219.45 - 194.24 = 25.21, \qquad \sqrt{\frac{s_1^2}{33} + \frac{s_2^2}{33}} = 1.264,$$

$h_1 = 0.675$, $h_1^2 = 0.456$, $h_2 = 0.923$, $h_2^2 = 0.853$, and

$$\text{DF} = \frac{(0.675 + 0.923)^2}{(0.456)/32 + (0.853)/32} = 62.4 = 62 \quad \text{(with the decimal portion deleted).}$$

The value of $T2$ for a 99% confidence coefficient with DF = 62 is $T2 = 2.66$. This provides a confidence coefficient of 99% for the single interval $\mu_1 - \mu_2$. The resulting confidence interval for $\mu_1 - \mu_2$ is L to U, where

$$L = 25.21 - 2.66(1.264) = 21.85, \qquad U = 25.21 + 2.66(1.264) = 28.57.$$

Confidence intervals for $\mu_1 - \mu_3$ and $\mu_2 - \mu_3$ are calculated in the same manner. The results for all three confidence intervals are shown in Table 11.2.2.

TABLE 11.2.2 Summary of Simultaneous Confidence Intervals

Quantity	Point estimate	L	U
$\mu_1 - \mu_2$	25.21	21.85	28.57
$\mu_1 - \mu_3$	13.03	9.85	16.21
$\mu_2 - \mu_3$	−12.18	−15.62	−8.74

Since we are at least 97% confident that all three intervals in Table 11.2.2 are simultaneously correct, we are at least 97% confident that *all* of the following statements are correct:

1. μ_1 is at least 21.85 mg larger than μ_2;
2. μ_1 is at least 9.85 mg larger than μ_3;
3. μ_3 is at least 8.74 mg larger than μ_2;
4. ration 1 results in the largest average cholesterol per egg by at least 9.85 mg;
5. ration 2 results in the smallest average cholesterol per egg by at least 8.74 mg;
6. ration 3 results in the second smallest average cholesterol per egg.

A Minitab Macro for Comparing More Than Two Population Means

We have written a Minitab macro that can be used to compute confidence intervals for all pairwise differences of $k > 2$ population means using the Bonferroni method. The macro is in the file **pairwise.mac** in the directory **macro** on the data disk. To use the macro, retrieve the file that contains the data and enter the following command in the Session window:

```
MTB > %a:\macro\pairwise ci cj ... ck  confidence
```

The k columns `i j ... k` contain the samples from k populations. The quantity `confidence` is the desired simultaneous confidence coefficient (in percent). We illustrate with an example.

EXAMPLE 11.2.3

CHOLESTEROL IN CHICKEN EGGS. We will use the macro **pairwise.mac** to compute the confidence intervals for all pairwise differences of means in Example 11.2.2 so that the confidence is at least 97% that all intervals are simultaneously correct. The samples are stored in columns 1, 2, and 3 in the file **eggyolk.mtp** on the data disk. Retrieve the file and enter the following command in the Session window:

```
MTB > %a:\macro\pairwise c1 c2 c3 97
```

The output is shown in Exhibit 11.2.2.

```
----------------------------------------------------------------------
Two Sample T-Test and Confidence Interval

Twosample T for ration1 vs ration2
          N      Mean     StDev   SE Mean
Ration1  33    219.45      4.72      0.82
Ration2  33    194.24      5.52      0.96

*99% C.I. for mu ration1 - mu ration2: ( 21.85,  28.57)
T-Test mu ration1 = mu ration2 (vs not =): T= 19.93  P=0.0000  DF=  62

Two Sample T-Test and Confidence Interval

Twosample T for ration1 vs ration3
          N      Mean     StDev   SE Mean
Ration1  33    219.45      4.72      0.82
Ration3  33    206.42      4.99      0.87

*99% C.I. for mu ration1 - mu ration3: ( 9.85,  16.21)
T-Test mu ration1 = mu ration3 (vs not =): T= 10.90  P=0.0000  DF=  63

Two Sample T-Test and Confidence Interval

Twosample T for ration2 vs ration3
          N      Mean     StDev   SE Mean
Ration2  33    194.24      5.52      0.96
Ration3  33    206.42      4.99      0.87

*99% C.I. for mu ration2 - mu ration3: ( -15.62,  -8.74)
T-Test mu ration2 = mu ration3 (vs not =): T= -9.40  P=0.0000  DF=  63
----------------------------------------------------------------------
```

EXHIBIT 11.2.2

The three confidence intervals are shown in Exhibit 11.2.2 in the three lines labeled "99% C.I. for mu . . . ." We have marked each of these lines with an asterisk ($*$). These lines report the 99% confidence interval for each individual pairwise difference of means. Each individual interval is a 99% confidence interval in order to provide a confidence of at least 97% that all three intervals are simultaneously correct.

Pooling Estimates of the Standard Deviations

As stated in Section 11.1, there may be situations when an investigator is comparing population means and knows that the population standard deviations are equal (or at least approximately equal). In such a situation, the k sample standard deviations can be pooled in the same manner as described in Section 11.1. For example, with $k = 3$ populations, the formula for the pooled standard deviation is

$$s_\text{p} = \sqrt{\frac{(n_1 - 1)s_1^2 + (n_2 - 1)s_2^2 + (n_3 - 1)s_3^2}{n_1 + n_2 + n_3 - 3}}. \qquad \textbf{Line 11.2.3}$$

If the decision is made to pool the sample standard deviations, then s_p replaces s_1, s_2, and s_3 in the confidence interval formulas. Minitab has a command and subcommand that can be used for computing confidence intervals for all pairwise differences of means when the σ_i are all equal and the pooling procedure is used. In this case, Tukey's HSD method may be used instead of the Bonferroni method, and the resulting confidence intervals may be shorter. We will not discuss Tukey's HSD procedure here. We recommend that the formulas in Box 11.1.2 be used (that is, do not pool) unless the investigator has reliable information indicating that the population standard deviations are all the same.

A Statistical Test to Determine if k Population Means Are Equal: The Analysis of Variance

When two or more population means are being compared, an investigator wants to know the differences between each pair. For example, suppose we are comparing three population means, μ_1, μ_2, and μ_3. Practically speaking, it would be virtually impossible for these three means to be *exactly* equal. Nevertheless, many investigators conduct a statistical test

$$H_0 : \mu_1 = \mu_2 = \mu_3 \quad \text{versus} \quad H_\text{a} : \mu_1,\ \mu_2,\ \mu_3 \text{ are not all equal} \qquad \textbf{Line 11.2.4}$$

to determine the amount of sample evidence in favor of concluding that the three means are not *exactly* equal. These hypotheses are typically tested using a P-value that summarizes the sample evidence in support of rejecting H_0. The P-value is generally computed assuming that the three populations are *normal* and that the standard deviations of the three populations are equal ($\sigma_1 = \sigma_2 = \sigma_3$).

If the P-value is small (say, 0.05 or smaller), an investigator would generally conclude that there are differences among the population means. In this case, confidence intervals for all pairwise differences of population means should be examined to see how large the differences are and to determine if the differences are large enough to be of practical importance in the problem under study.

If the P-value is not small (say, greater than 0.05), then an investigator would generally conclude that the sample data do not provide enough evidence to claim that the population means are different. In this case, confidence

200 customers in the sample, 33 owned brand 1, 42 owned brand 2, 40 owned brand 3, 38 owned brand 4, and 47 owned brand 5. One of the questions each customer was asked is "How many times have you had your refrigerator serviced in the past five years?" Let μ_1, μ_2, μ_3, μ_4, and μ_5 denote the respective population means of the frequency of repairs during the past five years. The data obtained from the sample of 200 customers are presented in Table 11.2.3 and stored in the file **refrig.mtp** on the data disk. You will notice that the variable in this problem is discrete rather than continuous so that the assumption of *normal* populations is not very reasonable. However, since the sample sizes all exceed 30, the formulas in Box 11.1.2 can be used to compute approximate confidence intervals for the differences between population means.

TABLE 11.2.3 Frequency of Repairs for Five Brands of Refrigerators

Brand 1	Brand 2	Brand 3	Brand 4	Brand 5
4	2	3	1	8
1	5	3	2	1
1	4	5	0	5
1	3	6	0	5
2	1	3	0	3
0	3	5	2	2
2	1	2	0	5
6	2	3	1	1
1	2	5	2	5
2	0	6	0	3
1	6	6	0	1
0	1	5	0	3
3	4	6	0	2
0	2	6	1	0
3	1	2	0	1
1	4	7	0	4
0	4	8	0	0
1	3	5	1	2
1	3	4	0	4
1	0	5	2	4
3	6	6	1	3
1	3	7	2	1
2	4	3	3	5
1	1	9	1	3
2	3	6	2	2
1	5	5	0	1
1	1	8	3	2
0	2	1	1	2
2	1	7	1	2
1	2	6	0	5
3	0	6	0	2
4	3	5	0	3
0	2	4	1	2

intervals for all pairwise differences of population mear
examined to see if any difference is small enough to cor
negligible for the problem under study.

Thus, decisions should be made based on confidence int
differences of population means, and not solely on the *I*

A large amount of computing is required to obtain the *P*-
potheses in Line 11.2.4. It has been standard practice over t
of the intermediate computations in a table called the **anal**
simply, **ANOVA** table). We present the ANOVA table for
in Exhibit 11.2.3.

```
One-Way Analysis of Variance

Analysis of Variance
Source      DF         SS          MS          F          p
Factor       2    10492.2      5246.1     202.56      0.000
Error       96     2486.3        25.9
Total       98    12978.5
```

EXHIBIT 11.2.3

In the ANOVA table, the *P*-value used to test the hyp
labeled p. In Exhibit 11.2.3 the *P*-value is 0.000 (accurate t
statistical test had been conducted using a value of α greater
H_0 would have been rejected. The numbers in the body of the
computations used to calculate the *P*-value. Their interpret
more advanced texts. Computational formulas for all these
in the appendix to this chapter (Section 11.8).

Using Minitab to Compute the ANOVA Table

To use Minitab to compute the ANOVA table shown in Exhi
file **eggyolk.mtp** and enter the following command in the Ses

```
MTB > aovoneway c1 c2 c3
```

The word aovoneway in the preceding Minitab command
"aov one way." The abbreviation aov is short for ANOVA
way mean that the entire set of data has been partitioned onl
the type of ration.

THINKING STATISTICALLY: SELF-TEST PROB

11.2.1 A company that sells extended service warranties fo
ducted a comparative study of the average frequer
ciated with five leading brands of refrigerators. W
as brand 1, brand 2, brand 3, brand 4, and brand 5.
sample of 200 customers was selected from a databa
who had purchased one of these refrigerators five

Brand 1	Brand 2	Brand 3	Brand 4	Brand 5
	1	5	0	4
	1	2	2	6
	4	5	1	2
	3	6	1	3
	2	7	1	5
	2	6		3
	5	7		1
	1			2
	5			1
				0
				2
				2
				2
				2

a. Use the macro **pairwise.mac** to compute confidence intervals for all pairwise differences of the five means such that you are at least 90% confident that all the intervals are correct.

b. The results from part (a) are summarized in Table 11.2.4.

TABLE 11.2.4 Summary of Simultaneous Confidence Intervals

Quantity	Point estimate	Margin of error	L	U
$\mu_1 - \mu_2$	−0.99	0.92	−1.91	−0.07
$\mu_1 - \mu_3$	−3.57	0.99	−4.56	−2.58
$\mu_1 - \mu_4$	0.74	0.75	−0.02	1.48
$\mu_1 - \mu_5$	−1.12	0.92	−2.04	−0.20
$\mu_2 - \mu_3$	−2.58	1.01	−3.59	−1.57
$\mu_2 - \mu_4$	1.73	0.77	0.96	2.50
$\mu_2 - \mu_5$	−0.13	0.93	−1.06	0.80
$\mu_3 - \mu_4$	4.31	0.86	3.45	5.17
$\mu_3 - \mu_5$	2.45	1.01	1.44	3.46
$\mu_4 - \mu_5$	−1.86	0.77	−2.63	−1.09

What are the conclusions about the average frequency of repairs for the five brands during the first five years after purchase?

c. Use Minitab to compute an ANOVA table. State the null and the alternative hypotheses corresponding to the statistical test in Line 11.2.4. What is the P-value for this test? If you chose $\alpha = 0.01$, what is your conclusion? What assumptions were made in order to obtain the P-value?

11.3 COMPARING TWO POPULATION MEANS USING PAIRED DIFFERENCES

In Section 11.1 we discussed confidence intervals for $\mu_1 - \mu_2$, the difference between two population means. This interval was computed using two simple random samples, one from population 1 and one from population 2. If every value in the confidence interval is positive, then we have evidence that μ_1 is greater than μ_2. If every value in the confidence interval is negative, then we have evidence that μ_2 is greater than μ_1. Thus, if zero is not included in the confidence interval, the sample data provide enough evidence to decide which mean is larger and which is smaller and by how much. If zero is included in the confidence interval, and the width is small enough, then one can conclude that the difference between μ_1 and μ_2 is negligible for practical purposes.

In some investigations the standard deviations of the populations may be too large to obtain a confidence interval short enough to decide which of the two means is greater. An obvious method for obtaining a shorter confidence interval is to increase the size of the sample taken from each population. However, this will generally increase the cost of the study. Statisticians have developed alternatives to increasing sample size that decrease the width of the confidence interval and do not increase the cost of the study. Example 11.3.1 illustrates one such approach.

EXAMPLE 11.3.1

COMPUTER KEYBOARDS. Suppose we want to compare two brands of computer keyboards, which we denote as keyboard 1 and keyboard 2. Keyboard 1 is a standard keyboard, while keyboard 2 is specially designed so that the keys need very little pressure to make them respond. The manufacturer of keyboard 2 would like to claim that typing can be done faster using keyboard 2. It is therefore of interest to the manufacturer to determine if keyboard 2 is better than keyboard 1 with regard to typing speed.

A study is undertaken to compare the two keyboards. A collection of 5,621 teachers attending a national high-school teachers' conference will be used as the sampled population of items. There are two populations of numbers associated with the sampled population, both of which are conceptual. In your mind's eye, imagine that each of the 5,621 teachers typed a prepared page of text using keyboard 1. The completion time (the time to correctly type the page of text) is measured in seconds; the collection of these 5,621 numbers is population 1. Let the mean of this population be denoted by μ_1 and the standard deviation by σ_1. Similarly, population 2 consists of the 5,621 numbers representing the completion times using keyboard 2. Let μ_2 and σ_2 denote the mean and the standard deviation, respectively, of population 2. The objective of the study is to determine $\mu_1 - \mu_2$, where μ_1 is the average completion time for the 5,621 teachers using keyboard 1 and μ_2 is the average completion time for the 5,621 teachers using keyboard 2.

Here are two methods that can be used to conduct the investigation in Example 11.3.1.

Method 1. Two Simple Random Samples. This method was considered in Section 11.1. A simple random sample of n_1 teachers is selected from the sampled population of 5,621 teachers and asked to type the page of text using keyboard 1. The completion times are measured to the nearest second for each of the n_1 teachers. Based on this sample, the estimates $\hat{\mu}_1$ and $\hat{\sigma}_1$ are computed using the formulas in Box 11.1.1. Another simple random sample, consisting of n_2 teachers, is then selected from the sampled population of teachers and asked to type the page of text using keyboard 2. The completion times for the second sample are used to compute the estimates $\hat{\mu}_2$ and $\hat{\sigma}_2$ using the formulas in Box 11.1.1. The formulas in Box 11.1.2 are used to compute L and U, the endpoints of a confidence interval for $\mu_1 - \mu_2$.

Method 2. One Simple Random Sample of Paired Differences. We now define a third conceptual population of numbers for the sampled population. Imagine that each of the 5,621 teachers types the page of text two times, once using keyboard 1 and once using keyboard 2. Let D_i = (time using keyboard 1) – (time using keyboard 2) for teacher i for $i = 1, \ldots, 5621$. The population of numbers $D_1, \ldots, D_{5621}$ is called the population of **paired differences**. The mean of this population is denoted μ_D and the standard deviation is denoted σ_D. You can imagine the three populations of numbers as schematically represented in Table 11.3.1, where Y_i is the completion time of the ith teacher using keyboard 1, X_i is the completion time of the ith teacher using keyboard 2, and $D_i = Y_i - X_i$.

TABLE 11.3.1 Three Conceptual Populations

Teacher (population item)	Population 1 (time in seconds using keyboard 1)	Population 2 (time in seconds using keyboard 2)	Population 3 (differences in time)
1	Y_1	X_1	$D_1 = Y_1 - X_1$
2	Y_2	X_2	$D_2 = Y_2 - X_2$
$\vdots$	$\vdots$	$\vdots$	$\vdots$
N	Y_N	X_N	$D_N = Y_N - X_N$
Mean	μ_1	μ_2	$\mu_D = \mu_1 - \mu_2$

It is important to note that $\mu_D = \mu_1 - \mu_2$. That is, the difference between the means of population 1 and population 2 is equal to the mean of the numbers in population 3. So we can obtain a confidence interval for $\mu_1 - \mu_2$ by using a simple random sample from the population of paired differences to compute a confidence interval for μ_D.

A simple random sample of n teachers is selected from the sampled population of 5,621 teachers. Each teacher types the page of text twice, once using keyboard 1 and once using keyboard 2. The order in which the keyboards are used is determined by a coin flip for each teacher. For the ith teacher, calculate

$$d_i = (\text{completion time using keyboard 1}) - (\text{completion time using keyboard 2}).$$

The values $d_1, \ldots, d_n$ represent a simple random sample from the population of paired differences (population 3). The sample mean and the sample standard deviation of $d_1, \ldots, d_n$ are, respectively,

$$\bar{d} = \frac{d_1 + d_2 + \cdots + d_n}{n}$$

$$s_d = \sqrt{\frac{(d_1 - \bar{d})^2 + (d_2 - \bar{d})^2 + \cdots + (d_n - \bar{d})^2}{(n-1)}}.$$

Thus, the estimate for μ_D is $\hat{\mu}_D = \bar{d}$ and the estimate for σ_D is $\hat{\sigma}_D = s_d$. A confidence interval for μ_D is obtained by computing L and U using the formulas in Box 6.3.1. These formulas are

$$L = \hat{\mu}_D - \text{ME} \qquad \text{and} \qquad U = \hat{\mu}_D + \text{ME},$$

where

$$\text{ME} = T2\frac{\hat{\sigma}_D}{\sqrt{n}}.$$

We have described two methods for conducting the keyboard study. In the first method we select two simple random samples, one from population 1 and one from population 2. In the second method we select one simple random sample from the population of paired differences. Which method should be used to conduct the study? To help make this decision, it seems reasonable to ask the following questions:

1. For a specified confidence coefficient, which method results in a shorter confidence interval?
2. How do the methods compare in terms of cost?

If we use method 1, then

$$\text{ME} = T2\sqrt{\frac{s_1^2}{n_1} + \frac{s_2^2}{n_2}},$$

where $T2$ is calculated with DF given in Line 11.1.1. If we use method 2, then

$$\text{ME} = T2\frac{\hat{\sigma}_D}{\sqrt{n}},$$

where $T2$ is calculated with DF $= n - 1$. To compare these two margins of error, assume for simplicity $n_1 = n_2 = n$. Unless n is very small, the values for $T2$ are very nearly equal. So, if we assume the values of $T2$ are equal, method 2 will provide a shorter confidence interval than method 1 if

$$\frac{\hat{\sigma}_D}{\sqrt{n}} < \sqrt{\frac{s_1^2}{n} + \frac{s_2^2}{n}}. \qquad \textbf{Line 11.3.1}$$

In Example 11.3.1 σ_1 characterizes the variability among the completion times of the 5,621 teachers when they all use keyboard 1. The standard deviation σ_2 characterizes the variability among the completion times of the 5,621 teachers when they all use keyboard 2. If there is a large variability in typing skills among the teachers, then both σ_1 and σ_2 will be relatively large. In contrast, even though typing skills among teachers may be highly variable, the population of paired differences may have relatively small variability, and σ_D may be considerably less than σ_1 and σ_2. For instance, suppose the completion time is 10 seconds less with keyboard 2 for every teacher in the sampled population. Then $D_i = 10$ for every teacher and $\sigma_D = 0$. In practice, not every teacher will obtain the same improvement, but σ_D should still be much smaller than either σ_1 or σ_2. This is because σ_D measures the variation in typing speed *differences* (of the same teacher using two different keyboards), whereas σ_1 and σ_2 measure the variation among typing speeds of teachers. Consequently, we expect $\hat{\sigma}_D$ to be smaller than the sample standard deviations s_1 and s_2. If this is the case, the inequality in Line 11.3.1 holds and method 2 will provide a shorter confidence interval than method 1.

Regarding cost considerations, method 1 requires a sample of $2n$ teachers, each of whom uses only one of the two keyboards. Method 2 requires a sample of n teachers, each of whom uses both keyboards. Thus, the number of completion times recorded is the same for each method ($2n$). There does not appear to be a substantial difference between the costs of the two methods. Therefore, it appears that the paired-differences approach (method 2) is preferred over the two-samples approach (method 1). We now provide two other examples where paired differences are used to compare two population means.

EXAMPLE 11.3.2

ASTHMA DRUGS. Two asthma drugs, say drug 1 and drug 2, are compared to determine if one drug provides a faster recovery rate. The sampled population of items is the set of all patients belonging to a health maintenance organization (HMO) who frequently suffer from asthma. There are two sampled populations of numbers, both of which are conceptual. Population 1 consists of the recovery times (in minutes) for the population of patients where each patient is administered drug 1 when suffering an asthma attack. Let μ_1 and σ_1 denote the mean and the standard deviation of this population. Population 2 consists of the recovery times (in minutes) for the population of patients where each patient is administered drug 2 when suffering an asthma attack. Let μ_2 and σ_2 denote the mean and the standard deviation of this population. We can also imagine a third population of numbers that consists of the differences between recovery times for drug 1 and drug 2 for each patient in the HMO when suffering an asthma attack. This is the population of paired differences. Let μ_D and σ_D denote the mean and the standard deviation for this population.

Method 1 for comparing the two drugs would be conducted using two simple random samples, one from population 1 and one from population 2. Alternatively, the study could be conducted using method 2 by selecting a simple random sample of n asthma sufferers who belong to the HMO. When a patient suffers an asthma attack, administer one drug and record the recovery time (in minutes). When this same patient suffers another asthma attack, administer the other drug and record the recovery time. It is advisable to randomize the order of drugs for each patient. This can be done by tossing a coin to determine which drug is administered for the first attack.

Because a great deal of variability is expected among recovery rates for different asthma patients, σ_1 and σ_2 are both likely to be larger than σ_D. This is because σ_D measures variability among differences between the two responses for the same patient. If σ_D is less than both σ_1 and σ_2, method 2 (the method of paired differences) should result in a shorter confidence interval for $\mu_D = \mu_1 - \mu_2$ than method 1 (the method using two simple random samples).

EXAMPLE 11.3.3

NOSE PATCHES FOR SNORING. A research group for a major pharmaceutical company wants to evaluate the effectiveness of a drug placed on a nose patch designed to prevent snoring during sleep. The sampled population of items is the collection of individuals who have sought help with a snoring problem from a sleep-disorder clinic during the past year. There are two populations of numbers in this study, and both are conceptual. Imagine that each member of the sampled population spends the night at the clinic wearing the nose patch treated with the drug. The time during which snoring was registered by a machine is recorded in minutes for each subject. This collection of numbers is population 1. We use μ_1 and σ_1 to denote the mean and the standard deviation of this population. Now imagine that each subject spends the night at the clinic wearing a **placebo** patch (a nose patch that contains no drug). The time during which snoring was registered by a machine is recorded in minutes for each subject. This collection of numbers is population 2. We use μ_2 and σ_2 to denote the mean and the standard deviation for this population. We can also imagine a third population of numbers that consists of the differences in snoring times between the two patches for each subject. We use μ_D and σ_D to denote the mean and the standard deviation of this population of paired differences. The study objective is to obtain an estimate and a confidence interval for $\mu_1 - \mu_2$.

We consider two methods for conducting the study to evaluate the effectiveness of the patch with the drug. Method 1 consists of selecting two simple random samples of patients from the clinic records, one sample from population 1 and the other from population 2. All patients in the sample from population 1 wear the nose patch with the drug, and all patients from population 2 wear the placebo patch. Using these two samples, an estimate and a confidence interval for $\mu_1 - \mu_2$ are obtained using the formulas in Boxes 11.1.1 and 11.1.2. Method 2 consists of selecting a simple random sample of n patients from clinic records. The snoring time is recorded twice for each subject, once with the nose patch with the drug, and once with the placebo patch. The patients do not know which patch they are wearing. This is called **blinding**. The time during which snoring was registered by a machine is recorded under both conditions, and the difference is calculated. This is done for each of the n patients, resulting in the simple random sample from the population of paired differences, $d_1, \ldots, d_n$.

It appears reasonable to expect σ_1 and σ_2 to be much larger than σ_D in this study. Thus, for a specified confidence coefficient, the method of paired differences is expected to yield shorter confidence intervals than the method with two simple random samples. You should think about whether a substantial difference in the cost between the two methods for this example is likely.

Point Estimation and Confidence Intervals Using Paired-Differences Data

Suppose we are interested in the difference between the means of two variables Y and X measured on each item in a sampled population. For the ith item, let Y_i denote the value of the variable Y and X_i denote the value of the variable X. These values form two populations of numbers, population 1 and population 2. Population 1 consists of the numbers $Y_1, \ldots, Y_N$ and population 2 consists of the numbers $X_1, \ldots, X_N$. Let μ_1, σ_1, μ_2, and σ_2 denote the means and the standard deviations for populations 1 and 2, respectively. The objective of the study is to obtain an estimate and a confidence interval for $\mu_1 - \mu_2$.

Define $D_1, D_2, \ldots, D_N$ as a population of paired differences where $D_i = Y_i - X_i$ for $i = 1, \ldots, N$. The mean of this population is $\mu_D = \mu_1 - \mu_2$, and the standard deviation is σ_D. To obtain an estimate and a confidence interval for μ_D, select a simple random sample of n items from the sampled population of items and record the values of Y and X for each item. Let $y_1, y_2, \ldots, y_n$ denote the sample values of Y and $x_1, x_2, \ldots, x_n$ denote the sample values of X. Let $d_i = y_i - x_i$ for $i = 1, \ldots, n$. Then, $d_1, d_2, \ldots, d_n$ is a simple random sample from the population $D_1, D_2, \ldots, D_N$. The data may be recorded as shown in Table 11.3.2.

TABLE 11.3.2 Paired-Differences Data

Pair	Sample from population 1	Sample from population 2	Difference
1	y_1	x_1	$d_1 = y_1 - x_1$
2	y_2	x_2	$d_2 = y_2 - x_2$
$\vdots$	$\vdots$	$\vdots$	$\vdots$
n	y_n	x_n	$d_n = y_n - x_n$
Sample means	$\bar{y}$	$\bar{x}$	$\bar{d} = \bar{y} - \bar{x}$

Note that the last column in Table 11.3.2 reports the values of the paired differences $d_i = y_i - x_i$. Point estimates for μ_1, μ_2, and $\mu_1 - \mu_2$ are given by the formulas

$$\hat{\mu}_1 = \bar{y}, \qquad \hat{\mu}_2 = \bar{x}, \qquad \hat{\mu}_D = \hat{\mu}_1 - \hat{\mu}_2 = \bar{y} - \bar{x} = \bar{d}. \qquad \textbf{Line 11.3.2}$$

The procedure described in Box 6.3.1 can be used to compute confidence intervals for μ_1 and μ_2. The procedure for computing a confidence interval for $\mu_D = \mu_1 - \mu_2$ with paired-differences data is presented in Box 11.3.1.

BOX 11.3.1: CONFIDENCE INTERVAL PROCEDURE FOR $\mu_D = \mu_1 - \mu_2$ WITH PAIRED-DIFFERENCES DATA

A simple random sample of size n is selected from a sampled population of items. Let y_i and x_i denote the values of the variables Y and X for the ith sample item. The following steps can be used to compute a confidence interval for $\mu_D = \mu_1 - \mu_2$:

1. Compute the difference $d_i = y_i - x_i$ for each pair, as shown in the last column of Table 11.3.2.
2. Compute

$$\hat{\mu}_D = \bar{d}.$$

3. Compute the sample standard deviation of the d_i. Namely,

$$\hat{\sigma}_D = s_d = \sqrt{\frac{(d_1 - \bar{d})^2 + (d_2 - \bar{d})^2 + \cdots + (d_n - \bar{d})^2}{(n-1)}}.$$

4. Specify a confidence coefficient and compute the margin of error ME, where

$$\text{ME} = (T2)\left(\frac{\hat{\sigma}_D}{\sqrt{n}}\right).$$

The value of $T2$ is obtained from Table T-2 using DF $= n - 1$.

5. Compute a confidence interval L to U for μ_D using the formulas

$$L = \hat{\mu}_D - \text{ME}, \qquad U = \hat{\mu}_D + \text{ME}.$$

The confidence interval procedure in Box 11.3.1 requires that the population D_1, $D_2, \ldots, D_N$ be *normal*, or that $n \geq 30$. Observe that the formulas for computing L and U in Box 11.3.1 are the same ones shown in Box 6.3.1 for computing a confidence interval for the mean of a single population. This is because $D_1, \ldots, D_N$ represents a single population of numbers (differences). We illustrate the procedure in the following example.

EXAMPLE 11.3.4

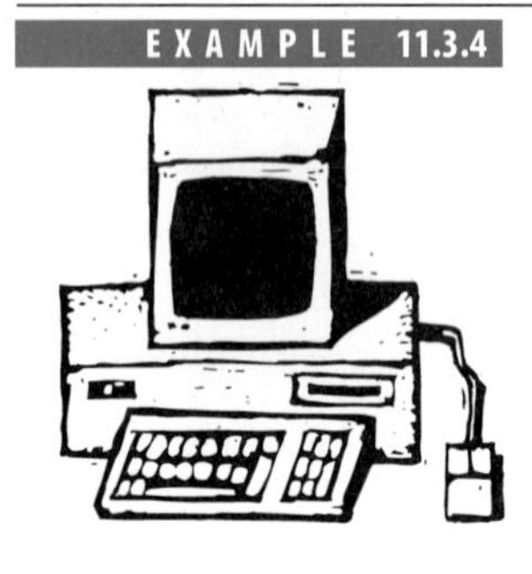

COMPUTER KEYBOARDS. Consider the keyboard study described in Example 11.3.1. The study objective is to compare two types of keyboards. A simple random sample of $n = 30$ teachers was selected from a population of high-school teachers attending a national conference. Each teacher typed the same page of text once using keyboard 1 and once using keyboard 2. For each teacher the order in which the keyboards were used was determined by the toss of a coin. For each teacher the variable measured was the time (in seconds) to correctly type the page of text. The data are reported in Table 11.3.3 and stored in the file **typing.mtp** on the data disk. Column 1 of the file contains the subject number, column 2 contains the time for keyboard 1 (Y), and column 3 contains the time for keyboard 2 (X).

TABLE 11.3.3 Completion Time (in seconds)

Subject	Keyboard 1 Y	Keyboard 2 X	Difference D
1	348	350	−2
2	435	442	−7
3	369	356	13
4	357	360	−3
5	376	373	3
6	412	405	7
7	396	376	20
8	317	314	3
9	366	366	0
10	340	337	3
11	347	352	−5
12	315	303	12
13	349	338	11
14	330	328	2
15	335	322	13
16	345	351	−6
17	374	361	13
18	374	370	4
19	380	375	5
20	319	318	1
21	387	382	5
22	313	317	−4
23	303	310	−7
24	404	393	11
25	355	362	−7
26	364	364	0
27	348	355	−7
28	361	368	−7
29	301	291	10
30	348	323	25

We follow the steps in Box 11.3.1 to compute a 95% confidence interval for μ_D.

1. The differences $d_i = y_i - x_i$ are shown in the last column of Table 11.3.3.
2. The average of the d_i is $\bar{d} = 3.53$ seconds. Thus, $\hat{\mu}_D = 3.53$ seconds.
3. The standard deviation of the d_i is $s_d = 8.56$ seconds. Thus, $\hat{\sigma}_D = 8.56$ seconds.
4. The value of $T2$ for a 95% confidence interval with DF $= n - 1 = 29$ is $T2 = 2.05$. The resulting margin of error is

$$\text{ME} = 2.05\left(\frac{8.56}{\sqrt{30}}\right) = 3.20.$$

5. The endpoints L and U of the 95% confidence interval for μ_D are

$$L = 3.53 - 3.20 = 0.33, \qquad U = 3.53 + 3.20 = 6.73.$$

With 95% confidence, we conclude that the average time to correctly type the page of text using keyboard 2 is between 0.33 and 6.73 seconds less than it is using keyboard 1. You should verify that $\hat{G}_0$ is much smaller than $\hat{G}_1$ and $\hat{G}_2$.

The paired-differences approach can be extended to compare more than two population means. This extension is referred to as **blocking** and is discussed in more-advanced textbooks.

Using Minitab to Compute a Confidence Interval for $\mu_1 - \mu_2$ with Paired-differences Data

The computations described in Box 11.3.1 can be performed using Minitab by entering the following commands in the Session window after the data have been retrieved:

```
MTB > let ci = cj-ck
MTB > tinterval confidence ci
```

The sample values $y_1, y_2, \ldots, y_n$ are in column `j` and the sample values $x_1, x_2, \ldots, x_n$ are in column `k`. The first command computes the $d_i = y_i - x_i$ and places them in column `i`. The second command computes a confidence interval for μ_D, where `confidence` is the specified confidence coefficient in percent (e.g., 90, 95, 99). This is the confidence interval shown in Box 11.3.1. To illustrate, we will compute a 95% confidence interval for μ_D in Example 11.3.4. Retrieve the data in the file **typing.mtp**. The data are in columns 2 and 3, so enter the following commands in the Session window:

```
MTB > let c4 = c2-c3
MTB > tinterval 95 c4
```

A portion of the Minitab output is shown in Exhibit 11.3.1.

```
-----------------------------------------------------------------------
Variable      N       Mean     StDev  SE Mean        95.0 % C.I.
C4           30       3.53      8.56     1.56  (     0.34,     6.73)
-----------------------------------------------------------------------
```

EXHIBIT 11.3.1

From this output we obtain $\hat{\mu}_D = 3.53$. The endpoints L and U of the 95% confidence interval reported in Exhibit 11.3.1 are $L = 0.34$ and $U = 6.73$.

THINKING STATISTICALLY: SELF-TEST PROBLEMS

11.3.1 A physical fitness program is designed to increase a person's upper-body strength. To determine the effectiveness of this program, a simple random sample of 35 members of a health club was selected to participate in the program. Before beginning the program, each participant was asked to do as many push-ups as possible in one minute. After one month in the program, each participant was again requested to complete as many push-ups as possible in one minute. The number of push-ups the 35 participants completed is shown in Table 11.3.4 and stored in the file **pushup.mtp** on the data disk. Column 1 of the file contains the subject number, column 2 contains the number of completed push-ups before the program (Y), column 3 contains the number of completed push-ups after the program (X), and column 4 contains the difference ($D = Y - X$).

TABLE 11.3.4 Number of Push-ups in One Minute

Subject	Before program Y	After one month X	Difference D
1	28	32	−4
2	34	32	2
3	28	42	−14
4	60	64	−4
5	20	41	−21
6	25	33	−8
7	32	49	−17
8	19	32	−13
9	29	50	−21
10	33	44	−11
11	27	34	−7
12	33	40	−7
13	30	35	−5
14	31	52	−21
15	32	39	−7
16	21	21	0
17	30	22	8
18	19	28	−9
19	36	42	−6
20	50	62	−12
21	40	55	−15
22	27	27	0
23	29	40	−11
24	38	39	−1
25	29	41	−12
26	30	62	−32
27	45	60	−15
28	30	30	0

Subject	Before program Y	After one month X	Difference D
29	37	44	−7
30	42	55	−13
31	36	52	−16
32	30	33	−3
33	32	34	−2
34	28	30	−2
35	34	49	−15

a. Describe the sampled population of items.
b. Describe the two conceptual populations of numbers for which it is of interest to compare the population means.
c. Describe the population of paired differences.
d. Use the paired-differences method to compute a 95% confidence interval for $\mu_D = \mu_1 - \mu_2$, where μ_1 is the mean of the first population of numbers in part (b) and μ_2 is the mean of the second population. Exhibit 11.3.2 reports some summary measures for the sample values $d_1, d_2, \ldots, d_n$. Interpret the confidence interval.

```
------------------------------------------------------------
Descriptive Statistics

Variable        N    Mean    Median  Tr Mean   StDev  SE Mean
D              35   -9.17     -8.00    -8.97    8.06     1.36

Variable      Min     Max       Q1       Q3
D          -32.00    8.00   -15.00    -3.00
------------------------------------------------------------
```

EXHIBIT 11.3.2

11.4 DISTRIBUTION-FREE (NONPARAMETRIC) METHODS

The confidence intervals for the difference between two population means presented in Boxes 11.1.2 and 11.3.1 should only be used if one (or both) of the following two conditions is satisfied:

1. the sample sizes are "sufficiently large"; or
2. the simple random samples are selected from *normal* populations.

In this section we describe two procedures for computing a confidence interval for the difference between two population means that can be used when neither of these conditions is satisfied. The **Mann-Whitney procedure** uses two simple random samples, and the **Wilcoxon procedure** uses paired differences. Since neither of these methods requires either condition listed above, they are called **distribution-free** or **nonparametric** methods.

The Mann-Whitney Confidence Interval for the Difference Between Two Population Means

Suppose an investigator wants to compare the means of two populations, population 1 and population 2. Let μ_1 denote the mean of population 1 and μ_2 denote the mean of population 2. Two samples are obtained by selecting a simple random sample of size n_1 from population 1 and a simple random sample of size n_2 from population 2. The sample data are represented in Table 11.4.1. It is not necessary for n_1 and n_2 to be equal.

TABLE 11.4.1 Simple Random Samples from Populations 1 and 2

Sample from population 1	Sample from population 2
y_1	x_1
y_2	x_2
$\vdots$	$\vdots$
$\vdots$	$\vdots$
$\vdots$	$\vdots$
y_{n_1}	$\vdots$
	$\vdots$
	$\vdots$
	$\vdots$
	x_{n_2}

An approximate confidence interval for $\mu_1 - \mu_2$ can be computed using the procedure described in Box 11.4.1. The approximation is satisfactory when both n_1 and n_2 are greater than or equal to 5.

BOX 11.4.1: THE MANN-WHITNEY PROCEDURE FOR OBTAINING A CONFIDENCE INTERVAL FOR $\mu_1 - \mu_2$

Let $Y_1, Y_2, \ldots, Y_{N_1}$ denote population 1 with mean μ_1 and $X_1, X_2, \ldots, X_{N_2}$ denote population 2 with mean μ_2. Obtain a simple random sample of size n_1 from population 1 and denote the sample values by $y_1, y_2, \ldots, y_{n_1}$. Obtain a simple random sample of size n_2 from population 2 and denote the sample values by $x_1, x_2, \ldots, x_{n_2}$. The following steps can be used to compute a confidence interval for $\mu_1 - \mu_2$.

1. Compute all possible differences $y_i - x_j$ for $i = 1, \ldots, n_1$, and $j = 1, \ldots, n_2$. In all, there will be $n_1 \times n_2$ such differences. For instance, if $y_1 = 3$, $y_2 = 5$, $x_1 = 2$, $x_2 = 3$, and $x_3 = 6$, then $n_1 = 2$, $n_2 = 3$, and the $n_1 \times n_2 = 6$ differences are shown in the following table:

$3 - 2 = 1$	$3 - 3 = 0$	$3 - 6 = -3$
$5 - 2 = 3$	$5 - 3 = 2$	$5 - 6 = -1$

2. Arrange the differences computed in step (1) in increasing order. For our example the ordered differences are

$$-3, \quad -1, \quad 0, \quad 1, \quad 2, \quad 3.$$

3. Calculate q, where

$$q = \frac{n_1 n_2}{2} - (T1)\sqrt{\frac{n_1 n_2 (n_1 + n_2 + 1)}{12}}.$$

The quantity $T1$ is a table value found in Table T-1 for the desired confidence coefficient. If q is not an integer, round it *down* to the nearest integer. If $q < 1$, then a confidence interval is not available for the selected confidence coefficient.

4. The lower endpoint L of the confidence interval for $\mu_1 - \mu_2$ is the ordered difference in step (2) located in position q when counting *up* from the smallest value. The upper endpoint U of the confidence interval for $\mu_1 - \mu_2$ is the ordered difference in step (2) located in position q when counting *down* from the largest value. The confidence that $\mu_1 - \mu_2$ is contained in the interval L to U is (approximately) equal to the specified confidence coefficient.

Assumption. Although the confidence interval presented in Box 11.4.1 does not require large sample sizes or *normal* populations, it does require that one condition be satisfied. In particular, it is necessary that the relative frequency histograms of the two populations have the same shape (at least approximately). Populations 1 and 2 have histograms with the same shape if population 2 results from adding a constant to each item in population 1. Figure 11.4.1 shows two histograms that have the same shape.

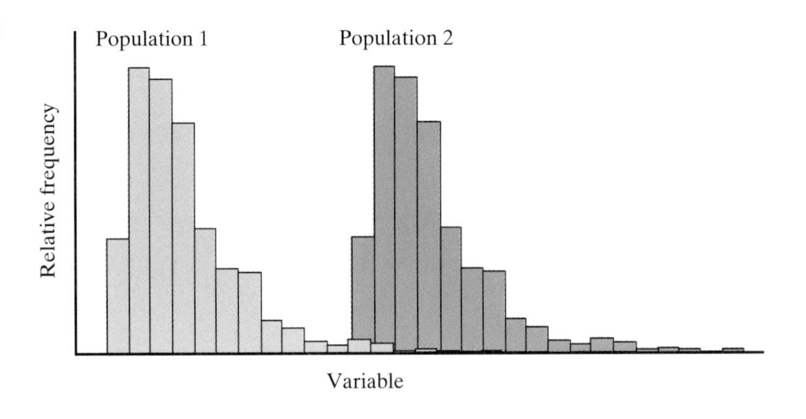

FIGURE 11.4.1
Two Population Histograms with the Same Shape

The following example illustrates the computations in Box 11.4.1.

EXAMPLE 11.4.1

MARATHON TRAINING. A sports scientist conducted a study to evaluate whether a newly proposed training method for marathon training is better than a traditional approach. She selected a simple random sample of 11 male marathon runners from a local running club to participate in the study. The runners were of roughly equal abilities. She selected a simple random sample of five runners from the 11 and instructed them to train for an upcoming marathon using the traditional method. The remaining six runners were instructed to train using the new method. The finishing times for all 11 runners (in minutes) are presented in Table 11.4.2.

TABLE 11.4.2 Marathon Training Data

Traditional Method		New Method	
Runner	Finishing time (minutes)	Runner	Finishing time (minutes)
1	145	1	152
2	135	2	147
3	139	3	138
4	148	4	133
5	142	5	137
		6	139

The two sampled populations are both conceptual. One population consists of all male runners in the running club where each runner trains using the traditional method (population 1). The second sampled population consists of all male runners in the running club where each runner trains using the new method (population 2). We will assume that the relative frequency histograms of these two populations have the same shape. The population means for these conceptual populations are denoted by μ_1 and μ_2, respectively. We now demonstrate how to use Box 11.4.1 to compute a 95% confidence interval for $\mu_1 - \mu_2$.

1. Compute the $n_1 \times n_2 = 5 \times 6 = 30$ possible differences $y_i - x_j$. These differences are shown in Table 11.4.3.

TABLE 11.4.3 Differences for Marathon Data

145 − 152 = −7	145 − 147 = −2	145 − 138 = 7	145 − 133 = 12	145 − 137 = 8	145 − 139 = 6
135 − 152 = −17	135 − 147 = −12	135 − 138 = −3	135 − 133 = 2	135 − 137 = −2	135 − 139 = −4
139 − 152 = −13	139 − 147 = −8	139 − 138 = 1	139 − 133 = 6	139 − 137 = 2	139 − 139 = 0
148 − 152 = −4	148 − 147 = 1	148 − 138 = 10	148 − 133 = 15	148 − 137 = 11	148 − 139 = 9
142 − 152 = −10	142 − 147 = −5	142 − 138 = 4	142 − 133 = 9	142 − 137 = 5	142 − 139 = 3

2. The differences computed in step (1) are arranged in increasing order and displayed in Table 11.4.4.

TABLE 11.4.4 Ordered Differences for Marathon Data

−17	−13	−12	−10	−8	−7
−5	−4	−4	−3	−2	−2
0	1	1	2	2	3
4	5	6	6	7	8
9	9	10	11	12	15

3. With $n_1 = 5, n_2 = 6$, and $T1 = 1.96$ for 95% confidence,

$$q = \frac{(5)(6)}{2} - (1.96)\sqrt{\frac{(5)(6)(5+6+1)}{12}} = 15 - (1.96)\sqrt{30} = 4.26,$$

which is rounded down to 4.

4. The lower bound L is located in position 4 of the ordered differences when counting up from the smallest value. This provides $L = -10$ minutes. The upper bound U is located in position 4 of the ordered differences when counting down from the largest value. This provides $U = 10$ minutes. We are at least 95% confident that $\mu_1 - \mu_2$ is between -10 and $+10$ minutes. Since both negative and positive values are contained in this interval, the data do not demonstrate that one method is better than the other.

Using Minitab to Compute the Mann-Whitney Confidence Interval

The Mann-Whitney confidence interval can be computed with Minitab using the command

```
MTB > mann whitney confidence ci cj
```

Here `confidence` represents the desired confidence coefficient expressed as a percentage (e.g., 90, 95, 99). The data for the two samples are contained in columns `i` and `j`. For the marathon data in Example 11.4.1, place the sample values from population 1 in column 1 of a Minitab worksheet and the sample values from population 2 in column 2. The Minitab command is

```
MTB > mann whitney 95 c1 c2
```

The relevant portion of the Minitab output is shown in Exhibit 11.4.1.

```
---------------------------------------------------------------
Mann-Whitney Confidence Interval and Test

Pop1        N =   5     Median =        142.00
Pop2        N =   6     Median =        138.50
Point estimate for ETA1-ETA2 is           1.50
96.4 Percent CI for ETA1-ETA2 is (-10.00,10.00)
---------------------------------------------------------------
```

EXHIBIT 11.4.1

The confidence interval reported on the last line of Exhibit 11.4.1 is from -10 minutes to 10 minutes, and the confidence coefficient is 96.4%. You might wonder why the confidence coefficient is 96.4% instead of the requested value of 95%. The reason is that it is not possible to construct an exact Mann-Whitney confidence interval for every possible confidence coefficient. Minitab uses the confidence coefficient that is closest to the one specified.

You might also wonder why Minitab labels the difference between population means as `ETA1-ETA2` instead of `MU1-MU2`. The term `ETA` is the Minitab designation for median. Minitab actually computes a confidence interval for the difference between the medians, M_1 and M_2, of the two populations. However, if two population histograms have the same shape, then $M_1 - M_2 = \mu_1 - \mu_2$. This is true even if $\mu_1 \neq M_1$ and $\mu_2 \neq M_2$. Thus, if two population histograms have the same shape, a confidence interval for the difference between the two population medians is the same as a confidence interval for the difference between the two population means.

Wilcoxon's Signed-Rank Confidence Interval Using Paired-Differences Data

A procedure due to Wilcoxon can be used to compute a confidence interval for $\mu_1 - \mu_2$ using paired differences. This confidence interval is valid even when the population of paired differences $(D_1, D_2, \ldots, D_N)$ is not *normal* and the sample size n is not "sufficiently large." Consider the sample of n paired differences displayed in Table 11.4.5.

TABLE 11.4.5 Paired-Differences Data from Populations 1 and 2

Pair number	Sample from population 1	Sample from population 2	Difference
1	y_1	x_1	$d_1 = y_1 - x_1$
2	y_2	x_2	$d_2 = y_2 - x_2$
⋮	⋮	⋮	⋮
⋮	⋮	⋮	⋮
⋮	⋮	⋮	⋮
n	y_n	x_n	$d_n = y_n - x_n$

A confidence interval for $\mu_1 - \mu_2$ can be obtained using the procedure outlined in Box 11.4.2.

BOX 11.4.2: THE WILCOXON CONFIDENCE INTERVAL FOR $\mu_1 - \mu_2$

A simple random sample of size n is selected from a sampled population of items. Let y_i and x_i denote the values of the variables Y and X for the ith sample item and $d_i = y_i - x_i$. The following steps can be used to compute a confidence interval for $\mu_D = \mu_1 - \mu_2$.

1. Calculate all possible pairwise averages of the d_i values (including pairs consisting of the same value). This provides $n(n+1)/2$ averages. For instance, if the values of the d_i are 3, 5, 6, 9, and 9, then the $5(6)/2 = 15$ pairwise averages are computed and displayed in the body of the following table. Each number in the body of the table is the average of the corresponding row and column headings.

	3	**5**	**6**	**9**	**9**
3	3				
5	4	5			
6	4.5	5.5	6		
9	6	7	7.5	9	
9	6	7	7.5	9	9

2. Next, arrange the pairwise averages in increasing order. For the illustration in step (1), we obtain

   3, 4, 4.5, 5, 5.5, 6, 6, 6, 7, 7, 7.5, 7.5, 9, 9, 9.

3. Calculate q, where

$$q = \frac{n(n+1)}{4} - (T1)\sqrt{\frac{n(n+1)(2n+1)}{24}}.$$

   The quantity $T1$ is a table value found in Table T-1 for the desired confidence coefficient. If q is not an integer, round it *down* to the nearest integer. If $q < 1$, then a confidence interval is not available for the selected confidence coefficient.

4. The lower endpoint L of the confidence interval for $\mu_1 - \mu_2$ is the ordered pairwise average in step (2) located in position q when counting *up* from the smallest average. The upper endpoint U of the confidence interval for $\mu_1 - \mu_2$ is the pairwise average in position q when counting *down* from the largest average. The confidence that $\mu_1 - \mu_2$ is contained in the interval L to U is approximately equal to the stated confidence coefficient.

Assumption. The confidence interval in Box 11.4.2 requires that the histogram for the population of paired differences is symmetric. If this condition is satisfied, an approximate confidence interval for $\mu_1 - \mu_2$ can be computed using the procedure in Box 11.4.2. The approximation is satisfactory when $n \geq 5$. We illustrate the computations in Box 11.4.2 using the following example.

EXAMPLE 11.4.2

SHOOTING FREE THROWS. A basketball coach has developed a new method for shooting free throws. To determine how well this method works, the coach selected a simple random sample of six players from the set of players participating in a women's state high-school basketball tournament. After warming up, each player was asked to shoot 60 free throws, and the number of successful attempts was recorded. The coach then taught all six players how to shoot free throws using the new method that she had developed. Each player then shot an additional 60 free throws using the new method. Table 11.4.6 reports the number of successful attempts for each set of 60 free throws. Assume the population of differences is symmetric.

TABLE 11.4.6 Number of Successful Free Throws

Sample number	Before instruction	After instruction	Difference
1	46	52	−6
2	32	47	−15
3	34	38	−4
4	38	33	5
5	32	47	−15
6	40	41	−1

We now follow the steps in Box 11.4.2 to compute the required confidence interval. The values for $d_i = y_i - x_i$ are shown in the last column of Table 11.4.6.

1. The $n(n+1)/2 = 6(7)/2 = 21$ pairwise averages of the d_i are shown in Table 11.4.7.

TABLE 11.4.7 Pairwise Averages for Free-Throw Data

	−6	**−15**	**−4**	**5**	**−15**	**−1**
−6	−6					
−15	−10.5	−15				
−4	−5	−9.5	−4			
5	−0.5	−5	0.5	5		
−15	−10.5	−15	−9.5	−5	−15	
−1	−3.5	−8	−2.5	2	−8	−1

2. Next we arrange the pairwise averages in increasing order as shown below:

$$-15.0,\ -15.0,\ -15.0,\ -10.5,\ -10.5,\ -9.5,\ -9.5,\ -8.0,\ -8.0,\ -6.0,$$
$$-5.0,\ -5.0,\ -5.0,\ -4.0,\ -3.5,\ -2.5,\ -1.0,\ -0.5,\ 0.5,\ 2.0,\ 5.0.$$

3. For a 90% confidence coefficient, the appropriate table value is $T1 = 1.65$. Compute

$$q = \frac{n(n+1)}{4} - (T1)\sqrt{\frac{n(n+1)(2n+1)}{24}} = \frac{(6)(7)}{4} - (1.65)\sqrt{\frac{(6)(7)(13)}{24}} = 2.63,$$

which we round down to 2.

4. The lower bound of the confidence interval is the pairwise average in position 2 as we count up from the smallest value. This provides the lower bound $L = -15.0$. The upper bound is in position 2 as we count down from the largest pairwise average. This provides the upper bound $U = 2$. The 90% confidence interval for $\mu_1 - \mu_2$ is -15 to 2. Since zero is contained in this confidence interval, the data provide inconclusive evidence for deciding if the new method will increase the average number of free throws made in 60 attempts.

Using Minitab to Compute the Wilcoxon Confidence Interval

There is a Minitab command that can be used to compute the Wilcoxon confidence interval. You must first place the differences $d_i = y_i - x_i$ in a column of a Minitab worksheet. The following Minitab command will compute the confidence interval:

```
MTB > winterval confidence ci
```

Here `confidence` is the desired confidence coefficient expressed as a percentage (e.g., 90, 95, 99). For the free-throw data in Example 11.4.2, the six differences $-6,\ -15,\ -4,\ 5,\ -15$, and -1 are placed in column 1 and the following command is entered in the Session window:

```
MTB > winterval  90  c1
```

The output is displayed in Exhibit 11.4.2.

```
------------------------------------------------------------------------
Wilcoxon Signed Rank Confidence Interval

                  Estimated     Achieved
              N      Median   Confidence   Confidence Interval
C1            6        -5.0         90.7   (     -15.0,       0.5)
------------------------------------------------------------------------
```

EXHIBIT 11.4.2

The computed confidence interval for $\mu_1 - \mu_2$ is given in the last line of Exhibit 11.4.2. This interval is -15 to 0.5. The interval computed in Example 11.4.2 is slightly wider than the interval reported in Exhibit 11.4.2 because the Minitab procedure uses the confidence coefficient closest to the chosen level for which an exact interval can be obtained.

THINKING STATISTICALLY: SELF-TEST PROBLEMS

11.4.1 A large utility company wants to determine the impact of insulation on energy costs for a house. A study was conducted in cooperation with a local home builder to determine the amount of energy saved by putting additional insulation in the ceiling of a house. Five pairs of recently built houses were sampled at random from a new housing development. The houses in each pair had identical floor plans and faced identical directions. One of the houses in each pair had additional insulation placed in the ceiling. Thus, apart from the amount of insulation placed in the ceiling, the two houses were very similar in each pair. At the end of one month, the amount of electricity used for the month was measured in kilowatt hours (kwh). The samples are shown in Table 11.4.8.

TABLE 11.4.8 Electricity Used (in kilowatt hours)

Sample number	Regular amount of insulation	Additional insulation	Difference
1	2,074	1,866	208
2	2,210	2,055	155
3	1,852	1,833	19
4	1,946	1,731	215
5	2,412	2,364	48

a. Compute the appropriate nonparametric confidence interval for $\mu_1 - \mu_2$, where μ_1 is the mean of the first population (regular insulation) and μ_2 is the mean of the second population (additional insulation). Assume the population of paired differences is symmetric and use a 90% confidence coefficient.

b. Compute a 90% confidence interval for $\mu_1 - \mu_2$ if it is known the populations are *normal*. How does this confidence interval compare to the one computed in part (a)?

11.4.2 Ticket prices for admission to games of a major league baseball team were lowered 20% in 1997 from the prices charged in 1996 in an effort to increase attendance. At the end of the 1997 season, attendances were recorded for five randomly chosen games from 1996 and four randomly chosen games from 1997. These data are reported in Table 11.4.9.

TABLE 11.4.9 Attendance at Baseball Games

1996		1997	
Game	Attendance (in thousands)	Game	Attendance (in thousands)
1	22	1	33
2	27	2	31
3	18	3	39
4	25	4	30
5	31		

a. Describe the two target populations and the two sampled populations.

b. Let μ_1 represent the average attendance for all games in 1996 and μ_2 represent the average attendance for all games in 1997. Construct a 90% confidence interval for $\mu_1 - \mu_2$ using the appropriate nonparametric confidence interval (assume the relative frequency histograms have the same shape). Do the data provide evidence that the average attendance increased in 1997?

c. Compute a 90% confidence interval for $\mu_1 - \mu_2$ if it is known the populations are *normal*. How does this confidence interval compare to the one computed in part (b)?

11.5 CHAPTER SUMMARY

KEY TERMS

1. pooling
2. distribution-free (nonparametric)
3. analysis of variance (ANOVA)
4. Bonferroni method
5. two separate simple random samples, one from each of two populations
6. paired differences
7. Mann-Whitney confidence interval
8. Wilcoxon confidence interval

SYMBOLS

1. $\mu_1 - \mu_2$—a difference between two population means
2. s_p—an estimate of σ based on pooling sample standard deviations

KEY CONCEPTS

1. If an investigator has reliable information indicating that the standard deviations of populations are equal, then for a specified confidence coefficient pooling may provide a shorter confidence interval for the difference between two population means.
2. The Bonferroni method enables one to specify the confidence that all confidence intervals in a set of *r* intervals simultaneously contain their respective parameters.
3. For a specified confidence coefficient, using paired differences will generally provide a shorter confidence interval for the difference between two population means than will separate simple random samples from each population.
4. Nonparametric confidence intervals can be computed for $\mu_1 - \mu_2$ if populations are not *normal* and sample sizes are small.

SKILLS

1. Construct a confidence interval for the difference between two population means using two simple random samples, one from each population.
2. Construct a confidence interval for the difference between two population means using paired-differences data.
3. Use the Bonferroni method to construct a set of *r* simultaneous confidence intervals for all pairwise differences of means for *k* populations.
4. Construct a confidence interval for the difference between two population means using the Mann-Whitney procedure.
5. Construct a confidence interval for the difference between two population means using the Wilcoxon procedure.

11.6 CHAPTER EXERCISES

11.1 An investigator is interested in studying the average number of days rats live when fed diets that contain different amounts of fat. Three populations were studied, where rats in population 1 were fed a high-fat diet, rats in population 2 were fed a medium-fat diet, and rats in population 3 were fed a low-fat diet. The variable of interest is "Days lived." Population 1 has a mean and a standard deviation denoted by μ_1 and σ_1, respectively. Population 2 has a mean and a standard deviation denoted by μ_2 and σ_2, respectively. Population 3 has a mean and a standard deviation denoted by μ_3 and σ_3, respectively. A simple random sample of 50 rats was obtained from each population, and the sample values are stored in columns 1, 2, and 3 in the file **fat.mtp** on the data disk. Exhibit 11.6.1 reports some descriptive measures for the sample data.

Variable	N	Mean	Median	TrMean	StDev	SEMean
Highfat	50	344.6	351.0	344.3	112.7	15.9
Medumfat	50	583.6	573.0	580.4	177.6	25.1
Lowfat	50	609.2	618.0	612.8	154.8	21.9

Variable	Min	Max	Q1	Q3
Highfat	98.0	587.0	254.3	421.0
Medumfat	191.0	1029.0	473.5	696.8
Lowfat	212.0	910.0	507.0	723.5

EXHIBIT 11.6.1

a. What are the estimates of μ_1, μ_2, $\mu_1 - \mu_2$, σ_1, and σ_2?

b. Estimate the average number of days rats lived in the population fed a medium-fat diet.

c. Estimate the average number of days rats lived in the population fed a high-fat diet.

d. On average, how much longer are rats estimated to live if they are fed a medium-fat diet as opposed to a high-fat diet?

e. Suppose that each population is a *normal* population. Minitab is used to produce Exhibit 11.6.2. With 95% confidence, how much longer (on average) will rats live if they are fed a medium-fat diet instead of a high-fat diet?

f. Is the confidence interval in part (e) correct if the populations are not *normal* populations?

g. The macro **pairwise.mac** was used to construct a set of confidence intervals for all pairwise differences of the population means such that the confidence is 95% that all three intervals are simultaneously correct. The results are shown in Exhibit 11.6.3. Explain these results in detail.

```
----------------------------------------------------------------
Twosample T for highfat vs medumfat
            N       Mean      StDev    SE Mean
Highfat    50        345        113         16
Medumfat   50        584        178         25

95% C.I. for mu highfat - mu medumfat: ( -298,  -180)
----------------------------------------------------------------
```

EXHIBIT 11.6.2

```
----------------------------------------------------------------
Twosample T for highfat vs medumfat
            N       Mean      StDev    SE Mean
Highfat    50        345        113         16
Medumfat   50        584        178         25

98% C.I. for mu highfat - mu medumfat: ( -312,  -166)
T-Test mu highfat = mu medumfat (vs not =): T= -8.04  P=0.0000  DF=  82

Twosample T for highfat vs lowfat
            N       Mean      StDev    SE Mean
Highfat    50        345        113         16
Lowfat     50        609        155         22

98% C.I. for mu highfat - mu lowfat: ( -331,  -199)
T-Test mu highfat = mu lowfat (vs not =): T= -9.78  P=0.0000  DF=  89

Twosample T for medumfat vs lowfat
            N       Mean      StDev    SE Mean
Medumfat   50        584        178         25
Lowfat     50        609        155         22

98% C.I. for mu medumfat - mu lowfat: ( -107,  56)
T-Test mu medumfat = mu lowfat (vs not =): T= -0.77  P=0.44  DF=  96
----------------------------------------------------------------
```

EXHIBIT 11.6.3

11.2 An experiment was conducted to compare the performance of two brands of golf balls, brand 1 and brand 2. A simple random sample of 36 golf balls was selected from a population of brand 1 golf balls in a sporting goods store. A simple random sample of 36 golf balls from a population of brand 2 golf balls was selected from the same store. The golf balls were taken to a driving range and hit by a golfer using a metal driver with a titanium shaft. Let μ_1 represent the average distance traveled for the sampled population of brand 1 golf balls and μ_2 represent the average distance traveled for the sampled population of brand 2 golf balls. Summary measures for the 72 balls used in the experiment are shown in Table 11.6.1. The distance a ball traveled was measured to the nearest yard.

TABLE 11.6.1 Summary of Distances Traveled (in yards) by Golf Balls

	Brand 1	Brand 2
Mean	251.75	248.92
Standard deviation	12.71	17.21

a. Estimate μ_1 and μ_2.

b. Compute a 95% confidence interval for $\mu_1 - \mu_2$ using the formulas in Box 11.1.2.

c. Using the confidence interval in (b), what is your conclusion?

11.3 A scientist in the research and development department of a breakfast cereal company is working to find a recipe that will yield a tasty new breakfast cereal. She has come up with four different recipes (1, 2, 3, and 4) that yield an acceptable product. An initial batch of cereal has been produced using each of the four recipes. The investigator decides to conduct a taste-panel experiment where several human subjects taste one of the four cereals and assign a score based on a rating scale of 1 to 10 (1 being the lowest score and 10 the highest). The taste panel consists of a simple random sample of 128 people chosen from the local community. The sample is randomly divided into four groups of size 32. The first group receives recipe 1, the second group receives recipe 2, and so on. The scores assigned by the taste-panel members are stored in columns 1, 2, 3, and 4 in the file **recipe.mtp** on the data disk. Table 11.6.2 shows the sample means and the sample standard deviations of the taste-panel scores. Let μ_1, μ_2, μ_3, and μ_4 denote the population means corresponding to the four recipes. Here the sampled population for recipe 1 is the conceptual population of ratings of all individuals in the local community after they have tasted recipe 1. The other three sampled populations are defined in the same manner.

a. Estimate μ_1, μ_2, μ_3, and μ_4.

b. Estimate all six pairwise differences of the population means.

c. The Minitab macro **pairwise.mac** was used with a simultaneous 95% confidence coefficient to compute confidence intervals for each of the $r = 6$ pairwise differences between the population means, $\mu_1 - \mu_2$, $\mu_1 - \mu_3$, $\mu_1 - \mu_4$, $\mu_2 - \mu_3$, $\mu_2 - \mu_4$, and $\mu_3 - \mu_4$. The output is shown in Exhibit 11.6.4 on page 360.
What can you conclude from these confidence intervals? What are your assumptions?

11.4 A state regulatory agency is studying the effects of secondhand smoke in the workplace. All companies in the state that employ more than 15 workers must file a report with the agency that describes the company's smoking policy. In particular, each company must report whether (1) smoking is allowed (no restrictions), (2) smoking is allowed only in restricted areas, or (3) smoking is banned. In order to determine the effect of secondhand smoke, the state agency needs to measure the nicotine level at the work site. It is not possible to measure the nicotine level for every company that reports to the agency, and so a simple random sample of 25 companies is selected from each category of smoking policy. Data collected by the agency are stored in columns 1, 2, and 3 in the file **smoke.mtp** on the data disk. The data represent the average of the nicotine levels measured at noon each workday for a week at each of the sampled work sites. Nicotine level is measured in micrograms per cubic meter of air. Summary statistics for all three samples are shown in Exhibit 11.6.5 on page 361.

a. Use this information to estimate the average nicotine level and the standard deviation for each of the three populations.

b. Use the formula in Box 11.1.2 to construct a 95% confidence interval for $\mu_1 - \mu_2$, the difference in the average nicotine level between population 1 (smoking allowed) and population 2 (smoking restricted). Assume that each population is *normal*.

TABLE 11.6.2 Summary of Taste-Panel Scores

	Recipe 1	Recipe 2	Recipe 3	Recipe 4
Mean	4.28	5.12	8.22	7.94
Standard deviation	1.17	1.10	1.16	1.11

```
------------------------------------------------------------------------------
TWOSAMPLE T FOR recipe1 VS recipe2
           N       MEAN      STDEV   SE MEAN
Recipe1   32       4.28       1.17      0.21
Recipe2   32       5.12       1.10      0.19

99 PCT CI FOR MU recipe1 - MU recipe2: (-1.62, -0.07)

TWOSAMPLE T FOR recipe1 VS recipe3
           N       MEAN      STDEV   SE MEAN
Recipe1   32       4.28       1.17      0.21
Recipe3   32       8.22       1.16      0.20

99 PCT CI FOR MU recipe1 - MU recipe3: (-4.73, -3.14)

TWOSAMPLE T FOR recipe1 VS recipe4
           N       MEAN      STDEV   SE MEAN
Recipe1   32       4.28       1.17      0.21
Recipe4   32       7.94       1.11      0.20

99 PCT CI FOR MU recipe1 - MU recipe4: (-4.43, -2.88)

TWOSAMPLE T FOR recipe2 VS recipe3
           N       MEAN      STDEV   SE MEAN
Recipe2   32       5.12       1.10      0.19
Recipe3   32       8.22       1.16      0.20

99 PCT CI FOR MU recipe2 - MU recipe3: (-3.86, -2.32)

TWOSAMPLE T FOR recipe2 VS recipe4
           N       MEAN      STDEV   SE MEAN
Recipe2   32       5.12       1.10      0.19
Recipe4   32       7.94       1.11      0.20

99 PCT CI FOR MU recipe2 - MU recipe4: (-3.56, -2.06)

TWOSAMPLE T FOR recipe3 VS recipe4
           N       MEAN      STDEV   SE MEAN
Recipe3   32       8.22       1.16      0.20
Recipe4   32       7.94       1.11      0.20

99 PCT CI FOR MU recipe3 - MU recipe4: (-0.49, 1.05)
------------------------------------------------------------------------------
```

EXHIBIT 11.6.4

c. The macro **pairwise.mac** is used to construct a set of confidence intervals for all pairwise differences of the population average nicotine levels using a 95% simultaneous confidence coefficient. The results are shown in Exhibit 11.6.6. Does it appear that population 2 (smoking restricted) has a lower average nicotine level than population 1 (smoking allowed)? Is the average nicotine level less in population 3 (smoking banned) than in population 2 (smoking restricted)? Assume that each population is *normal.*

```
Descriptive Statistics

Variable       N      Mean    Median    TrMean     StDev    SEMean
Allow         25     8.148     7.710     8.071     1.785     0.357
Restrict      25    1.2416    1.2500    1.2552    0.2185    0.0437
Banned        25   0.32040   0.32000   0.32000   0.04748   0.00950
```

EXHIBIT 11.6.5

```
Two Sample T-Test and Confidence Interval

Twosample T for allow vs restrict
           N       Mean       StDev    SE Mean
Allow     25       8.15        1.78       0.36
Restrict  25      1.242       0.218      0.044

98% C.I. for mu allow - mu restrict: ( 5.98,  7.832)

Two Sample T-Test and Confidence Interval

Twosample T for allow vs banned
         N       Mean       StDev    SE Mean
Allow   25       8.15        1.78       0.36
Banned  25     0.3204      0.0475    0.00915

98% C.I. for mu allow - mu banned: ( 6.91,  8.7472)

Two Sample T-Test and Confidence Interval

Twosample T for restrict vs banned
           N       Mean       StDev    SE Mean
Restrict  25      1.242       0.218      0.044
Banned    25     0.3204      0.0475     0.0093

98% C.I. for mu restrict - mu banned: ( 0.807,  1.0356)
```

EXHIBIT 11.6.6

11.5 A national fast-food restaurant chain plans to open a new store in one of three Arizona cities: Tempe, Chandler, or Mesa. In order to decide which city to select, the company wants to determine the average amount of money spent at fast-food restaurants each month in each city. To do this, the company selects a simple random sample of residents from each city using the city telephone directories. The sampled residents were then telephoned and asked, "How much money did your household spend at fast-food restaurants last month?"

a. Describe the three sampled populations for this example.

b. The responses to the question posed in the study are contained in the first three columns of the file **food.mtp** on the data disk. Summary statistics for the data are shown in Exhibit 11.6.7.

```
-----------------------------------------------------------------------
Variable        N      Mean    Median    TrMean     StDev    SEMean
Tempe          50    203.12    198.50    202.86     53.25      7.53
Chandler       50    231.96    241.10    232.36     53.53      7.57
Mesa           50    180.78    184.00    180.82     50.88      7.20
-----------------------------------------------------------------------
```

EXHIBIT 11.6.7

Based on this information, estimate the average amount of money spent per household last month at fast-food restaurants in each of the three cities.

c. Use the formula in Box 11.1.2 to compute a 95% confidence interval for the difference between the average money spent per household last month at fast-food restaurants for the cities of Chandler and Tempe.

d. The macro **pairwise.mac** is used to compute a set of confidence intervals for all pairwise differences of the population means using a 95% simultaneous confidence coefficient. The results are shown in Exhibit 11.6.8. Does it appear that there is any pairwise difference that exceeds $10 per month?

11.6 An airline provides meals for passengers on flights from Los Angeles to New York. Meals must be ordered 24 hours before each flight when the exact passenger count is unknown. The airline would like to determine if the time of day that a flight departs is useful

```
-----------------------------------------------------------------------
Two Sample T-Test and Confidence Interval

Twosample T for Tempe vs Chandler
             N      Mean     StDev   SE Mean
Tempe       50     203.1      53.3       7.5
Chandler    50     232.0      53.5       7.6

98% C.I. for mu Tempe - mu Chandler: ( -54.9,  -2.8)
T-Test mu Tempe = mu Chandler (vs not =): T= -2.70  P=0.0082  DF=  97

Two Sample T-Test and Confidence Interval

Twosample T for Tempe vs Mesa
          N      Mean     StDev   SE Mean
Tempe    50     203.1      53.3       7.5
Mesa     50     180.8      50.9       7.2

98% C.I. for mu Tempe - mu Mesa: ( -3.0,  47.7)
T-Test mu Tempe = mu Mesa (vs not =): T= 2.14  P=0.034  DF=  97

Two Sample T-Test and Confidence Interval

Twosample T for Chandler vs Mesa
             N      Mean     StDev   SE Mean
Chandler    50     232.0      53.5       7.6
Mesa        50     180.8      50.9       7.2

98% C.I. for mu Chandler - mu Mesa: ( 25.7,  76.6)
T-Test mu Chandler = mu Mesa (vs not =): T= 4.90  P=0.0000  DF=  97
-----------------------------------------------------------------------
```

EXHIBIT 11.6.8

for predicting the number of meals to order. The analyst assigned to work on this problem selects a simple random sample of 36 flights from a list of over 600 scheduled flights from Los Angeles to New York departing between the hours of 6:00 a.m. and 9:00 a.m. Pacific Standard Time (PST) during a three-month period. The airline serves breakfast to the passengers during these flights. A simple random sample of 36 flights is also selected from Los Angeles to New York flights that are scheduled to depart between the dinner hours of 5:00 p.m. and 8:00 p.m. PST during the same three-month period. After the completion of each sampled flight, the number of meals served was recorded. The sample means and the sample standard deviations for the two samples are reported in Table 11.6.3.

TABLE 11.6.3 Summary of Meal Counts

	Morning flights	**Evening flights**
Mean	102.06	113.75
Standard deviation	27.13	23.43

a. Define the two sampled populations.

b. Let μ_1 represent the average number of meals served for the population of morning flights and μ_2 represent the average number of meals served for the population of evening flights. Compute a 95% confidence interval for $\mu_1 - \mu_2$ using the formulas in Box 11.1.2. Interpret the result.

11.7 Faced with the problem of increasing student enrollment and decreasing funding for course support, the department chair of a statistics department must find new alternatives for teaching the introductory statistics class. There are presently 20 sections of size 40 scheduled each semester. The chair wants to compare two alternative formats to the present structure. The first alternative is to teach five sections of size 160. The second alternative is to teach five sections of size 160 and offer a weekly "help session." In order to compare these formats, the chair scheduled one section of each format for the fall semester during the class period from 9:15 a.m. to 10:30 a.m. on Tuesdays and Thursdays. Students were randomly assigned to one of the three formats, and three instructors with comparable teaching abilities were each randomly assigned to teach one of the sections. At the end of the semester a common 100-point final exam was given to all sections. The exam results are contained in the file **examscor.mtp** on the data disk. The number of students taking the exam was less than the total number enrolled because some students dropped out of the course during the semester.

a. Define the three sampled populations for this problem.

b. Summary measures for the three samples are shown in Exhibit 11.6.9. The scores for students in the section of size 40 are identified with the label `Small`. The scores for students in the section of 160 without the help session are identified with the label `Large`. The scores for students in the section of 160 with the help session are identified with the label `Help`.

Use this information to estimate the average test score for each of the three populations.

c. The macro **pairwise.mac** is used to compute a set of confidence intervals for all pairwise differences of the population means using a 95% simultaneous confidence coefficient. The results are shown in Exhibit 11.6.10. Interpret the results in this exhibit.

11.8 An experiment was conducted by the research and development department of a hair shampoo manufacturing company in which a new shampoo (shampoo 1) was compared with a leading brand of shampoo (shampoo 2). The variable of interest is the softness of the hair after washing the hair with a shampoo. Because the type of hair varies so widely among individuals, it was decided to use paired differences. Each sampled subject washed her hair once with each shampoo. The time between washings was one week, and the order of shampoo brand was determined by flipping a coin. After washing the hair, the degree of hair softness was evaluated by a trained hair specialist who assigned a score from 0 to 10, with 10 being the most soft and 0 the least soft. Forty subjects were randomly chosen from a population of consumers who had volunteered to participate in marketing research studies. The data are shown in Table 11.6.4 on page 365 and stored in the file **shampoo.mtp**. The subject number is in column 1, the score with shampoo 1 (Y) is in column 2, the score with shampoo 2 (X) is in column 3, and the differences ($D = Y - X$) are in column 4.

```
Population      N      Mean    Median    TrMean     StDev    SEMean
Small          35     70.17     71.00     70.35     12.79      2.16
Large         139     72.48     72.00     72.50     13.00      1.10
Help          142    82.831     83.00    82.891     8.025     0.673
```

EXHIBIT 11.6.9

```
------------------------------------------------------------------------
Two Sample T-Test and Confidence Interval

Twosample T for small vs large
          N      Mean     StDev   SE Mean
Small    35      70.2      12.8       2.2
Large   139      72.5      13.0       1.1

98% C.I. for mu small - mu large: ( -8.3,  3.7)
T-Test mu small = mu large (vs not =): T= -0.95  P=0.35  DF=  53

Two Sample T-Test and Confidence Interval

Twosample T for small vs help
          N      Mean     StDev   SE Mean
Small    35      70.2      12.8       2.2
Help    142     82.83      8.02      0.67

98% C.I. for mu small - mu help: ( -18.3,  -7.00)
T-Test mu small = mu help (vs not =): T= -5.59  P=0.0000  DF=  40

Two Sample T-Test and Confidence Interval

Twosample T for large vs help
          N      Mean     StDev   SE Mean
Large   139      72.5      13.0       1.1
Help    142     82.83      8.02      0.67

98% C.I. for mu large - mu help: ( -13.5,  -7.23)
T-Test mu large = mu help (vs not =): T= -8.01  P=0.0000 DF=  228
------------------------------------------------------------------------
```

EXHIBIT 11.6.10

a. Define the sampled population of items for this problem.

b. Let μ_1 represent the average softness score for the sampled population when all consumers are treated with shampoo 1 and μ_2 represent the average softness score for the sampled population when all consumers are treated with shampoo 2. Estimate $\mu_1 - \mu_2$.

c. Compute s_d, the standard deviation of the d_i values.

d. Compute a 95% confidence interval for $\mu_D = \mu_1 - \mu_2$.

e. What conclusions are suggested by the confidence interval in part (d)?

11.9 A utility company provides energy audits for its customers in order to help reduce utility bills. An energy audit is performed by inspecting a customer's residence and making recommendations on how to conserve energy better. To determine the effectiveness of this program, a simple random sample of 30 households was selected from the population of residential customers. The kilowatt hours (kwh) for the month of July 1996 were measured for each of the 30 households in the sample. An energy audit was then performed, and the average kwh were measured again in July 1997. The weather conditions in July 1996 and July 1997 were very similar. The differences between the kwh measurements in 1996 and 1997 were computed and summarized in Exhibit 11.6.11 on page 366. A positive difference means the 1996 measurement is greater than the 1997 measurement. A negative difference means the 1996 measurement is less than the 1997 measurement.

a. Define the sampled population of items in this problem.

b. Let μ_1 represent the average kwh for the population of residential customers in July 1996 and μ_2 represent the average kwh for the same population of residential customers in July 1997. Estimate $\mu_D = \mu_1 - \mu_2$ using Exhibit 11.6.11.

TABLE 11.6.4 Softness Scores for Two Hair Shampoos

Subject	Shampoo 1 Y	Shampoo 2 X	Difference D
1	8	7	1
2	7	7	0
3	7	7	0
4	7	7	0
5	6	6	0
6	4	3	1
7	7	7	0
8	7	7	0
9	7	7	0
10	6	6	0
11	5	5	0
12	6	6	0
13	7	7	0
14	8	7	1
15	7	6	1
16	8	9	−1
17	7	7	0
18	7	6	1
19	7	6	1
20	8	8	0
21	7	7	0
22	6	6	0
23	7	7	0
24	8	7	1
25	7	7	0
26	5	5	0
27	5	6	−1
28	8	7	1
29	6	6	0
30	6	6	0
31	7	7	0
32	9	9	0
33	8	8	0
34	9	10	−1
35	7	7	0
36	8	9	−1
37	6	6	0
38	7	7	0
39	6	6	0
40	5	5	0

c. Determine s_d using Exhibit 11.6.11 on page 366.

d. Compute a 95% confidence interval for μ. Caution: Read Box 7.2.1 concerning sample size.

e. What effect does the energy audit program have on electrical usage?

11.10 Clothing made of heat-resistant material is used by people who work near high-temperature ovens and furnaces in manufacturing plants. A company that manufactures heat-resistant fabrics is testing three different formulas for making heat-resistant suits that can withstand very high temperatures. Several batches of these fabrics are made using each of three formulas (formula 1, formula 2, and formula 3). Sample pieces are cut out and tested in a laboratory to determine the maximum temperature (in degrees Fahrenheit) they can withstand. The resulting data are stored in the file **heat.mtp**. The temperatures for formula 1 are in column 1, the temperatures for formula 2 are in column 2, and the temperatures for formula 3 are in column 3.

a. Define the three sampled populations for this problem.

b. The macro **pairwise.mac** is used to compute a set of confidence intervals for all pairwise differences of the population means using a 95% simultaneous confidence coefficient. The results are shown in Exhibit 11.6.12 on page 366. What conclusions do you draw from these data?

11.11 Coronary artery bypass surgery is routinely performed using veins in the body to replace clogged coronary arteries. These vein grafts can also become blocked with fat and cholesterol accumulation. To prevent blockage, some doctors recommend an LDL cholesterol-lowering drug for their patients. To investigate the potential benefits of this drug, a study was conducted using male patients who had undergone bypass surgery. A simple random sample of 133 patients was selected from all male patients who had undergone bypass surgery in a large hospital. These patients were randomly assigned to one of three treatment groups. Group I patients were put on a low-fat low-cholesterol diet with no medication. Group II patients were put on the same diet and asked to take 20 mg of a cholesterol-lowering drug each day. Group III patients were placed on the diet and asked to take 40 mg of the cholesterol-lowering drug each day. Five years after the date of the bypass surgery, doctors used angiograms to measure the percentage of blockage of the graft for each patient. The data for these 133 patients are stored in the file **bypass.mtp** on the data disk.

It is desired to compare the population means for three conceptual populations. The numbers in population 1 are the percentages of blockage of the grafts for all patients in the sampled population if they all had been put on a low-fat low-cholesterol diet with no medication. Populations 2 and 3 are defined in a similar manner. Let μ_i and σ_i denote the mean and the standard deviation of population i, for $i = 1, 2, 3$.

a. Compute confidence intervals for $\mu_1 - \mu_2$, $\mu_1 - \mu_3$, and $\mu_2 - \mu_3$ to provide a 97% confidence that all three intervals are simultaneously correct.

b. Do the data demonstrate that, on average, the aggressive treatment (diet plus 40 mg of drug each day) is superior to the other two treatments?

c. Do the data demonstrate that, on average, the moderate treatment (diet plus 20 mg of drug each day) is superior to the diet-only treatment?

```
---------------------------------------------------------------------
Variable        N     Mean    Median    TrMean     StDev    SEMean
Diff           30    297.3     264.5     289.5     302.1      55.1

Variable      Min      Max        Q1        Q3
Diff       -370.0   1062.0     162.0     434.8
---------------------------------------------------------------------
```

EXHIBIT 11.6.11

```
---------------------------------------------------------------------
Two Sample T-Test and Confidence Interval

Two sample T for formula1 vs formula2
            N      Mean      StDev    SE Mean
Formula1   32     424.5       10.4        1.8
Formula2   36     447.4       11.2        1.9

*98% CI for mu formula1 - mu formula2: ( -29.4,  -16.5)
T-Test mu formula1 = mu formula2 (vs not =): T= -8.73  P=0.0000  DF=  65

Two sample T for formula1 vs formula3
            N      Mean      StDev    SE Mean
Formula1   32     424.5       10.4        1.8
Formula3   30     425.9       11.0        2.0

*98% CI for mu formula1 - mu formula3: ( -8.1,  5.3)
T-Test mu formula1 = mu formula3 (vs not =): T= -0.52  P=0.61  DF=  59

Two sample T for formula2 vs formula3
            N      Mean      StDev    SE Mean
Formula2   36     447.4       11.2        1.9
Formula3   30     425.9       11.0        2.0

*98% CI for mu formula2 - mu formula3: ( 14.8,  28.2)
T-Test mu formula2 = mu formula3 (vs not =): T= 7.86  P=0.0000  DF=  62
---------------------------------------------------------------------
```

EXHIBIT 11.6.12

11.12 After witnessing a steady increase in traffic accident rates, a city council passed an ordinance requiring the speed limit on all major streets to be lowered by five miles per hour. One year after the ordinance took effect, accident statistics were compiled for six randomly chosen intersections of major streets. The total number of accidents at these intersections are given in Table 11.6.5 for the periods covering one year prior to and one year following the passage of the ordinance.

a. Let μ_1 denote the average number of accidents over all intersections of major streets in this city for the year prior to the ordinance and μ_2 denote the average number of accidents over all intersections of major streets in this city after the passage of the ordinance. Use a 90% confidence coefficient, and compute the appropriate nonparametric confidence interval for $\mu_1 - \mu_2$.

b. Do the data provide evidence that the average number of accidents has decreased since the ordinance was passed?

TABLE 11.6.5 Number of Accidents

Intersection	Before the ordinance	After the ordinance	Difference
Main and Broadway	12	9	3
Jefferson and Sheridan	8	7	1
Laurel and Maple	6	4	2
Busch and Plum	11	5	6
Peashway and Hiawatha	5	5	0
Princeton and Stover	10	4	6

11.7 SOLUTIONS FOR SELF-TEST PROBLEMS IN CHAPTER 11

11.1.4

a. The items in target population 1 are all male programmers who work for this corporation. Target population 2 consists of all female programmers who work for this corporation. Since the corporation has payroll records for all its employees, these records will be used to obtain the samples. Thus, any programmer who is presently employed can appear in the sample. Hence, the two sampled populations are identical to the two target populations.

b. The variable of interest is "Salary."

c. From Table 11.1.2, $n_1 = 12$, $\hat{\mu}_1 = \bar{y}_1 = 44.83$, $s_1 = 7.79$, $n_2 = 8$, $\hat{\mu}_2 = \bar{y}_2 = 41.87$, $s_2 = 7.16$, and $\hat{\mu}_1 - \hat{\mu}_2 = 2.96$. We estimate the average salary for male programmers is \$2,960 more than the average salary for female programmers.

d. The normal rankit-plots are shown in Figures 11.7.1 and 11.7.2.

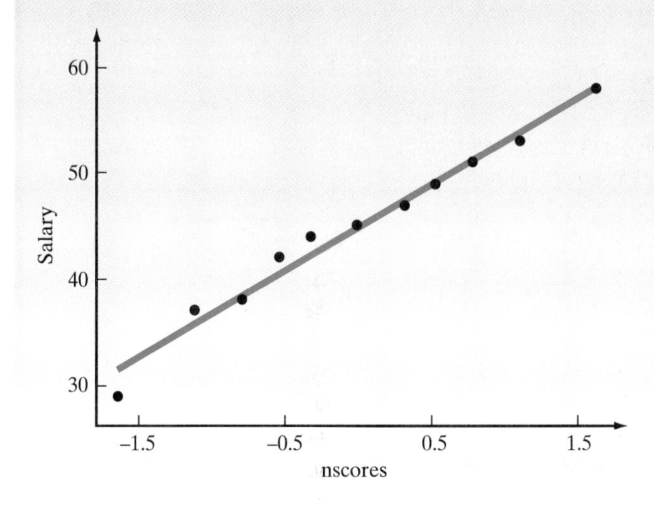

FIGURE 11.7.1
Normal Rankit-plot for Salaries of Male Programmers (in thousands of dollars)

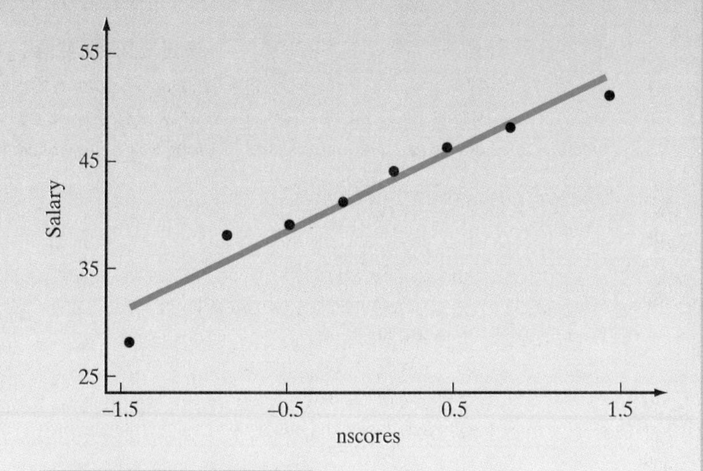

FIGURE 11.7.2
Normal Rankit-plot for Salaries of Female Programmers (in thousands of dollars)

Based on these plots, it seems reasonable to conclude that the two samples were selected from *normal* populations.

e. Using Line 11.1.1, we compute the degrees of freedom with $h_1 = (7.79)^2/12 = 5.06$ and $h_2 = (7.16)^2/8 = 6.41$. Using these values, we obtain

$$\text{DF} = \frac{(5.06 + 6.41)^2}{(5.06)^2/11 + (6.41)^2/7} = 16.04.$$

The table value from Table T-2 with a 95% confidence coefficient and DF = 16 is $T2 = 2.12$. Using Box 11.1.2, we obtain

$$L = 2.96 - 2.12\sqrt{5.06 + 6.41} = -4.22,$$

$$U = 2.96 + 2.12\sqrt{5.06 + 6.41} = 10.14.$$

The confidence is 95% that $\mu_1 - \mu_2$ is in the interval -4.22 to 10.14. Since values in this interval are both positive (men have a higher average salary than women) and negative (women have a higher average salary than men), the data do not provide sufficient evidence for us to decide whether or not the average salary in the population of men is larger than the average salary in the population of women. The confidence interval for the difference between two means can also be obtained using the following Minitab command after retrieving the data in the file **programr.mtp**:

```
MTB > twosample 95 c1 c2
```

The Minitab output is shown in Exhibit 11.7.1.

f. The value for s_p is $\sqrt{[(11)(7.79)^2 + (7)(7.16)^2]/18} = 7.55$. The confidence interval using Minitab is obtained using the commands and subcommand

```
MTB > twosample 95 c1 c2;
SUBC> pooled.
```

The output is shown in Exhibit 11.7.2.

The 95% confidence interval for $\mu_1 - \mu_2$ is from −4.3 to 10.2 (thousand dollars).

11.2.1

a. The Minitab command is

```
MTB > %a:\macro\pairwise  c1-c5  90
```

b. These confidence intervals tell us the following with an overall confidence of at least 90%:

(1) On average, brand 1 refrigerators require between 0.07 and 1.91 fewer repairs than brand 2, between 2.58 and 4.56 fewer repairs than brand 3, and between 0.20 and 2.04 fewer repairs than brand 5. Since the confidence interval for $\mu_1 - \mu_4$ contains both positive and negative values, it cannot be stated that these two means differ.

(2) On average, brand 2 requires between 1.57 and 3.59 fewer repairs than brand 3 and between 0.96 and 2.50 more repairs than brand 4. The data do not provide sufficient evidence for us to decide if μ_5 is larger than μ_2, equal to μ_2, or smaller than μ_2 since $\mu_2 - \mu_5$ contains negative values as well as positive values. More data will be needed to resolve this issue.

(3) Brand 3, on average, requires between 3.45 and 5.17 more repairs than brand 4 and between 1.44 and 3.46 more repairs than brand 5.

(4) On average, brand 4 requires between 1.09 and 2.63 fewer repairs than brand 5.

Combining all this information, we conclude that brands 1 and 4 require the fewest number of repairs on average. The worst brand is brand 3. The data are insufficient to decide if there is any difference between brand 2 and brand 5 or between brand 1 and brand 4.

```
---------------------------------------------------------------------
TWOSAMPLE T FOR C1 VS C2
          N        MEAN       STDEV    SE MEAN
Male     12       44.83        7.79        2.2
Female    8       41.87        7.16        2.5

95%  CI  mu Male - mu Female: (-4.2, 10.1)
---------------------------------------------------------------------
```

EXHIBIT 11.7.1

```
---------------------------------------------------------------------
Twosample T for male vs female
          N       Mean      StDev    SE Mean
Male     12      44.83       7.79        2.2
Female    8      41.87       7.16        2.5

95% C.I. for mu Male - mu Female: ( -4.3,  10.2)
T-Test mu male = mu female (vs not =): T= 0.86  P=0.40  DF=  18
Both use Pooled StDev = 7.55
---------------------------------------------------------------------
```

EXHIBIT 11.7.2

```
ANALYSIS OF VARIANCE
SOURCE     DF        SS        MS        F         p
FACTOR      4    412.55    103.14    43.31     0.000
ERROR     195    464.33      2.38
TOTAL     199    876.88
```

EXHIBIT 11.7.3

c. The ANOVA table in Exhibit 11.7.3 is obtained from Minitab.

The null and the alternative hypotheses are as follows:

$$H_0: \mu_1 = \mu_2 = \mu_3 = \mu_4 = \mu_5$$

versus

$$H_a: \mu_1, \mu_2, \mu_3, \mu_4, \mu_5 \text{ are not all equal.}$$

The P-value is 0.000 (to three decimal places). Since it is less than $\alpha = 0.01$, we reject the null hypothesis that all the means are equal. Thus, the data indicate (using $\alpha = 0.01$) that the five brands of refrigerators do not have equal population means. We came to this same conclusion using the simultaneous confidence intervals, but the intervals provided much more information than the test. All we can conclude from the statistical test is that differences exist among the population means. The confidence intervals give us information to determine which brand is the best, which brand is the worst, and the magnitude of the differences among all brand averages. The computations for the P-value assume all five populations are *normal* and that the standard deviations of the five populations are equal.

11.3.1

a. The sampled population consists of all members of the health club.

b. In your mind's eye, suppose that every member of the health club did as many push-ups as possible for one minute. The population consisting of the number of push-ups completed by each member is population 1. Now assume each member of the health club participates in the program for one month. The population consisting of the number of completed push-ups by each member after one month in the program is population 2.

c. The population of paired differences is the set of numbers that reports the difference in completed push-ups before the program and after one month in the program for each member of the health club. Notationally, if Y_i represents the completed push-ups in one minute before the program for member i and X_i represents the completed push-ups in one minute after one month of the program for the ith member, then the population of paired differences is $D_1, D_2, \ldots, D_N$, where $D_i = Y_i - X_i$.

d. To compute a 95% confidence interval with DF $= n - 1 = 34$, use $T2 = 2.03$. This provides the 95% confidence interval

$$L = -9.17 - (2.03)(8.06/\sqrt{35}) = -11.94,$$
$$U = -9.17 + (2.03)(8.06/\sqrt{35}) = -6.40.$$

We conclude (with 95% confidence) that the average number of completed push-ups after the program is from 6.40 to 11.94 more than before the program.

11.4.1

a. Since the houses are paired, the appropriate confidence interval is the Wilcoxon interval described in Box 11.4.2. The values for $d_i = y_i - x_i$ are shown in the last column of Table 11.4.8. The following steps are used to compute the interval.

(1) The $n(n + 1)/2 = 5(6)/2 = 15$ pairwise averages of the d_i are shown in Table 11.7.1.

TABLE 11.7.1 Pairwise Averages for Insulation Data

	208	**155**	**19**	**215**	**48**
208	208				
155	181.5	155			
19	113.5	87	19		
215	211.5	185	117	215	
48	128	101.5	33.5	131.5	48

(2) Next we arrange the pairwise averages in increasing order as shown below:

19, 33.5, 48, 87, 101.5, 113.5, 117, 128, 131.5, 155, 181.5, 185, 208, 211.5, 215.

(3) For a 90% confidence coefficient we use $T1 = 1.65$ and compute

$$q = \frac{n(n+1)}{4} - (T1)\sqrt{\frac{n(n+1)(2n+1)}{24}}$$

$$= \frac{(5)(6)}{4} - (1.65)\sqrt{\frac{(5)(6)(11)}{24}} = 1.38,$$

which we round down to 1.

(4) The lower bound of the confidence interval is the pairwise average in position 1 as we count up from the minimum value. This provides the lower bound $L = 19$ kilowatt hours. The upper bound is in position 1 as we count down from the maximum pairwise average. This provides the upper bound $U = 215$ kilowatt hours.

b. If it is assumed that the sampled populations are *normal*, then one can use the confidence interval for paired-differences data in Box 11.3.1. To compute this interval, $\bar{d} = 129$, $s_d = 90.79$, and $T2 = 2.13$ for DF = 4 and a 90% confidence coefficient. This provides the confidence interval from $L = 42.51$ to $U = 215.49$ kilowatt hours. This interval is shorter than the interval computed in part (a). This is generally the case since the additional information that the populations are *normal* reduces uncertainty.

11.4.2

a. The first target population consists of the attendance figures for all ball games in 1996. The second target population consists of the attendance figures for all ball games in 1997. Since attendance figures are available for all games in both years, the sampled populations are identical to the target populations.

b. This problem uses two simple random samples, so the appropriate confidence interval is the Mann-Whitney interval described in Box 11.4.1. The following steps are used to compute the interval.

(1) Compute the $n_1 \times n_2 = 5 \times 4 = 20$ possible differences $y_i - x_j$. These differences are shown in Table 11.7.2.

(2) The differences computed in step (1) are arranged in increasing order and displayed in Table 11.7.3.

TABLE 11.7.3 Ordered Differences for Attendance Data

−21	−17	−15	−14	−13
−12	−12	−11	−9	−8
−8	−8	−6	−6	−5
−4	−3	−2	0	1

(3) With $n_1 = 5$, $n_2 = 4$, and $T1 = 1.65$ for 90% confidence,

$$q = \frac{(5)(4)}{2} - (1.65)\sqrt{\frac{(5)(4)(5+4+1)}{12}}$$

$$= 10 - (1.65)\sqrt{16.67} = 3.26,$$

which is rounded down to 3.

(4) The lower bound is in position 3 counting up from the smallest difference. This value is $L = -15$. The upper bound is in position 3 counting down from the largest difference. This value is $U = -2$. With a confidence level of at least 90%, we can say that $\mu_1 - \mu_2$ is in the interval from $L = -15$ to $U = -2$. Since all values in the interval are negative, the evidence suggests the average attendance in 1997 is greater than it was in 1996.

c. If it is assumed that the sampled populations are *normal*, then one can use the confidence interval in Box 11.1.2. To compute this interval, $\bar{y}_1 = 24.60$, $\bar{y}_2 = 33.25$, $s_1 = 4.93$, $s_2 = 4.03$, $h_1 = 4.86$, $h_2 = 4.06$, and $T2 = 1.94$ for 90% confidence coefficient with DF = 6 (rounded down). This provides the confidence interval from $L = -14.5$ to $U = -2.8$. This interval is shorter than the interval computed in part (b). This is generally the case since the additional information that the populations are *normal* reduces uncertainty.

TABLE 11.7.2 Differences for Attendance Data

22 − 33 = −11	22 − 31 = −9	22 − 39 = −17	22 − 30 = −8
27 − 33 = −6	27 − 31 = −4	27 − 39 = −12	27 − 30 = −3
18 − 33 = −15	18 − 31 = −13	18 − 39 = −21	18 − 30 = −12
25 − 33 = −8	25 − 31 = −6	25 − 39 = −14	25 − 30 = −5
31 − 33 = −2	31 − 31 = 0	31 − 39 = −8	31 − 30 = 1

11.8 APPENDIX: ANALYSIS OF VARIANCE

In Section 11.2 we compared the means of $k = 3$ populations and introduced the analysis of variance (ANOVA) table. We discussed how the *P*-value in the table is used to test

$$H_0 : \mu_1 = \mu_2 = \mu_3 \quad \text{versus} \quad H_a : \mu_1,\ \mu_2,\ \mu_3 \text{ are not all equal.} \qquad \textbf{Line 11.8.1}$$

The *P*-value summarizes the evidence contained in the sample in support of rejecting H_0. This test is generally conducted assuming that the three populations are *normal* and that the standard deviations of the three populations are equal ($\sigma_1 = \sigma_2 = \sigma_3$).

A large number of calculations are needed to obtain the *P*-value for the test in Line 11.8.1 and computer programs are generally used for this purpose. It has been standard practice over the years to display some of the intermediate computations in the ANOVA table. Since *P*-values are generally no longer computed by hand, the ANOVA table no longer holds the central place in statistical computing that it once did. Yet, for the sake of tradition, ANOVA tables are still commonly exhibited in books, in computer output, and in technical publications. For this reason, we demonstrate how the numbers in the ANOVA table are computed using the small set of data in Table 11.8.1. Although this example contains only $k = 3$ populations, the method can be used for any number of populations. The data in Table 11.8.1 represent three simple random samples of size 5, 3, and 4 from populations 1, 2, and 3, respectively, where the *i*th population has mean μ_i and standard deviation σ_i. These data are stored in columns 1, 2, and 3 in the file **easyprob.mtp** on the data disk.

TABLE 11.8.1 Small Data Set

Sample from population 1	Sample from population 2	Sample from population 3
6	1	1
8	5	3
9	3	7
5		5
7		

We used the Minitab command below to obtain the ANOVA table shown in Exhibit 11.8.1. Only a portion of the Minitab output is shown in the exhibit.

```
MTB > aovoneway c1 c2 c3
```

```
-----------------------------------------------------------------------
Analysis of Variance

Source      DF        SS        MS        F        p
Factor       2     36.00     18.00     4.26    0.050
Error        9     38.00      4.22
Total       11     74.00
-----------------------------------------------------------------------
```

EXHIBIT 11.8.1

Notice that the ANOVA table has six columns, labeled `Source`, `DF`, `SS`, `MS`, `F`, and `p`, respectively. There are three rows, labeled `Factor`, `Error`, and `Total`, respectively.

The word `Factor` refers to population. The most useful numbers in the ANOVA table are the P-value, which appears in the last column (0.050), and the number 4.22 that appears in the row labeled `Error` and the column labeled `MS`. This number is the value of s_p^2, the pooled estimate of σ^2.

We now show how to compute the quantities in Exhibit 11.8.1 using the data in Table 11.8.1. Table 11.8.1 is reproduced in Table 11.8.2 with computations in lines (1) to (4) added to the bottom of the table. The computations shown in these lines are as follows:

1. n_i, the sample size of the ith column;
2. s_i^2, the variance of the numbers in the ith column (this is the estimate of σ_i^2);
3. s_i, the standard deviation of the numbers in the ith column (this is the estimate of σ_i);
4. $\text{DF}_i = n_i - 1$, the degrees of freedom of the ith column.

TABLE 11.8.2 Data for Easy Computing

	Sample from population 1	Sample from population 2	Sample from population 3
	6	1	1
	8	5	3
	9	3	7
	5		5
	7		
(1) n_i	5	3	4
(2) s_i^2	2.50	4.0	6.66
(3) s_i	1.58	2.00	2.58
(4) $\text{DF}_i = n_i - 1$	4	2	3

The total number of observations is $n_T = n_1 + n_2 + n_3 = 12$. As stated above, the test of Line 11.8.1 is conducted assuming that the three population standard deviations are equal to a common value σ. That is, $\sigma_1 = \sigma_2 = \sigma_3 = \sigma$. You might think it is odd to conduct a test concerning the equality of three population means by using a method called "the analysis of *variance*." Perhaps it would seem more reasonable to use the phrase "an analysis of means." However, the test of Line 11.8.1 is conducted by comparing two estimates of σ^2, denoted by $\hat{\sigma}_T^2$ and $\hat{\sigma}_E^2$. In Boxes 11.8.1 and 11.8.2, we explain how to compute $\hat{\sigma}_T^2$ and $\hat{\sigma}_E^2$.

BOX 11.8.1: HOW TO COMPUTE $\hat{\sigma}_T^2$

To compute $\hat{\sigma}_T^2$, combine the n_T observations into one sample and compute

$$\text{SSD}y = (y_1 - \bar{y})^2 + (y_2 - \bar{y})^2 + \cdots + (y_{n_T} - \bar{y})^2$$

where $\bar{y} = (y_1 + y_2 + \ldots + y_{n_T})/n_T$. The term SSD$y$ is called the *total* sum of squares of the deviation of the y_i values from the combined sample mean $\bar{y}$. The value of SSDy is located in Exhibit 11.8.1 in the row labeled `Total` and the column labeled `SS`. We compute

$$\hat{\sigma}_T^2 = \frac{\text{SSD}y}{n_T - 1}.$$

The denominator in $\hat{\sigma}_{\mathrm{T}}^2$, $n_{\mathrm{T}} - 1$, is called the "degrees of freedom for total." It is located in Exhibit 11.8.1 in the row labeled `Total` and the column labeled `DF`.

The value of $\bar{y}$ for the data in Table 11.8.2 is $\bar{y} = 5$, and the value for SSDy is SSD$y = (6-5)^2 + (8-5)^2 + \cdots + (5-5)^2 = 74.00$. Since $n_{\mathrm{T}} = 12$, $\hat{\sigma}_{\mathrm{T}}^2 = 74/(12-1) = 6.7273$.

BOX 11.8.2: HOW TO COMPUTE $\hat{\sigma}_{\mathrm{E}}^2$

The estimate $\hat{\sigma}_{\mathrm{E}}^2$ is equal to s_{p}^2 where s_{p} is defined in Line 11.2.3. That is,

$$\hat{\sigma}_{\mathrm{E}}^2 = \frac{(n_1 - 1)s_1^2 + (n_2 - 1)s_2^2 + (n_3 - 1)s_3^2}{n_1 + n_2 + n_3 - 3}.$$

This estimate is located in Exhibit 11.8.1 in the row labeled `Error` and the column labeled `MS`. The estimate $\hat{\sigma}_{\mathrm{E}}^2$ is often called the "error mean square." The numerator of $\hat{\sigma}_{\mathrm{E}}^2$ is called the "sum of squares for error" and is located in the row labeled `Error` and the column labeled `SS`. The denominator of $\hat{\sigma}_{\mathrm{E}}^2$ is called the "degrees of freedom for error" and is located in the row labeled `Error` and the column labeled `DF`. The degrees of freedom for error with k populations is equal to $n_1 + n_2 + \cdots + n_k - k$.

The value of the sum of squares for error in Exhibit 11.8.1 is 38.00, and the degrees of freedom for error is $12 - 3 = 9$. Thus, $\hat{\sigma}_{\mathrm{E}}^2 = 38.00/9 = 4.2222$.

If H_0 is true, then $\hat{\sigma}_{\mathrm{T}}^2$ is also an estimate of the common variance σ^2. If H_0 is not true, then $\hat{\sigma}_{\mathrm{T}}^2$ is an estimate of the common variance σ^2 plus a nonnegative quantity that depends on the means μ_1, μ_2, and μ_3. The quantity $\hat{\sigma}_{\mathrm{E}}^2$ is an estimate of the common variance σ^2 regardless of whether H_0 is true. Thus, if H_0 is true, we would expect the ratio

$$\frac{\hat{\sigma}_{\mathrm{T}}^2}{\hat{\sigma}_{\mathrm{E}}^2}$$

to vary around 1. If $\hat{\sigma}_{\mathrm{T}}^2/\hat{\sigma}_{\mathrm{E}}^2$ is "large," one would conclude that the numerator estimates something larger than σ^2 and hence that H_0 is not true. However, the question of importance is, "How large must $\hat{\sigma}_{\mathrm{T}}^2/\hat{\sigma}_{\mathrm{E}}^2$ be to reject H_0?" To answer the question, it is customary to examine the function of $\hat{\sigma}_{\mathrm{T}}^2/\hat{\sigma}_{\mathrm{E}}^2$ shown in Line 11.8.2.

$$F = \frac{\hat{\sigma}_{\mathrm{T}}^2}{\hat{\sigma}_{\mathrm{E}}^2}\left(\frac{n_{\mathrm{T}} - 1}{k - 1}\right) - \left(\frac{n_{\mathrm{T}} - 3}{k - 1}\right) \qquad \textbf{Line 11.8.2}$$

Line 11.8.2 is shown in Exhibit 11.8.1 in the row labeled `Factor` and the column labeled `F`. Although it is not obvious from the computations we have presented, the value of F (called F-value) in Line 11.8.2 is obtained by dividing the number in the row labeled `Factor` in the column labeled `MS` (the factor mean square) by the number in the row labeled `Error` in the column labeled `MS` (the error mean square).

The term $k-1$ is the degrees of freedom for `Factor`. The F-value in Line 11.8.2 for our example is

$$F = \frac{6.7273}{4.2222}\left(\frac{11}{2}\right) - \left(\frac{9}{2}\right) = 4.26.$$

Box 11.8.3 discusses the F-value and describes how it is used to obtain the P-value.

BOX 11.8.3: COMPUTATIONS OF F AND p

The fifth column of the ANOVA table is labeled `F`, which stands for `F` in Line 11.8.2. It is obtained by dividing the factor mean square by the error mean square. The symbol `F` is used in honor of one of the founding fathers of modern statistics, R. A. Fisher. The quantity F and the degrees of freedom for `Factor` and `Error` are used to determine the P-value for the test in Line 11.8.1. The degrees of freedom for factor is also called the **numerator degrees of freedom**, and the degrees of freedom for error is also called the **denominator degrees of freedom**. This terminology is used because the factor mean square appears in the numerator of F and the error mean square appears in the denominator. Using the F-value, the following Minitab commands can be used to obtain the P-value:

```
MTB > cdf F-value k3;
SUBC> F k1 k2.
MTB>  let k4=1-k3
MTB>  print k4
```

where the `F-value` is the computed value of F in Line 11.8.2, `k1` is the degrees of freedom for factor, and `k2` is the degrees of freedom for error. The function `cdf` computes 1 – (P-value) and places this value in `k3`. The P-value is placed in `k4`.

An approximate P-value can be obtained from Table T-5, which is on the last page of this book. For the appropriate degrees of freedom for the numerator (DFn) and the denominator (DFd), see if the computed value of F is greater than or equal to the table value corresponding to $\alpha = 0.01$. If so, then the P-value is less than or equal to 0.01. If the computed value of F is less than or equal to the table value corresponding to $\alpha = 0.05$, then the P-value is greater than or equal to 0.05. Otherwise the P-value is between 0.01 and 0.05.

For the data in Table 11.8.1, the Minitab commands are

```
MTB > cdf 4.26 k3;
SUBJ> F 2 9.
MTB> let k4=1-k3
MTB> print k4
```

From these commands, the P-value of 0.05 is obtained. Thus, H_0 would be rejected if the selected value of α is greater than this value.

CHAPTER 12

MULTIPLE REGRESSION

12.1 INTRODUCTION

In Chapter 9 we discussed simple linear regression and showed how a prediction function can be used to predict values of a variable Y using a single predictor variable X. In many problems a single predictor variable will not adequately predict Y and it is therefore necessary to employ a prediction function based on several predictor variables. For example, a physician may be interested in predicting a patient's blood pressure using two predictor variables: (1) the amount of a medicine administered, and (2) the patient's weight loss in a two-month period. The manager of a coal-burning power plant may want to predict the amount of sulfur dioxide that will be present tomorrow in a national park located 25 miles from the plant based on two predictor variables: (1) the amount of sulfur dioxide emitted at the plant today, and (2) today's relative humidity. The director of admissions at a university may want to predict a student's grade point average for the first year at the university based on four predictor variables: (1) the student's grade point average in high-school mathematics courses; (2) the student's grade point average in high-school English courses; (3) the student's score on the mathematics section of the Scholastic Aptitude Test; and (4) the student's score on the verbal section of the Scholastic Aptitude Test. You can think of other situations where two or more variables might be used to predict the value of a variable Y. This topic will be discussed in this chapter, and we begin with an example.

EXAMPLE 12.1.1

SCHOLARSHIP. An educational foundation at a major university awards scholarships to high-school graduates to help defray expenses during their first year at the university. The foundation would like to award scholarships to students who will earn a grade point average (GPA) of at least 2.80 the first year at the university. Since the scholarship is awarded before the student enters the university, the GPA will be predicted using two predictor variables: (1) high-school grade point average, and (2) score on an achievement test taken by all applicants. We will denote the predicted variable GPA with the uppercase letter Y and the two predictor variables as X_1 and X_2, where X_1 is the high-school grade point average and X_2 is the score on the achievement test. The Y value for a randomly chosen student with a high-school grade point average equal to X_1 and a score on the achievement test equal to X_2 is denoted by $Y(X_1, X_2)$. A formula used to predict Y using X_1 and X_2 is called a **prediction function** and denoted by $\widehat{Y}(X_1, X_2)$. Suppose that the prediction function for this problem is

$$\widehat{Y}(X_1, X_2) = -0.740 + 0.426X_1 + 0.0306X_2$$

"for X_1 from 1.50 to 4.00 and X_2 from 60 to 100."

Using this function, the predicted first-year GPA of a student who has a high-school grade point average of 2.78 and a score of 75 on the achievement test is

$$\widehat{Y}(2.78, 75) = -0.740 + 0.426(2.78) + 0.0306(75) = 2.74.$$

The terminology, notation, and concepts discussed in this chapter are similar to those used in Chapter 9. When more than one predictor variable is used to predict Y, the procedure is called **multiple regression** instead of simple regression. Throughout most of this chapter we will use two predictor variables for illustration. However, the results are easily generalized to include more than two predictor variables. As in simple linear regression, we say that $Y(X_1, X_2)$ is a value of Y to be chosen at random from the subpopulation of Y values for specified values of X_1 and X_2. $\widehat{Y}(X_1, X_2)$ is the predicted value of Y. The quantity $\mu_Y(X_1, X_2)$ is the average of the subpopulation of Y values for specified values of X_1 and X_2.

In your mind's eye, you can imagine any three-variable population of N items as displayed in Table 12.1.1.

TABLE 12.1.1 Representation of a Sampled Population of Three Variables, Y, X_1, and X_2

Population item number	Variable Y	Variable X_1	Variable X_2
1	Y_1	$X_{1,1}$	$X_{1,2}$
2	Y_2	$X_{2,1}$	$X_{2,2}$
3	Y_3	$X_{3,1}$	$X_{3,2}$
⋮	⋮	⋮	⋮
N	Y_N	$X_{N,1}$	$X_{N,2}$

The subscript i on the term Y_i means it is the Y value of the ith population item. For example, Y_6 is the Y value of the sixth population item. The subscripts i, j on the term $X_{i,j}$ denote the X value of the jth predictor variable for the ith population item. For example, $X_{4,2}$ denotes the X value of the second predictor variable (X_2) for the fourth population item. The population regression function is generally unknown, but linear functions of the form

$$\mu_Y(X_1, X_2) = \beta_0 + \beta_1 X_1 + \beta_2 X_2 \quad \text{“for specified values of } X_1 \text{ and } X_2\text{”} \qquad \textbf{Line 12.1.1}$$

often provide useful approximations in many applied problems. A function of this form is called a **population multiple linear regression function** or simply a **population regression function**. The value of $\mu_Y(X_1, X_2)$ in Line 12.1.1 changes β_1 units when X_1 increases one unit and X_2 is held fixed. Also, the value of $\mu_Y(X_1, X_2)$ changes β_2 units when X_2 increases one unit and X_1 is held fixed.

Since the values of β_0, β_1, and β_2 in Line 12.1.1 are typically unknown, they must be estimated from sample data. A schematic representation of a sample of size n from the population displayed in Table 12.1.1 is shown in Table 12.1.2. Lowercase letters are used to represent sample values.

TABLE 12.1.2 Schematic Representation of a Sample of Size n from a Three-Variable Population

Sample item	Variable Y	Variable X_1	Variable X_2
1	y_1	$x_{1,1}$	$x_{1,2}$
2	y_2	$x_{2,1}$	$x_{2,2}$
3	y_3	$x_{3,1}$	$x_{3,2}$
$\vdots$	$\vdots$	$\vdots$	$\vdots$
n	y_n	$x_{n,1}$	$x_{n,2}$

Using sample data, we obtain estimates of β_0, β_1, and β_2 and use them to estimate $\mu_Y(X_1, X_2)$. The estimates of β_0, β_1, and β_2 are denoted $\widehat{\beta}_0$, $\widehat{\beta}_1$, and $\widehat{\beta}_2$, respectively. The estimate of $\mu_Y(X_1, X_2)$ is

$$\widehat{\mu}_Y(X_1, X_2) = \widehat{\beta}_0 + \widehat{\beta}_1 X_1 + \widehat{\beta}_2 X_2.$$

The estimate $\widehat{\mu}_Y(X_1, X_2)$ is called the sample regression function. As in simple linear regression, we use the sample regression function as the prediction function. That is,

$$\widehat{Y}(X_1, X_2) = \widehat{\mu}_Y(X_1, X_2) = \widehat{\beta}_0 + \widehat{\beta}_1 X_1 + \widehat{\beta}_2 X_2.$$

EXAMPLE 12.1.2

SCHOLARSHIP. This is a continuation of Example 12.1.1, where a foundation awards scholarships to graduating high-school students. Suppose the population regression function is given by

$$\mu_Y(X_1, X_2) = \beta_0 + \beta_1 X_1 + \beta_2 X_2 \quad \text{“for } X_1 \text{ from 1.50 to 4.00 and } X_2 \text{ from 60 to 100.”}$$

A simple random sample of 35 students was obtained from the sampled population, and the values of Y, X_1, and X_2 were recorded for each student. The data are stored in the file **35scholr.mtp** on the data disk. This sample was used to compute the sample regression function

$$\widehat{\mu}_Y(X_1, X_2) = -0.740 + 0.426 X_1 + 0.0306 X_2.$$

This sample regression function is the estimate of the population regression function. Suppose a in the sampled population has a high-school grade point average of 3.00 and received a score of 86 on the achievement test. Since $\widehat{Y}(3.00, 86) = \widehat{\mu}_Y(3.00, 86)$, the first-year GPA of this student is predicted to be

$$\widehat{Y}(3.00, 86) = -0.740 + 0.426(3.00) + 0.0306(86) = 3.17.$$

THINKING STATISTICALLY: SELF-TEST PROBLEMS

12.1.1 An investigator wants to study the relationship of Y with the two predictor variables X_1 and X_2. It is assumed that the population regression function has the form

$$\mu_Y(X_1, X_2) = \beta_0 + \beta_1 X_1 + \beta_2 X_2$$

"for X_1 from 1.0 to 3.0 and X_2 from 45 to 70."

A simple random sample from the population was used to compute the sample regression function

$$\widehat{\mu}_Y(X_1, X_2) = 3.2 - 9.7X_1 + 0.15X_2.$$

a. What is the value of $\widehat{\mu}_Y(1.6, 62)$? Describe in words what this number represents.
b. Estimate the average of the subpopulation of Y values, where $X_1 = 1.5$ and $X_2 = 54$.
c. Suppose an item is to be chosen at random from the subpopulation where $X_1 = 1.5$ and $X_2 = 54$. What symbol represents the predicted value of Y for this item?
d. What is the predicted value of Y for a randomly chosen item from the subpopulation where $X_1 = 1.5$ and $X_2 = 54$?
e. What symbol represents the average value of the subpopulation of Y values where $X_1 = 2.8$ and $X_2 = 49$?
f. Estimate the change in $\mu_Y(X_1, X_2)$ when X_1 increases one unit and X_2 is held fixed.
g. Estimate $\mu_Y(2.0, 80)$.

12.2 INFERENCE IN MULTIPLE REGRESSION

In this section we discuss prediction functions and prediction intervals for

$$Y(X_1, X_2) \qquad \textbf{Line 12.2.1}$$

and point estimates and confidence intervals for

$$\beta_0, \quad \beta_1, \quad \beta_2, \quad \text{and} \quad \mu_Y(X_1, X_2). \qquad \textbf{Line 12.2.2}$$

The computations required in multiple regression are impractical to perform without a computer. In this section we use Minitab to perform the calculations. Other software packages will perform the same computations and provide output similar in format to the Minitab output.

The assumptions used in this chapter for statistical inference in multiple linear regression are described in Box 12.2.1. These assumptions are extensions of those for simple linear regression described in Box 9.3.1. As described in Section 9.5, sample data can be used to help decide if these assumptions are reasonable.

BOX 12.2.1: ASSUMPTIONS FOR STATISTICAL INFERENCE IN MULTIPLE LINEAR REGRESSION

Suppose an investigator is interested in studying the regression of Y on X_1 and X_2 in a three-variable population

$$(Y_1, X_{1,1}, X_{1,2}), (Y_2, X_{2,1}, X_{2,2}), \ldots, (Y_N, X_{N,1}, X_{N,2}).$$

To make valid statistical inferences using point estimates, prediction functions, confidence intervals, and prediction intervals, the three-variable population must satisfy assumptions (1), (2), and (3). In addition, the sample must be obtained by either method (4a) or (4b).

1. The subpopulation of Y values for each distinct value of the pair (X_1, X_2) has a mean denoted by $\mu_Y(X_1, X_2)$, where

$$\mu_Y(X_1, X_2) = \beta_0 + \beta_1 X_1 + \beta_2 X_2 \quad \text{"for specified values of } X_1 \text{ and } X_2\text{."}$$

The term "for specified values of X_1 and X_2" defines the values of X_1 and X_2 for which this linear regression function is valid.

2. For each distinct value of the pair (X_1, X_2), the subpopulation of Y values must have the same standard deviation. This common standard deviation is denoted by σ (that is, the standard deviation of every subpopulation is equal to σ).

3. The subpopulation of Y values for each distinct value of the pair (X_1, X_2) is a *normal* population. This assumption is not needed for point estimation but is required for confidence intervals and prediction intervals.

4a. A simple random sample of n items is selected from the sampled population, and Y, X_1, and X_2 are observed for each sampled item. The sample values are denoted as $(y_1, x_{1,1}, x_{1,2})$, $(y_2, x_{2,1}, x_{2,2})$, $\ldots$, $(y_n, x_{n,1}, x_{n,2})$. The sample values are schematically represented in Table 12.1.2.

4b. An investigator chooses any desired set of values for (X_1, X_2). A simple random sample of one or more Y values is then selected from each subpopulation determined by the selected values of (X_1, X_2) to obtain a sample size of n. As in (4a), the sample values are denoted as $(y_1, x_{1,1}, x_{1,2})$, $(y_2, x_{2,1}, x_{2,2})$, $\ldots$, $(y_n, x_{n,1}, x_{n,2})$.

In practice, it would be rare, if ever, that we encountered a problem where the population regression function has the *exact* form shown in (1) of Box 12.2.1. For this reason, the equation in (1) serves as a regression **model**. A regression model can be viewed as an idealized relationship between $\mu_Y(X_1, X_2)$ and X_1, X_2. Although the regression model in Box 12.2.1 may only be approximate, in many applications it can provide useful results. In a similar manner, assumptions (2) and (3) in Box

12.2.1 are unlikely to be satisfied *exactly*. They represent an idealized situation that might provide useful results in practical applications.

Minitab can be used to compute the sample regression function by entering the command

```
MTB > regress ci 2  cj ck
```

where Y is in column `i`, X_1 is in column `j`, and X_2 is in column `k`. The number `2` that appears between `ci` and `cj` means that there are two predictor variables. You might think of this command as stating "regress the predicted variable Y, which is in column `i`, on the two (`2`) predictor variables X_1 and X_2, which are in columns `j` and `k`, respectively." If there are m predictor variables (in columns `j`, . . . , `q`), the Minitab command is

```
MTB > regress ci m cj ...cq
```

Minitab subcommands can be used to compute confidence intervals and prediction intervals as we demonstrate later in this section. We now present an example.

EXAMPLE 12.2.1

SCHOLARSHIP. This is a continuation of Example 12.1.1, where an educational foundation awards scholarships to graduating high-school seniors. The foundation wants to predict first-year college GPA (Y) based on the following two predictors: (1) high-school grade point average (X_1), and (2) score on an achievement test (X_2). Suppose that the assumptions in Box 12.2.1 are satisfied. Hence, the model

$$\mu_Y(X_1, X_2) = \beta_0 + \beta_1 X_1 + \beta_2 X_2$$

"for X_1 from 1.50 to 4.00 and X_2 from 60 to 100"

is used to represent the population regression function. A simple random sample of 35 students was obtained from the population, and values of the three variables Y, X_1, and X_2 were obtained for each student. The sample data are stored in the file **35scholr.mtp**, where column 1 contains the sample number, and columns 2, 3, and 4 contain sample values of Y, X_1, and X_2, respectively. To obtain the prediction function and point estimates of the quantities in Line 12.2.2, retrieve the data and enter the following command in the Session window:

```
MTB > regress c2 2 c3 c4
```

The output is shown in Exhibit 12.2.1.

```
--------------------------------------------------------------------
Regression Analysis

The regression equation is
Y = - 0.740 + 0.426 X1 + 0.0306 X2

Predictor        Coef       Stdev    t-ratio        p
Constant      -0.7398      0.2692      -2.75    0.010
X1            0.42563     0.04798       8.87    0.000
X2           0.030582    0.003326       9.20    0.000

S = 0.1898       R-Sq = 87.2%     R-Sq(adj) = 86.4%
--------------------------------------------------------------------
```

EXHIBIT 12.2.1

The third line in Exhibit 12.2.1 contains the prediction function

$$\widehat{Y}(X_1, X_2) = -0.740 + 0.426X_1 + 0.0306X_2.$$

From this we obtain $\widehat{\beta}_0 = -0.740$, $\widehat{\beta}_1 = 0.426$, and $\widehat{\beta}_2 = 0.0306$. These estimates appear with greater numerical accuracy in the column labeled `Coef`. The first number in this column (in the row labeled `Constant`) is $\widehat{\beta}_0$; the number in the row labeled `X1` is $\widehat{\beta}_1$; and the number in the row labeled `X2` is $\widehat{\beta}_2$. The prediction function is also the sample regression function

$$\widehat{\mu}_Y(X_1, X_2) = -0.740 + 0.426X_1 + 0.0306X_2.$$

The number labeled `S` in the last line of Exhibit 12.2.1 is $\widehat{\sigma}$, the estimate of σ. This estimate is $\widehat{\sigma} = 0.1898$. The remaining quantities in Exhibit 12.2.1 will be explained as they are needed.

Using Minitab to Compute Prediction Intervals and Confidence Intervals in Multiple Regression

We now discuss how to use Minitab to compute prediction intervals for

$$Y(X_1, X_2) \qquad \textbf{Line 12.2.3}$$

and confidence intervals for

$$\beta_0, \quad \beta_1, \quad \beta_2, \quad \text{and} \quad \mu_Y(X_1, X_2). \qquad \textbf{Line 12.2.4}$$

Retrieve the data where the Y variable is in column `i` and the X_1 and X_2 variables are in columns `j` and `k`, respectively. Enter the following Minitab command and subcommands in the Session window:

```
MTB > regress ci 2 cj ck;
SUBC> predict a b;
SUBC> confidence = p.
```

The quantity `a` is the value of X_1 and `b` is the value of X_2 you want to use in $\widehat{Y}(X_1, X_2)$ and $\widehat{\mu}_Y(X_1, X_2)$. The quantity `p` is the desired confidence coefficient expressed as a percent. We illustrate with an example.

EXAMPLE 12.2.2

SCHOLARSHIP. We continue Example 12.2.1, where a foundation awards scholarships to high-school seniors. Suppose the foundation wants a 95% confidence interval for $\mu_Y(X_1, X_2)$ when $X_1 = 3.00$ and $X_2 = 86$. The sample data are stored in the file **35scholr.mtp** on the data disk; column 1 contains the sample number, column 2 contains Y, column 3 contains X_1, and column 4 contains X_2. Retrieve the data and enter the following command and subcommands in the Session window:

```
MTB > regress c2 2 c3 c4;
SUBC> predict 3.00 86;
SUBC> confidence = 95.
```

A portion of the output is shown in Exhibit 12.2.2.

```
Regression Analysis

The regression equation is
Y = - 0.740 + 0.426 X1 + 0.0306 X2

Predictor        Coef        Stdev     t-ratio       p
Constant      -0.7398       0.2692       -2.75   0.010
X1            0.42563      0.04798        8.87   0.000
X2           0.030582     0.003326        9.20   0.000

S = 0.1898        R-sq = 87.2%      R-sq(adj) = 86.4%

     Fit  Stdev.Fit          95.0% C.I.             95.0% P.I.
  3.1672     0.0379   ( 3.0899,  3.2444)   ( 2.7729,  3.5615)
```

EXHIBIT 12.2.2

Exhibit 12.2.2 is the same output shown in Exhibit 12.2.1 except that two additional lines have been added to the bottom of Exhibit 12.2.2. In the column labeled `Fit` is the number 3.1672. This is the value of both $\widehat{\mu}_Y(3.00, 86)$ and $\widehat{Y}(3.00, 86)$. In the column labeled `95.0% C.I.` is the 95% confidence interval for $\mu_Y(3.00, 86)$. From this we can state that

The confidence is 95% that $\mu_Y(3.00, 86)$ is between 3.0899 and 3.2444.

In the column labeled `95.0% P.I.` is the 95% prediction interval for $Y(3.00, 86)$. From this we can state that

The confidence is 95% that $Y(3.00, 86)$ is between 2.7729 and 3.5615.

Notice that the 95% prediction interval for $Y(3.00, 86)$ is wider than the 95% confidence interval for $\mu_Y(3.00, 86)$. This is because $Y(3.00, 86)$ is a value to be selected at random from the subpopulation where $X_1 = 3.00$ and $X_2 = 86$ and could be *any* value in that subpopulation. In contrast, $\mu_Y(3.00, 86)$ is a *single* value, namely the mean of the subpopulation.

Box 9.3.2 showed how the standard error for $\widehat{\beta}_1$ reported in Minitab can be used to compute a confidence interval for β_1 in simple linear regression. In a similar manner, Minitab reports the standard errors for $\widehat{\beta}_0, \widehat{\beta}_1$, and $\widehat{\beta}_2$ in a multiple regression analysis. These standard errors are found in the column labeled `Stdev` in Exhibit 12.2.2. A formula to compute a confidence interval for β_0, β_1, or β_2 using the Minitab output is

$$L = \texttt{Coef} - (T2)\texttt{Stdev}, \quad U = \texttt{Coef} + (T2)\texttt{Stdev}, \qquad \textbf{Line 12.2.5}$$

where `Coef` and `Stdev` are obtained from Exhibit 12.2.2 for the β parameter of interest, and $T2$ is obtained from Table T-2 with

$$\text{DF} = n - (\text{number of } \beta \text{ parameters in the population regression function}).$$

In Example 12.2.2 the number of β parameter is $= 3$, $n = 35$, and DF $= 35 - 3 = 32$. For a 95% confidence interval, $T2 = 2.04$. The 95% confidence interval for β_1 is

$$L = 0.42563 - (2.04)(0.04798) = 0.32775,$$

$$U = 0.42563 + (2.04)(0.04798) = 0.52351.$$

The 95% confidence interval for β_2 is

$$L = 0.030582 - (2.04)(0.003326) = 0.023796,$$

$$U = 0.030582 + (2.04)(0.003326) = 0.037368.$$

EXPLORATION 12.2.1

An electric utility company is investigating the possibility of charging a fixed monthly rate for household electricity. To study this problem, it is necessary to determine how the annual cost of electricity (Y) is related to three predictor variables: the size of the dwelling in square feet (X_1); the monthly household income in dollars (X_2); and the number of people living in the household (X_3). A simple random sample of 40 households was obtained from the population of all residential customers. The sample is stored in the file **electric.mtp** on the data disk. Column 1 contains the sample number, column 2 contains Y, column 3 contains X_1, column 4 contains X_2, and column 5 contains X_3. If the assumptions in Box 12.2.1 are satisfied (for three predictor variables), this implies that the population regression function is of the form

$$\mu_Y(X_1, X_2, X_3) = \beta_0 + \beta_1 X_1 + \beta_2 X_2 + \beta_3 X_3, \qquad \text{Line 12.2.6}$$

where X_1 is from 850 to 3,300 square feet, X_2 is from \$2,000 to \$6,000, and X_3 is from 1 to 8 people. The data were retrieved and the following Minitab command was entered in the Session window to produce the output in Exhibit 12.2.3:

```
MTB > regress c2 3 c3 c4 c5

---------------------------------------------------------

Regression Analysis

The regression equation is
Y = 379 + 0.248 X1 + 0.110 X2 + 28.0 X3

Predictor         Coef        Stdev     t-ratio        p
Constant        379.11        44.32        8.55    0.000
X1             0.24794      0.07169        3.46    0.001
X2             0.11031      0.03956        2.79    0.008
X3               27.96        11.69        2.39    0.022

S = 36.18        R-sq = 98.9%      R-sq(adj) = 98.8%
---------------------------------------------------------
```

EXHIBIT 12.2.3

1. Write the sample regression function.
   ▶**Answer:** *Using Exhibit 12.2.3, the sample regression function is*

$$\widehat{\mu}_Y(X_1, X_2, X_3) = 379 + 0.248X_1 + 0.110X_2 + 28X_3.$$

2. Interpret the estimate of β_2.
   ▶**Answer:** *For a given size of residence and a given number of residents, it is estimated that the average annual electric bill will increase \$0.11 for every \$1.00 increase in monthly household income.*

3. Locate the estimate of σ in Exhibit 12.2.3.
   ▶**Answer:** *The estimate of the standard deviation of any subpopulation is denoted by* `S` *in Exhibit 12.2.3. So* $\widehat{\sigma} =$ `S` $=$ \$36.18.

4. Estimate the average annual electric bill for the subpopulation of households that contain 950 square feet, have a monthly income of \$2,000, and have three people in residence.
   ▶**Answer:** *The estimate of* $\mu_Y(950, 2000, 3)$ *is*

$$\widehat{\mu}_Y(950, 2000, 3) = 379 + 0.248(950) + 0.110(2000) + 28.0(3) = \$918.60.$$

5. A family of four in the sampled population has a monthly income of \$3,600 and a 1,750-square-foot house. Predict this family's annual electric bill.
   ▶**Answer:** *The predicted value is* $\widehat{Y}(1750, 3600, 4) = \$1,321.00$.

6. Since the number of people living in a household changes from time to time (children move out or relatives visit), the utility company wants to determine the average annual electric bill if the predictor variable X_3 is not considered. In this case, suppose the population regression function is given by

$$\mu_Y(X_1, X_2) = \beta_0 + \beta_1 X_1 + \beta_2 X_2. \qquad \textbf{Line 12.2.7}$$

The values of β_0, β_1, and β_2 in Line 12.2.7 are generally not equal to the values of β_0, β_1, and β_2 in Line 12.2.6. However, to keep the notation simple, we will use the same symbols for β_0 and β_1 in each regression model. The following Minitab command was used with **electric.mtp** to produce Exhibit 12.2.4:

```
MTB > regress c2 2 c3 c4
```

```
-----------------------------------------------------------------

Regression Analysis

The regression equation is
Y = 359 + 0.338 X1 + 0.0921 X2

Predictor        Coef        Stdev     t-ratio         p
Constant       358.94        46.20        7.77     0.000
X1            0.33802      0.06477        5.22     0.000
X2            0.09212      0.04122        2.23     0.032

S = 38.42         R-sq = 98.7%      R-sq(adj) = 98.7%
-----------------------------------------------------------------
```

EXHIBIT 12.2.4

Write the sample regression function.

▶**Answer:** *The sample regression function is*

$$\widehat{\mu}_Y(X_1, X_2) = 359 + 0.338X_1 + 0.0921X_2.$$

7. Estimate the average annual electric bill for the subpopulation of households that contain 950 square feet and have a monthly income of \$2,000.

▶**Answer:** *The estimate is*

$$\widehat{\mu}_Y(950, 2000) = 359 + 0.338(950) + 0.0921(2000) = \$864.30.$$

8. The following Minitab commands were used to produce Exhibit 12.2.5:

```
MTB > regress c2 2 c3 c4;
SUBC> predict 950 2000;
SUBC> confidence=95.
```

```
-------------------------------------------------------------
Regression Analysis

The regression equation is
Y = 359 + 0.338 X1 + 0.0921 X2

Predictor        Coef       Stdev     t-ratio        p
Constant       358.94       46.20        7.77    0.000
X1            0.33802     0.06477        5.22    0.000
X2            0.09212     0.04122        2.23    0.032

S = 38.42          R-sq = 98.7%      R-sq(adj) = 98.7%

     Fit  Stdev.Fit    95.0% C.I.            95.0% P.I.
  864.29     23.46    (816.76, 911.83)  (773.06, 955.53)
-------------------------------------------------------------
```

EXHIBIT 12.2.5

What is the 95% confidence interval for the average annual electric bill for the subpopulation of households with 950 square feet and having a monthly income of \$2,000?

▶**Answer:** *The answer is contained in the last line of Exhibit 12.2.5. The 95% confidence interval for* $\mu_Y(950, 2000)$ *is \$816.76 to \$911.83.*

Variable Selection

When k predictor variables are available for use in a regression analysis, it is often of interest to determine how well a subset of m of the variables predicts Y. For example, an investigator might have $k = 4$ predictor variables available to perform a regression analysis. If a subset of two or three of these variables predicts Y about as well as the full set of variables, the investigator may want to use the subset of predictors for two reasons:

1. A regression model with fewer predictor variables is simpler and easier to interpret.
2. The cost of collecting and processing a sample decreases as the number of predictor variables decreases.

We illustrate with an example.

EXAMPLE 12.2.3

GPA SCORES. A director of admissions at a university wants to admit only those students who will obtain a grade point average (GPA) of at least 2.00 during their first academic year. Of course, the decision to admit a student must be made before the first-year GPA is known, and so it is necessary to predict GPA. The director knows from experience that four predictor variables do a good job of predicting the first-year GPA (Y). These four variables are

X_1 = the grade point average of all high-school mathematics courses,
X_2 = the grade point average of all high-school English courses,
X_3 = the score on the mathematics section of the Scholastic Aptitude Test, and
X_4 = the score on the verbal section of the Scholastic Aptitude Test.

The director would like to know how well a prediction function that uses only X_1 and X_2 predicts Y. If this two-variable prediction function can predict Y about as well as the prediction function based on all four predictor variables, then the Scholastic Aptitude Test would not be needed to predict first-year GPA.

In Section 9.4 it was stated that the standard deviation of the subpopulations in a regression analysis measures how well a prediction function predicts Y. Thus, to compare a prediction function based on a subset of m predictor variables with one that includes all k predictors, one often compares the standard deviations associated with each prediction function. We demonstrate this situation in the next example.

EXAMPLE 12.2.4

SCHOLARSHIP. This is a continuation of Example 12.2.1, where an educational foundation awards scholarships to graduating high-school seniors. In this problem it is desired to predict GPA (Y) for the first year in college using data collected for the $k = 2$ predictor variables: (1) high-school grade point average (X_1), and (2) score on an achievement test (X_2). Suppose that the assumptions in Box 12.2.1 are satisfied. Hence, the regression model is

$$\mu_Y(X_1, X_2) = \beta_0 + \beta_1 X_1 + \beta_2 X_2$$

"for X_1 from 1.50 to 4.00 and X_2 from 60 to 100."

A simple random sample of 35 students was obtained from the population. For convenience, the Minitab output in Exhibit 12.2.1 is reported again in Exhibit 12.2.6.

```
Regression Analysis

The regression equation is
Y = - 0.740 + 0.426 X1 + 0.0306 X2

Predictor        Coef       Stdev    t-ratio       p
Constant      -0.7398      0.2692      -2.75   0.010
X1            0.42563     0.04798       8.87   0.000
X2           0.030582    0.003326       9.20   0.000

S = 0.1898       R-Sq = 87.2%     R-Sq(adj) = 86.4%
```

EXHIBIT 12.2.6

The estimate of σ is labeled S in the last line of Exhibit 12.2.6. Thus, $\widehat{\sigma} = 0.1898$. Now suppose we want to determine how well a prediction function based only on X_1 compares with the prediction function in Exhibit 12.2.6. To do this, we need to estimate σ when the population regression function is represented by the model $\mu_Y(X_1) = \beta_0 + \beta_1 X_1$. Suppose the assumptions in Box 9.3.1 are satisfied. Exhibit 12.2.7 shows the sample regression function when X_1 is the only predictor variable.

```
Regression Analysis

The regression equation is
Y = 1.44 + 0.537 X1

Predictor        Coef       Stdev    t-ratio       p
Constant       1.4395      0.2399       6.00   0.000
X1            0.53690     0.08726       6.15   0.000

S = 0.3567       R-sq = 53.4%     R-sq(adj) = 52.0%
```

EXHIBIT 12.2.7

The estimate of σ in Exhibit 12.2.7 is $\widehat{\sigma} = 0.3567$. Since this value is greater than the value in Exhibit 12.2.6, the prediction intervals based on the prediction function in Exhibit 12.2.7 will generally be wider than those based on the prediction function in Exhibit 12.2.6.

The decision of which prediction function to select is based on several considerations. In general, it is desirable to use the fewest number of variables necessary to provide prediction intervals that are short enough to be useful. To make this decision, an investigator must rely on his or her practical knowledge of the subject matter and on other insights concerning the interrelationships of the predictor variables. Many statistical tools are available to help in this process. Such tools are discussed in textbooks on regression analysis under topics called "subset selection" or "variable selection."

THINKING STATISTICALLY: SELF-TEST PROBLEMS

12.2.1 Consider the population regression function $\mu_Y(X_1, X_2, X_3) = \beta_0 + \beta_1 X_1 + \beta_2 X_2 + \beta_3 X_3$ employed in Exploration 12.2.1, where Y is the annual cost of electricity, X_1 is the size of the dwelling in square feet, X_2 is the monthly household income in dollars, and X_3 is the number of people living in the household. The following Minitab command and subcommands are used to produce Exhibit 12.2.8:

```
MTB > regress c2 3 c3 c4 c5;
SUBC> predict 950 2000 3;
SUBC> predict 1000 3200 5.
```

```
---------------------------------------------------------------
Regression Analysis

The regression equation is
Y = 379 + 0.248 X1 + 0.110 X2 + 28.0 X3

Predictor        Coef        Stdev     t-ratio        p
Constant       379.11        44.32        8.55    0.000
X1            0.24794      0.07169        3.46    0.001
X2            0.11031      0.03956        2.79    0.008
X3              27.96        11.69        2.39    0.022

S = 36.18        R-sq = 98.9%      R-sq(adj) = 98.8%

     Fit  Stdev.Fit          95.0% C.I.           95.0% P.I.
  919.15      31.85   (854.54, 983.76)  (821.37, 1016.94)

     Fit  Stdev.Fit          95.0% C.I.           95.0% P.I.
 1119.85      59.27  (999.62, 1240.07) (978.99, 1260.71)
---------------------------------------------------------------
```

EXHIBIT 12.2.8

a. What is the estimate of β_3?
b. Find the 95% confidence interval for average annual electric bill for the subpopulation of households that contain 950 square feet, have a monthly income of $2,000, and have three people.
c. What is the predicted value and the 95% prediction interval for the annual electric bill for a household with 1,000 square feet, a monthly income of $3,200, and five people?

12.3 CHAPTER SUMMARY

KEY TERM

1. multiple regression

SYMBOL

1. $\mu_Y(X_1, X_2)$—population regression function with two predictor variables

KEY CONCEPTS

1. In the population regression function $\mu_Y(X_1, X_2) = \beta_0 + \beta_1 X_1 + \beta_2 X_2$, the value of $\mu_Y(X_1, X_2)$ changes β_1 units when X_1 increases one unit and X_2 is held fixed, and it changes β_2 units when X_2 increases one unit and X_1 is held fixed.
2. It is desirable to use the fewest number of predictor variables necessary in order to facilitate interpretation and reduce costs of a prediction function.

SKILLS

1. Use a computer software package to compute a prediction function using a sample with two or more predictor variables.
2. Use a computer software package to compute prediction intervals and confidence intervals for parameters in a multiple regression analysis.

12.4 CHAPTER EXERCISES

12.1 The Master of Business Administration (MBA) degree has become increasingly popular since the late 1970s. Admissions to these programs are based on information submitted by the applicant. Faculty members of a large business school questioned whether the criteria used to make admission decisions were good predictors of success in the program. In order to examine this situation, a faculty committee selected a sample of recent graduates and used regression analysis to predict performance based on information contained in the application form for admission. Program success is measured by a student's grade point average at graduation from the MBA program (Y). Although grade point average is not the only measure of success, it could be obtained easily from student records. In order to select the sample, a list was obtained that contained the names of all graduating MBA students for the past two years. A simple random sample of 49 students was selected from this list. A staff member obtained the application for each sampled student and recorded the values of Y and two predictor variables. The predictor variables are

X_1 = the applicant's undergraduate grade point average, and

X_2 = the applicant's score on the Graduate Management Admissions Test (GMAT).

The data are stored in the file **mbagrad.mtp** on the data disk. Column 1 contains the sample number, column 2 contains Y, column 3 contains X_1, and column 4 contains X_2. The population regression function is assumed to be of the form $\mu_Y(X_1, X_2) = \beta_0 + \beta_1 X_1 + \beta_2 X_2$ for X_1 from 2.00 to 4.00 and for X_2 from 300 to 750. The sample of 49 students produced the Minitab output shown in Exhibit 12.4.1.

```
Regression Analysis


The regression equation is
Y = 1.10 + 0.406 X1 + 0.00190 X2

Predictor        Coef        Stdev     t-ratio       p
Constant       1.0993       0.2532        4.34   0.000
X1            0.40627      0.08253        4.92   0.000
X2          0.0018966    0.0002908        6.52   0.000

S = 0.1624       R-sq = 71.2%      R-sq(adj) = 70.0%
```

EXHIBIT 12.4.1

a. What is the sample regression function?

b. Interpret the meaning of the number 0.40627 in the column labeled `Coef` in Exhibit 12.4.1.

c. Predict Y for a student selected at random from the sampled population if the student had an undergraduate grade point average of 3.60 and a GMAT score of 600.

d. Use Minitab or another software package to compute a 95% prediction interval for part (c).

e. Estimate the average grade point average at graduation for the subpopulation of students with an undergraduate grade point average of 3.60 and a GMAT score of 600.

f. Use Minitab or another software package to compute a 95% confidence interval for the parameter described in part (e).

g. Explain why the interval in part (f) is shorter than the interval in part (d).

h. If the predicted value of Y is greater than or equal to 3.60, the student is admitted to the MBA program. If a student earned an undergraduate grade point average of 3.40, compute the minimum score required on the GMAT exam to be admitted to the program.

i. Estimate the standard deviation of the subpopulation of students with an undergraduate grade point average of 3.60 and a GMAT score of 600.

12.2 An economics professor wants to investigate the relationship among several variables related to unemployment. One question of interest is whether a person's wage after a period of unemployment can be predicted by the person's wage immediately before unemployment and by the duration of the unemployment period. If so, such a prediction function would be useful in advising unemployed workers. The professor contacted the state unemployment insurance office and collected a simple random sample of employment records for 163 workers who had been unemployed and eventually found employment. The data are stored in the file **uiwage.mtp** on the data disk. The sample number is in column 1, the worker's hourly wage after the unemployment period (Y) is stored in column 2, the hourly wage before unemployment (X_1) is stored in column 3, and the number of days of the unemployment period (X_2) is stored in column 4. Suppose the assumptions in Box 12.2.1 are satisfied. Among other things, this implies that the population regression function is $\mu_Y(X_1, X_2) = \beta_0 + \beta_1 X_1 + \beta_2 X_2$ for X_1 from \$4.00 to \$18.00 per hour and for X_2 from 5 days to 365 days. The Minitab output for the regression analysis is shown in Exhibit 12.4.2.

a. What is the sample regression function?

b. Predict Y for a worker who had been earning \$5.00 an hour and was unemployed for 30 days.

c. Use Minitab or another software package to compute a 95% prediction interval for the predicted value in part (b).

d. Estimate the standard deviation that measures how well the two predictor variables X_1 and X_2 predict Y.

e. The Minitab output in Exhibit 12.4.3 reports the simple linear regressions of Y on X_1 and Y on X_2, respectively. If you could use only one of the two predictor variables to predict Y, which one would you select?

12.3 What factors determine the ability of a university professor to perform well in the classroom? Is it true that faculty who are strong in research are poor classroom instructors? These questions are often asked by students, parents, taxpayers, and governing boards of many state universities. The president of a large state university appointed a faculty committee to help answer these questions. The committee decided to use regression analysis to determine if it is possible to predict classroom teaching performance. All courses in the university are evaluated by students at the end of each term using the same evaluation form. Students evaluate faculty on many dimensions, but the committee decided to use the overall evaluation as the measure of teaching performance. The overall evaluation is a score based on a scale from 1 (poor) to 5

```
Regression Analysis

The regression equation is
Y = 0.444 + 0.836 X1 - 0.00403 X2

Predictor        Coef       Stdev    t-ratio       p
Constant       0.4437      0.3836       1.16   0.249
X1            0.83574     0.03581      23.34   0.000
X2          -0.004030    0.001759      -2.29   0.023

S = 1.568        R-sq = 78.0%      R-sq(adj) = 77.7%
```

EXHIBIT 12.4.2

```
Regression Analysis of Y on X1

The regression equation is
Y = 0.003 + 0.844 X1

Predictor        Coef       Stdev     t-ratio        p
Constant       0.0032      0.3362        0.01    0.993
X1            0.84414     0.03609       23.39    0.000

S = 1.588        R-sq = 77.3%      R-sq(adj) = 77.1%


Regression Analysis of Y on X2

The regression equation is
Y = 8.06 - 0.00823 X2

Predictor        Coef       Stdev     t-ratio        p
Constant       8.0603      0.4214       19.13    0.000
X2          -0.008233    0.003661       -2.25    0.026

S = 3.279        R-sq = 3.0%       R-sq(adj) = 2.4%
```

EXHIBIT 12.4.3

(outstanding). The predicted variable Y is the average overall evaluation for all courses taught by a faculty member during the past five years. Three predictor variables were used in the prediction function:

X_1 = the average number of journal articles published per year during the past five years;

X_2 = self-reported hours per week devoted to teaching activities during the past five years;

X_3 = self-reported hours per week devoted to research activities during the past five years.

Assumptions in Box 12.2.1 appear to be satisfied, so the population regression function is represented as $\mu_Y(X_1, X_2, X_3) = \beta_0 + \beta_1 X_1 + \beta_2 X_2 + \beta_3 X_3$ for X_1 from 0 to 5, for X_2 from 0 to 40 hours, and for X_3 from 0 to 40 hours. Data from a simple random sample of 63 faculty members were used to produce the Minitab output in Exhibit 12.4.4.

a. What is the sample regression function?

b. Predict Y for a faculty member who published an average of one journal article per year, spent 20 hours a week on teaching activities, and spent 15 hours a week on research activities.

c. Construct a 95% confidence interval for β_3.

d. If you were to predict Y for two professors who have the same values of X_1 and X_2, will the professor who spends more hours on research be predicted to have lower teaching evaluations?

e. What is the estimated standard deviation of the subpopulation of faculty with one journal article, 20 hours of teaching activities per week, and 20 hours of research activities per week?

12.4 A computer benchmarking group wants to study the relationship between the price of a Pentium personal computer (Y) and variables that describe the computer's key features. In particular, the three predictor variables are

X_1 = the number of megabytes (Mb) of RAM (memory),

X_2 = the speed of the processor in megahertz (MHz), and

X_3 = the number of gigabytes (Gb) of hard disk space.

All the computers are equipped with a state-of-the-art CD-ROM drive and a fast modem. The price of the monitor is not included. The group selects a simple random sample of 40 computer retail stores. From each selected store, the group chooses the most popular computer configuration sold by that store and obtains the selling price in dollars (Y). It also records the memory (X_1), speed (X_2), and disk space (X_3). The data

```
Regression Analysis

The regression equation is
Y = 3.50 + 0.172 X1 + 0.0118 X2 + 0.0103 X3

Predictor        Coef        Stdev     t-ratio        p
Constant       3.4961       0.1589       22.01    0.000
X1            0.17193      0.03589        4.79    0.000
X2           0.011840     0.005354        2.21    0.031
X3           0.010350     0.004501        2.30    0.025

S = 0.2351       R-sq = 54.5%      R-sq(adj) = 52.2%
```

EXHIBIT 12.4.4

are stored in the file **computer.mtp**, where the sample number is in column 1, Y is in column 2, X_1 is in column 3, X_2 is in column 4, and X_3 is in column 5. Suppose the assumptions in Box 12.2.1 are satisfied. Among other things, this implies that the regression of Y on X_1, X_2, and X_3 has the form

$$\mu_Y(X_1, X_2, X_3) = \beta_0 + \beta_1 X_1 + \beta_2 X_2 + \beta_3 X_3$$

for X_1 from 16 Mb to 64 Mb, for X_2 from 120 MHz to 200 MHz, and for X_3 from 1.2 Gb to 4.2 Gb. The Minitab output for the regression analysis is shown in Exhibit 12.4.5.

a. Estimate $\mu_Y(X_1, X_2, X_3)$.

b. Estimate the average price for the subpopulation of Pentium computers with 32 Mb of RAM, 166 MHz processor speed, and 3.6 Gb of hard disk space.

c. Use Minitab or another software package to compute a 95% confidence interval for the parameter estimated in part (b).

d. Estimate the additional cost, on average, for each additional gigabyte of disk space if X_1 and X_2 remain fixed. Construct a 95% confidence interval for this parameter.

e. Predict the price for a computer with 64 Mb of RAM, 200 MHz processor speed, and 4.2 Gb of hard disk space.

f. Use Minitab or another software package to compute a 95% prediction interval for part (e).

g. What is the interpretation of β_0?

12.5 An agency administers proficiency tests for those seeking jobs as office professionals. Two types of tests are administered. Test 1 consists of typing business correspondence. Test 2 concerns basic accounting principles. The agency director wants to determine if the scores on these tests can be used to predict starting salary for an office professional. To do this, she selects a simple random sample of 36 individuals who took the proficiency tests during the past two years and were hired as office professionals. For each sampled subject, she recorded the starting annual salary in thousands of

```
Regression Analysis

The regression equation is
Y = 792 + 5.54 X1 + 3.02 X2 + 35.4 X3

Predictor        Coef        Stdev     t-ratio        p
Constant       792.12        67.15       11.80    0.000
X1             5.5395       0.7981        6.94    0.000
X2             3.0178       0.3819        7.90    0.000
X3              35.39        14.89        2.38    0.023

S = 79.07        R-sq = 78.9%      R-sq(adj) = 77.1%
```

EXHIBIT 12.4.5

dollars (Y), the score on test 1 (X_1), and the score on test 2 (X_2). The data are stored in **proftest.mtp**, where the sample number is in column 1, Y is in column 2, X_1 is in column 3, and X_2 is in column 4. Suppose the assumptions in Box 12.2.1 are satisfied. This implies that the regression function of Y on X_1 and X_2 is of the form

$$\mu_Y(X_1, X_2) = \beta_0 + \beta_1 X_1 + \beta_2 X_2$$

for X_1 from 30 to 100 and for X_2 from 30 to 100. The Minitab output for the regression analysis is shown in Exhibit 12.4.6.

a. What is the sampled population?

b. Estimate the population regression function.

c. Predict the starting salary for an individual who scored 95 on test 1 and 98 on test 2.

d. Use Minitab or another software package to compute a 95% prediction interval for part (c).

e. Estimate the average salary for the subpopulation of individuals who scored 75 on both tests.

f. Use Minitab or another software package to compute a 95% confidence interval for the parameter in part (e).

```
Regression Analysis

The regression equation is
Y = 20.6 + 0.0429 X1 + 0.0988 X2

Predictor        Coef       Stdev    t-ratio       p
Constant      20.6234      0.9954      20.72   0.000
X1            0.04294     0.01360       3.16   0.003
X2           0.098763    0.009203      10.73   0.000

S = 1.116        R-sq = 82.7%     R-sq(adj) = 81.6%
```

EXHIBIT 12.4.6

12.5 SOLUTIONS FOR SELF-TEST PROBLEMS

12.1.1 a. $\widehat{\mu}_Y(1.6, 62) = -3.02$. This is the estimate of the average of the subpopulation of Y values where $X_1 = 1.6$ and $X_2 = 62$.

b. The estimate is $\widehat{\mu}_Y(1.5, 54) = -3.25$.

c. The symbol is $\widehat{Y}(1.5, 54)$.

d. The predicted value is $\widehat{Y}(1.5, 54) = -3.25$. This is the same value used to estimate the average of the subpopulation of Y values where $X_1 = 1.5$ and $X_2 = 54$.

e. The symbol is $\mu_Y(2.8, 49)$.

f. Suppose that X_1 changes from "a" to "$a + 1$" and X_2 remains constant. Then the estimated change in $\mu_Y(X_1, X_2)$ is

$$\begin{aligned}
&\widehat{\mu}_Y(a+1, X_2) - \widehat{\mu}_Y(a, X_2) \\
&= \widehat{\beta}_0 + \widehat{\beta}_1(a+1) + \widehat{\beta}_2 X_2 - [\widehat{\beta}_0 + a\widehat{\beta}_1 + \widehat{\beta}_2 X_2] \\
&= \widehat{\beta}_1 = -9.7.
\end{aligned}$$

g. Since $\mu_Y(X_1, X_2)$ is defined for X_1 from 1.0 to 3.0 and X_2 from 45 to 70, this estimate cannot be obtained from the given prediction function.

12.2.1 a. $\widehat{\beta}_3 = \$27.96$ per person.

b. We must obtain a 95% confidence interval for $\mu_Y(950, 2000, 3)$. The confidence interval reported in Exhibit 12.2.8 is from \$854.54 to \$983.76.

c. We must obtain the predicted value and a 95% prediction interval for $Y(1000, 3200, 5)$. From the last line of Exhibit 12.2.8, we obtain $\widehat{Y}(1000, 3200, 5) = \$1,119.85$. The 95% prediction interval is from \$978.99 to \$1,260.71.

CHAPTER 13

PROCESS IMPROVEMENT

13.1 INTRODUCTION

In previous chapters we discussed inferences concerning populations of items. In some situations the items in the population were produced sequentially by some well-defined series of actions. For instance, in Example 1.3.1 a manager wants to determine the proportion of aspirin bottles with broken pills. The bottles of aspirin were produced one at a time using a well-defined sequence of operations. In Example 2.5.3 a manager wants to determine if some recently manufactured truck tires satisfy certain technical specifications. The tires were produced one at a time following a particular set of operations. In both of these examples, the population of interest consists of a collection of items that were produced using a set of operations over a fixed period of time. This set of operations is called a **process**. If the process used to make bottles of aspirin is not working properly, then these bottles of aspirin might contain too many broken pills. If the process used to make tires does not operate properly, then the tires might not meet the desired level of quality.

In this chapter we discuss inferences concerning processes that produce items of a population. Inferences about a process are needed to determine if the process is operating properly. A process is an ongoing operation that converts **inputs** into **outputs**. The outputs of the process for any fixed period of time can be viewed as a population of items. We formalize this concept in Definition 13.1.1.

DEFINITION 13.1.1: PROCESS

A process is a series of actions or operations used to convert a set of inputs into a set of outputs. The values of a variable of interest measured on every output item from the process for a fixed-time interval form a population of numbers.

Process output can be examined to determine if the process is operating as it is designed to operate. This is done by periodically sampling output items and measuring variables that describe process performance. Periodic sampling is required so that problems can be detected as soon as they occur. The following example serves as an illustration.

EXAMPLE 13.1.1

DAMAGED ASPIRIN. In Example 1.3.1 the quality assurance manager of a company that manufactures medical products wants to determine the proportion of aspirin bottles in last week's output of 120,000 bottles that contains damaged tablets. The sampled population (and the target population) is the set of 120,000 aspirin bottles produced last week and stored in a warehouse. If 2% or more of the 120,000 bottles contain damaged tablets, all bottles will be sold under a different brand name at a reduced price. A simple random sample of 200 bottles is obtained from the warehouse and used to estimate the proportion of bottles in the population that contain damaged tablets.

Suppose that 1% of the first 60,000 aspirin bottles produced contained damaged tablets. At this point in time, something went wrong with the process and 4% of the next 60,000 bottles produced contained damaged tablets. The proportion of the 120,000 bottles in the warehouse that contains damaged tablets is therefore 2.5%. A simple random sample of 200 bottles is to be selected and used to estimate π, the proportion of bottles that contains damaged tablets. If the lower bound of the confidence interval for π exceeds 2%, the company will sell all 120,000 bottles in the warehouse at a reduced price. The disadvantage of this sampling procedure is that if the confidence interval suggests that the proportion of bottles with damaged tablets exceeds 2%, then the first 60,000 bottles will be sold at a reduced price even though the proportion of defective bottles in this group is less than 2%. To avoid this situation, a sampling procedure is needed that selects bottles periodically as they are produced. Such a procedure will detect a change in the production process and provide an opportunity to fix the problem more quickly. As an example, suppose that a simple random sample of one bottle was obtained from every 600 bottles as they were produced. In this case, we would expect the samples to indicate that the proportion of bottles with damaged aspirin for the first 60,000 bottles is less than 2%. However, beginning with bottle number 60,001, we would expect the samples to indicate that something has changed since 4% of all bottles now contain damaged tablets.

The procedure of sampling items periodically is better than waiting until all items have been produced. Not only would the company have been able to sell the first 60,000 bottles at the regular price, but it also would have received an indication when the proportion of bottles with damaged tablets changed from 1% to 4%. This would have allowed it to stop the production process and correct the problem. Using this sampling procedure, the same number of bottles would be examined (200), but each bottle would be examined at a different point in time.

The production of aspirin bottles in Example 13.1.1 can be viewed as a process in which a set of operations is performed to convert inputs (plastic bottles, cotton, aspirin tablets) into outputs (bottles of aspirin). Figure 13.1.1 represents the aspirin bottling process. Not all processes concern manufacturing operations, as we demonstrate in the next two examples.

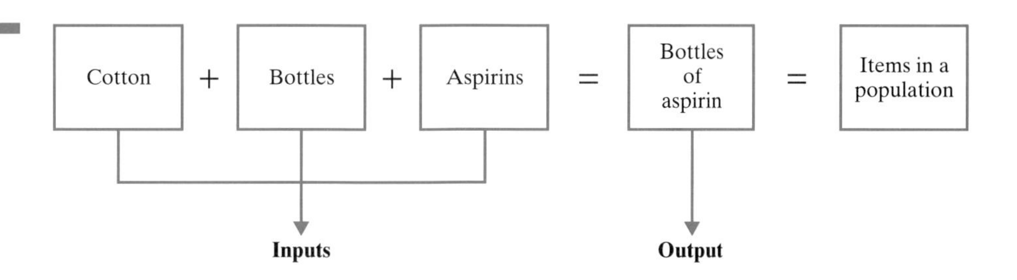

FIGURE 13.1.1 Representation of Aspirin Bottling Process

EXAMPLE 13.1.2

REPAIR OF COPIER MACHINES. A company that repairs copier machines has established the following process for servicing customers. When a customer calls to report a service problem, the dispatcher calls the repair person who is geographically closest to the customer. Once the repair person reaches the site of the broken copier, the problem is diagnosed and (hopefully) fixed. The output of the process is a fixed copier. Inputs include the service personnel, the parts used to repair the copier, and the broken copier. A variable of interest is the amount of time it takes to fix the copier (measured in minutes from the time of the customer's initial call).

EXAMPLE 13.1.3

BLOOD PRESSURE. A person with a certain disease will be given a new medication that is known to have seriously altered the systolic blood pressure of some patients. A person's blood pressure can be thought of as a measure that results from many inputs. Inputs include a person's diet, physical activity, and type of job. We view the blood pressure readings of the person who is to receive the drug as measurements of a process over time. By studying this process, we can determine if it changes when the new medication is administered. In order to do this, the person's blood pressure is taken at 7:00 a.m., 1:00 p.m., and 7:00 p.m. each day for 30 days before the drug is administered. This is done to benchmark the person's blood pressure without the drug. These three daily readings are averaged, and the average is recorded. After the drug is given to the patient, the blood pressure is monitored in the same manner each day to determine if the process measurements (blood pressure readings) have changed dramatically. If the blood pressure becomes too high or too low, the amount of the drug administered can be altered.

The goal of a process is to produce items that meet certain specifications. To ensure this goal is realized, the process must be periodically monitored. This is done by routinely collecting samples of process output and measuring variables that reflect process performance. By monitoring how values of these variables change from sample to sample, one can make inferences about the process performance. In the next two sections we describe several methods that are used to monitor a process.

THINKING STATISTICALLY: SELF-TEST PROBLEMS

13.1.1 Describe each of the following common activities as a process. List possible inputs and outputs. List variables that might be measured for the process output.

a. Taking a final examination in a course
b. Making popcorn in a microwave oven
c. Hitting a golf ball
d. Taking a course in statistics

13.2 PROCESS VARIABILITY AND ITS CAUSES

As stated in the previous section, a process is a series of actions or operations taken to convert a set of inputs into outputs. The performance of a process determines the cost and quality of process output. Process improvement is a key goal of any successful venture. This goal was fueled by the philosophy espoused by W. E. Deming, P. Crosby, K. Ishikawa, J. M. Juran, G. Taguchi, and others.

One of the main ideas Deming promoted is that process costs can be reduced and the quality of output improved by reducing process variability. For instance, consider a process used to manufacture ball bearings. The process is designed to produce ball bearings that are 2 millimeters in diameter. However, not every ball bearing will be *exactly* 2 millimeters in diameter. Some will be slightly larger and some slightly smaller. Thus, there is variability in the output of the process. The smaller the variability, the smaller the uncertainty about the performance of any single ball bearing, and the better the product quality. This suggests that the ability to improve a process depends on the ability to measure and explain variability. Consequently, statistical methods play a critical role in the successful implementation of quality improvement procedures. In the following two examples we demonstrate how the reduction of process variability leads to process improvement.

EXAMPLE 13.2.1

FILLING BOXES OF CEREAL. Suppose you purchase a box of dry cereal that has a stated net weight of 425 grams. You probably don't believe that the weight of the cereal is *exactly* 425 grams, but you certainly expect it to be close to this amount. The box of cereal you purchase is filled by a machine designed to package a specified weight of cereal in each box. Let's view this filling operation as a process. The inputs consist of empty cardboard boxes, plastic lining, cereal, a filling machine, and an operator. The output is a filled box of cereal. Because of variation in the process, not all net weights will be exactly the same, nor will they all be exactly equal to the net weight stated on the box.

The company that packages the cereal is legally required to be within some specified limit of the advertised net weight. Thus, the company must ensure that most of the boxes contain a weight of cereal that is at least as great as the legal limit. Let's assume the net weights of the 100,000 boxes of cereal produced in a given week (the process output) form a *normal* population with mean $\mu = 425$ grams and standard deviation $\sigma = 5$ grams. A relative frequency histogram of the population of net weights is shown in Figure 13.2.1.

If the advertised net weight is 425 grams, do you think the company would have trouble justifying this advertised net weight to a regulatory agency? To answer this question, we note that if the population of net weights is *normal* with mean

FIGURE 13.2.1
Histogram of Net Weights for 100,000 Boxes of Cereal

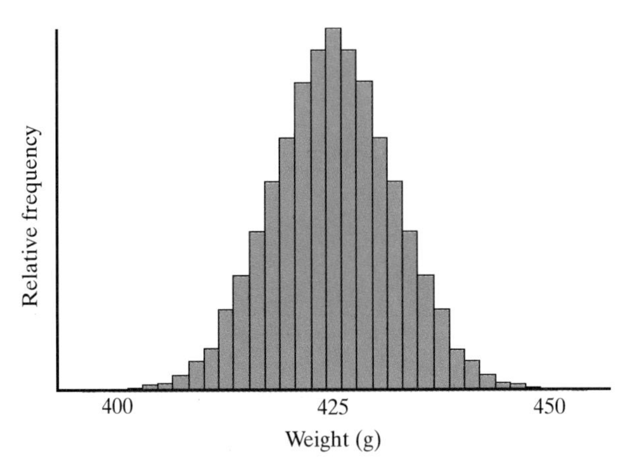

$\mu = 425$ grams, then 50% of the net weights will be less than the advertised weight of 425 grams. There would be serious legal problems if half of the filled boxes contained less than the advertised net weight. Suppose the legal requirement is that at least 99% of the boxes must contain 425 or more grams of cereal (425 grams is the advertised weight). The company will want to adjust the machine so that μ will be as small as possible and still meet the legal requirement. What is the smallest value of μ that will ensure that the legal requirement is satisfied? To answer this question, we must find the value of μ such that exactly 99% of the boxes will contain at least 425 grams of cereal (that is, exactly 1% of the boxes will contain less than 425 grams).

We can use the results of Chapter 6 to determine the required value of μ. Let Y represent the weight of a box of cereal in the population of 100,000 boxes. Since $\sigma = 5$ grams, we must determine the value of μ so that $\Pr(Y < 425) = 0.01$. This means that a proportion 0.01 of weights is less than 425 grams. Or equivalently, if one box is drawn at random from the population of boxes, the probability is 0.01 that it contains less than 425 grams of cereal. From line (21) in Table 6.2.1, it follows that 1% of the population values are less than $\mu - 2.326\sigma$. That is, $\Pr(Y < 425) = \Pr(Y < \mu - 2.326\sigma) = 0.01$. We now set $\mu - 2.326\sigma = 425$ and solve for μ. With $\sigma = 5$, this results in the equation $\mu - 2.326(5.0) = 425$, and the solution is $\mu = 436.63$ grams. Thus, if the machine is adjusted to dispense an average of 436.63 grams of cereal, then only 1% of the net weights will be less than the advertised weight of 425 grams, and the legal requirement will be satisfied.

Earlier in this section we stated that process costs can be reduced if process variability is reduced. Let's examine how this might occur in this example. Suppose the company purchases a new filling machine that has a standard deviation for net weights of $\sigma = 2$ grams. What value of μ is required so that only 1% of the boxes are underfilled? Performing the same computations as before, we let $\sigma = 2$ and solve for μ in the equation $\mu - 2.326(2) = 425$. This solution is $\mu = 429.65$ grams. Notice that μ is less for the machine with $\sigma = 2$ grams than it is for the machine with $\sigma = 5$ grams. Since the company's objective is to keep the average net weight (μ) as low as possible while still maintaining the advertised net weight in 99% of the boxes, the process with the smaller value of σ is better.

We can estimate the economic benefit that would result from using the new filling machine in the following manner. Observe that if $\mu = 429.65$ grams, then the amount of cereal needed to fill 100,000 boxes is 42,965,000 grams, or 42,965 kilograms. With the machine currently in use, the amount of cereal needed to fill 100,000 boxes is 43,663 kilograms. So the new machine will require 698 kilograms less cereal than the present machine for every 100,000 filled boxes. Most likely, the money saved on cereal will pay for the new machine in a short period of time and help reduce costs on a permanent basis.

EXAMPLE 13.2.2

PRODUCTION OF CRYSTALS. Crystals are used in many different types of electronic devices. The process used to produce a crystal is to combine various chemicals (inputs) and allow the crystal to grow for a given amount of time. The output of this process is a crystal. A variable of interest is the weight of the crystal. Suppose a customer orders 100 crystals and states he will only accept crystals that weigh between 11 and 13 grams. The weights of the crystals produced by a standard process can be viewed as a *normal* population with mean $\mu = 12$ grams and a standard deviation $\sigma = 1$ gram. What proportion of the total crystals produced by the process will be acceptable to the customer? That is, what proportion of crystals in the population are between 11 grams and 13 grams? Since the population of crystal weights is *normal*, we need to determine $\Pr(11 \leq Y \leq 13)$, where $Y_1, Y_2, \ldots, Y_N$ is the population of crystal weights resulting from the production process for a given period of time. Since $\mu = 12$ and $\sigma = 1$, we need to know the proportion of crystals that are within one standard deviation of the mean. From Box 6.2.1 we know that about 68.26% of the items in a *normal* population are within 1 standard deviation of the mean. Thus, we would expect 68.26% of the crystals to be within the required range of 11 grams to 13 grams.

Now suppose that a new process for growing the crystals provides a standard deviation of only $\sigma = 0.5$ grams. In this case, the values 11 grams and 13 grams are each two standard deviations from the mean of $\mu = 12$ grams. From Box 6.2.1 we know that 95.45% of the items in a *normal* population are within two standard deviations of the population mean. Thus, 95.45% of the crystals produced by the new process will be acceptable to the customer. As this problem demonstrates, reduction of the standard deviation makes the outcome of the process more predictable, so that fewer items are unacceptable to the customer.

The performance of a process is often described by the mean μ and the standard deviation σ of values for a variable measured on a population of output items. In fact, it is customary to refer to μ as the process mean and to σ as the process standard deviation. If a process is not working properly, it might have a mean that is not acceptable, a standard deviation that is too large, or both.

Typically, several factors affect the performance of a process. For instance, in Example 13.2.1 factors that cause net weight to vary from box to box include the moisture content of the cereal, the quality of the filling machine, and the ability of the machine operator. These factors cause variations of net weight from one box to another and are inherent in the process. It might be possible to reduce the effect of any of these factors by changing the process, but it will not be possible to eliminate them entirely. There may also be more unusual reasons for net weights to differ from box to box. For example, a part on the filling machine might be broken, the machine might be out of adjustment, or the operator might not be attentive.

The sources of variability in any process are classified as either **chance causes** or **assignable causes**. Chance causes of variability are inherent in any process and occur routinely but in an unpredictable or random manner. For instance, in Example 13.1.2 the time to repair a copier is affected by the variability in the technical abilities of the repair personnel. Similarly, the nature of the repair problem varies from site to site. The amount of variability attributed to chance causes can only be reduced by changing the process. In Example 13.1.2 the process could be changed by using only repair personnel with at least 10 years of experience. This would reduce the process variability attributed to inexperienced personnel. Alternatively, the company could decide that each repairperson will work on only one brand of copier so that the type of problems confronting the repairperson would be more similar from site to site.

Assignable causes are the second source of process variability. In contrast to chance causes, assignable causes are not inherent in the process and are not expected to occur on a regular basis. Suppose a repairperson in Example 13.1.2 has to wait an hour to get into a locked copier room while the person with the key is at lunch. This is not a common situation that occurs on a regular basis. Although assignable causes of variability occur on an infrequent basis, they can have a significant effect on process output.

BOX 13.2.1: CAUSES OF PROCESS VARIABILITY

Measurements of process output vary from item to item. Sources of variability are categorized as one of two types:

1. **Chance causes**. Chance causes of variability are inherent in any process. They are expected and occur routinely but in an unpredictable or random manner.
2. **Assignable causes**. Assignable causes of variability are due to unexpected events that may affect the process.

It is important to identify and classify the sources of variability in a process since different actions are required for each type in order to improve the process. If an assignable cause of variability is detected, the source of this variability must be identified and eliminated. One also needs to determine what can be done to prevent the situation from recurring. In our copier example, it might now be required for the dispatcher to confirm that the repairperson will have access to the copier before being sent to the site. In contrast, since chance causes of variability are expected, there should be no reaction to variability in output due to chance causes. Those who monitor a process need to recognize that the process is working as designed.

A process in which all variability is attributed to chance causes is called a **stable process**. A stable process is said to be **in control**. A process that also includes an assignable cause of variation is referred to as an **unstable process**, or a process that is **out of control**. In Section 13.3 we provide guidelines for deciding whether a process is stable or unstable.

BOX 13.2.2: TYPES OF PROCESSES

A process is categorized as one of two types:

1. a **stable process** consists of only chance causes of variation;
2. an **unstable process** has both assignable causes and chance causes of variation.

It should again be noted that every process has variability and that every process can be improved. If a process is unstable, it will be improved as soon as the assignable cause is identified and eliminated. A stable process can be improved by redesigning the process so that the magnitude of variability attributed to chance causes is reduced. In order to improve a process, it is necessary to first determine if a process is stable or unstable. In Section 13.3 we present methods that are used for this purpose.

THINKING STATISTICALLY: SELF-TEST PROBLEMS

13.2.1 Consider Example 13.2.1, where a company is required to have at least 99% of the net weights of cereal boxes greater than or equal to 425 grams. Suppose that the net weights of a population of boxes from filling machine 1 is a *normal* population with a standard deviation of $\sigma = 5$ grams. The company purchases another machine (machine 2) to replace machine 1. The net weights of boxes filled with machine 2 is a *normal* population with a standard deviation of $\sigma = 2$ grams. The cost of machine 2 is $5,000. Assume that all operating costs for machines 1 and 2 are equal and that the cost of a gram of cereal is $0.005.

a. How many boxes of cereal need to be filled using machine 2 before the cost savings in cereal will equal the $5,000 spent for machine 2?

b. Rework part (a) if the standard deviation of machine 2 is 1 gram instead of 2 grams.

13.2.2 Consider the daily process of driving to work or school. The amount of time it takes you to get from your house to your destination varies from day to day. What are some possible chance causes to explain this variability? What are some possible assignable causes?

13.3 SOME TOOLS FOR MONITORING A PROCESS

In this section we discuss some graphical tools that are used for monitoring processes. If you have not read Section 7.7, you might want to do so now because some of the material in this section uses results from Section 7.7. We first consider the task of collecting data from a process. The first step is to determine the variables that should be measured on the process output. For example, the variables of interest might be the net weight of a box of cereal, or the length of a bolt, or the number of damaged aspirin tablets in a bottle. The device used to measure a variable is called a **gauge**. It is imperative that a gauge be both accurate and precise. A gauge is accurate if it reports values that are neither consistently lower nor consistently higher than the true value. A gauge is precise if repeated measurements of the same item produce similar results. Data collected from a process will be useful only if variables can be measured accurately and precisely.

Once the variables are defined and suitable gauges are obtained, data can be collected. For monitoring the performance of a process, the usual practice is to select a sample of the output at regular intervals and measure the variables of interest. In previous chapters we have used simple random sampling to collect items from a population. In order to select a simple random sample, one must follow the procedure described in Box 2.5.1. However, practicalities do not always make it possible to collect observations from a process using simple random sampling. The following approach is sometimes used for selecting samples for monitoring a process. During each scheduled interval of time (such as each hour, or each day, or each shift), a sample of size n is selected. These may be n consecutive items at a given time point, or a simple random sample of n items from the collection of all items produced during the time interval, or n items selected during the time interval by some other method. Each such sample of n items is called a **rational subgroup**. Although rational subgroups are not strictly simple random samples as defined in Box 2.5.1, they are treated as such when performing data analyses. However, in order for this to be appropriate, someone familiar with the process must determine the interval at which samples are selected, and the sample size selected at each point in time.

To aid in monitoring a process, sample data are used to construct charts that display process performance. Some of the most commonly used charts are ***x*-bar ($\bar{x}$) charts**, **p-charts**, and **Pareto charts**. We now present a brief description of each of these charts.

x-bar charts

x-bar charts are used to monitor quantitative variables. An x-bar chart is constructed by first selecting random samples (rational subgroups) of size n from the process output at each of k different points in time. The values of n and k and the frequency with which samples are collected should be determined in consultation with someone familiar with the process. The k sample means are then plotted against time on a line plot, as described in Section 4.4. The pattern exhibited by the sample means is used to determine if a process is in control, as the following example describes.

EXAMPLE 13.3.1

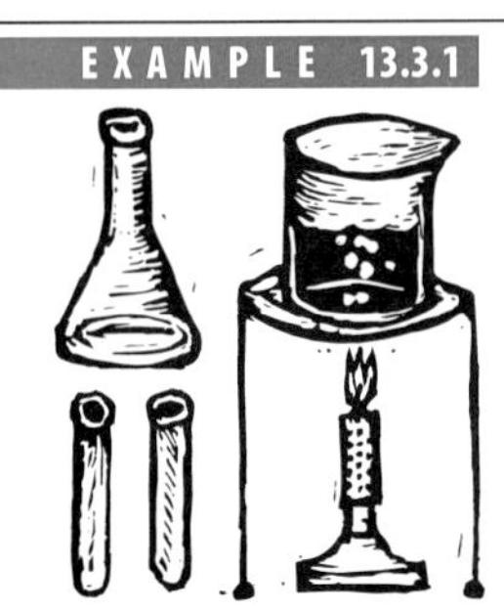

MONITORING A LIQUID CHEMICAL. A company that manufactures circuit boards uses a liquid chemical to wash the boards. The liquid chemical must be monitored periodically to ensure that the amount of acid contained in the mixture is within a specified range. Once each day an operator collects four beakers of liquid from random locations in a tank containing the chemical. The acid concentration of the liquid in each beaker is measured in micrograms per liter. Table 13.3.1 reports the measurements for a recent 30-day period. The $k = 30$ sample means are plotted against time in the line plot shown in Figure 13.3.1.

TABLE 13.3.1 Acid Concentrations (in micrograms per liter) for Samples of Size $n = 4$ for $k = 30$ Days

Day	Four observations of acid concentration	Day	Four observations of acid concentration
1	0.68, 0.73, 0.70, 0.69	16	0.71, 0.73, 0.69, 0.71
2	0.68, 0.68, 0.68, 0.68	17	0.70, 0.69, 0.71, 0.70
3	0.66, 0.66, 0.70, 0.70	18	0.66, 0.69, 0.69, 0.72
4	0.65, 0.68, 0.67, 0.68	19	0.70, 0.71, 0.71, 0.68
5	0.72, 0.70, 0.68, 0.70	20	0.70, 0.70, 0.69, 0.71
6	0.66, 0.67, 0.70, 0.65	21	0.70, 0.72, 0.68, 0.70
7	0.69, 0.69, 0.66, 0.72	22	0.68, 0.67, 0.65, 0.60
8	0.62, 0.64, 0.66, 0.64	23	0.66, 0.67, 0.66, 0.65
9	0.72, 0.70, 0.70, 0.72	24	0.66, 0.66, 0.66, 0.66
10	0.68, 0.70, 0.66, 0.68	25	0.68, 0.68, 0.66, 0.66
11	0.68, 0.68, 0.72, 0.72	26	0.72, 0.74, 0.73, 0.69
12	0.71, 0.69, 0.71, 0.69	27	0.71, 0.69, 0.73, 0.71
13	0.66, 0.66, 0.69, 0.71	28	0.63, 0.62, 0.64, 0.63
14	0.65, 0.67, 0.69, 0.67	29	0.67, 0.68, 0.66, 0.67
15	0.67, 0.67, 0.66, 0.68	30	0.66, 0.68, 0.68, 0.62

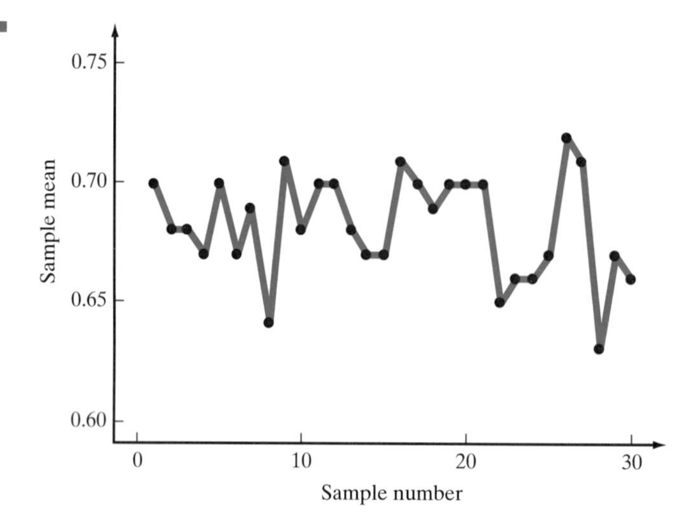

FIGURE 13.3.1
Line Plot of $k = 30$ Sample Means for Acid Concentrations (in micrograms per liter)

Figure 13.3.1 shows the daily variability in the acid concentration of the liquid chemical. To monitor this process properly, it is necessary to determine if the variability is due to chance causes or assignable causes. To do this, one needs to know the expected behavior when the process is in control (stable).

To determine if the process in Example 13.3.1 is in control, it is necessary to know how the chart appears when the process is stable and all variability is due to chance causes. Suppose that the process in Example 13.3.1 was operating satisfactorily for the 30-day period that preceded the time when the data in Table 13.3.1 were collected. Additionally, suppose that the population of measurements for this period was a *normal* population with an average concentration of $\mu = 0.68$ micrograms per liter and a standard deviation of $\sigma = 0.04$ micrograms per liter (for these 30 days, μ and σ are known). If the process is in control, we expect the population of acid concentrations for the next 30 days also to be a *normal* population with mean $\mu = 0.68$ and $\sigma = 0.04$. Thus, if the process is in control, the $k = 30$ samples in Table 13.3.1 can be viewed as a set of 30 simple random samples of size $n = 4$ from a *normal* population with mean $\mu = 0.68$ and $\sigma = 0.04$. As such, the 30 sample means in Figure 13.3.1 can be viewed as a simple random sample of size $k = 30$ from a derived population of sample means based on $n = 4$ observations (see Section 7.7). From Box 7.7.2 we know that a derived population of sample means has a mean of $\mu_{\text{SM}} = \mu$ and a standard deviation of $\sigma_{\text{SM}} = \sigma/\sqrt{n}$ (with $f = 1$ in Line 7.7.3). Furthermore, if the sampled population is *normal*, then the derived population of sample means is *normal*. The mean of the derived population when the process is in control in Example 13.3.1 is $\mu_{\text{SM}} = \mu = 0.68$, and the standard deviation is $\sigma_{\text{SM}} = \sigma/\sqrt{n} = 0.04/\sqrt{4} = 0.02$.

One way to determine if a process is in control is to see if the appropriate proportion of sample means fall within $3\sigma_{\text{SM}}$ units of the population mean μ_{SM}. If the derived population of sample means is *normal*, Box 6.2.1 states that 99.73% of the sample means are between $\mu_{\text{SM}} - 3\sigma_{\text{SM}}$ and $\mu_{\text{SM}} + 3\sigma_{\text{SM}}$. Thus, if the process is in control, we expect 99.73% of the 30 sample means in Figure 13.3.1 to be no more than $3\sigma_{\text{SM}}$ units from μ_{SM}. Figure 13.3.2 shows the line plot in Figure 13.3.1 with a horizontal line at the value of $\mu_{\text{SM}} = 0.68$. Lines are also drawn at $\mu_{\text{SM}} - 3\sigma_{\text{SM}} = 0.68 - 3(0.04/\sqrt{4}) = 0.62$ and $\mu_{\text{SM}} + 3\sigma_{\text{SM}} = 0.68 + 3(0.04/\sqrt{4}) = 0.74$.

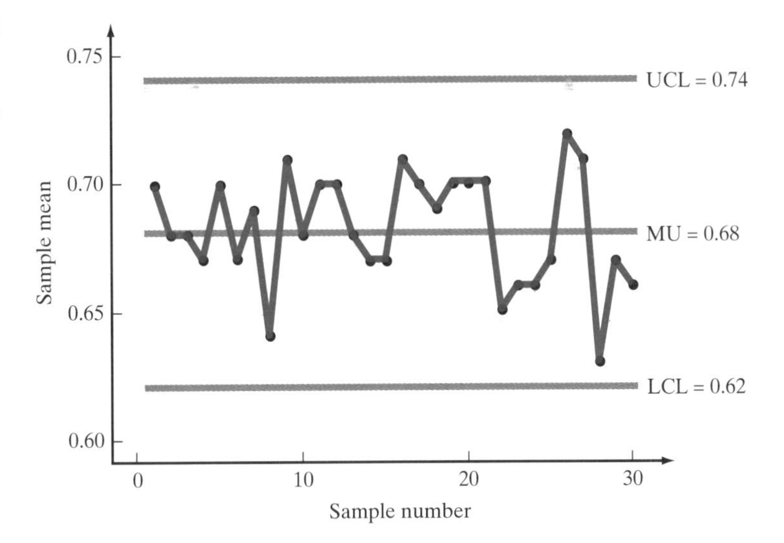

FIGURE 13.3.2
x-bar Control Chart for Acid Concentration (in micrograms per liter)

Figure 13.3.2 is called an ***x*-bar control chart**. The line at $\mu_{SM} - 3\sigma_{SM}$ is labeled LCL and called the **lower control limit**. The line at $\mu_{SM} + 3\sigma_{SM}$ is labeled UCL and called the **upper control limit**. The line at μ_{SM} is labeled MU (which stands for μ) and called the **center line**. The formulas for the control lines on an x-bar chart are presented in Box 13.3.1.

BOX 13.3.1: CONTROL LIMITS FOR AN x-BAR CHART

Suppose that when a process is in control the values of the variable of interest form a *normal* population with mean μ and standard deviation σ. A set of k random samples of size n is selected from the process at each of k different points in time. The set of k sample means represents a simple random sample of k observations from a derived population of sample means with mean $\mu_{SM} = \mu$ and standard deviation $\sigma_{SM} = \sigma/\sqrt{n}$. The k sample means are plotted on an x-bar control chart. Horizontal lines are drawn on the chart at the center line (MU), the upper control limit (UCL), and the lower control limit (LCL). Formulas for these limits are

1. $\text{MU} = \mu_{SM} = \mu$,
2. $\text{UCL} = \mu_{SM} + 3\sigma_{SM} = \mu + 3\sigma/\sqrt{n}$, and
3. $\text{LCL} = \mu_{SM} - 3\sigma_{SM} = \mu - 3\sigma/\sqrt{n}$.

Additional lines are sometimes drawn at $\mu_{SM} - \sigma_{SM}$, $\mu_{SM} + \sigma_{SM}$, $\mu_{SM} - 2\sigma_{SM}$, and $\mu_{SM} + 2\sigma_{SM}$.

As long as a process is stable, we expect 99.73% of the sample means to be within $3\sigma_{SM}$ units of μ_{SM}. Besides determining if any sample means are outside the upper and lower control limits, one should also look for unusual patterns such as a persistent upward or downward trend. If sample means appear outside of either control limit or display a trend, it is a signal that the process may no longer be stable. When such a situation arises, the process should be stopped and the underlying assignable cause determined. Figure 13.3.2 suggests that the process being monitored in Example 13.3.1 is in control (stable).

EXAMPLE 13.3.2

MONITORING A LIQUID CHEMICAL. We continue the problem introduced in Example 13.3.1 by reporting data for a different set of $k = 30$ samples of size $n = 4$. The sample means of the 30 samples are reported in Table 13.3.2. An x-bar control chart with $\mu_{\text{SM}} = 0.68$ and $\sigma_{\text{SM}} = 0.02$ is shown in Figure 13.3.3.

TABLE 13.3.2 Sample Means of Acid Concentration (in micrograms per liter) for a New Set of $k = 30$ Samples of Size $n = 4$

Time period	Acid concentration	Time period	Acid concentration
1	0.67	16	0.67
2	0.68	17	0.68
3	0.75	18	0.69
4	0.66	19	0.66
5	0.66	20	0.75
6	0.68	21	0.66
7	0.68	22	0.68
8	0.67	23	0.67
9	0.68	24	0.67
10	0.67	25	0.67
11	0.67	26	0.66
12	0.67	27	0.66
13	0.68	28	0.73
14	0.67	29	0.67
15	0.67	30	0.68

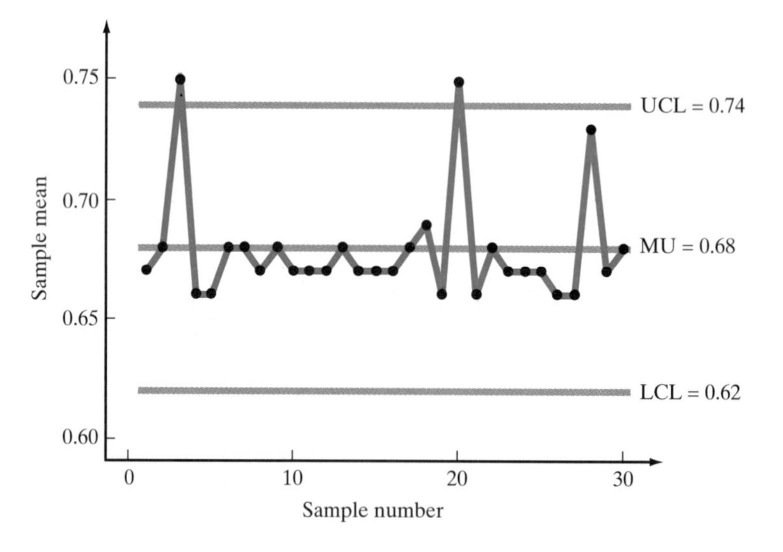

FIGURE 13.3.3
Control Chart of Sample Means in Table 13.3.2

The center line in Figure 13.3.3 is located at $\mu_{\text{SM}} = 0.68$. The upper and lower control limits are $\text{UCL} = 0.68 + 3(0.02) = 0.74$ and $\text{LCL} = 0.68 - 3(0.02) = 0.62$. Two points (points 3 and 20) are above the UCL. Since we do not expect points above the UCL, it appears that this process is unstable.

In situations where a process is well established, or when the process parameters are specified by the investigator, the values of μ and σ will be known. In situations where the parameters μ and σ are not specified, these values are estimated using samples from the process. Modifications to the control limits in Box 13.3.1 are required in this situation. This topic is discussed in texts on statistical quality control.

p-charts

As noted earlier, the x-bar chart is used to monitor a quantitative variable. Often, we wish to monitor a qualitative variable that is measured on process output. For example, we may wish to monitor the proportion of defective parts in a manufacturing process. The variable "Part quality" has two possible values, *defective* and *not defective*. To monitor a qualitative variable, one plots sample proportions computed from samples collected over time. The resulting chart with accompanying control limits is called a **p-chart**. We now present an example where a p-chart is used to monitor a service process.

EXAMPLE 13.3.3

Customer satisfaction. A utility company conducts a monthly survey to determine the level of customer satisfaction. Each month a simple random sample of 100 customers is asked several questions concerning the customer's level of satisfaction with the company's performance. Customers are asked to assign an overall rating on a scale from 1 to 5, where 1 is extremely unsatisfied and 5 is extremely satisfied. A p-chart is used to monitor the proportion of sampled customers who assign an overall rating of either 4 or 5. For example, if 30 customers in the sample of 100 assign an overall rating of either 4 or 5, then the sample proportion plotted on the p-chart is 0.30. The number of customers who assigned a rating of 4 or 5 for each of the past 12 months is shown in Table 13.3.3.

TABLE 13.3.3 Number of Customers Assigning a Rating of Either 4 or 5 for the Past 12 Months

Time period	Number assigning 4 or 5	Sample size	Time period	Number assigning 4 or 5	Sample size
1	64	100	7	57	100
2	59	100	8	58	100
3	59	100	9	64	100
4	57	100	10	60	100
5	60	100	11	61	100
6	61	100	12	59	100

Suppose that the average proportion of customers who assigned a rating of either 4 or 5 during the past 48 months was 0.60. We select this value to represent the population proportion when the process is stable. The formulas used to compute the control limits for a p-chart are based on the central limit theorem for zero–one populations described in Box 7.7.4. Let π represent the mean of the zero–one population of outputs when the process is in control. Box 7.7.4 states that if n is "sufficiently large," then the derived population of all possible sample proportions is a *normal* population with mean π and standard deviation $\sqrt{\pi(1-\pi)/n}$. So, when the process is in control, 99.73% of the sample proportions should be within $3\sqrt{\pi(1-\pi)/n}$ units of π. Thus, the control limits are

$$\text{UCL} = \pi + 3\sqrt{\frac{\pi(1-\pi)}{n}},$$

$$\text{LCL} = \pi - 3\sqrt{\frac{\pi(1-\pi)}{n}}. \qquad \textbf{Line 13.3.1}$$

For our example, $\text{UCL} = 0.60 - 3\sqrt{0.60(1 - 0.60)/100} = 0.75$, and $\text{LCL} = 0.60 - 3\sqrt{0.60(1 - 0.60)/100} = 0.45$. The p-chart for the data in Table 13.3.3 is shown in Figure 13.3.4.

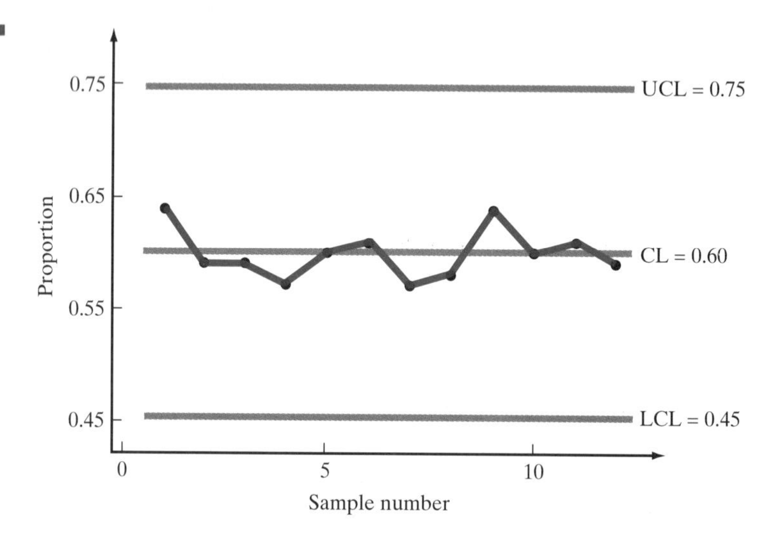

FIGURE 13.3.4
p-Chart of Sample Proportions in Table 13.3.3

Since all sample proportions are within the control limits in Figure 13.3.4, the process appears to be stable.

Pareto charts

A useful tool for identifying causes of process variability is the **Pareto chart**. A Pareto chart is a bar chart that reports the frequency of each identifiable cause of process variability. The bars in the chart are arranged so that the greatest frequency is at the left, the next-greatest frequency is second from the left, and so on. The idea behind the Pareto chart is that typically only a few causes provide the majority of variability in a process. This notion was first espoused by the Italian economist Vilfredo Pareto, who stated that 85% of the world's wealth was owned by 15% of the world's people. In a similar manner, the majority of variation in a process can be attributed to only a few causes. The Pareto chart is a graphical method of separating the important causes from the less important ones. We present an example where a Pareto chart is used to study variability in a service process.

EXAMPLE 13.3.4

COPIER REPAIRS. An important process variable in Example 13.1.2 is the amount of time taken to fix a copier. This variable can have a very large value when repairs cannot be made on the first visit thus requiring a return trip to complete the repairs. In order to improve service, it is necessary to decrease the number of calls that require more than one visit to repair a machine. Over a period of two months, records were kept that listed reasons why repairs were not completed in one visit. The collected data are shown in Table 13.3.4. This table lists reasons why service calls were not completed on the first visit and the frequency for each reason. The Pareto chart for these data is shown in Figure 13.3.5.

TABLE 13.3.4 Reasons for Not Repairing Copier on First Visit

Reason	Frequency	Relative frequency
1—Part not available	13	0.52
2—Personnel not properly trained	6	0.24
3—Documentation not clear	3	0.12
4—Tools not available	2	0.08
5—Other	1	0.04

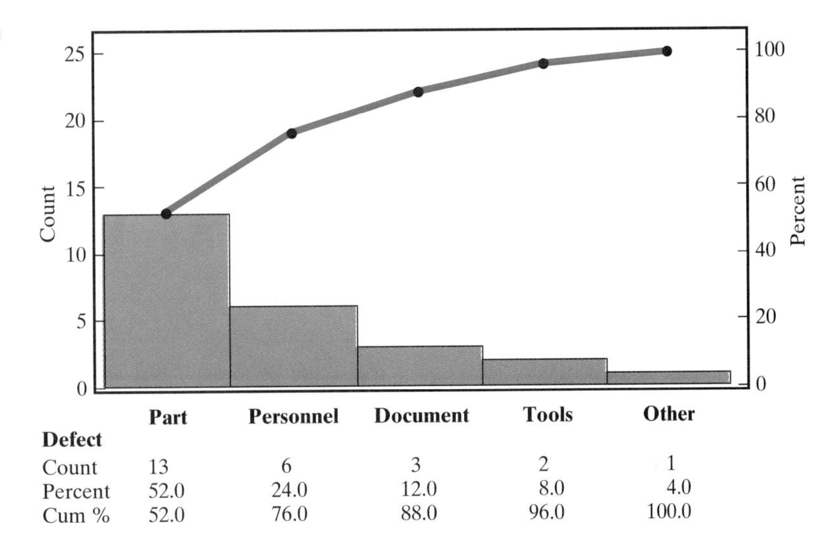

FIGURE 13.3.5
Pareto Chart for Reasons Copier Not Fixed on First Visit

From the Pareto chart we see that the major reason for return trips is that the part needed to fix the copier was not available. The process could be improved if the repair trucks had a better inventory of parts. By keeping records on the frequency of parts that are used, the service trucks could be better equipped, and the number of return trips could be reduced. The second major reason for not repairing a copier on the first trip is that the repair personnel do not have the technical expertise required to solve the problem. This indicates that additional training of personnel might yield fewer return trips to repair a copier. The line above the bars in Figure 13.3.5 reports the cumulative frequency. The value of the cumulative frequency is read from the vertical axis on the right and is reported in the last line of the chart. From this line we note that 76% of all return visits are because of either reason 1 (*Part*) or reason 2 (*Personnel*).

Using Minitab to Construct Process Control Charts

Minitab was used to produce all the figures in this section. To construct an x-bar control chart using k samples of size n, place all the observations in one column of a Minitab worksheet in the order in which they were collected. The following commands are then entered in the Session window:

```
MTB > xbarchart ci n;
SUBC> mu a;
SUBC> sigma b.
```

The command in the first line instructs Minitab to construct an x-bar control chart for the data in column `i`. The number `n` tells Minitab that each of the k samples has n observations. Thus, there are $k \times n$ observations in column `i`. The values `a` and `b` are the process mean and standard deviation, respectively, when the process is in control. This information is needed so that the center line and control limits can be computed. If these subcommands are omitted, Minitab will use the data to estimate the process mean and standard deviation.

To demonstrate these commands, we use the data in Table 13.3.1 that are stored in the file **acid.mtp** on the data disk. Column 1 contains the observation number (from 1 to 120), column 2 contains the sample number (from 1 to 30), and column 3 contains the acid concentration in micrograms per liter.

To use Minitab to construct this chart, retrieve the data and enter the following command and subcommands in the Session window:

```
MTB > xbarchart c3 4;
SUBC> mu 0.68;
SUBC> sigma 0.04.
```

The resulting x-bar control chart is shown in Figure 13.3.2.

To construct a p-chart using Minitab, suppose you have k samples, where each sample is of size n and q_i is number of the n items that have the characteristic of interest. Place the q_i in one column of a Minitab worksheet and the sample sizes n in another column. Enter the following command and subcommand in the Session window:

```
MTB > pchart ci cj;
SUBC> p a.
```

The command instructs Minitab to construct a p-chart where n, the number of items in the ith sample, is in column `j`, and q_i, the number of the n items in the ith sample that has the characteristic of interest, is in column `i`. The `p` in the subcommand tells Minitab that the population proportion is known, and `a` tells Minitab that the known value is a. Minitab uses the value a to compute the lower control limit, the center line, and the upper control limit.

To demonstrate these commands, we use the data in Table 13.3.3 which are stored in the file **satisfid.mtp** on the data disk. There are 12 samples, each of size 100, and column 1 contains the sample number, which goes from 1 to 12. Column 2 contains q_i, the number of the 100 customers in sample i who assigned a rating of 4 or 5. Column 3 contains 100, the size of each sample. The proportion of customers who mark the questionnaire with a 4 or 5 is known for the past 48 months (when the process was in control) and is 0.60. So the subcommand is `p 0.60`. To use Minitab to construct this chart, retrieve the data and enter the following command and subcommand in the Session window:

```
MTB > pchart c2 c3;
SUBC> p 0.60.
```

The output from the Minitab commands are shown in Figure 13.3.4.

To use Minitab to construct a Pareto chart, place the categories of the qualitative variable in a column of a Minitab worksheet. Place the frequencies of each category in an adjacent column. Enter the following command and subcommand in the Session window:

```
MTB > %pareto ci;
SUBC> counts cj.
```

The command in the first line instructs Minitab to construct a Pareto chart using the variable categories in column `i`. The counts are contained in column `j`.

To demonstrate these commands, we use the data in Table 13.3.4 that are stored in the file **copier.mtp** on the data disk. Column 1 contains the categories for the qualitative variable "Reason," and column 2 contains the frequencies for each value. To use Minitab to construct the Pareto chart, retrieve the data and enter the following command and subcommand in the Session window:

```
MTB > %pareto c1;
SUBC> counts c2.
```

The resulting Pareto chart is shown in Figure 13.3.5.

THINKING STATISTICALLY: SELF-TEST PROBLEMS

13.3.1 Recall Example 13.3.3 in which the proportion of 100 customers who assigned a rating of either 4 or 5 in a satisfaction survey are plotted on a p-chart. Table 13.3.5 reports sample proportions for the next 12-month period. Construct a p-chart for these values using the same center line and control limits shown in Figure 13.3.4. Does it appear that customer satisfaction changed during this 12-month period? If so, how has it changed?

TABLE 13.3.5 Sample Proportions of 100 Customers Scoring a Rating of Either 4 or 5 for Past 12 Months

Time period	Satisfaction proportion	Time period	Satisfaction proportion
1	0.68	7	0.74
2	0.67	8	0.69
3	0.73	9	0.75
4	0.72	10	0.73
5	0.66	11	0.76
6	0.72	12	0.79

13.3.2 Compute the control limits for an x-bar chart used to monitor the length of bolts produced in a manufacturing process. When the process is in control, a population of bolts forms a *normal* population with mean $\mu = 5$ inches and $\sigma = 0.05$ inches. The x-bar chart is based on $k = 50$ samples of size $n = 9$.

13.4 CHAPTER SUMMARY

KEY TERMS

1. process
2. chance causes of variation
3. assignable causes of variation
4. stable process
5. unstable process
6. rational subgroups
7. x-bar control chart
8. p-chart
9. lower control limit
10. upper control limit
11. center line
12. Pareto chart

SYMBOLS

1. μ_{SM}—mean of a derived population of sample means
2. σ_{SM}—standard deviation of a derived population of sample means
3. LCL—lower control limit
4. UCL—upper control limit
5. MU—center line

KEY CONCEPTS

1. Process costs can be reduced and the quality of output improved by reducing process variability.
2. In order to improve a process, it is first necessary to determine if the process is stable.
3. If an assignable cause of variability is detected, it must be identified and eliminated.
4. The upper and lower control limits on an x-bar chart define the range of values that contain 99.73% of the sample means in a derived population of sample means when the process is stable.
5. The majority of variation in a process can often be attributed to only a few causes.

SKILLS

1. Use a software package to construct an x-bar chart and a p-chart.
2. Use a software package to construct a Pareto chart.

13.5 CHAPTER EXERCISES

13.1 A company that manufactures machine bolts must produce bolts that have an average length of 6 centimeters (cm). In addition, 99.73% of the bolts must be within 0.3 cm of this average. What is the maximum allowable value for the standard deviation of a population of bolt lengths manufactured by this process if this specification is satisfied? Assume the population of bolt lengths is a *normal* population.

13.2 A 24-hour fast-food restaurant must ensure product quality in order to meet customer expectations. Based on company specifications, a small order of french fries must weigh an average of 68 grams of french fries. Since it is not possible to weigh each box before it is sold, the manager of the restaurant monitors this process in the following manner. At the end of each eight-hour shift, the manager selects one box of french fries from the rack of prepared food and weighs the amount of fries placed in the box. This provides $n = 3$ measurements each day. The $n = 3$ measurements are averaged and plotted on an x-bar control chart. If the control limits are LCL = 66 grams and UCL = 70 grams, what is the standard deviation of a population of weights resulting from this process when it is in control? Assume the population of weights is *normal*.

13.3 Consider the process used to grow crystals in Example 13.2.2. Assume that when a crystal growing process is in control, a resulting population of crystal weights is a *normal* population with mean $\mu = 12$ grams and standard deviation $\sigma = 1$ gram.

a. What proportion of crystal weights are between 11 grams and 13 grams when the process is in control?

b. A simple random sample of 100 crystals contained 60 crystals that weighed between 11 and 13 grams. Use this sample and Box 3.2.1 to compute a 95% confidence interval for π, the proportion of crystal weights between 11 and 13 grams. Based solely on this confidence interval, does it appear that the process is in control?

13.4 A filling machine is used to place cola in aluminum cans that are advertised to contain 12 fluid ounces. Assume that when the process is in control, the population of cans filled during a 24-hour period is a *normal* population with an average of $\mu = 12.1$ fluid ounces per can.

a. What is the standard deviation of the population (σ) if it is known that 95% of the cans are greater than or equal to 12 fluid ounces?

b. An x-bar control chart is used to monitor the filling process. Each hour a random sample of 9 cans is selected and measured to determine the fluid ounces contained in each can. If the process is in control, the population of volumes of filled cans from the process is *normal* with a mean of $\mu = 12.1$ fluid ounces and a standard deviation of $\sigma = 0.06$ fluid ounces. Compute the values of the center line, the upper control limit, and the lower control limit for this chart.

13.5 The manufacturing of circuit boards involves a process in which gold plating is placed on each board. When the process is in control, the population of gold-plating thickness values is a *normal* population with a mean of 30 microinches and a standard deviation of 0.50 microinches. If a board has more than 31 microinches or less than 29 microinches of plating, it will not work properly and must be rejected.

a. When the process is in control, what proportion of the circuit boards will be rejected?

b. Compute the center line, the upper control limit, and the lower control limit for an x-bar control chart based on samples of size 4 when the process mean is $\mu = 30$ and the standard deviation is $\sigma = 0.50$.

c. Construct an x-bar control chart for the data set **gold.mtp** on the data disk. This data set consists of measurements of gold-plating thickness on circuit boards from a process designed to have a mean of 30 microinches and a standard deviation of 0.5 microinches. The measurements appear in the third column of the file. The data represent samples of size $n = 4$ selected hourly over a 10-hour period. Thus, the first four measurements represent the sample of $n = 4$ selected during the first hour, the next four measurements represent the sample of $n = 4$ during the second hour, and so on. Does the process appear to be in control?

13.6 A newspaper in a large metropolitan area contacts every person who has canceled his or her subscription during the past month to determine the reason for the cancellation. Table 13.5.1 reports the reasons and the frequency of each reason. The data are stored in **paper.mtp** on the data disk. Construct a Pareto chart for these data. What do you conclude?

TABLE 13.5.1 Reasons for Canceling Subscription

Reason	Frequency
1—Too expensive	312
2—Not delivered on time	830
3—Switching to another paper	142
4—Moving out of the area	123
5—Never have time to read it	68

13.7 An owner of a gasoline station with 12 pumps wants to monitor his pumps to ensure they are properly calibrated. Once every week the owner draws gasoline into a five-gallon container from each of the 12 pumps. The volume shown on the pump gauge is then recorded and plotted on an x-bar control chart. There is one x-bar control chart for each pump. An x-bar control chart based on samples of size $n = 1$ is called an **individual chart**, or an **I-chart**. Suppose that the volumes measured from a properly calibrated pump over the period of one year can be viewed as a *normal* population with $\mu = 5$ gallons and $\sigma = 0.1$ gallons. What are the values of the center line, the upper control limit, and the lower control limit for the I-chart?

13.6 SOLUTIONS TO SELF-TEST PROBLEMS IN CHAPTER 13

13.1.1

a. Possible inputs are study time, study method, previous knowledge of the subject being tested, and interest in the subject. The output is a completed final exam. A variable of interest might be the percentage of correct responses on the exam. Another variable of interest might be the amount of time taken to complete the exam.

b. Inputs include the microwave oven, the bag of popcorn, and the cooking time. The output is a bag of cooked popcorn. Possible variables of interest are the number of unpopped kernels of corn and a taste score measured on a scale from 1 to 10.

c. Possible inputs are the player, the club, the ball, and the length of the back swing. The output is the ball after it has been struck. Possible variables of interest are the distance and the direction the ball traveled.

d. Possible inputs are the instructor, the course materials, the course format, and interest in the topic. The output is the knowledge gained about the field of statistics. Possible variables of interest are the grade earned in the course, the amount of enjoyment in taking the course, or the ability to apply statistical methods.

13.2.1

a. Using the computations in Example 13.2.1, the required average net weight for machine 1 is 436.63 grams and for machine 2 is 429.65 grams. Thus, for every box filled using machine 2 there is an average savings of \$0.005/gram × (436.63 − 429.65) grams = \$0.0349 per box. Thus, to recover the cost of \$5,000 for machine 2, the company must fill \$5,000/\$0.0349= 143,267 boxes.

b. Solve for μ in the equation $\mu - 2.326(1) = 425$ to obtain $\mu = 427.33$ grams. Thus, the savings per box is \$0.005/gram × (436.63 − 427.33) grams = \$0.0465. To recover the \$5,000 cost for machine 2, the company must fill \$5,000/\$0.0465 = 107,527 boxes.

13.2.2 Chance causes include the traffic flow, the frequency with which you hit red lights, or your level of aggressiveness on a particular day. An assignable cause would be an unexpected event such as an accident or a flat tire.

13.3.1 The control chart is shown in Figure 13.6.1. From Figure 13.6.1 it appears that the sample proportions are not centered about the center line. Rather, there is a steady increase, which suggests the process has changed. In particular, it appears that the proportion of customers who assigned a score of 4 or 5 has increased.

13.3.2 Using the formulas in Box 13.3.1, we obtain $LCL = 5 - 3(0.05)/\sqrt{9} = 4.95$ and $UCL = 5 + 3(0.05)/\sqrt{9} = 5.05$.

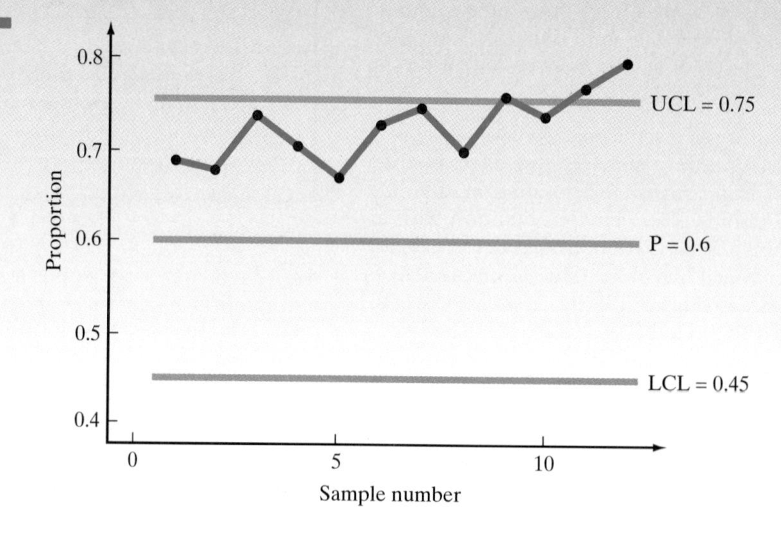

FIGURE 13.6.1
p-Chart for Proportions in Table 13.3.5

CHAPTER 14

SAMPLE SURVEYS

14.1 INTRODUCTION

In Section 2.5 we introduced the technique of simple random sampling. We noted that simple random sampling is a statistically valid sampling procedure that enables us to make inferences about a sampled population. However, in many applications, it is neither convenient nor efficient to use simple random sampling. Fortunately, there are many cost-effective, convenient, and statistically valid alternative sampling procedures that can be used in these situations. Many of these alternative procedures also provide smaller margins of error for estimating population parameters than simple random sampling. In this chapter we introduce a few of these alternative sampling procedures. To start, we examine the steps involved in planning and conducting a sample survey.

14.2 CONDUCTING A SAMPLE SURVEY

Box 14.2.1 lists steps that should be followed to conduct a sample survey.

BOX 14.2.1: STEPS FOR CONDUCTING A SAMPLE SURVEY

Step 1. State the survey objectives.

Step 2. Plan the sample selection process.

Step 3. Plan the data analysis.
Step 4. Conduct the survey.
Step 5. Analyze the data.
Step 6. Report the results of the survey.

Each of these steps is now discussed in more detail.

Step 1. State the Survey Objectives

The starting point of every sample survey is a statement of the objectives. Objectives may be rather vague at first, but ultimately they must be clear enough to determine the information that needs to be collected. We illustrate with an example.

EXAMPLE 14.2.1

CUSTOMER SATISFACTION SURVEY. At a recent sales meeting, the sales director of a used-car dealership asked, "I wonder what our customers think about our dealership?" In order to answer this question, it was decided to conduct a sample survey. The first step in this process is to translate the sales director's question into specific survey objectives. One objective might be to determine customer satisfaction with past performance. To accomplish this objective, customers would be asked to evaluate the dealership on several performance measures, such as knowledge of the sales staff, convenience of shopping hours, and price competitiveness. Another objective might be to determine customer requirements for future services. This information would help the dealership remain competitive in the future.

In defining the objectives of a survey, it is necessary to clearly define the target and sampled populations. Also, the required margin of error must be established. In order to help formulate the objectives, it is useful to ask the following two questions:

1. What actions are being considered that require information from a sample survey?
2. How will the survey results be used to select the most appropriate action?

If these two questions cannot be answered, more work is needed to establish the survey objectives.

Step 2. Plan the Sample Selection Process

Once objectives have been established, the next step is to determine how the data will be collected. Possible methods for data collection include (1) telephone surveys, (2) mail surveys, (3) personal interviews, (4) direct input into a computer, and (5) direct observation. In selecting a particular collection method, one must consider the required sample size, the relative costs of each method, the time available for completing the survey, and the likelihood of collecting accurate and complete information. Telephone surveys are useful for assessing attitudes and opinions because much information can be collected in a very short period of time. Mail surveys are useful if the questionnaire is too long to administer in a telephone interview. Personal interviews are useful in situations where detailed information is required. Computers are being used more frequently to collect information. You may have provided information in this manner by answering a questionnaire on the Internet.

Direct observation is typically used when one collects data from nonhuman items. For example, an inspector on an assembly line uses direct observation to identify manufactured items that do not meet specifications.

After the method for collecting data has been determined, a sampling procedure must be selected. In Section 14.3 we discuss several alternatives to simple random sampling. To select the most appropriate method, one must consider the physical structure of the population and any information that is presently known about the population. The sample size must be determined by making tradeoffs between the desired margin of error and budget constraints. If data are collected by telephone or personal interviews, interviewers must be trained to ask questions in a consistent manner. Questionnaires must be designed and pretested using a small group of subjects from the sampled population. Pretesting provides an opportunity to determine if all the questions are correctly interpreted and if they elicit the desired information.

Step 3. Plan the Data Analysis

Many surveys fail because no consideration is given to the data analysis until the data have already been collected. Planning the data analysis is necessary to prevent a situation where data are collected in such a manner that it is impossible to perform a valid statistical analysis. For example, if the formulas in Box 6.3.1 will be used to compute a confidence interval for μ and it is not known whether the population is *normal*, then a sample of size 30 or more will be required to assure that the confidence coefficient will be close to the stated value.

A helpful practice in this regard is to analyze the data collected during pretesting of the questionnaire. This provides an opportunity to discover data characteristics that suggest particular analysis procedures. It also helps to select the computer software for analyzing the data.

Step 4. Conduct the Survey

Data are now collected by carefully following the established sampling protocol. Efforts must be made to collect information from members of the selected sample who do not respond to the survey. **Nonrespondents** often have attitudes and characteristics that are quite different from subjects who respond to a survey. For this reason, follow-up visits and telephone calls are needed to collect information from this group. The following example provides an illustration.

EXAMPLE 14.2.2

CUSTOMER SATISFACTION SURVEY. Consider the customer satisfaction survey conducted by the used-car dealership in Example 14.2.1. In order to select a sample of customers, the sales director compiled a list of all customers who had purchased a car from the dealership during the past three months. This group of customers represents the sampled population. A simple random sample of 500 customers was selected from the list, and a short questionnaire was mailed to each sampled customer. One question asked the customer to rate the dealership's overall performance on a scale from 1 to 7, where 1 is "very poor" and 7 is "very good." A parameter of interest is the proportion of customers in the sampled population who would assign a score of 5 or higher. After two weeks the sales director had received responses from only 100 of the 500 sampled customers. Of the 100 respondents, 45 customers assigned a rating of 5 or higher. The sales director was concerned that 80% of the sampled customers had not answered the survey, and so he sent a second questionnaire to each of the 400 nonrespondents. At the end of the next two-week period, the sales director had received 40 questionnaires from the group of 400 nonrespondents. Of the 40 respondents who replied to the second mailing, 34 assigned a rating of 5 or higher. The sales director noted that the proportion of customers who assigned a rating of 5 or higher was greater in the second group ($34/40 = 85\%$) than it was in the first group of 100 responses ($45/100 = 45\%$). The fact that these proportions are so different suggests that the attitudes of the original 400 nonrespondents are different than those of the 100 customers who responded during the first two weeks of the survey. In particular, the data suggest that customers who were not satisfied with the dealership were more likely to respond to the survey.

The previous example illustrates a problem encountered with the presence of nonrespondents. The customers who responded to the first mailing do not comprise a simple random sample from the sampled population. Rather, they represent a simple random sample from the subset of the sampled population who would have responded to the survey if they had been selected. As noted in the example, this subset of the sampled population might have different attitudes toward the car dealership than other members of the sampled population. The second set of responses in Example 14.2.2 suggests that customers who would respond to the survey are more likely to rate the dealership lower than would members of the sampled population who would not respond to the survey.

Step 5. Analyze the Data

After the data are collected, they must be analyzed. Except in very small studies, a computer will be used to store the data and perform computations. Thus, the data must be entered into a computer database. This often involves coding some of the variables. For instance, the variable "Gender" has two values, *male* and *female*. It might be convenient to use numerical codes to denote these values, such as 1 for male and 2 for female. Decisions regarding coding should be made based on the statistical software used to analyze the data. Some software packages accept only numerical values for variables, while others do not have this restriction.

Care must be taken to ensure that editing and coding errors are minimized. Data must be examined for obvious typing errors or other mistakes. After the quality of the data has been assured, one conducts the appropriate data analysis.

Step 6. Report the Results of the Survey

The final step is to write a report that summarizes all aspects of the survey. The report should include the study objectives, the data collection process, and conclusions based on the data analysis. It is advisable to supplement numerical summaries with judiciously chosen graphs and charts, as the audience will comprehend visual information more readily than numerical information.

The steps outlined in Box 14.2.1 often overlap. For example, it is not possible to determine how data will be collected in step 2 without some knowledge of the how the data will be analyzed in step 5. As another example, it is not possible to state the desired margin of error in step 1 without considering the budget constraints and the sampling procedure selected in step 2.

The Section on Survey Research Methods of the American Statistical Association (ASA) has published a series of pamphlets that describe the sample survey process. This series is entitled "What Is a Survey?" and can be ordered from the American Statistical Association, 1429 Duke Street, Alexandria, VA 22314-3402.

THINKING STATISTICALLY: SELF-TEST PROBLEMS

14.2.1 Bob Marshall is the editor of a regional magazine in Arizona. Although the majority of the magazine's subscribers live in Arizona, Bob believes there is a market for subscribers who live outside the state. In order to determine if such a market exists, Bob obtained lists of registered participants from several national conferences held in Phoenix during the past month. From these lists he constructed a list of people who live outside Arizona. Current subscribers were dropped from the list, and then a simple random sample of 1,200 people was selected from the remaining names on the list. Each person in the sample was sent a copy of the magazine. Enclosed with each magazine was a stamped postcard that requested demographic information about the reader. The last question on the postcard was, "Will you subscribe to this magazine?"
a. State the target population and the sampled population.
b. Only 400 of the 1,200 sampled convention participants returned the postcard. Of these 400 respondents, 100 stated they would subscribe to the magazine. Do you expect the proportion of the 800 nonrespondents who would subscribe to the magazine to be close in value to the proportion of the 400 respondents who would subscribe to the magazine?

14.3 SOME ALTERNATIVES TO SIMPLE RANDOM SAMPLING

In Chapter 2 we stated that a sample estimate is expected to be a good estimate of a population parameter if the sample is representative of the population. A sample is representative of a population if it possesses characteristics similar to the population.

Judgment Sampling

It might seem that the best way to ensure that a sample is representative is to have a person with expert knowledge about the population select items that appear to be most representative of the population. Such a procedure is called **judgment sampling**. For example, the sales manager in Example 14.2.1 might handpick the customers that he believes are the most "typical" based on information reported by the customer at the time of the sale. Although judgment sampling appears to have merit, in general it cannot be recommended for two reasons. First, it is virtually impossible for any expert to be free of bias. Different people look at situations from different perspectives, and an item that appears representative to one person may not appear so to another person. The second problem with judgment sampling is that it cannot be used for making valid statistical inferences. This is because statistical inferences require a random sample. There is no well-defined random process used in judgment sampling. Thus, even if a judgment sample is representative of a population, it cannot be used to construct valid confidence intervals or perform valid statistical tests.

Simple Random Sampling

In order to make a statistical inference, it is necessary that a well-defined random process be used to select the sample. Simple random sampling is one such random process. Let's review the procedure for selecting a simple random sample. The simple random sampling procedure presented in Box 2.5.1 is repeated in Box 14.3.1.

BOX 14.3.1: A PROCEDURE FOR SELECTING A SIMPLE RANDOM SAMPLE OF SIZE n FROM A POPULATION OF SIZE N

In your mind's eye, envision the following. Number the population items from 1 to N. Obtain N small plastic balls that are indistinguishable except that painted on one ball is the number 1, on another ball is painted the number 2, on another ball is painted the number 3, and so forth until finally on the last ball the number N is painted. There is one number, from 1 to N, painted on each ball. Put the balls in a bin and shake the bin until the balls are thoroughly mixed. Then select n balls from the bin. The numbers on these n balls define the population items that belong in the sample. The values of the variable corresponding to these items are the sample values obtained by the simple random sampling procedure.

If a simple random sample of n items is selected from a population of N items by the procedure explained in Box 14.3.1, then every group of n items has the same chance of being selected. No preference is shown for selecting one group of n items over any other group of n items.

We now describe three alternative random sampling procedures that may be used instead of simple random sampling to provide a valid statistical inference. These methods are **systematic random sampling**, **two-stage** or **multi-stage random sampling**, and **stratified random sampling**.

Systematic Random Sampling

Unless an entire population can be thoroughly mixed like the balls described in Box 14.3.1, the implementation of simple random sampling requires that every item in the population be identified with a number. Once a number is selected, the item in the population associated with the number can be identified and included in the sample. A list of all population items with the associated identification number is called the sampling **frame**. In many situations it is either impossible or inconvenient to obtain a sampling frame. Systematic sampling is a method of sampling that can be used when no frame is available. With this method samples are obtained at regular intervals of time or space. Consider the following two examples.

EXAMPLE 14.3.1

CARROT PLANT DAMAGE. A hail storm has damaged the carrot plants on a vegetable farm. An insurance adjustor is sent to the farm to estimate the proportion of the carrot plants that was damaged by the hail storm. In order to select a simple random sample of carrot plants, it would be necessary for the adjustor to obtain a sampling frame that identifies every carrot plant on the farm. It would be practically impossible to identify and number each plant, so the adjustor decides to use the following sampling plan. The carrots are planted in one row that is about one-half mile long (2,640 feet). There are about 10 carrot plants per foot, so there are about 26,400 plants in the row. He selects a simple random sample of size 1 from the integers between 1 and 30, inclusive. Suppose the selected number is 4. A carrot plant will be selected four feet from the end of the row and inspected for damage. The adjustor then continues down the row, stopping every 30 feet to select another carrot plant. The sample will consist of carrot plants taken at 4 feet, at 34 feet, at 64 feet, and so on, until the adjustor reaches the end of the row. The resulting sample size will be $2{,}640/30 = 88$ plants. This sampling procedure is random because a simple random sample of one integer was selected to start the process. However, it is not a simple random sampling procedure as described in Box 14.3.1 because not every possible group of $n = 88$ plants has the same chance of being selected. For example, it would not be possible using this method to select any sample that contains two plants that are four feet apart.

EXAMPLE 14.3.2

QUALITY CONTROL INSPECTION. A manufacturer of potato chips monitors the weights of bags of chips to ensure that they satisfy the advertised net weight printed on each package. Bags of chips are packed in boxes with 24 bags in a box. In a typical day, 1,000 boxes of chips (24,000 bags) are prepared for shipping. A quality control manager wants to sample 50 of the boxes produced each day and weigh all the bags in each box. To do this, the manager could number each box from 1 to 1,000 at the end of each day and select 50 random integers between 1 and 1,000, inclusive, using a random number generator. The boxes corresponding to the sampled numbers could be located and all bags in each box weighed. This process would provide a simple random sample of boxes. However, the manager doesn't want to wait until the end of a day to begin weighing the bags. If the bag weights are not in conformance, he wants to discover the problem as early as possible. Thus, he decides to randomly select one integer from 1 to 20, inclusive. Suppose the number selected is 17. The manager then selects boxes as they are produced throughout the day by first selecting the 17th box and then selecting every 20th box thereafter. The sample consists of the 50 boxes numbered 17, 37, 57, . . . , 997. Notice that this sample is *not* a simple random sample of boxes, because not every group of 50 boxes has an equal chance of being selected. For example, boxes 17 and 18 cannot appear in the same sample.

The process described in Examples 14.3.1 and 14.3.2, in which every kth item of a population is sampled, is called **systematic random sampling**. Systematic random sampling is similar to simple random sampling in that it is a random process that shows "no preference" for selecting any single item in the population. The difference between these two sampling procedures is that with systematic random sampling, not every group of n items has the same chance of being selected. There is also a danger in using systematic random sampling if the variable of interest is systematically related to the order in which items are selected from the population. As an example of such a situation, suppose 20 filling machines are used to fill the bags in Example 14.3.2. If the boxes are filled with bags from machines in an alternating order, and the 20th machine is underfilling every bag, there will be underfilled bags in boxes 20, 40, 60, . . . , 1,000. However, the selected sample of boxes 17, 37, 57, . . . , 997 will not contain any of the boxes with the underfilled bags. Because of this possible problem, it is important that someone familiar with the population be consulted before selecting a systematic random sample.

Typically, systematic random samples are analyzed as if they are simple random samples. That is, sample estimates based on systematic samples are computed using the formulas for simple random sampling. This is proper only if the variable of interest is not related to the order in which the items are selected. Such is the case in Example 14.3.1, where the hail storm would not be expected to damage the carrot plants in any systematic manner. However, this may not be the case in Example 14.3.2, where the possibility of a systematic malfunctioning of one of the filling machines exists.

Two-Stage or Multi-Stage Sampling

In some applications, population items are naturally grouped into larger units. Such populations are described as either **nested** or **hierarchical** populations. For instance, in Example 14.3.2 the items of interest (bags of chips) are grouped in larger units (boxes). Samples from nested populations may be obtained by selecting simple random samples in stages. This type of sampling is called **two-stage** or **multi-stage** sampling. The following two examples describe this sampling method.

EXAMPLE 14.3.3

OPINIONS ON FINANCING A STADIUM. A member of the city council plans to conduct a survey to determine the opinions of adults in the city concerning a proposed sales tax to finance a new baseball stadium. The target population is all adults aged 21 or over living in the city. The council member used the following random sampling procedure. A list of every block in the city was obtained from the office of the city manager, and a simple random sample of blocks was selected from this list. For each of the selected blocks, a list of residential housing units was obtained and a simple random sample of housing units was selected from each block. Interviewers were sent to the selected housing units, and every adult aged 21 or older living in the unit was asked a set of questions concerning the proposed sales tax.

EXAMPLE 14.3.4

EXAMINING FOR DAMAGE TO POTATO CHIPS. A company that manufactures potato chips has just discovered that a machine malfunction has damaged some of the bags of chips produced last week. To determine the proportion of bags that were damaged, it is necessary to select a sample of bags from the warehouse that contains the bags produced last week. The bags of chips are packaged in cartons of 200 bags. Cartons are stored on platforms that each hold 1,000 cartons. The platforms are stored in the warehouse in 20 rows of 10 platforms each. This population of potato chip bags can be viewed as a nested population. The bags are grouped by cartons, cartons are grouped by platforms, and platforms are grouped by rows. The following procedure could be used to select a sample of 100 cartons. The 200 platforms could be numbered from 1 to 200, and the procedure in Box 14.3.1 used to select a simple random sample of 100 platforms. One could then select a simple random sample of one carton from each selected platform and examine all bags in the carton for damage.

Formulas for computing point estimates and confidence intervals using multi-stage sampling can be found in textbooks on sampling theory.

Stratified Random Sampling

Stratification is the process of grouping population items into subpopulations prior to sampling. The subpopulations formed in this manner are called **strata**. Once items have been separated into strata, a simple random sample is selected from *every* stratum in the population. In situations where the variable of interest displays greater variability among strata than within strata, stratified random sampling will provide a smaller margin of error than simple random sampling. The reason why the margin of error is reduced is because the investigator is using additional information, namely, knowledge of an item's stratum membership. The next example provides an illustration of a stratified random sampling application.

EXAMPLE 14.3.5 **ESTIMATING BUILDING COSTS.** A builder of single-family homes must estimate the cost of buildings in progress at the end of each fiscal year. The builder typically has about 2,000 houses under construction at any given time, and it is necessary to estimate the average cost for this population of houses. It is very time consuming to obtain the necessary cost information for each house so a sample of 100 houses will be selected. The builder has 10 different house models under construction. Houses of the same model have relatively equal costs, but there is great variability in costs among houses of different models. Since cost is the variable of interest, stratification on model type will reduce the margin of error for estimating average cost. The builder obtains a list of all 2,000 houses that reports the model type of each house. The population of 2,000 houses is then separated into 10 strata based on model type, and a simple random sample is selected from each stratum. The sample size selected from each stratum depends on (1) the standard deviation of the variable of interest (building cost) within each stratum, (2) the number of items in each stratum, and (3) the cost of measuring an item in each stratum. Rules for deciding the sample size selected from each stratum can be found in textbooks on sampling theory. A stratified random sample of 100 houses will provide a smaller margin of error for estimating the population average cost than a simple random sample of 100 houses.

The calculations for estimating population parameters based on a stratified random sample are more complicated than those based on a simple random sample. This is because individual strata estimates must be combined to estimate a population parameter. Formulas for computing point estimates and confidence intervals using a stratified random sample can be found in textbooks on sampling theory.

THINKING STATISTICALLY: SELF-TEST PROBLEMS

14.3.1 Discuss the two reasons why judgment sampling is not recommended for selecting a sample.

14.3.2 Explain how simple random sampling differs from systematic random sampling. Explain how the two sampling procedures are similar.

14.3.3 In Example 14.3.1 an adjustor wanted to determine the amount of damage a hail storm caused to carrot plants on a vegetable farm. Suppose that in addition to carrots, there were five other vegetables that needed to be examined for hail damage. Each type of vegetable has a different tolerance to hail, and so the proportion of plants damaged by the hail is likely to be different for each type of vegetable. Describe how stratification could be used to estimate the proportion of all plants on the farm that were damaged by hail.

14.4 CHAPTER SUMMARY

KEY TERMS

1. judgment sampling
2. simple random sampling
3. sampling frame
4. systematic random sampling
5. multi-stage random sampling
6. nested population
7. stratified random sampling

KEY CONCEPTS

1. A sample is representative of a population if it possesses similar characteristics to the population.
2. Statistical inferences cannot be based on a judgment sample.
3. Stratified random sampling can provide a smaller margin of error than simple random sampling.
4. Systematic random sampling does not require a sampling frame.

SKILLS

1. Design a sample survey following the steps outlined in Box 14.2.1.
2. Determine when it is appropriate to use an alternative to simple random sampling.

14.5 CHAPTER EXERCISES

14.1 A marketing research firm was hired by a long-distance phone company to determine the proportion of residential phone customers listed in a city phone book who are enrolled in a competitor's long-distance program. The residential section of the phone book consists of 1,000 pages. For the sake of simplicity, suppose that each page has 4 columns of 100 phone numbers. Thus, the sampled population has 400,000 phone numbers. The research firm collected a sample of 100 phone numbers in the following manner. A random integer from 1 to 1,000 was selected, and the phone directory was opened to the page corresponding to the selected number. A random integer from 1 to 4 was selected to identify one of the four columns on the page. Finally, a random integer from 1 to 100 was selected to identify one phone number from the selected column. This entire process was repeated until a sample of 100 phone numbers was obtained.

a. Was the sample collected by the research firm a simple random sample?

b. Describe a systematic random sampling procedure for selecting the sample of 100 phone numbers.

14.2 The owner of a professional baseball team wants to survey fans to determine if they would like to have designated alcohol-free sections in the stadium. The owner has a list of all season-ticket holders and plans to send each one a questionnaire concerning this issue with the ticket renewal forms. However, he would also like to collect information from the population of fans who attend games but are not season-ticket holders. No list of these fans is available. Describe a sampling method that might be employed to select a random sample of fans who attend the games but are not season-ticket holders.

14.3 The proportion of sample items responding to a survey is called the **response rate**. For example, if a sample of 1,000 people is selected, and responses are obtained for 600 of the 1,000 people, the response rate is 60%. Consider the following three methods for data collection: mail survey, telephone survey, and personal interviews.

a. Which method do you think generally provides the highest response rate?

b. Which method do you think generally provides the lowest response rate?

c. Which method do you think is the most expensive?

14.4 The dean of a liberal arts college wants to determine the proportion of undergraduate students in the college who plan to register for a "college survival" course next semester. This course teaches study skills, provides academic advising, and helps students with career planning. The dean obtains a list of all undergraduate students enrolled in courses during the current semester. This list also reports each student's mailing address, age, and department major. Describe a sampling procedure that could be used to conduct a mail survey of this population of students.

14.5 An accountant must perform an audit for a department store. One of the tasks is to estimate the proportion of billing statements that show an incorrect balance. After billing statements are printed, they are placed alphabetically in a long metal tray and prepared for mailing. In a typical month about 4,000 statements are in the tray. The accountant wants to select a sample of statements so he can estimate the proportion of statements that report an incorrect balance. Each billing statement must be compared to the original bill of sale to determine if the balance is correct. In order to achieve the desired margin of error, a sample of 400 statements is needed.

a. Describe a simple random sampling procedure for collecting the sample of 400 billing statements.

b. Describe a systematic random sampling procedure for collecting the sample of 400 billing statements.

c. Which of the two procedures do you think is better?

14.6 SOLUTIONS TO SELF-TEST PROBLEMS IN CHAPTER 14

14.2.1

a. The target population is the set of all people who are not residents of Arizona and who do not currently subscribe to the magazine. The sampled population is the set of all people who (1) do not live in Arizona, (2) attended a national conference in Phoenix during the past month, and (3) do not currently subscribe to the magazine.

b. No. It is reasonable to expect that a smaller proportion of the nonrespondents would subscribe to the magazine since they did not bother to return the postcard.

14.3.1 The items selected will be dependent on the expert used to select the sample. Also, a judgment sample cannot be used to make valid statistical inferences.

14.3.2 Simple random sampling ensures that every possible sample of n items has the same chance of being selected. This is not true with systematic sampling. Simple random sampling and systematic random sampling are similar in that both procedures use a random process to ensure that "no preference" is shown for any single item.

14.3.3 Since the proportion of damaged plants is expected to vary by the type of vegetable, the plants should be stratified based on the type of vegetable. Separate estimates could be made for each type of vegetable, and then these estimates could be combined to estimate the proportion of all plants on the farm that was damaged by hail.

APPENDIX A

ANSWERS TO ODD-NUMBERED EXERCISES

Chapter 1

1.1(1) the number of pages in the book **(2)** whether the book has a hard or a soft cover **(3)** the number of copies the company expects to sell **(4)** the prices of similar books **(5)** the desired profit margin **1.3** They can use the amount of money they spent for groceries last month. **1.5** My automobile's miles per gallon and the cost per gallon of gasoline. **1.7a.** 0.12 **b.** No, since the entire population of purchases is contained in the database. **c.** No, since the price paid for each car is contained in the database. **d.** 0.62 **e.** no **f.** no **g.** Yes. The statement is based on a sample, so there is uncertainty. **1.9a.** The population of items is the set of 100,000 radios in the warehouse. **b.** The sample is 50 radios selected from the population of 100,000. **c.** The variable of interest is "Condition of radio," and the possible values are *damaged* and *not damaged*. **d.** The parameter of interest is the proportion of radios in the warehouse that is damaged. **1.11** Yes. You can determine the exact value of a parameter if you are able to measure the variable of interest for every item in the population. **1.13a.** all registered voters in the state **b.** The variable of interest is "Tax cut preference." This variable has three values—*state income tax*, *state property tax*, and *no preference*. **c.** 0.452. There is uncertainty in this estimate since only 489 registered voters in the state were sampled. **1.15a.** all households in the United States **b.** the proportion of households in the population with more than one television set **c.** $\widehat{\pi} = 0.680$; no **1.17a.** the set of the 30 most recent league-sanctioned games **b.** the score for a game, which we label "Score" **c.** the bowler's average score for the 30 games **d.** The Greek letter μ is used to represent the population parameter since it is the symbol for a population average.

Chapter 2

2.1a. The target population is all registered voters in the city. The sampled population is all senior citizens in the city who are registered to vote. **b.** Yes. The opinion of senior citizens concerning taxation is often quite different than the opinion of younger voters. Thus, it may not be good judgment to think that the sampled population closely resembles the target population. **2.3a.** The target population is the set of all customers who shopped in the store during the past 12 months. The sampled population is the set of all customers who have a store credit card. **b.** No. Since a simple random sample was obtained, 23% is a valid estimate of the proportion of credit card owners (the sampled population) who regularly listen to the station. It is not known if 23% is a valid estimate of the proportion of all customers who regularly listen to the station. **c.** The manager must determine whether the sampled population has the same listening habits as the target population. **2.5** quantitative **2.7a.** quantitative **b.** quantitative **2.9** qualitative with two possible categories—*yes* and *no* **2.11** quantitative **2.13a.** The target population of items is the set of all pens manufactured last month. **b.** The variable of interest is whether a pen is defective, which we label "Condition of pen." **c.** The variable is qualitative with two possible categories—*defective* and *not defective*. **d.** The parameter of interest is π, the proportion of defective pens. **2.15a.** The target population is the collection of all households in the United States that have television sets. The sampled population is the set of households in Miami that have television sets. **b.** If the viewing habits of the people in the households in Miami are similar to those of the entire United States, one can make a good judgment inference from the sampled population to the target population. However, viewing habits often vary across different regions of the country, which could make it difficult to make a good judgment inference. **c.** The variable of interest is "Watched the Super Bowl." **d.** This is a qualitative variable consisting of two categories—*yes* and *no*. **e.** The parameter of interest in the sampled population is π, the proportion of households in Miami with television sets that were tuned to the Super Bowl. **f.** $\widehat{\pi} = 0.40$. **2.17a.** The target population is all pet owners in the United States. **b.** The variable of interest is "How often do you feed human food to your pet?" **c.** qualitative

Chapter 3

3.1a. Pr(ball will be green) = 0.40. **b.** Pr(ball will be either green or red) = 0.50. **3.3** 920 balls **3.5** 0.99 **3.7** 0.26 **3.9** $L = 0.796$ and $U = 0.904$ **3.11a.** $n = 2{,}401$ (rounded up) **b.** No. Even though the point estimate exceeds 0.50, there are values in the confidence interval that are less than 0.50. **3.13a.** $\widehat{\pi} = 0.500$. **b.** $L = 0.443$ and $U = 0.557$ **c.** $n = 463$ (rounded up). **3.15a.** $k = 162$. **b.** $L = 0.738$ and $U = 0.882$ **c.** $n = 1{,}041$.

Chapter 4

4.1a. 7,423 **b.** 0.012, 0.043, 0.109, 0.198, 0.244, 0.206, 0.126, 0.047, 0.015 **c.** 0.362 **d.** 0.821 **e.** 0.117 **f.** 1.00 **g.** 0.198 **4.3a.** 88 **b.** 44 **c.** 0.596 **4.5** There is an upward trend beginning at the eighth quarter. **4.7** As "Appldgpa" increases, "Gradgpa" tends to increase.

Chapter 5

5.1a. $Y_{10} = 52$ and $Y_{21} = 67$. There is no Y_{42} in the data. **b.** No, $Y_{15} = 94$. **5.3** $M = 209$ pounds. **5.5** The set of numbers less than or equal to M is 179, 194, 198, 245, 249, 251, 255, and 260. This set constitutes a proportion 0.533 of the values. **5.7a.** $\mu = 28.00$ (\$28,000). **b.** $\sigma = 20.41$ (\$20,410). **c.** $M = 23$ (\$23,000). **5.9** 72.18 **5.11** At least 75% of the students will receive a grade of C. **5.13a.** The number of years each patient in the population lives after being diagnosed with this type of cancer is recorded. These values are averaged for all patients. **b.** the median **5.15** It would be more accurate if the newspaper report read, "The average number of bus rides per citizen last year was 13.7." The city keeps a record of the number of passengers that ride every bus each day. By adding up all of these numbers for the entire year and dividing by the city's population, the average number of rides per citizen can be determined. **5.17** From step (3) in Box 5.3.2, we see that SSDY is the sum of squares of N numbers. Since the square of a number is greater than or equal to 0, the sum of squares of these N numbers is greater than or equal to 0. Hence, SSDY ≥ 0, and $\sigma = \sqrt{\text{SSDY}/N} \geq 0$.

Chapter 6

6.1a. yes, since $\mu - 2.326\sigma = 454 > 453$ grams **b.** 456 grams **6.3a.** all cars that were driven in the United States last year **b.** between 0.95 and 0.975 **c.** 0.818 **6.5a.** $\widehat{\mu} = 8.95$ miles. **b.** $L = 8.24$ miles and $U = 9.66$ miles **6.7a.** $\widehat{\mu} = \$1.19$ and

$\widehat{\sigma} = \$0.024$. **b.** $L = \$1.17$ and $U = \$1.21$ **6.9a.** $\widehat{\mu} = 3.72$ pounds. **b.** $L = 3.45$ pounds and $U = 3.99$ pounds **6.11a.** $L = -1.14$ inches and $U = 0.14$ inches **b.** The evidence of the cream's effectiveness is not conclusive. **6.13** $n = 150$ (rounded up) **6.15a.** yes **b.** $L = 0.27$ microinches and $U = 0.43$ microinches **c.** $n = 46$ boards (rounded up). **6.17** sample 3

Chapter 7

7.1 the larger sample size of 45 **7.3a.** the mean, since consumers are interested in the total money they spend on gasoline for any given time period **b.** $L = \$1.199$ and $U = \$1.253$ **7.5** $\widehat{M} = \$57{,}000$. The 95% confidence interval is $L = \$49{,}000$ to $U = \$61{,}000$. **7.7a.** $\widehat{M} = 2{,}234.5$ hours. **b.** $L = 1{,}974$ and $U = 2{,}342$. The purchasing officer should not purchase the light bulbs.

Chapter 8

8.1a. $H_0 : \pi \leq 0.50$ versus $H_a : \pi > 0.50$ **b.** $L = 0.62$. H_0 is rejected. **8.3a.** $H_0 : \mu \leq 18.6$ versus $H_a : \mu > 18.6$ **b.** Since α is less than the P-value, H_0 is not rejected. The farmer will continue to use the standard brand of fertilizer. **8.5a.** $H_0 : \pi \leq 0.35$ versus $H_a : \pi > 0.35$ **b.** $L = 0.36 > 0.35$, and there is sufficient sample evidence to conclude that $\pi > 0.35$. The legislator will introduce the bill. **8.7a.** $H_0 : \pi \geq 0.21$ versus $H_a : \pi < 0.21$ **b.** Since $U = 0.136 < 0.21$, there is sufficient sample evidence to conclude that $\pi < 0.21$. **c.** The P-value is less than $\alpha = 0.005$. **8.9a.** $H_0 : M \geq 45{,}000$ versus $H_a : M < 45{,}000$ **b.** Since $U = 46{,}000 > 45{,}000$, the null hypothesis cannot be rejected. **8.11a.** $H_0 : \mu = 10$ versus $H_a : \mu \neq 10$ **b.** Since $L = 9.99 < 10$ and $U = 10.05 > 10$, there is not sufficient sample evidence to conclude that $\mu \neq 10$. **c.** The P-value is greater than $\alpha = 0.01$.

Chapter 9

9.1a. $\widehat{Y}(42) = -1{,}768.5$. **b.** $\widehat{Y}(90) = -4{,}034.1$. **c.** The prediction function is not defined for $X = 30$. **9.3a.** $\widehat{Y}(79) = \widehat{\mu}_Y(79) = 39.69$. **b.** The answer is not available since $\widehat{\mu}_Y(X)$ is defined only for X between 50 degrees and 90 degrees. **c.** In the range 50 degrees to 90 degrees, the amount of trash is expected to increase in a linear fashion as the temperature increases. However, if the temperature is too high, people may stay home and the amount of trash will no longer increase in this manner. **9.5a.** Yes; based on the scatter plot this seems reasonable. **b.** $\widehat{\beta}_0 = -223.73$ and $\widehat{\beta}_1 = 61.4428$. **c.** The sample regression function is $\widehat{\mu}_Y(X) = \widehat{\beta}_0 + \widehat{\beta}_1 X = -224 + 61.4X$. **d.** The population parameter is β_1. The estimate is $\widehat{\beta}_1 = 61.4428$ (thousand dollars/thousand feet). **e.** The predicted cost for a well drilled 8,000 feet is 267.812 (thousand dollars). The 95% prediction interval is from 250.296 to 285.328 (thousand dollars). **f.** The estimate of the average cost of all wells that were drilled to 8,000 feet is 267.812 (thousand dollars). **g.** $\widehat{\rho}_{Y,X} = \sqrt{0.992} = 0.996$. **h.** The proportion is $\rho^2_{Y,X}$. The estimate is $\widehat{\rho}^2_{Y,X} = 99.2\%$(R = sq). **9.7a.** Yes; based on the scatter plot this seems reasonable. The population regression function is

$$\mu_Y(X) = \beta_0 + \beta_1 X \quad \text{“for } X \text{ from 0 to 40.”}$$

b. The sample regression function is

$$\widehat{\mu}_Y(X) = \widehat{\beta}_0 + \widehat{\beta}_1 X = 3.61 - 0.0261X.$$

c. The 95% confidence interval for β_1 is $L = -0.0318$ to $U = -0.0204$. Since all values in the confidence interval are negative, this supports the argument that work has a negative impact on GPA. **d.** The predicted GPA is $\widehat{Y}(20) = 3.09$. The 95% prediction interval is from 2.48 to 3.69. **9.9a.** Yes. The scatter plot suggests this is a reasonable assumption. **b.** Yes. The simple random sample of students was obtained using sampling method (4a). **c.** No. There are no standardized residuals that have a magnitude greater than 3. **d.** Yes. The assumptions in Box 9.3.1 appear to be satisfied.

Chapter 10

10.1a. The first sampled population consists of all the bolts in the most recently purchased lot from supplier 1. The second sampled population consists of all the bolts in the most recently purchased lot from supplier 2. **b.** $\widehat{\pi}_1 = 0.069$ and $\widehat{\pi}_2 = 0.029$. **c.** $\widehat{\pi}_1 - \widehat{\pi}_2 = 0.040$. **d.** $L = 0.010$ and $U = 0.070$. **10.3a.** $\widehat{\pi}_1 = 0.530$. **b.** $\widehat{\pi}_2 = 0.417$. **c.** $L = 0.049$ and $U = 0.177$. **10.5a.** $\widehat{\pi}_1 = 0.422$. **b.** $\widehat{\pi}_2 = 0.291$. **c.** $\widehat{\pi}_1 - \widehat{\pi}_2 = 0.131$. **d.** $L = 0.099$ and $U = 0.163$. **10.7a.** There are two sampled populations, and both are conceptual. Sampled population 1 consists of all members of the health club where each person completes program 1. Sampled population 2 consists of all members of the health club where each person completes program 2. **b.** $\widehat{\pi}_1 - \widehat{\pi}_2 = 0.220$. **c.** $L = 0.088$ and $U = 0.352$. Since all values in the confidence interval are positive, we can conclude that $\pi_1 > \pi_2$. Hence, program 1 appears to be a better program. **10.9a.** The chi-squared statistic is 112.203 and DF = 2. Using Table 10.3.4, we see that the P-value is less than 0.001. Since the P-value is less than α, there is sufficient evidence to conclude that support of the propositions and ownership of a handgun are related. **b.** $L_1 = -0.514$, $U_1 = -0.286$, $L_2 = 0.389$, $U_2 = 0.603$, $L_3 = -0.180$, and $U_3 = -0.012$.

Chapter 11

11.1a. $\widehat{\mu}_1 = 344.6$, $\widehat{\sigma}_1 = 112.7$, $\widehat{\mu}_2 = 583.6$, $\widehat{\sigma}_2 = 177.6$, and $\widehat{\mu}_1 - \widehat{\mu}_2 = -239.0$. **b.** 583.6 days. **c.** 344.6 days. **d.** Rats that were fed a medium-fat diet were estimated to live an average of 239 days longer than rats that were fed a high-fat diet. **e.** With 95% confidence, rats that are fed a medium-fat diet live 180 to 298 days longer on average than rats fed a high-fat diet. **f.** Since each sample size is larger than 30, the confidence interval is approximately correct even if the populations are not *normal.* **g.** With at least 95% confidence, we arrive at the following conclusions: (1) Rats fed a medium-fat diet lived an average of 166 to 312 days longer than rats fed a high-fat diet. (2) Rats fed a low-fat diet lived an average of 199 to 331 days longer than rats fed a high-fat diet. (3) Since the confidence interval for $\mu_2 - \mu_3$ contains zero, the results are inconclusive concerning the comparison of the low-fat diet and the medium-fat diet. **11.3a.** $\widehat{\mu}_1 = 4.28$, $\widehat{\mu}_2 = 5.12$, $\widehat{\mu}_3 = 8.22$, and $\widehat{\mu}_4 = 7.94$. **b.** The estimates are shown in the following table.

Quantity	Point estimate
$\mu_1 - \mu_2$	−0.84
$\mu_1 - \mu_3$	−3.94
$\mu_1 - \mu_4$	−3.66
$\mu_2 - \mu_3$	−3.10
$\mu_2 - \mu_4$	−2.82
$\mu_3 - \mu_4$	0.28

c. From these confidence intervals we can conclude, with confidence at least 95%, that μ_3 and μ_4 are both greater than μ_1 and μ_2 and that μ_2 is greater than μ_1. However, we cannot decide whether μ_3 is greater than μ_4 or μ_4 is greater than μ_3. This is because the confidence interval for $\mu_3 - \mu_4$ contains both positive and negative values. **11.5a.** Population 1 consists of all households listed in the Tempe telephone directory. Population 2 consists of all households listed in the Chandler telephone directory. Population 3 consists of all households listed in the Mesa telephone directory. **b.** $\hat{\mu}_1 = \$203.12$, $\hat{\mu}_2 = \$231.96$, and $\hat{\mu}_3 = \$180.78$. **c.** $L = -\$49.91$ and $U = -\$7.69$. **d.** Yes. The average for Chandler exceeds the average for Mesa by at least $25.70 (with a confidence of at least 95%). **11.7a.** The sampled populations in this problem are conceptual. The first sampled population consists of the test scores of all students assuming they were assigned to classes of size 40 and that they completed the course. The second sampled population consists of the test scores of all students assuming they were assigned to classes of size 160 and that they completed the course. The third sampled population consists of the test scores of all students assuming they were assigned to classes of size 160 with the "help" session and that they completed the course. **b.** $\hat{\mu}_{\text{small}} = 70.17$, $\hat{\mu}_{\text{large}} = 72.48$, and $\hat{\mu}_{\text{help}} = 82.83$. **c.** Based on the set of confidence intervals in Exhibit 11.6.10, the class with the help session has a greater average test score than either of the other two class formats. Since the confidence interval that compares small and large sections contains both negative and positive values, it cannot be determined if there is any difference between these two formats. **11.9a.** The sampled population of items is all residential customers in July 1996. **b.** $\widehat{\mu}_1 - \widehat{\mu}_2 = 297.3$ kwh. **c.** $s_{\text{d}} = 302.1$ kwh. **d.** $L = 184.2$ kwh and $U = 410.4$ kwh. **e.** Based on this confidence interval, the average kwh was from 184.2 kwh to 410.4 kwh less in July 1997 than it was in July 1996. If weather conditions are comparable for these two months, it appears the energy audit program is useful in decreasing electrical usage. **11.11a.** $\widehat{\mu}_1 = 61.0$, $\widehat{\mu}_2 = 52.5$, and $\widehat{\mu}_3 = 25.14$. The 99% confidence interval for $\mu_1 - \mu_2$ is from $L = 2.8$ to $U = 14.2$; the 99% confidence interval for $\mu_1 - \mu_3$ is from $L = 30.4$ to $U = 41.3$; and the 99% confidence interval for $\mu_2 - \mu_3$ is from $L = 21.8$ to $U = 33.0$. The confidence that all three intervals are simultaneously correct is 97%. **b.** Yes. The proportion of blockage with the aggressive treatment is at least 30.4% less than treatment 1 and at least 21.8% less than treatment 2. **c.** Yes. The proportion of blockage with the moderate treatment is at least 2.8% less than treatment 1.

Chapter 12

12.1a. $\widehat{\mu}_Y(X_1, X_2) = 1.10 + 0.406X_1 + 0.00190X_2$. **b.** This number is $\widehat{\beta}_1$, the estimate of β_1. It is the estimated change in $\mu_Y(X_1, X_2)$ when X_1 increases one unit and X_2 is fixed. **c.** $\widehat{Y}(3.60, 600) = 3.70$. **d.** The 95% prediction interval is $L = 3.36$ to $U = 4.04$. Since a grade point average cannot exceed 4.00, we can write the upper bound as $U = 4.00$. **e.** $\widehat{\mu}_Y(3.60, 600) = 3.70$. **f.** The 95% confidence interval is $L = 3.63$ to $U = 3.77$. **g.** The confidence interval in part (f) is shorter than the prediction interval in part (d) because there is less uncertainty when estimating the mean of a subpopulation than there is when predicting the value of an item to be chosen at random from the subpopulation. **h.** $X_2 = 590$ (rounding up). **i.** $\widehat{\sigma} = 0.1624$. **12.3a.** $\widehat{\mu}_Y(X_1, X_2, X_3) = 3.50 + 0.172X_1 + 0.0118X_2 + 0.0103X_3$. **b.** $\widehat{Y}(1, 20, 15) = 4.06$. **c.** The 95% confidence interval for β_3 is $L = 0.00134$ to $U = 0.01936$. **d.** No. Since $\widehat{\beta}_3$ is positive, the professor with the greater value of X_3 will be predicted to have the greater value of Y. **e.** $\widehat{\sigma} = 0.2351$. **12.5a.** The sampled population consists of all individuals who took the proficiency tests during the past two years and were hired as office professionals. **b.** $\widehat{\mu}_Y(X_1, X_2) = 20.6 + 0.0429X_1 + 0.0988X_2$. **c.** $\widehat{Y}(95, 98) = 34.4$ (thousand dollars). **d.** The 95% prediction interval is $L = 31.956$ to $U = 36.807$ (thousand dollars). **e.** $\widehat{\mu}_Y(75, 75) = 31.2$ (thousand dollars). **f.** The 95% confidence interval is $L = 30.834$ to $U = 31.668$ (thousand dollars).

Chapter 13

13.1 $\sigma = 0.1$ cm. **13.3a.** The proportion of crystal weights between 11 grams and 13 grams is 0.6826. **b.** $\widehat{\pi} = 0.60$, $L = 0.50$, and $U = 0.70$. Since 0.6826 is included in this interval, there is no evidence to suggest the process is out of control. **13.5a.** 4.55% of the boards will be rejected. **b.** The center line is 30, UCL = 30.75, and LCL = 29.25. **c.** There are no points beyond the control limits, and it appears the process is in control. **13.7** The center line is 5 gallons, UCL = 5.3, and LCL = 4.7.

Chapter 14

14.1a. Yes, every group of 100 phone numbers has the same chance of being selected. **b.** This could be done in several ways. Since there are 1,000 pages in the book, one could select a random integer from 1 to 10 to identify the first sampled page and then sample every tenth page thereafter. For example, if the random integer selected is 6, then the systematic sample consists of pages 6, 16, 26, . . . , 996. For each sampled page, one could select the first name on the page. The systematic procedure seems appropriate since it is unlikely the response variable of interest (subscription to a long-distance phone program) would be related to the order of the phone numbers in the book. **14.3a.** The highest response rate would be expected from personal interviews. **b.** The lowest response rate would be expected from a mail survey. **c.** It is expected that personal interviews would be the most expensive method. **14.5a.** In order to select a simple random sample, the accountant could randomly select 400 integers between 1 and 4,000. The accountant would then count through the entire tray and pull out the statements that correspond to the sampled numbers. **b.** In order to select a systematic sample, the accountant could pick one random integer from 1 to 10. Suppose the number 3 is selected. The accountant would select the third statement and continue through the tray, selecting every tenth statement that follows. For example, the accountant would pick statements 3, 13, 23, . . . , 3,993. **c.** Both of these procedures would be relatively easy to perform, but the systematic approach would probably be faster. Since the accounts are ordered alphabetically, there is no reason to believe that the probability of finding a statement with an incorrect balance is related to its position in the tray.

APPENDIX B

CALCULUS SCORES DATA

Item number	Score	Major	Item number	Score	Major	Item number	Score	Major	Item number	Score	Major
1	64	1	41	47	1	81	67	1	121	68	1
2	86	1	42	64	1	82	75	1	122	70	1
3	85	1	43	88	1	83	67	1	123	70	1
4	57	1	44	67	1	84	97	1	124	89	1
5	63	0	45	89	1	85	78	1	125	41	0
6	94	1	46	84	0	86	67	0	126	72	1
7	71	0	47	81	1	87	87	1	127	68	1
8	87	1	48	90	1	88	71	1	128	68	1
9	69	1	49	87	1	89	79	1	129	75	1
10	73	0	50	74	1	90	71	0	130	73	1
11	77	1	51	77	1	91	64	1	131	84	1
12	70	0	52	92	1	92	96	1	132	79	1
13	76	1	53	71	1	93	57	0	133	98	1
14	86	1	54	71	1	94	73	1	134	98	1
15	63	1	55	59	0	95	78	1	135	80	1
16	78	1	56	81	1	96	68	0	136	81	1
17	88	1	57	62	1	97	76	1	137	70	0
18	85	1	58	81	1	98	85	0	138	82	1
19	54	0	59	51	0	99	74	1	139	78	1
20	66	1	60	69	0	100	91	1	140	55	1
21	73	1	61	70	0	101	88	1	141	57	0
22	67	1	62	53	1	102	63	0	142	78	1
23	64	1	63	74	1	103	82	0	143	72	1
24	58	1	64	69	0	104	83	1	144	89	1
25	91	1	65	90	1	105	88	1	145	72	0
26	94	0	66	84	1	106	57	0	146	59	1
27	82	0	67	64	1	107	45	0	147	83	1
28	81	0	68	69	0	108	72	1	148	86	0
29	71	0	69	48	0	109	82	1	149	63	0
30	74	1	70	63	0	110	91	1	150	73	0
31	65	1	71	84	1	111	79	1	151	82	0
32	77	1	72	68	1	112	70	0	152	88	1
33	57	1	73	55	1	113	64	1	153	93	1
34	71	0	74	85	1	114	77	0	154	65	1
35	83	1	75	74	0	115	68	0	155	72	1
36	83	0	76	83	1	116	61	0	156	86	1
37	65	0	77	78	1	117	86	1	157	66	1
38	84	1	78	63	1	118	76	1	158	64	1
39	92	1	79	79	0	119	98	1	159	67	1
40	87	1	80	85	0	120	66	1	160	68	0

Calculus Scores Data (Continued)

Item number	Score	Major	Item number	Score	Major	Item number	Score	Major	Item number	Score	Major
161	78	0	211	79	1	261	49	0	311	71	1
162	74	1	212	75	1	262	64	1	312	91	0
163	81	1	213	82	1	263	63	1	313	96	1
164	82	0	214	83	1	264	78	1	314	69	1
165	87	0	215	45	1	265	75	1	315	73	1
166	58	0	216	73	1	266	85	1	316	89	1
167	55	1	217	60	0	267	80	1	317	62	0
168	63	1	218	76	1	268	87	1	318	59	1
169	69	1	219	73	1	269	66	1	319	72	0
170	71	0	220	60	0	270	79	1	320	52	0
171	64	1	221	76	1	271	67	1	321	68	1
172	45	0	222	95	1	272	64	1	322	72	0
173	89	1	223	66	1	273	67	1	323	68	0
174	94	1	224	66	1	274	75	1	324	54	0
175	83	1	225	57	1	275	81	1	325	55	1
176	77	1	226	89	1	276	74	1	326	69	1
177	75	1	227	66	1	277	73	1	327	68	1
178	92	1	228	79	1	278	71	1	328	83	1
179	59	0	229	64	0	279	69	0	329	86	1
180	67	0	230	88	1	280	84	1	330	89	1
181	76	1	231	51	1	281	68	0	331	77	1
182	83	0	232	71	1	282	73	1	332	58	1
183	79	1	233	90	1	283	66	0	333	63	0
184	77	1	234	76	0	284	90	1	334	81	1
185	47	0	235	66	0	285	73	1	335	72	1
186	73	0	236	80	1	286	93	1	336	75	1
187	80	1	237	88	1	287	79	1	337	78	1
188	71	0	238	72	1	288	89	1	338	69	1
189	75	0	239	70	1	289	76	0	339	74	1
190	80	1	240	75	1	290	84	1	340	74	0
191	80	0	241	64	1	291	62	1	341	86	0
192	86	1	242	79	1	292	86	1	342	72	0
193	84	1	243	91	0	293	81	0	343	58	1
194	61	1	244	60	0	294	67	1	344	76	0
195	81	1	245	74	0	295	84	1	345	68	1
196	85	1	246	60	0	296	94	1	346	83	1
197	73	1	247	67	0	297	69	1	347	59	0
198	74	1	248	58	0	298	64	1	348	65	1
199	57	0	249	71	1	299	67	1	349	75	1
200	71	0	250	82	1	300	78	1	350	70	1
201	88	0	251	81	1	301	64	1	351	67	1
202	82	0	252	64	1	302	73	1	352	82	1
203	90	1	253	77	0	303	79	1	353	86	1
204	82	0	254	56	1	304	90	1	354	69	0
205	88	0	255	88	0	305	64	1	355	79	0
206	78	1	256	71	1	306	77	1	356	86	0
207	88	1	257	96	1	307	92	1	357	94	1
208	76	0	258	57	0	308	74	1	358	61	0
209	93	1	259	81	1	309	80	1	359	83	1
210	69	1	260	71	1	310	63	0	360	78	1

Calculus Scores Data (Continued)

Item number	Score	Major	Item number	Score	Major	Item number	Score	Major	Item number	Score	Major
361	57	0	411	95	0	461	94	1	511	80	1
362	62	1	412	87	1	462	84	0	512	81	0
363	73	0	413	72	1	463	74	1	513	89	1
364	65	0	414	75	1	464	57	1	514	82	1
365	81	1	415	85	1	465	81	1	515	57	0
366	91	1	416	61	1	466	68	0	516	61	0
367	69	1	417	70	0	467	66	0	517	70	1
368	82	1	418	86	1	468	87	1	518	61	1
369	65	0	419	86	1	469	62	0	519	95	1
370	86	1	420	68	1	470	87	1	520	60	1
371	91	0	421	67	1	471	99	1	521	64	1
372	67	1	422	60	0	472	82	1	522	62	1
373	65	0	423	68	0	473	80	1	523	83	1
374	75	1	424	67	1	474	78	1	524	80	0
375	91	1	425	64	0	475	70	0	525	95	1
376	89	1	426	90	1	476	65	0	526	78	1
377	88	1	427	73	1	477	92	1	527	76	0
378	80	1	428	54	1	478	75	0	528	81	1
379	84	0	429	89	0	479	83	1	529	100	1
380	75	0	430	69	1	480	93	1	530	88	1
381	74	1	431	77	0	481	83	0	531	71	1
382	59	1	432	68	1	482	85	1	532	78	1
383	48	0	433	81	0	483	56	0	533	71	0
384	75	1	434	74	0	484	75	0	534	66	1
385	82	1	435	77	1	485	90	1	535	87	1
386	63	1	436	82	1	486	59	1	536	80	1
387	54	0	437	64	0	487	80	1	537	96	1
388	52	0	438	60	0	488	89	1	538	41	0
389	88	1	439	55	0	489	81	0	539	85	0
390	61	1	440	65	1	490	71	1	540	87	1
391	72	1	441	64	1	491	86	1	541	84	1
392	69	1	442	76	0	492	66	1	542	80	1
393	66	1	443	75	1	493	86	1	543	69	1
394	93	0	444	66	0	494	69	1	544	64	0
395	75	1	445	74	0	495	61	1	545	58	1
396	70	1	446	54	1	496	69	1	546	56	1
397	67	1	447	67	1	497	87	1	547	59	1
398	68	1	448	71	0	498	84	1	548	83	1
399	72	1	449	68	1	499	86	1	549	78	0
400	80	1	450	72	1	500	81	1	550	62	0
401	71	0	451	84	1	501	64	1	551	50	1
402	75	0	452	85	0	502	75	1	552	86	1
403	75	0	453	55	1	503	69	1	553	74	1
404	63	0	454	58	0	504	63	0	554	85	1
405	81	1	455	80	1	505	52	1	555	91	1
406	95	1	456	75	1	506	84	1	556	47	1
407	73	1	457	84	1	507	77	1	557	62	1
408	84	1	458	55	0	508	82	1	558	67	1
409	73	1	459	82	0	509	70	1	559	91	1
410	60	0	460	79	1	510	57	0	560	76	1

Calculus Scores Data (Continued)

Item number	Score	Major	Item number	Score	Major	Item number	Score	Major	Item number	Score	Major
561	93	1	611	63	1	661	93	1	711	41	0
562	92	1	612	96	1	662	96	0	712	93	1
563	62	1	613	68	1	663	75	0	713	79	1
564	89	1	614	72	0	664	67	1	714	92	1
565	60	1	615	68	1	665	67	1	715	84	0
566	76	1	616	86	0	666	73	0	716	64	0
567	66	1	617	72	1	667	90	1	717	65	1
568	70	1	618	75	1	668	80	1	718	83	1
569	91	1	619	70	1	669	58	0	719	88	1
570	88	0	620	74	1	670	57	1	720	89	0
571	72	1	621	57	0	671	79	1	721	87	1
572	75	1	622	85	1	672	88	1	722	84	1
573	59	0	623	84	1	673	87	0	723	65	0
574	85	0	624	67	1	674	69	1	724	74	1
575	63	0	625	68	0	675	57	1	725	71	0
576	79	1	626	83	1	676	58	1	726	60	0
577	68	1	627	78	1	677	71	1	727	83	1
578	83	1	628	83	1	678	64	0	728	89	1
579	66	0	629	83	1	679	73	0	729	73	1
580	74	0	630	71	0	680	78	0	730	99	1
581	69	1	631	70	1	681	80	0	731	63	0
582	78	1	632	77	1	682	78	1	732	75	1
583	68	1	633	79	1	683	83	1	733	75	0
584	70	0	634	91	0	684	53	0	734	76	1
585	60	0	635	56	1	685	90	1	735	72	0
586	84	1	636	59	0	686	67	1	736	93	1
587	80	1	637	75	1	687	76	0	737	85	1
588	72	0	638	89	1	688	79	1	738	87	1
589	69	1	639	68	1	689	90	1	739	83	0
590	84	1	640	89	1	690	82	0	740	82	1
591	67	1	641	63	0	691	92	1	741	85	0
592	70	1	642	75	0	692	82	1	742	75	0
593	91	1	643	86	1	693	82	1	743	41	1
594	78	1	644	80	1	694	60	1	744	74	1
595	67	1	645	62	1	695	77	1	745	98	1
596	77	1	646	80	1	696	88	1	746	61	1
597	79	0	647	85	1	697	42	1	747	79	1
598	76	1	648	67	0	698	56	0	748	76	1
599	83	0	649	60	1	699	72	1	749	72	1
600	86	1	650	80	1	700	81	1	750	78	1
601	83	1	651	73	1	701	76	1	751	71	1
602	78	0	652	85	1	702	95	1	752	61	1
603	75	0	653	69	1	703	82	1	753	90	1
604	77	1	654	66	0	704	81	0	754	82	1
605	62	1	655	77	1	705	67	0	755	85	0
606	77	0	656	72	1	706	62	1	756	83	1
607	81	1	657	71	1	707	71	1	757	88	0
608	90	0	658	86	1	708	88	1	758	70	1
609	95	1	659	79	0	709	75	1	759	90	0
610	87	0	660	72	0	710	85	1	760	61	1

Calculus Scores Data (Continued)

Item number	Score	Major	Item number	Score	Major	Item number	Score	Major	Item number	Score	Major
761	78	1	811	69	1	861	71	0	911	80	1
762	59	1	812	74	1	862	74	1	912	79	1
763	77	1	813	75	1	863	84	1	913	59	1
764	83	1	814	92	0	864	72	1	914	62	0
765	59	0	815	74	1	865	74	1	915	76	1
766	68	0	816	77	1	866	75	0	916	70	1
767	61	0	817	63	1	867	91	0	917	77	1
768	82	0	818	92	1	868	52	0	918	82	1
769	82	1	819	63	1	869	81	1	919	76	1
770	60	1	820	93	1	870	92	1	920	56	1
771	58	1	821	68	0	871	79	1	921	72	1
772	65	1	822	78	1	872	78	1	922	90	1
773	64	1	823	75	1	873	52	1	923	72	1
774	65	0	824	76	0	874	83	1	924	73	1
775	87	1	825	76	1	875	58	0	925	89	1
776	72	1	826	70	1	876	70	1	926	83	1
777	68	0	827	57	1	877	61	1	927	70	1
778	59	1	828	79	1	878	69	0	928	85	1
779	74	0	829	87	1	879	72	0	929	82	0
780	78	1	830	83	1	880	69	1	930	52	0
781	94	1	831	56	1	881	95	1	931	64	0
782	59	0	832	79	1	882	66	1	932	71	1
783	72	0	833	61	1	883	61	0	933	55	0
784	74	1	834	60	1	884	68	1	934	94	1
785	66	0	835	65	1	885	78	1	935	77	1
786	94	0	836	78	1	886	78	1	936	80	1
787	84	1	837	87	1	887	77	1	937	79	1
788	72	1	838	79	1	888	74	1	938	85	0
789	62	1	839	90	1	889	76	1	939	83	1
790	65	1	840	69	0	890	81	0	940	79	0
791	76	1	841	79	1	891	75	1	941	49	0
792	85	1	842	84	1	892	62	1	942	76	0
793	69	0	843	54	1	893	82	1	943	63	0
794	79	1	844	75	1	894	67	0	944	78	1
795	82	1	845	86	1	895	60	1	945	64	0
796	82	0	846	53	1	896	90	1	946	81	0
797	97	1	847	85	1	897	93	1	947	41	0
798	73	1	848	89	1	898	81	0	948	83	1
799	84	1	849	64	0	899	80	1	949	67	1
800	68	1	850	58	1	900	82	1	950	81	0
801	58	0	851	82	1	901	83	1	951	79	1
802	80	1	852	70	1	902	52	0	952	63	0
803	83	1	853	71	0	903	92	1	953	78	0
804	52	1	854	61	1	904	78	0	954	45	1
805	77	1	855	70	1	905	84	0	955	92	1
806	66	1	856	81	1	906	75	0	956	72	1
807	63	0	857	49	0	907	74	1	957	72	0
808	74	1	858	76	1	908	61	1	958	63	1
809	74	1	859	86	1	909	71	1	959	80	0
810	78	1	860	88	1	910	87	0	960	90	0

Calculus Scores Data (Continued)

Item number	Score	Major	Item number	Score	Major	Item number	Score	Major	Item number	Score	Major
961	59	0	1011	94	1	1061	71	1	1111	85	0
962	69	0	1012	74	0	1062	89	0	1112	74	0
963	72	0	1013	67	0	1063	60	0	1113	64	1
964	76	0	1014	82	0	1064	60	1	1114	47	0
965	76	0	1015	70	1	1065	84	1	1115	71	1
966	78	0	1016	65	1	1066	72	1	1116	67	1
967	50	1	1017	81	0	1067	61	1	1117	88	1
968	59	0	1018	90	1	1068	80	1	1118	83	0
969	93	1	1019	68	0	1069	92	1	1119	82	1
970	80	1	1020	73	0	1070	78	0	1120	87	0
971	89	1	1021	100	1	1071	73	1	1121	82	0
972	82	0	1022	77	0	1072	86	1	1122	78	1
973	94	1	1023	84	1	1073	77	1	1123	69	1
974	94	1	1024	68	1	1074	66	0	1124	65	1
975	76	0	1025	55	0	1075	69	0	1125	75	0
976	79	1	1026	88	0	1076	91	1	1126	84	0
977	66	0	1027	71	0	1077	67	1	1127	81	1
978	86	1	1028	88	0	1078	69	1	1128	62	1
979	93	1	1029	86	0	1079	92	1	1129	59	0
980	72	1	1030	75	1	1080	70	0	1130	71	1
981	83	1	1031	83	1	1081	88	1	1131	78	1
982	67	0	1032	59	1	1082	69	1	1132	62	1
983	51	1	1033	65	1	1083	69	1	1133	79	1
984	76	0	1034	58	0	1084	78	0	1134	82	1
985	77	1	1035	68	1	1085	56	0	1135	89	0
986	87	1	1036	78	1	1086	62	1	1136	70	1
987	70	0	1037	75	0	1087	61	0	1137	82	1
988	79	0	1038	70	1	1088	98	1	1138	77	0
989	61	1	1039	74	1	1089	72	1	1139	73	0
990	79	0	1040	60	1	1090	79	0	1140	88	1
991	77	0	1041	59	0	1091	68	1	1141	81	0
992	90	1	1042	80	1	1092	83	1	1142	84	1
993	53	0	1043	83	1	1093	44	0	1143	64	0
994	80	1	1044	77	1	1094	78	1	1144	85	0
995	63	0	1045	80	1	1095	88	1	1145	82	1
996	78	1	1046	71	1	1096	82	1	1146	97	1
997	67	0	1047	89	1	1097	75	1	1147	71	1
998	71	1	1048	89	1	1098	75	1	1148	71	0
999	74	1	1049	93	1	1099	67	1	1149	71	1
1000	84	0	1050	74	1	1100	76	1	1150	65	1
1001	79	1	1051	77	1	1101	70	1	1151	80	1
1002	88	0	1052	83	1	1102	61	1	1152	61	1
1003	86	1	1053	76	1	1103	92	1	1153	75	0
1004	86	1	1054	63	0	1104	69	1	1154	70	1
1005	52	1	1055	66	1	1105	79	0	1155	81	1
1006	71	1	1056	58	1	1106	80	1	1156	85	0
1007	60	1	1057	70	1	1107	49	1	1157	77	0
1008	71	0	1058	73	0	1108	52	0	1158	79	0
1009	67	1	1059	78	1	1109	78	0	1159	68	1
1010	84	1	1060	86	1	1110	63	1	1160	74	1

Calculus Scores Data (Continued)

Item number	Score	Major	Item number	Score	Major	Item number	Score	Major	Item number	Score	Major
1161	74	1	1211	72	1	1261	69	0	1311	75	1
1162	95	1	1212	93	1	1262	74	1	1312	74	1
1163	69	0	1213	59	0	1263	71	1	1313	72	0
1164	69	1	1214	74	1	1264	75	0	1314	67	1
1165	77	0	1215	60	1	1265	59	1	1315	81	0
1166	87	1	1216	83	0	1266	77	0	1316	71	1
1167	65	1	1217	74	1	1267	64	1	1317	96	1
1168	66	1	1218	53	0	1268	69	1	1318	91	1
1169	78	1	1219	76	0	1269	78	1	1319	73	1
1170	53	0	1220	74	1	1270	72	0	1320	84	1
1171	92	1	1221	78	1	1271	68	1	1321	76	0
1172	70	0	1222	58	0	1272	73	1	1322	78	0
1173	74	1	1223	78	0	1273	84	0	1323	75	1
1174	73	1	1224	87	0	1274	69	1	1324	71	1
1175	68	1	1225	85	1	1275	73	0	1325	54	0
1176	82	1	1226	67	1	1276	99	0	1326	71	0
1177	67	1	1227	65	0	1277	71	1	1327	87	1
1178	73	0	1228	60	0	1278	89	1	1328	66	1
1179	82	0	1229	70	1	1279	63	1	1329	74	0
1180	69	1	1230	87	1	1280	64	1	1330	73	1
1181	61	0	1231	51	0	1281	72	1	1331	58	1
1182	92	0	1232	73	0	1282	63	0	1332	80	1
1183	93	1	1233	88	1	1283	88	1	1333	62	0
1184	69	1	1234	63	0	1284	88	1	1334	87	0
1185	63	0	1235	84	1	1285	78	0	1335	69	1
1186	74	1	1236	75	1	1286	75	1	1336	77	1
1187	65	1	1237	67	0	1287	75	1	1337	64	0
1188	55	0	1238	54	1	1288	63	0	1338	75	1
1189	81	0	1239	77	0	1289	71	0	1339	74	1
1190	69	1	1240	64	1	1290	73	0	1340	70	1
1191	71	1	1241	78	1	1291	85	0	1341	71	1
1192	59	1	1242	79	1	1292	52	0	1342	72	0
1193	68	0	1243	69	1	1293	79	0	1343	75	0
1194	63	0	1244	61	1	1294	68	1	1344	94	1
1195	81	1	1245	74	1	1295	68	1	1345	88	1
1196	78	1	1246	75	1	1296	74	0	1346	73	1
1197	81	1	1247	66	0	1297	71	1	1347	77	1
1198	72	0	1248	63	0	1298	59	0	1348	67	1
1199	65	1	1249	85	1	1299	77	1	1349	91	1
1200	71	1	1250	67	0	1300	76	0	1350	95	1
1201	68	1	1251	72	1	1301	64	0	1351	54	1
1202	77	0	1252	71	0	1302	75	1	1352	69	1
1203	79	1	1253	79	1	1303	77	1	1353	99	1
1204	74	0	1254	95	1	1304	64	0	1354	62	1
1205	81	0	1255	78	1	1305	66	0	1355	85	0
1206	60	1	1256	77	1	1306	74	1	1356	69	1
1207	90	1	1257	71	1	1307	93	1	1357	96	0
1208	63	1	1258	74	1	1308	95	1	1358	58	1
1209	87	1	1259	52	1	1309	74	0	1359	63	0
1210	82	1	1260	69	0	1310	68	1	1360	86	1

Calculus Scores Data (Continued)

Item number	Score	Major	Item number	Score	Major	Item number	Score	Major	Item number	Score	Major
1361	71	1	1411	83	1	1461	63	0	1511	69	1
1362	74	1	1412	77	1	1462	79	1	1512	91	1
1363	71	1	1413	93	1	1463	71	1	1513	59	0
1364	70	1	1414	84	1	1464	78	1	1514	73	1
1365	61	0	1415	98	1	1465	69	1	1515	66	1
1366	70	1	1416	91	1	1466	71	1	1516	56	1
1367	77	1	1417	81	0	1467	77	1	1517	67	1
1368	82	1	1418	83	1	1468	71	0	1518	65	1
1369	90	1	1419	55	0	1469	58	0	1519	83	1
1370	90	1	1420	100	1	1470	97	1	1520	86	0
1371	84	1	1421	61	1	1471	70	1	1521	73	0
1372	79	1	1422	92	1	1472	68	0	1522	69	1
1373	65	1	1423	94	1	1473	75	1	1523	83	1
1374	72	1	1424	58	0	1474	75	1	1524	70	0
1375	74	1	1425	55	1	1475	58	0	1525	81	1
1376	73	1	1426	86	1	1476	76	1	1526	77	1
1377	71	1	1427	80	1	1477	55	0	1527	80	1
1378	80	1	1428	72	1	1478	87	1	1528	47	1
1379	74	1	1429	75	1	1479	96	1	1529	63	1
1380	65	1	1430	69	1	1480	62	0	1530	92	1
1381	72	0	1431	82	1	1481	66	1	1531	74	0
1382	69	1	1432	91	1	1482	73	1	1532	93	1
1383	80	0	1433	56	1	1483	77	0	1533	70	1
1384	82	1	1434	66	0	1484	84	1	1534	90	1
1385	78	1	1435	62	1	1485	89	1	1535	65	0
1386	72	0	1436	93	1	1486	81	1	1536	66	1
1387	80	1	1437	81	1	1487	74	0	1537	75	0
1388	73	1	1438	70	0	1488	72	1	1538	90	0
1389	71	0	1439	75	0	1489	68	1	1539	89	1
1390	87	1	1440	65	1	1490	88	0	1540	97	1
1391	66	1	1441	88	1	1491	53	1	1541	64	1
1392	78	1	1442	70	1	1492	76	1	1542	76	1
1393	60	1	1443	66	0	1493	56	1	1543	67	1
1394	63	1	1444	79	1	1494	69	0	1544	57	1
1395	67	1	1445	80	1	1495	67	0	1545	81	1
1396	65	1	1446	92	1	1496	73	1	1546	80	1
1397	77	1	1447	65	1	1497	77	1	1547	68	0
1398	72	1	1448	60	1	1498	67	0	1548	60	0
1399	75	1	1449	76	0	1499	74	1	1549	75	1
1400	64	1	1450	71	1	1500	84	1	1550	73	1
1401	57	1	1451	58	0	1501	72	0	1551	74	1
1402	64	0	1452	57	0	1502	68	1	1552	95	1
1403	69	0	1453	91	0	1503	84	1	1553	87	0
1404	84	1	1454	79	0	1504	58	1	1554	79	0
1405	99	1	1455	76	1	1505	84	0	1555	98	1
1406	71	1	1456	68	1	1506	64	0	1556	70	1
1407	70	1	1457	84	0	1507	81	1	1557	54	0
1408	75	0	1458	66	0	1508	86	1	1558	78	0
1409	69	0	1459	89	1	1509	71	1	1559	94	1
1410	56	1	1460	77	1	1510	77	1	1560	80	1

Calculus Scores Data (Continued)

Item number	Score	Major	Item number	Score	Major	Item number	Score	Major	Item number	Score	Major
1561	67	0	1611	65	1	1661	60	1	1711	57	1
1562	77	1	1612	66	0	1662	80	1	1712	92	1
1563	65	0	1613	55	0	1663	65	1	1713	84	1
1564	76	1	1614	84	1	1664	78	1	1714	96	1
1565	69	1	1615	76	0	1665	63	1	1715	75	0
1566	85	0	1616	59	1	1666	83	1	1716	57	0
1567	71	1	1617	67	1	1667	72	1	1717	83	1
1568	69	1	1618	74	1	1668	76	1	1718	71	1
1569	79	1	1619	58	0	1669	80	1	1719	82	1
1570	84	1	1620	72	0	1670	100	1	1720	90	0
1571	66	0	1621	68	1	1671	96	1	1721	84	0
1572	71	1	1622	68	0	1672	76	1	1722	71	0
1573	58	1	1623	81	0	1673	65	0	1723	88	1
1574	82	0	1624	78	1	1674	74	0	1724	75	1
1575	67	1	1625	84	1	1675	64	1	1725	78	0
1576	78	1	1626	75	0	1676	63	0	1726	55	0
1577	74	1	1627	71	1	1677	69	1	1727	73	0
1578	62	1	1628	83	0	1678	94	1	1728	73	1
1579	80	0	1629	72	1	1679	53	0	1729	95	1
1580	78	1	1630	86	1	1680	78	0	1730	77	1
1581	94	1	1631	75	1	1681	82	1	1731	61	0
1582	68	0	1632	69	1	1682	75	1	1732	72	0
1583	84	0	1633	67	1	1683	99	1	1733	71	1
1584	74	0	1634	83	1	1684	75	1	1734	71	1
1585	81	1	1635	81	0	1685	62	0	1735	62	0
1586	98	0	1636	73	1	1686	62	1	1736	75	1
1587	70	1	1637	90	1	1687	85	1	1737	83	1
1588	79	1	1638	61	1	1688	76	1	1738	83	0
1589	80	1	1639	69	1	1689	83	1	1739	57	1
1590	70	1	1640	65	1	1690	90	0	1740	48	1
1591	73	0	1641	75	1	1691	79	1	1741	95	0
1592	84	1	1642	91	1	1692	73	0	1742	87	1
1593	80	1	1643	70	1	1693	52	1	1743	86	1
1594	90	1	1644	90	0	1694	67	0	1744	85	1
1595	75	1	1645	73	1	1695	87	1	1745	79	1
1596	61	0	1646	88	1	1696	57	0	1746	81	0
1597	87	1	1647	81	1	1697	97	1	1747	78	1
1598	75	0	1648	77	1	1698	88	1	1748	61	0
1599	60	1	1649	75	1	1699	59	1	1749	75	1
1600	61	0	1650	81	1	1700	78	0	1750	58	1
1601	79	1	1651	77	1	1701	80	1	1751	70	1
1602	73	1	1652	68	1	1702	63	0	1752	72	0
1603	58	0	1653	86	1	1703	74	1	1753	99	0
1604	63	1	1654	80	0	1704	75	1	1754	76	1
1605	87	1	1655	85	0	1705	71	1	1755	77	1
1606	76	1	1656	83	1	1706	86	1	1756	75	1
1607	64	1	1657	77	1	1707	86	1	1757	72	1
1608	71	0	1658	52	0	1708	78	1	1758	79	1
1609	68	0	1659	71	0	1709	72	1	1759	76	1
1610	66	1	1660	88	0	1710	82	0	1760	58	0

Calculus Scores Data (Continued)

Item number	Score	Major	Item number	Score	Major	Item number	Score	Major	Item number	Score	Major
1761	71	1	1811	72	1	1861	92	1	1911	89	1
1762	84	1	1812	55	0	1862	53	0	1912	85	1
1763	80	0	1813	66	0	1863	98	1	1913	74	1
1764	87	1	1814	78	1	1864	78	1	1914	69	0
1765	67	1	1815	75	1	1865	74	1	1915	92	1
1766	82	1	1816	71	0	1866	78	1	1916	85	0
1767	63	1	1817	74	0	1867	64	1	1917	80	1
1768	87	1	1818	67	1	1868	95	1	1918	78	1
1769	90	1	1819	62	0	1869	71	1	1919	68	1
1770	82	1	1820	61	0	1870	66	0	1920	88	1
1771	79	1	1821	63	1	1871	62	1	1921	60	0
1772	80	1	1822	77	0	1872	65	1	1922	59	1
1773	63	1	1823	62	0	1873	79	1	1923	72	1
1774	76	1	1824	73	0	1874	63	0	1924	97	1
1775	65	0	1825	85	1	1875	90	1	1925	70	0
1776	61	0	1826	65	0	1876	83	1	1926	81	1
1777	89	1	1827	85	1	1877	63	1	1927	54	0
1778	64	1	1828	56	0	1878	78	1	1928	84	1
1779	87	1	1829	70	1	1879	78	1	1929	61	0
1780	74	1	1830	69	1	1880	77	1	1930	78	1
1781	76	1	1831	63	1	1881	83	1	1931	74	1
1782	71	1	1832	84	1	1882	74	1	1932	72	1
1783	91	1	1833	87	1	1883	62	1	1933	87	1
1784	63	1	1834	83	1	1884	69	1	1934	87	1
1785	49	0	1835	86	1	1885	76	0	1935	78	1
1786	81	1	1836	63	1	1886	46	0	1936	85	1
1787	72	1	1837	86	1	1887	73	1	1937	61	1
1788	74	1	1838	63	1	1888	56	1	1938	88	1
1789	77	0	1839	68	1	1889	72	0	1939	59	0
1790	77	1	1840	69	0	1890	85	1	1940	83	1
1791	79	1	1841	73	0	1891	81	1	1941	82	1
1792	68	1	1842	64	1	1892	57	0	1942	65	0
1793	74	1	1843	70	1	1893	71	1	1943	58	1
1794	77	1	1844	75	1	1894	72	1	1944	75	0
1795	86	1	1845	83	1	1895	94	1	1945	74	1
1796	68	1	1846	76	1	1896	92	1	1946	59	0
1797	93	1	1847	68	1	1897	97	0	1947	78	1
1798	64	0	1848	72	0	1898	82	1	1948	66	0
1799	52	1	1849	61	0	1899	90	1	1949	96	1
1800	63	1	1850	68	1	1900	71	1	1950	57	1
1801	76	1	1851	75	0	1901	76	1	1951	61	0
1802	76	1	1852	84	1	1902	79	1	1952	74	1
1803	72	1	1853	75	1	1903	85	0	1953	100	1
1804	72	1	1854	73	0	1904	93	1	1954	72	0
1805	66	1	1855	76	1	1905	85	0	1955	59	1
1806	61	0	1856	44	0	1906	68	1	1956	59	1
1807	75	1	1857	84	1	1907	96	1	1957	93	1
1808	64	0	1858	68	0	1908	78	1	1958	65	1
1809	78	1	1859	68	1	1909	75	0	1959	82	1
1810	63	0	1860	62	1	1910	70	1	1960	59	0

Calculus Scores Data (Continued)

Item number	Score	Major	Item number	Score	Major	Item number	Score	Major	Item number	Score	Major
1961	71	0	2011	79	0	2061	58	0	2111	64	1
1962	83	1	2012	67	1	2062	64	1	2112	75	1
1963	81	1	2013	84	0	2063	62	0	2113	67	1
1964	50	0	2014	61	1	2064	66	1	2114	76	0
1965	83	1	2015	52	0	2065	97	1	2115	82	1
1966	91	0	2016	70	1	2066	69	1	2116	93	1
1967	72	0	2017	45	1	2067	83	1	2117	77	1
1968	80	1	2018	77	1	2068	76	1	2118	85	1
1969	59	0	2019	74	1	2069	80	1	2119	79	1
1970	79	0	2020	73	0	2070	76	1	2120	75	1
1971	69	1	2021	68	1	2071	62	1	2121	82	1
1972	91	1	2022	65	0	2072	95	1	2122	83	1
1973	66	0	2023	98	1	2073	93	1	2123	91	1
1974	75	0	2024	62	0	2074	65	0	2124	69	1
1975	81	1	2025	74	1	2075	70	1	2125	77	1
1976	98	1	2026	58	0	2076	83	1	2126	68	1
1977	79	0	2027	76	1	2077	70	1	2127	83	1
1978	86	1	2028	80	1	2078	64	0	2128	91	1
1979	73	1	2029	90	1	2079	68	1	2129	75	1
1980	59	1	2030	89	1	2080	71	1	2130	52	1
1981	88	1	2031	99	1	2081	87	1	2131	75	1
1982	62	0	2032	86	1	2082	71	1	2132	73	1
1983	68	1	2033	78	1	2083	71	0	2133	88	1
1984	68	0	2034	64	1	2084	65	0	2134	52	0
1985	84	1	2035	68	1	2085	97	1	2135	53	1
1986	81	0	2036	79	1	2086	78	1	2136	70	1
1987	62	1	2037	85	0	2087	71	1	2137	86	0
1988	68	1	2038	88	1	2088	41	1	2138	72	1
1989	66	1	2039	73	1	2089	81	1	2139	69	1
1990	67	1	2040	71	0	2090	76	1	2140	83	1
1991	75	0	2041	86	0	2091	45	1	2141	72	1
1992	65	0	2042	85	0	2092	64	1	2142	72	1
1993	63	1	2043	57	0	2093	83	1	2143	53	0
1994	78	1	2044	70	1	2094	69	0	2144	70	0
1995	81	1	2045	88	1	2095	88	1	2145	74	1
1996	80	0	2046	78	1	2096	78	0	2146	71	1
1997	97	1	2047	81	1	2097	82	1	2147	89	1
1998	91	1	2048	81	0	2098	75	1	2148	78	1
1999	52	1	2049	72	0	2099	77	1	2149	70	1
2000	96	1	2050	84	1	2100	80	0	2150	79	1
2001	59	1	2051	63	0	2101	83	0	2151	67	0
2002	50	0	2052	84	1	2102	86	1	2152	68	1
2003	64	0	2053	69	1	2103	70	1	2153	59	1
2004	81	1	2054	79	1	2104	81	1	2154	67	1
2005	71	1	2055	78	0	2105	79	1	2155	82	0
2006	76	1	2056	87	1	2106	84	1	2156	73	0
2007	81	1	2057	72	1	2107	78	1	2157	74	1
2008	57	0	2058	64	1	2108	72	1	2158	81	1
2009	79	1	2059	59	1	2109	98	1	2159	86	1
2010	58	1	2060	69	1	2110	79	1	2160	68	1

Calculus Scores Data (Continued)

Item number	Score	Major	Item number	Score	Major	Item number	Score	Major	Item number	Score	Major
2161	82	0	2211	59	1	2261	81	1	2311	82	1
2162	74	1	2212	77	0	2262	75	1	2312	90	0
2163	73	1	2213	78	0	2263	72	0	2313	68	1
2164	65	0	2214	75	1	2264	99	1	2314	64	0
2165	55	0	2215	55	0	2265	98	1	2315	80	1
2166	81	1	2216	86	0	2266	66	0	2316	71	1
2167	73	1	2217	60	0	2267	79	0	2317	71	1
2168	82	0	2218	71	1	2268	60	0	2318	49	1
2169	77	1	2219	69	0	2269	69	0	2319	72	1
2170	83	1	2220	70	0	2270	62	0	2320	75	1
2171	71	0	2221	72	1	2271	73	0	2321	53	1
2172	62	1	2222	70	1	2272	67	1	2322	80	1
2173	71	0	2223	82	1	2273	66	0	2323	72	1
2174	60	0	2224	89	1	2274	86	1	2324	76	1
2175	91	1	2225	79	1	2275	75	1	2325	76	1
2176	92	1	2226	95	1	2276	87	1	2326	58	1
2177	73	0	2227	57	1	2277	58	1	2327	90	0
2178	65	1	2228	63	1	2278	76	1	2328	98	1
2179	78	1	2229	63	0	2279	67	1	2329	73	1
2180	66	1	2230	81	1	2280	61	1	2330	87	1
2181	93	1	2231	92	1	2281	76	1	2331	69	1
2182	64	1	2232	81	1	2282	82	1	2332	42	0
2183	84	1	2233	71	0	2283	65	1	2333	76	1
2184	94	1	2234	81	1	2284	72	1	2334	78	1
2185	61	1	2235	85	1	2285	79	0	2335	61	0
2186	86	1	2236	74	1	2286	56	1	2336	73	0
2187	85	1	2237	59	1	2287	72	0	2337	85	1
2188	89	1	2238	60	0	2288	72	1	2338	75	1
2189	79	0	2239	74	1	2289	78	0	2339	75	1
2190	87	0	2240	60	1	2290	72	0	2340	60	0
2191	64	1	2241	74	1	2291	72	0	2341	79	1
2192	63	1	2242	71	0	2292	86	1	2342	78	0
2193	66	1	2243	62	0	2293	73	0	2343	90	1
2194	71	1	2244	50	1	2294	67	0	2344	80	0
2195	55	0	2245	100	1	2295	74	0	2345	61	1
2196	73	1	2246	78	1	2296	71	1	2346	66	0
2197	74	1	2247	76	1	2297	69	0	2347	72	1
2198	65	0	2248	74	1	2298	75	0	2348	62	1
2199	79	0	2249	76	1	2299	69	1	2349	94	1
2200	85	1	2250	53	0	2300	80	1	2350	79	1
2201	76	0	2251	82	1	2301	83	1	2351	85	0
2202	88	1	2252	72	1	2302	85	1	2352	55	0
2203	63	0	2253	83	1	2303	74	1	2353	74	1
2204	69	1	2254	75	1	2304	84	0	2354	74	1
2205	84	1	2255	72	1	2305	70	0	2355	82	1
2206	76	1	2256	77	1	2306	87	0	2356	67	0
2207	76	1	2257	82	1	2307	83	1	2357	86	1
2208	87	1	2258	92	1	2308	66	1	2358	80	1
2209	51	1	2259	84	1	2309	81	1	2359	76	0
2210	71	1	2260	66	1	2310	73	1	2360	70	0

Calculus Scores Data (Continued)

Item number	Score	Major	Item number	Score	Major	Item number	Score	Major	Item number	Score	Major
2361	55	1	2411	88	1	2461	88	0	2511	75	0
2362	70	1	2412	66	1	2462	70	0	2512	82	1
2363	81	1	2413	54	0	2463	68	1	2513	78	1
2364	76	0	2414	61	1	2464	65	1	2514	74	1
2365	76	1	2415	61	1	2465	49	0	2515	78	1
2366	91	1	2416	63	0	2466	78	1	2516	79	0
2367	57	1	2417	83	1	2467	58	1	2517	80	0
2368	84	1	2418	84	1	2468	92	0	2518	82	1
2369	74	0	2419	72	0	2469	72	1	2519	71	1
2370	74	1	2420	75	0	2470	72	1	2520	84	0
2371	85	1	2421	54	1	2471	81	1	2521	68	1
2372	95	1	2422	70	1	2472	67	0	2522	71	1
2373	70	0	2423	89	1	2473	73	1	2523	70	1
2374	69	1	2424	97	1	2474	74	0	2524	81	1
2375	75	0	2425	77	1	2475	90	1	2525	83	0
2376	71	1	2426	66	1	2476	87	1	2526	68	1
2377	64	1	2427	70	1	2477	75	1	2527	76	0
2378	96	1	2428	59	0	2478	72	1	2528	66	0
2379	79	1	2429	73	1	2479	79	0	2529	85	1
2380	68	1	2430	78	1	2480	56	1	2530	76	0
2381	58	0	2431	65	0	2481	84	1	2531	75	1
2382	72	1	2432	73	1	2482	73	1	2532	55	1
2383	71	1	2433	55	1	2483	86	1	2533	86	1
2384	94	0	2434	79	1	2484	85	1	2534	87	1
2385	65	0	2435	77	1	2485	78	1	2535	97	1
2386	96	0	2436	84	1	2486	80	1	2536	78	1
2387	82	1	2437	80	1	2487	60	0	2537	92	1
2388	51	1	2438	74	0	2488	69	1	2538	72	0
2389	81	1	2439	73	1	2489	78	1	2539	87	1
2390	81	1	2440	84	1	2490	68	1	2540	61	0
2391	77	0	2441	91	0	2491	65	0	2541	74	1
2392	64	1	2442	91	0	2492	74	1	2542	99	1
2393	57	1	2443	67	1	2493	59	0	2543	73	0
2394	86	1	2444	56	1	2494	72	1	2544	70	0
2395	70	1	2445	80	1	2495	86	1	2545	63	0
2396	82	1	2446	67	1	2496	70	0	2546	81	1
2397	88	1	2447	88	1	2497	74	0	2547	72	1
2398	79	1	2448	78	0	2498	93	1	2548	68	1
2399	89	0	2449	81	1	2499	68	1	2549	72	1
2400	100	1	2450	63	1	2500	77	1	2550	79	0
2401	76	1	2451	84	1	2501	72	1	2551	74	1
2402	68	0	2452	76	1	2502	52	0	2552	66	0
2403	84	1	2453	78	1	2503	84	1	2553	63	1
2404	79	1	2454	86	1	2504	70	1	2554	66	1
2405	75	0	2455	88	1	2505	66	1	2555	65	1
2406	73	1	2456	75	1	2506	78	1	2556	81	1
2407	88	1	2457	72	1	2507	55	1	2557	75	1
2408	75	1	2458	80	1	2508	74	1	2558	74	0
2409	68	1	2459	78	1	2509	62	0	2559	68	1
2410	78	1	2460	68	0	2510	69	1	2560	92	1

Calculus Scores Data (Continued)

Item number	Score	Major	Item number	Score	Major	Item number	Score	Major	Item number	Score	Major
2561	85	0	2571	68	0	2581	70	0	2591	84	1
2562	78	1	2572	75	1	2582	73	1	2592	83	1
2563	71	0	2573	79	1	2583	72	0	2593	85	1
2564	70	1	2574	81	0	2584	86	1	2594	90	0
2565	73	1	2575	89	1	2585	78	1	2595	71	1
2566	86	0	2576	56	0	2586	74	1	2596	70	1
2567	86	0	2577	72	0	2587	90	1	2597	73	0
2568	71	1	2578	74	0	2588	88	1	2598	61	1
2569	86	1	2579	67	1	2589	72	1	2599	53	1
2570	68	0	2580	80	1	2590	75	0	2600	84	1

APPENDIX C

A BRIEF INTRODUCTION TO MINITAB

INTRODUCTION

Minitab is a powerful interactive statistical computing package that can be used to perform a variety of data analyses ranging from the calculation of simple descriptive statistics to very complex univariate and multivariate modeling. The early versions of Minitab were command-driven. The more recent versions offer both options—a command language as well as a mouse-and-menu approach. In this appendix we give a very brief introduction to Minitab so that you can use it to perform computations presented in this text. We will explain the command language only, because it is very simple for all of the activities you will be required to do in this book. For brevity, our descriptions are based on the assumption that you are using a PC, but most of our command-language descriptions will apply verbatim to mainframe, workstation, and Macintosh versions.

As you read this text, you will see many Minitab commands that are used to perform calculations and construct graphs. In this appendix we merely introduce you to the Minitab computing system. We presume that you have used computers and know some of the basic operations involved.

MINITAB USER INTERFACE

Recent versions of Minitab for personal computers operate under the Windows operating system, whereas earlier versions operate under DOS. When you use Minitab for Windows, the display consists of several windows. The three main windows are (a) the Session window, (b) the Data window, and (c) the Graph window. The Session window is where you can enter data, enter commands, and observe the results of the calculations you instruct Minitab to perform. The Graph window is where high-resolution graphs appear when appropriate commands are issued. You can enter the data values directly into the cells of the Data window as you would in any spreadsheet program.

In the early DOS versions and the mainframe versions of Minitab, there is only one window, the Session window. The later DOS versions do offer a Data window, but only one of the two windows can be viewed at any given time.

FOUR MAIN ACTIVITIES

To use Minitab to process a set of data, you must

1. start Minitab,
2. enter data into Minitab,
3. perform the desired computations or construct the required graphs, and
4. save the results in a file for later use.

We will discuss each of these topics in sequence.

1. START MINITAB

Under the Windows operating system, Minitab can be invoked by clicking the left mouse button on the icon corresponding to Minitab (ask your local systems expert if there is no obvious icon for Minitab). This will open the Session window and/or the Data window. The Graph window appears only when a command is issued to construct a graph. Activate the Session window, and you will notice the following Minitab prompt:

```
MTB >
```

Under DOS, Minitab can usually be invoked by typing the command `minitab` at the DOS prompt. For instance, at the prompt

```
C:\>
```

type `minitab` and press ENTER.

Minitab commands are typed on the same line as the Minitab prompt `MTB >`. After a command is typed, press ENTER; this instructs Minitab to execute the command. Alternatively, you can use the menu items on the menu bar if you want to use the point-and-click approach with the mouse.

2. ENTER DATA INTO MINITAB

Before you can process data or construct graphs, you must make your data accessible to Minitab. This can be done in several ways, three of which are as follows:

- manual data entry using the keyboard;
- reading the data from an ASCII data file;
- reading the data from a Minitab worksheet file.

In Minitab, data are organized in a rectangular table (spreadsheet) containing several rows and columns. The number of rows and columns available for data storage depends on the amount of memory available on your computer. Each data value belongs to a cell identified by its row and column. By default, the columns are labeled C1, C2, C3, . . . , and so forth. Generally, columns represent variables and rows represent items. Two data types are available in Minitab—numeric and nonnumeric. A simple example of nonnumeric data arises when you want to record the gender of a subject and the categories are *male* and *female*.

a. ***Manual entry of numeric data using the keyboard***

We will show you how to enter data into the Session window using the keyboard. The command `set` can be used to enter data values into a single column of the worksheet. Suppose you want to enter the following two columns of data into two columns of a Minitab worksheet:

```
 6    21
28    39
11    63
23
20
```

Suppose you elect to enter the first column into column 1 and the second column into column 34 of the worksheet. The commands to do this are

```
MTB > set c1
DATA> 6 28 11 23 20
DATA> end
MTB > set c34
DATA> 21 39 63
DATA> end
```

You must leave at least one blank space between each number. The prompts `MTB >` and `DATA>` are printed by Minitab. The rest of the information is typed by the user. Remember that after each line, you must press ENTER. Minitab is not case sensitive, so you may use uppercase letters, lowercase letters, or a mixture of the two. Observe that Minitab displays the `MTB >` prompt after you type the command `end`. If you wanted to enter the numbers in column 3 instead of column 1, you would type `set c3` instead of `set c1`. It is always advisable to check the data after they are entered to make sure there are no errors. This can be done by instructing Minitab to print the columns of data just created. The command for this is

```
MTB > print c1 c34
```

Minitab will respond as follows (the format for the printout may differ slightly depending on your version of Minitab):

```
Data Display

 Row    C1    C34

   1     6     21
   2    28     39
   3    11     63
   4    23
   5    20
```

b. ***Commands and subcommands***

Often a command allows options, called `subcommands`. In situations where you want to follow a command with a subcommand, you must end the command with a semicolon (`;`). If more than one subcommand is used for a command, each subcommand except the last must end with a semicolon. The last subcommand must end with a period.

c. ***Manual entry of nonnumeric data***

If you want to enter nonnumeric data into a column, you must include a subcommand called `format` with the `set` command. Suppose you want column 1 in the worksheet to contain the following list of names:

```
John Lavoy
Susan Jensen
Matthew Stevensen
Mary Meager
```

The longest of the names has 17 characters (including spaces). The following command and subcommand may be used to put these names into column 1:

```
MTB > set c1;
SUBC> format(a17).
DATA> John Lavoy
DATA> Susan Jensen
DATA> Matthew Stevensen
DATA> Mary Meager
DATA> end
```

Notice the semicolon after `set c1` and the period after `format(a17)`. When entering nonnumeric data using this approach, enter each value on a separate line because spaces are considered part of the nonnumeric data and cannot be used to separate entries. The specification `a17` in the `format` subcommand tells Minitab that each data item is at most 17 characters long. You can specify a value greater than 17 if you wish, but if you specify a value less than 17, some names in this example will be truncated. To view the list of names just entered in column 1, use the following command:

```
MTB> print c1
```

Minitab will respond as follows:

```
C1
  John Lavoy          Susan Jensen          Matthew Stevensen
  Mary Meager
```

d. *Entering data from files on the data disk*

The files on the data disk accompanying this book contain most of the data used in the book. The data are stored on the disk in two forms: (1) in ASCII (American Standard Code for Information Interchange) form; and (2) in Minitab-portable (`mtp`) form. Data stored in ASCII form have the extension `.dat` and can be read into any computer. Data stored in Minitab-portable form have the extension `.mtp` and can be read into the Minitab worksheet by many different computer systems. Since we use Minitab in this book, we will not use the ASCII form of the data on the disk. However, each data set on the disk is in ASCII form in case you need it for your computer system. To read a data file from a portable Minitab worksheet file into the Session window, first start Minitab and then execute the following command and subcommand (the prompt `SUBC` is printed by Minitab, asking for a subcommand):

```
MTB > retrieve 'a:\filename';
SUBC> portable.
```

To illustrate, the first place in the text where we encounter data that are on the disk is Table 4.2.1. These data are stored in the file `radon.mtp` on the data disk. Insert the disk into drive `a:`, start Minitab, and enter the following command and subcommand in the Session window:

```
MTB > retrieve 'a:\radon';
SUBC> portable.
```

Minitab responds with

```
Retrieving worksheet from file: a:\radon.mtp
Worksheet was saved on  4/15/1997
```

With some versions of Minitab you might receive the message "Unknown object skipped." You may ignore this message. The first thing you may want to do is examine the contents of the file you just retrieved. You can do this by entering the command `information` at the Minitab prompt. The output from this command is as follows:

```
Information on the Worksheet

Column  Count  Name
C1         51  Building
C2         51  Radoncon
```

From this output you can learn which columns contain data and the names given to the variables in these columns. It is wise to print these data and check them. This is accomplished with the command

```
MTB > print c1 c2
```

The output is as follows (see Table 4.2.1):

```
 Row  Building  Radoncon

   1         1   13.6000
   2         2    2.8000
   3         3    2.9000
   4         4    3.8000
   5         5   15.9000
   6         6    1.7000
   7         7    3.4000
   8         8   13.7000
   9         9    6.1000
  10        10   16.8000
  11        11    7.9000
  12        12    3.5000
  13        13    2.2000
  14        14    4.1000
  15        15    3.2000
  16        16    2.9000
  17        17    3.7000
  18        18    2.9000
  19        19    2.0000
  20        20    2.9000
  21        21   11.2000
  22        22    1.9000
  23        23    2.0000
  24        24    6.0000
  25        25    2.9000
  26        26    7.7000
  27        27    5.1000
```

```
28          28    13.2000
29          29     3.8000
30          30    13.9000
31          31     2.4000
32          32     7.9000
33          33     1.4000
34          34     5.9000
35          35     6.5000
36          36    11.8000
37          37    13.2000
38          38     2.8000
39          39     6.9000
40          40     0.7000
41          41    12.9000
42          42     3.6000
43          43     3.6000
44          44     8.1000
45          45    17.0000
46          46     8.2000
47          47     9.8000
48          48    13.0000
49          49    11.3000
50          50     4.0000
51          51     6.0000
```

You will notice that the "name" attached to each column of data is rather interesting. This is because in early releases of Minitab, a column "name" must have no more than eight characters; so we must sometimes use a meaningful abbreviation of a full name. In later releases of Minitab, a column "name" can have up to 31 characters, but we will continue to use a maximum of 8 characters.

3. PERFORM COMPUTATIONS OR CONSTRUCT REQUIRED GRAPHS

Throughout this book Minitab commands are shown when needed to perform the computations required. The output from each set of commands is also shown. Thus, if you do not have the Minitab computing package, you can still use the information in the output. If you have a computing package other than Minitab, you can use it and the output you obtain will be similar to the Minitab output shown in the text. For many of the graphs we show the Minitab commands that can be used to construct them.

For some of the computations required in this book, there are no simple commands. In these situations we have written programs, called **macros**, which in essence allow us to define new Minitab commands that can be used for these computations. The macros are in the subdirectory `macro` on the data disk that accompanies this book. When macros are introduced and used in the text, we show you the commands and subcommands needed.

4. SAVE THE RESULTS IN A FILE FOR LATER USE

Suppose you want to save the contents of a Minitab worksheet for later use. You can save the data in a portable Minitab worksheet file named `filename.mtp`, where you can use any name instead of `filename` as long as it contains eight characters or less. If you want to store this file in the directory

```
c:\text>
```

the command and subcommand are

```
MTB > save 'c:\text\filename.mtp';
SUBC> portable.
```

Of course, instead of the directory `text`, you can use any other valid directory.

The entire contents of the Session window can also be saved in a file. The file may be edited using your favorite editor or word processor to delete unwanted information or include additional information explaining the results. You can then print the contents of the file using standard print commands. A report may be quickly created in this manner. To do this, *before* you begin entering commands you must tell Minitab that you want to save the material in the Session window. Suppose you want to save the Session window in a file named `report.txt` in your working directory. For example, if your working directory is

```
c:\text>
```

you would then issue the command

```
MTB > outfile 'c:\text\report.txt'
```

to tell Minitab to start saving the contents of the Session window. When you want to stop recording the contents of the Session window, enter the command

```
MTB > nooutfile
```

When you exit Minitab, you will find that your working directory has a file named `report.txt`. This file will contain commands issued and the computations reported in the Session window and will include any low-resolution line-printer plots that were created. High-resolution graphics will not be saved in this file. You can use this output in your word processor to write a report.

EXITING MINITAB

To exit Minitab, enter `stop` at the `MTB >` prompt. Of course, make sure that you have saved any relevant information to files.

This appendix is by no means a complete description of everything available in Minitab. You are encouraged to read the Minitab manual and experiment with all the options. Some of you may be more comfortable using the point-and-click procedures with the mouse. Try that approach after you have become familiar with the basic commands and concepts discussed here. The online help available in newer versions of Minitab for Windows can be very useful in learning these commands.

INDEX

TABLE T-4 χ^2 Table Values for Confidence Intervals for σ. Entries are Values of χ_1^2 and χ_2^2.

	0.90		0.95		0.99			0.90		0.95		0.99	
DF	χ_1^2	χ_2^2	χ_1^2	χ_2^2	χ_1^2	χ_2^2	**DF**	χ_1^2	χ_2^2	χ_1^2	χ_2^2	χ_1^2	χ_2^2
3	0.352	7.81	0.216	9.35	0.0717	12.8	32	20.1	46.2	18.3	49.5	15.1	56.3
4	0.711	9.49	0.484	11.1	0.207	14.9	33	20.9	47.4	19.0	50.7	15.8	57.6
5	1.15	11.1	0.831	12.8	0.412	16.7	34	21.7	48.6	19.8	52.0	16.5	59.0
6	1.64	12.6	1.24	14.4	0.676	18.5	35	22.5	49.8	20.6	53.2	17.2	60.3
7	2.17	14.1	1.69	16.0	0.989	20.3	36	23.3	51.0	21.3	54.4	17.9	61.6
8	2.73	15.5	2.18	17.5	1.34	22.0	37	24.1	52.2	22.1	55.7	18.6	62.9
9	3.33	16.9	2.70	19.0	1.73	23.6	38	24.9	53.4	22.9	56.9	19.3	64.2
10	3.94	18.3	3.25	20.5	2.16	25.2	39	25.7	54.6	23.7	58.1	20.0	65.5
11	4.57	19.7	3.82	21.9	2.60	26.8	40	26.5	55.8	24.4	59.3	20.7	66.8
12	5.23	21.0	4.40	23.3	3.07	28.3	41	27.3	56.9	25.2	60.6	21.4	68.1
13	5.89	22.4	5.01	24.7	3.57	29.8	42	28.1	58.1	26.0	61.8	22.1	69.3
14	6.57	23.7	5.63	26.1	4.07	31.3	43	29.0	59.3	26.8	63.0	22.9	70.6
15	7.26	25.0	6.26	27.5	4.60	32.8	44	29.8	60.5	27.6	64.2	23.6	71.9
16	7.96	26.3	6.91	28.8	5.14	34.3	45	30.6	61.7	28.4	65.4	24.3	73.2
17	8.67	27.6	7.56	30.2	5.70	35.7	46	31.4	62.8	29.2	66.6	25.0	74.4
18	9.39	28.9	8.23	31.5	6.26	37.2	47	32.3	64.0	30.0	67.8	25.8	75.7
19	10.1	30.1	8.91	32.9	6.84	38.6	48	33.1	65.2	30.8	69.0	26.5	77.0
20	10.9	31.4	9.59	34.2	7.43	40.0	49	33.9	66.3	31.6	70.2	27.2	78.2
21	11.6	32.7	10.3	35.5	8.03	41.4	50	34.8	67.5	32.4	71.4	28.0	79.5
22	12.3	33.9	11.0	36.8	8.64	42.8	55	39.0	73.3	36.4	77.4	31.7	85.7
23	13.1	35.2	11.7	38.1	9.26	44.2	60	43.2	79.1	40.5	83.3	35.5	92.0
24	13.8	36.4	12.4	39.4	9.89	45.6	65	47.4	84.8	44.6	89.2	39.4	98.1
25	14.6	37.7	13.1	40.6	10.5	46.9	70	51.7	90.5	48.8	95.0	43.3	104
26	15.4	38.9	13.8	41.9	11.2	48.3	75	56.1	96.2	52.9	101	47.2	110
27	16.2	40.1	14.6	43.2	11.8	49.6	80	60.4	102	57.2	107	51.2	116
28	16.9	41.3	15.3	44.5	12.5	51.0	85	64.7	108	61.4	112	55.2	122
29	17.7	42.6	16.0	45.7	13.1	52.3	90	69.1	113	65.6	118	59.2	128
30	18.5	43.8	16.8	47.0	13.8	53.7	95	73.5	119	69.9	124	63.2	134
31	19.3	45.0	17.5	48.2	14.5	55.0	100	77.9	124	74.2	130	67.3	140

TABLE T-5 Each Value of α (.05 or .01) Is the Proportion of the F Population That Is Greater Than the Corresponding Table Entry.*

		DF*n*														
DF*d*	α	**1**	**2**	**3**	**4**	**5**	**6**	**7**	**8**	**9**	**10**	**12**	**14**	**16**	**18**	**20**
2	.05	18.51	19.00	19.16	19.25	19.30	19.33	19.35	19.37	19.38	19.40	19.41	19.42	19.43	19.44	19.45
2	.01	98.50	99.00	99.17	99.25	99.30	99.33	99.36	99.37	99.39	99.40	99.42	99.43	99.44	99.44	99.45
3	.05	10.13	9.55	9.28	9.12	9.01	8.94	8.89	8.85	8.81	8.79	8.74	8.72	8.69	8.67	8.66
3	.01	34.12	30.82	29.46	28.71	28.24	27.91	27.67	27.50	27.34	27.22	27.03	26.92	26.85	26.77	26.67
4	.05	7.71	6.94	6.59	6.39	6.26	6.16	6.09	6.04	6.00	5.97	5.91	5.87	5.84	5.82	5.80
4	.01	21.20	18.00	16.69	15.98	15.52	15.21	14.98	14.80	14.66	14.55	14.37	14.25	14.15	14.08	14.02
5	.05	6.61	5.79	5.41	5.19	5.05	4.95	4.88	4.82	4.77	4.73	4.68	4.64	4.60	4.58	4.56
5	.01	16.26	13.27	12.06	11.39	10.97	10.67	10.46	10.29	10.16	10.05	9.89	9.77	9.68	9.61	9.55
6	.05	5.99	5.14	4.76	4.53	4.39	4.28	4.21	4.15	4.10	4.06	4.00	3.96	3.92	3.90	3.87
6	.01	13.75	10.92	9.78	9.15	8.75	8.47	8.26	8.10	7.98	7.87	7.72	7.60	7.52	7.45	7.40
7	.05	5.59	4.74	4.35	4.12	3.97	3.87	3.79	3.73	3.68	3.64	3.57	3.53	3.49	3.47	3.44
7	.01	12.25	9.55	8.45	7.85	7.46	7.19	6.99	6.84	6.72	6.62	6.47	6.36	6.27	6.21	6.16
8	.05	5.32	4.46	4.07	3.84	3.69	3.58	3.50	3.44	3.39	3.35	3.28	3.24	3.20	3.17	3.15
8	.01	11.26	8.65	7.59	7.01	6.63	6.37	6.18	6.03	5.91	5.81	5.67	5.56	5.48	5.41	5.36
9	.05	5.12	4.26	3.86	3.63	3.48	3.37	3.29	3.23	3.18	3.14	3.07	3.03	2.99	2.96	2.94
9	.01	10.56	8.02	6.99	6.42	6.06	5.80	5.61	5.47	5.35	5.26	5.11	5.01	4.92	4.86	4.81
10	.05	4.96	4.10	3.71	3.48	3.33	3.22	3.14	3.07	3.02	2.98	2.91	2.86	2.83	2.80	2.77
10	.01	10.04	7.56	6.55	5.99	5.64	5.39	5.20	5.06	4.94	4.85	4.71	4.60	4.52	4.46	4.41
11	.05	4.84	3.98	3.59	3.36	3.20	3.09	3.01	2.95	2.90	2.85	2.79	2.74	2.70	2.67	2.65
11	.01	9.65	7.21	6.22	5.67	5.32	5.07	4.89	4.74	4.63	4.54	4.40	4.29	4.21	4.15	4.10
12	.05	4.75	3.89	3.49	3.26	3.11	3.00	2.91	2.85	2.80	2.75	2.69	2.64	2.60	2.57	2.54
12	.01	9.33	6.93	5.95	5.41	5.06	4.82	4.64	4.50	4.39	4.30	4.16	4.05	3.97	3.91	3.86
13	.05	4.67	3.81	3.41	3.18	3.03	2.92	2.83	2.77	2.71	2.67	2.60	2.55	2.51	2.48	2.46
13	.01	9.07	6.70	5.74	5.21	4.86	4.62	4.44	4.30	4.19	4.10	3.96	3.86	3.78	3.72	3.66
14	.05	4.60	3.74	3.34	3.11	2.96	2.85	2.76	2.70	2.65	2.60	2.53	2.48	2.44	2.41	2.39
14	.01	8.86	6.51	5.56	5.04	4.69	4.46	4.28	4.14	4.03	3.94	3.80	3.70	3.62	3.56	3.51
15	.05	4.54	3.68	3.29	3.06	2.90	2.79	2.71	2.64	2.59	2.54	2.48	2.42	2.38	2.35	2.33
15	.01	8.68	6.36	5.42	4.89	4.56	4.32	4.14	4.00	3.89	3.80	3.67	3.56	3.49	3.42	3.37
16	.05	4.49	3.63	3.24	3.01	2.85	2.74	2.66	2.59	2.54	2.49	2.42	2.37	2.33	2.30	2.28
16	.01	8.53	6.23	5.29	4.77	4.44	4.20	4.03	3.89	3.78	3.69	3.55	3.45	3.37	3.31	3.26
17	.05	4.45	3.59	3.20	2.96	2.81	2.70	2.61	2.55	2.49	2.45	2.38	2.33	2.29	2.26	2.23
17	.01	8.40	6.11	5.18	4.67	4.34	4.10	3.93	3.79	3.68	3.59	3.46	3.35	3.27	3.21	3.16
18	.05	4.41	3.55	3.16	2.93	2.77	2.66	2.58	2.51	2.46	2.41	2.34	2.29	2.25	2.22	2.19
18	.01	8.29	6.01	5.09	4.58	4.25	4.01	3.84	3.71	3.60	3.51	3.37	3.27	3.19	3.13	3.08
19	.05	4.38	3.52	3.13	2.90	2.74	2.63	2.54	2.48	2.42	2.38	2.31	2.26	2.21	2.18	2.16
19	.01	8.18	5.93	5.01	4.50	4.17	3.94	3.77	3.63	3.52	3.43	3.30	3.19	3.12	3.05	3.00
20	.05	4.35	3.49	3.10	2.87	2.71	2.60	2.51	2.45	2.39	2.35	2.28	2.22	2.18	2.15	2.12
20	.01	8.10	5.85	4.94	4.43	4.10	3.87	3.70	3.56	3.46	3.37	3.23	3.13	3.05	2.99	2.94
30	.05	4.17	3.32	2.92	2.69	2.53	2.42	2.33	2.27	2.21	2.16	2.09	2.04	1.99	1.96	1.93
30	.01	7.56	5.39	4.51	4.02	3.70	3.47	3.30	3.17	3.07	2.98	2.84	2.74	2.66	2.60	2.55
40	.05	4.08	3.23	2.84	2.61	2.45	2.34	2.25	2.18	2.12	2.08	2.00	1.95	1.90	1.87	1.84
40	.01	7.31	5.18	4.31	3.83	3.51	3.29	3.12	2.99	2.89	2.80	2.66	2.56	2.48	2.42	2.37
60	.05	4.00	3.15	2.76	2.53	2.37	2.25	2.17	2.10	2.04	1.99	1.92	1.86	1.82	1.78	1.75
60	.01	7.08	4.98	4.13	3.65	3.34	3.12	2.95	2.82	2.72	2.63	2.50	2.39	2.31	2.25	2.20
80	.05	3.96	3.11	2.72	2.49	2.33	2.21	2.13	2.06	2.00	1.95	1.88	1.82	1.77	1.73	1.70
80	.01	6.96	4.88	4.04	3.56	3.26	3.04	2.87	2.74	2.64	2.55	2.42	2.31	2.23	2.17	2.12
100	.05	3.94	3.09	2.70	2.46	2.31	2.19	2.10	2.03	1.97	1.93	1.85	1.79	1.75	1.71	1.68
100	.01	6.90	4.82	3.98	3.51	3.21	2.99	2.82	2.69	2.59	2.50	2.37	2.27	2.19	2.12	2.07

DF*n* = degrees of freedom in the numerator.
DF*d* = degrees of freedom in the denominator.